IET ENERGY ENGINEERING SERIES 236

Green HV Switching Technologies for Modern Power Networks

Other volumes in this series:

Green HV Switching Technologies for Modern Power Networks

Edited by
Kaveh Niayesh

The Institution of Engineering and Technology

Published by The Institution of Engineering and Technology, London, United Kingdom

The Institution of Engineering and Technology is registered as a Charity in England & Wales (no. 211014) and Scotland (no. SC038698).

The Institution of Engineering and Technology
Futures Place
Kingsway, Stevenage
Hertfordshire SG1 2UA, United Kingdom

www.theiet.org

British Library Cataloguing in Publication Data
A catalogue record for this product is available from the British Library

ISBN 978-1-83953-710-3 (hardback)
ISBN 978-1-83953-711-0 (PDF)

Typeset in India by MPS Limited

Cover Image: Jose A. Bernat Bacete via Getty Images

Contents

About the editor

Kaveh Niayesh is a professor at the Norwegian University of Science and Technology. He earned his PhD at the RWTH-Aachen University of Technology, Germany. In the last 20 years, he held various academic and industrial positions in four different countries. His research interests focus on high voltage and switchgear technology, including current interruption, insulation materials, and diagnostic and condition assessment. He has (co)authored the book *"Power Switching Components"*, more than 150 papers, and 14 patents.

Chapter 1

Introduction

Kaveh Niayesh[1]

In this chapter, the background and motivation for a new book in the field of high-voltage switching technology are elaborated in the light of new developments and requirements in power networks, mainly caused by a paradigm shift towards perceiving the impact of human-made systems and technologies over their whole lifetime and even beyond.

1.1 Sustainable power network

1.1.1 Sustainable development goals: increased environmental awareness

Affordable energy is one of the important requirements for sustaining the nowadays high living standard in our modern societies. Electrical energy has been increasingly dominating our society for many years. This is mainly because of its superiority compared to all other energy sources including its efficient transmission, flexibility of operation and control, as well as its versatile conversion to all other energy forms necessary for different residential and industrial consumers.

In the early development of electrical energy systems, the focus has been put on addressing the technical issues related to the design of the power equipment to fulfil the desired functionalities like reliable insulation at high operating voltages or interruption of fault and load currents, in order to make them more cost competitive. The other societal aspects, like their environmental impact, have not been on the main agenda for electrical engineers unless they faced some conflicts with other stakeholders.

This way of thinking has changed over the last few decades, as the adverse impact of human-related pollution (and emission) became tangible for almost everybody in the world. Even though the environmental movement started as early as in the mid of twentieth century, but the focus at that time was more on local problems like industrial smog, chemical pesticides, and water and air quality. Over time, some more global phenomena, such as global warming and climate change, made themselves noticeable for the human beings. The first scientific studies

[1]Norwegian University of Science and Technology (NTNU), Trondheim, Norway

suggesting an increase in the globe temperature date back to the end of the nineteenth century, and many scientific efforts have been made on the estimation and simulation of different scenarios. The politics, however, became aware of this problem not before the 1980s. The main impact of the emissions caused by some materials used in the industrial systems is linked to their high global warming potential, especially when it is combined with high atmospheric lifetimes, and, in some cases, to their ozone depletion potential. The main source of these gases has been in the industry related to refrigerators but also some of the gases used in high voltage power equipment, e.g., SF_6, have been considered as major pollutants.

The first significant action, the so-called Montreal protocol [1], targeted limiting of ozone-depleting gases in 1987. It was followed by the adoption of an international treaty in 1992; the so-called United Nations convention on climate change, which was supposed to serve as a framework for international cooperation to strive to limit the average global temperature increase, and the resulting climate change. After many years of negotiation, the first legally binding protocol for developed countries, the so-called Kyoto protocol [2], came up in 1997. The first commitment period was from 2008 to 2012. The targets for the first commitment period covered the emissions of the six main greenhouse gases, namely, carbon dioxide (CO_2), methane (CH_4), nitrous oxide (N_2O), hydrofluorocarbons (HFCs), perfluorocarbons (PFCs), and sulphur hexafluoride (SF_6). Different emission reduction targets were foreseen for different countries, and, even for a few countries including Norway, Australia, and Iceland, an increase in emissions was agreed, leading to possibilities for trading of the emissions under the joint implementation mechanism of the Kyoto protocol, allowing a country with an emission reduction or limitation commitment to earn emission reduction units from an emission reduction or emission removal project in another country with similar commitments. This first step was apparently not far reached and therefore further actions needed to follow.

Another important milestone has been the Paris Agreement [3], which was reached by the world leaders at the UN climate change conference (COP21) in Paris in December 2015. The Agreement is a legally binding international treaty, now between 193 parties (192 countries and European Union), which entered into force on the 4th of November 2016, and sets long-term goals to guide all nations to substantially reduce global greenhouse gas emissions to limit the global temperature increase in this century to 2°C while pursuing efforts to limit the increase even further to 1.5°; to review countries' commitments every 5 years; and to provide financing to developing countries to mitigate climate change, strengthen resilience, and enhance abilities to adapt to climate impacts.

Despite all these actions, the human-made (anthropogenic) emission of all greenhouse gases (GHGs) has been steadily increasing as the recent report of intergovernmental panel on climate change (IPCC) [4] shows, see Table 1.1. The contributions of different emission sources to the equivalent CO_2 emissions are not the same: fossil fuel and industries account for more than 60% of global anthropogenic emissions, followed by methane (CH_4) with almost 18%, and land use, land use change and forestry with almost 11% of net anthropogenic emissions in

2019. It can further be observed that the net global GHG emissions have been increased by 21 Gt CO2-eq over a period of 30 years, interestingly the major increase has been caused by fossil fuels and industries.

As shown in Figure 1.1, emissions from different sources have been increased quite differently from 1990 to 2019. Of special interest is the rapid rise of fluorinated greenhouse gases (F-gases) emissions by almost 250% from 1990 to 2019, resulting in an increased share of the global net anthropogenic greenhouse gas emissions for the F-gas emissions from 1.1% to 2.4% in the period of 1990–2019. The data presented in Table 1.1 and Figure 1.1 justify targeting the GHG emission reduction caused by the fossil fuels and industries as well as fluorinated gases.

Table 1.1   Global net anthropogenic greenhouse gas emissions by different emission sources in 2019 based on data provided in [4]

Emission sources	2019 emissions (Gt CO_2 – eq)	Emission increase from 1990 to 2019 (Gt CO_2 – eq)
Fossil fuel and industry	38 ± 3	15
Methane	11 ± 3.2	2.4
Land use, forestry	6.6 ± 4.6	1.0
Nitrous oxide	2.7 ± 1.6	0.65
Fluorinated gases	1.4 ± 0.41	0.95
Total	59 ± 6.6	21

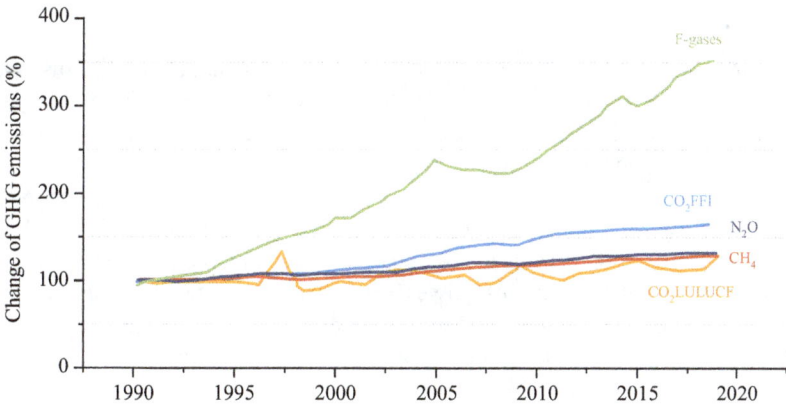

Figure 1.1   Change of global net anthropogenic greenhouse gas emissions caused by different sources including CO_2 from fossil fuel combustion and industrial processes (CO_2-FFI); net CO_2 from land use, land use change and forestry (CO_2-LULUCF); methane (CH_4); nitrous oxide (N_2O); fluorinated gases (HFCs; PFCs, SF_6, NF_3). This figure is made based on the information provided in [4].

The Paris Agreement was considered as the first step towards the net-zero emission. Parties of the agreement began submitting climate action plans known as nationally determined contributions (NDCs). Initial commitments, even if fully implemented, would only be enough to slow warming to 3°. By 2030, countries must cut emissions by at least 45% compared to 2010 levels, and the transition to net zero emissions must be fully completed by 2050.

Besides the endeavours of the United Nation climate action, there are also a number of activities and decisions on other levels such as on the European Union level. The most recent move in this context is related to the new proposal of regulations [5] for limiting F-gas usage in different products and in the EU, the so-called F-gas regulation, which was released in April 2022, and most likely will be the basis for the next revision of F-gas regulations, which is going to replace the earlier EU regulation No. 517/2014 from 2014 [6]. The proposal suggests strict limitations with a timetable to prohibit the use of high GWP gases for different products and equipment. In particular, installation and replacement of medium and high-voltage switchgear, with insulating or breaking medium using, or whose functioning relies upon, gases with GWP of 10 or more, or with GWP of 2,000 or more will be prohibited after the first of January 2026, the first of January 2028, the first of January 2030, or the first of January 2031, depending on their rated voltage level, unless evidence is provided that no suitable alternative is available based on technical grounds within the lower GWP ranges referred to above. This proposal is made in response to the increased climate ambition of the EU through the regulation 2021/1119 (the European Climate Law) which targets at least 55% reduction of greenhouse gas emissions by 2030, and EU climate neutrality at the latest by 2050 [7]. Similar limitations are also expected to be introduced in some other territories like the state of California in the US [8].

To achieve a prosperous future for human beings while protecting our planet, many further actions besides climate action must be taken. At the UN sustainable development summit in September 2015, the so-called 2030 Agenda for sustainable development with 17 sustainable development goals and 169 targets was adopted. These goals are shown in Figure 1.2, highlighting the most relevant ones in the context of the present book. The core of this agenda is a plan of action for people, the planet and prosperity to strengthen peace and end poverty in all its forms. Among the sustainable development goals, the goal no. 7 "affordable and clean energy" is very relevant in the context of power networks. This should ensure access to affordable, reliable, sustainable, and modern energy for all by targeting a substantial increase in the share of renewable energy in the global energy mix, doubling the global rate of improvement in energy efficiency, enhancing international cooperation to facilitate access to clean energy research and technology, including renewable energy, energy efficiency and advanced and cleaner fossil-fuel technology, and promoting investment in energy infrastructure and clean energy technology, as well as expanding infrastructure and upgrading technology for supplying modern and sustainable energy services for all in developing countries, in particular least developed countries, small island developing states, and land-locked developing countries, in accordance with their respective programs of support, by 2030 [9].

Figure 1.2 Seventeen sustainable development goals of the United Nations, highlighting the most relevant goals in the context of this book, namely goal no. 7 "affordable and clean energy" and goal no. 13 "climate action"

The targets related to the sustainable development goal no. 7 are formulated in a rather general form without giving many quantitative measures; but some stress on increased efficiency and renewable energy can be identified. These may be better understood as some means to facilitate the net-zero emission, explained earlier, while keeping energy accessible for everyone. The trend to install more renewable energy resources is facilitated by a rapid decrease in the unit cost of some forms of renewable energy during the last two decades as the recent report of IPCC [4] indicates. It seems that at least three technologies, namely photovoltaic (PV), onshore wind, and offshore wind become highly cost-competitive compared to the new fossil fuel (coal and gas) power in 2020. As the levelised cost of energy (LCOE) for these two technologies lies in the lower band of the cost for new fossil fuel power of 55–148 USD per MWh in 2020, see Figure 1.3. It must be noted that the costs of grid integration, and environmental and social externalities that may modify the overall (monetary and non-monetary) costs of technologies and alter their deployment are not considered in this comparison. Installed renewable energy sources have also significantly increased during the last two decades, so that it is, by 2020, about 700 GW for photovoltaic and onshore wind, each, contributing to 3% and 6% of electricity produced, respectively, as shown in Figure 1.4.

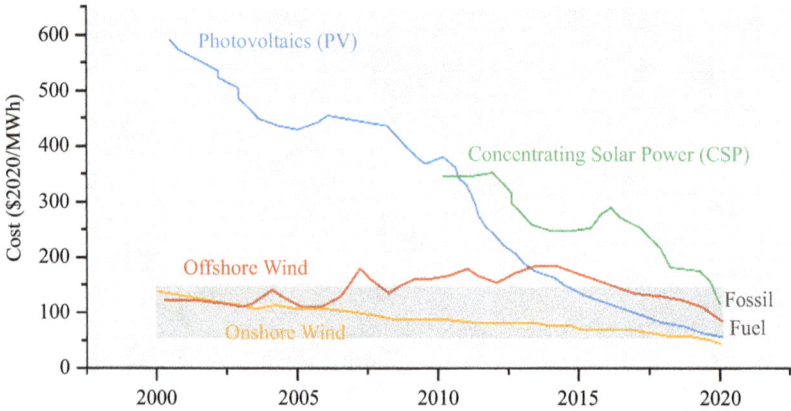

Figure 1.3   *Cost development of different renewable energy sources over two decades (2000–2020). The grey zone shows the cost range for new fossil fuel power in 2020. This figure is made based on the information provided in [4].*

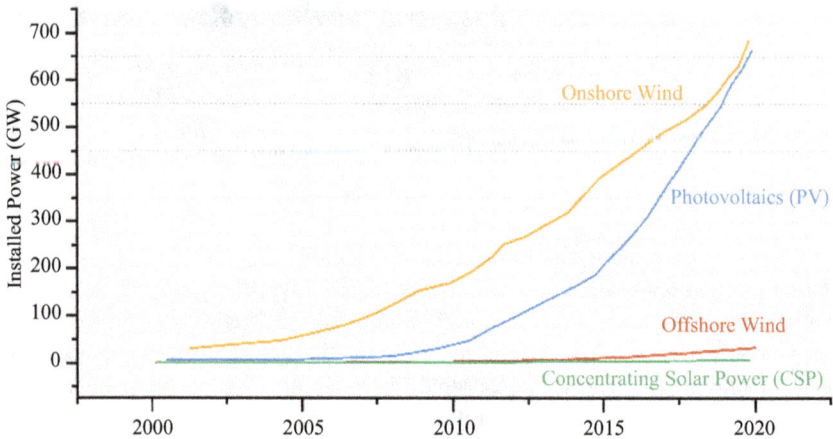

Figure 1.4   *Development of worldwide installed power for four different renewable energy sources over two decades (2000–2020). This figure is made based on the information provided in [4].*

There is no reason to believe that this trend will be changed. So, an even larger share of renewable energy sources in the power network within the near future is to be expected. A recent study [10] on global energy perspectives based on World Energy Council's (WEC) World Energy Scenarios 2019 estimates the share of renewable sources including solar or photovoltaic, wind, geothermal and bio-energies, on global power generation from 41% to 60% for three different future scenarios, by 2060, where wind will account for 10–18%, and solar from 9 to 20%. In all three future scenarios, it is assumed that the world population reached 10 billion by 2060, the differentiation is however on the approaches taken in different scenarios ranging from a government-driven approach, where global convergence implements climate-based policies (the best in terms of global emission reduction) to market-driven approach, where open economics follows the target of providing affordable energy for all, or even an approach with strong commitment to national unilateralism, where fragmented economics follow energy security policies (the worst in terms of global emission reduction).

1.1.2 Impact on power network development trends

The increased environmental awareness has impacted the generation, transmission, as well as consumption of energy. This has been mainly following the goal of reduction of greenhouse gas emissions as discussed in the last section, which is also known as *decarbonisation*. The installation of renewable energy sources as a key element to reduce the fossil fuel emissions by energy generation poses new challenges and requirements for the power networks. In conventional power networks, the generation of energy is performed in large generator units, powered by fossil fuels or in some cases nuclear power plants. The power ratings of thermal power plants or nuclear power plants consisting of many generator units could be many hundreds to a few thousands of MVA. The power ratings of the renewable generating units are rather low, and their number could be very large. So, the trend in power networks employing renewables could be in some cases to have many energy production units distributed all over the power network. This is called the *decentralisation* of the power-generating units resulting in bi-directional power flow.

In some cases, many of the renewable energy units are gathered in the form of an energy farm with power ratings comparable to conventional power plants. The energy generated by the renewables is, however, only possible where the energy generating agent, e.g., wind or sunshine, is available. In many cases, this is not given in the regions where the energy is required, and therefore some measures for transmission of the energy produced by the renewables to the regions with a high demand of energy should be realised. Conventional AC power networks show deficiencies for this task. To illustrate the problem with the utilisation of AC transmission systems, two examples are considered here. The first example is related to the renewable energy generation by the wind turbines placed on the high seas with long distances to the coasts, the so-called offshore wind. To bridge long distances using the cables, high-voltage direct current (HVDC) based systems show clear advantages, as the cable charging currents, which significantly limit the transmission capacity of the AC cables, can be avoided. An increasing activity is observed in many different countries, e.g., in the North Sea region, see Figure 1.5, as well as in the east coast of the United States to install offshore wind parks, artificial islands and sub-sea interconnections.

In addition, there are several plans to establish HVDC grids to better utilise the renewable energy sources by connecting them through an HVDC offshore grid. One

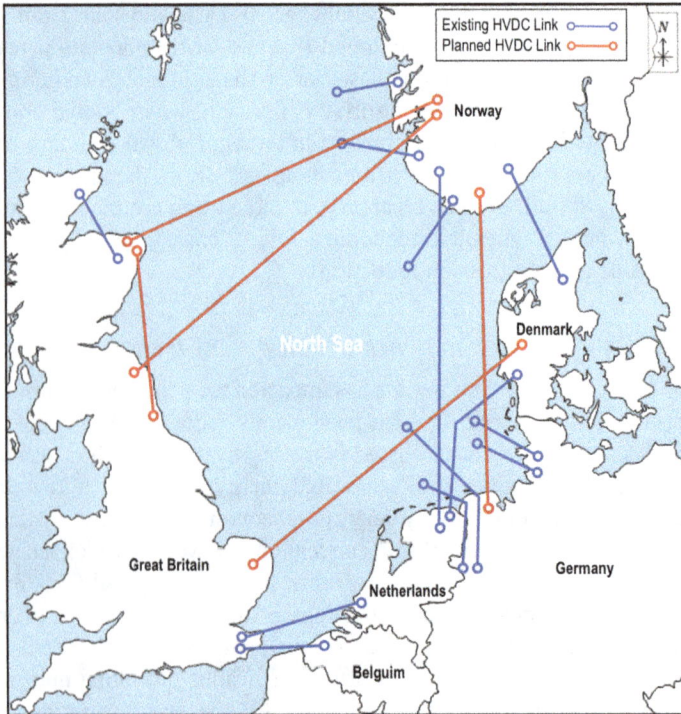

Figure 1.5   An overview of HVDC interconnections in the North Sea, data provided in [11] extended and updated

example is the development of the grid in the Europe over the next 30 years. This is illustrated very well in the comparison between the European grid in 2020 and the expected European grid in 2050 [12], as shown in Figure 1.6. In this figure, an increased renewable generation, especially wind in the northern part of Europe, and photovoltaics in the southern part of Europe by 2050 are shown. Furthermore, the grid development connecting many the renewable resources together using HVDC is foreseen.

The second example is related to the transfer of large amounts of energy over long distances, e.g., between the renewable energy generation in the northern part of Germany and the highly industrial region in the south part of Germany with large energy demands, where the HVDC systems show clear advantage, see Figure 1.7. These HVDC corridors are also considered as an important part of the so-called energy transition.

Another problem with the renewable sources like wind and photovoltaics is their intermittence (uncertainty of the energy production), as the presence and intensity of the energy generating agent, e.g., wind, is hardly predictable. There have been many approaches to combat this disadvantage, one of them is the application of energy storage units in the power system, including batteries, fly-wheels, superconducting magnetic energy storage. The renewable energy can also

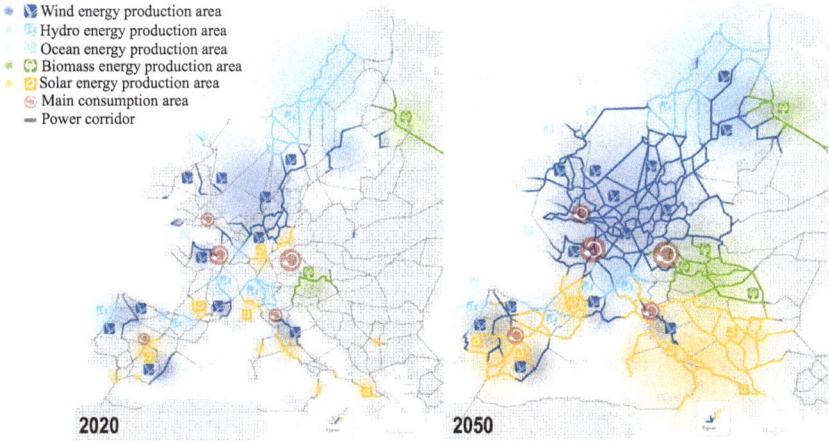

* Wind energy production area
* Hydro energy production area
* Ocean energy production area
* Biomass energy production area
* Solar energy production area
* Main consumption area
* Power corridor

2020   2050

Figure 1.6   European grid development 2020–2050, adapted based on the information provided in [12].

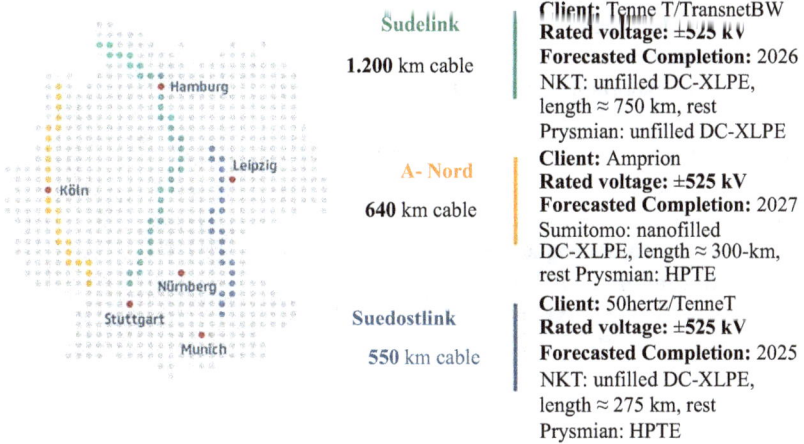

Hamburg

Leipzig

Köln

Nürnberg

Stuttgart

Munich

Südelink	**Client:** Tenne T/TransnetBW	
1.200 km cable	**Rated voltage: ±525 kV**	
	Forecasted Completion: 2026	
	NKT: unfilled DC-XLPE, length ≈ 750 km, rest Prysmian: unfilled DC-XLPE	
A- Nord	**Client:** Amprion	
640 km cable	**Rated voltage: ±525 kV**	
	Forecasted Completion: 2027	
	Sumitomo: nanofilled DC-XLPE, length ≈ 300-km, rest Prysmian: HPTE	
Suedostlink	**Client:** 50hertz/TenneT	
550 km cable	**Rated voltage: ±525 kV**	
	Forecasted Completion: 2025	
	NKT: unfilled DC-XLPE, length ≈ 275 km, rest Prysmian: HPTE	

Figure 1.7   HVDC corridors in Germany from [13]

be used for hydrogen production through electrolysis, which can be used as a clean energy source to replace fossil fuels, or be converted back to the electricity, whenever necessary, using fuel cells. The hydrogen generation from renewable sources is considered as one of the most promising technologies to reduce the CO_2 emission and is integrated into some of the newer offshore wind installations. Unpredictability of renewable energy generation and fluctuations in electricity price have an undermined role in hydrogen production.

Renewable energy sources are connected to the grid through power electronic inverters. In case of any fault in the grid, the power electronic inverter detects a voltage dip and disconnects the energy source from the grid. This results, however, in the stability problem in the grid. To avoid this problem, new grid codes [14] established for the steady operation of power systems require that the renewable energy sources, like wind or photovoltaic parks, remain connected to the grid for a specified time depending on the level of voltage drop at the common connection point, which is affected by the severity and location of the fault. This peculiar requirement related to the integration of renewable energy source to the grid is the so-called *fault ride through*. The flow of fault current through the renewable generating units and their associated power electronic circuitry for long time is challenging. In some studies [15], fault current limiters are proposed to be used in conjunction with renewable energy source grid integrations to improve their fault ride-through capabilities. Furthermore, the fault current limiters are considered as a way for integrating renewable sources into existing distribution substations without investing in new transformers [16]. So, fault current limiters found new applications in the context of integration of renewable sources to the grid, in addition to their other applications in power networks linked to the increased short circuit current levels of the power networks caused by increasing interconnections (meshing) in order to improve the reliability and availability of the energy supply.

On the energy consumption side, a clear shift towards electrical energy for many different applications is observed, i.e., the so-called electrification trend in many different sectors. One of the sectors significantly affected by the electrification wave is transportation. Not only the electric vehicles (EVs) are increasingly replacing the conventional combustion-engine-based vehicles thanks to the rapid decrease in the cost of batteries, as shown in Figure 1.8, but also other transportation methods, like aviation and ships, are also following the same trend to become all (or more)-electric-ships and all (or more)-electric-airplanes.

The mentioned consequences of decarbonisation change the power networks from the conventional form with large concentrated (mostly fossil fuel driven)

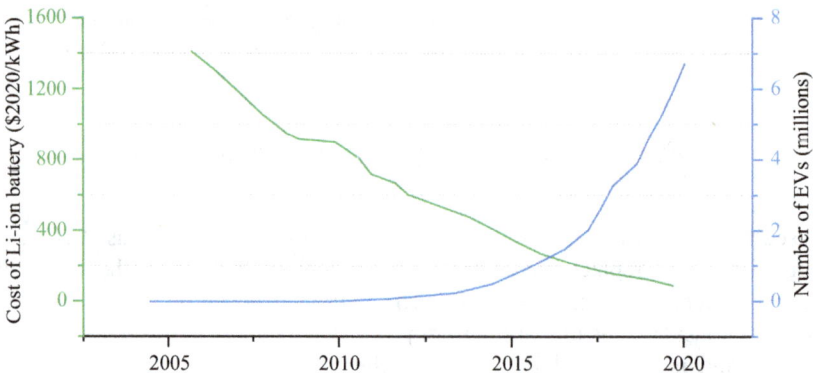

Figure 1.8   Development of Li-ion batteries' cost and worldwide number of personal electric vehicles from 2005 to 2020. This figure is made based on the information provided in [4].

energy production units connected to AC transmission systems, which transfer power unidirectionally from the energy sources to passive energy consumers, to a modern grid, where many renewable sources with power electronics units, most likely fault current limiters as well as energy storage units, are integrated to different voltage levels and at different places in power systems and a mixture of AC and DC grids is used to transfer bidirectionally the power to and from the energy consumers, which actively interact with the power system, as shown in Figure 1.9.

Moving towards more clean energy sources is not the only drive that impacts the power networks. Trends towards liberalisation of electricity markets, as well as increased application of digital transformative technologies, such as Internet of Things, co-contribute to the shaping of future modern networks. The rationale for liberalisation is that by making companies compete fairly with one another, efficiency is encouraged, quality and innovation increase, prices decrease, and consumers have an overall broader choice, apart from a more secure supply [18]. It has proven to be challenging to reconcile liberalisation, on the one hand, while meeting climate targets and decarbonising the electricity grid, on the other hand [19]. The

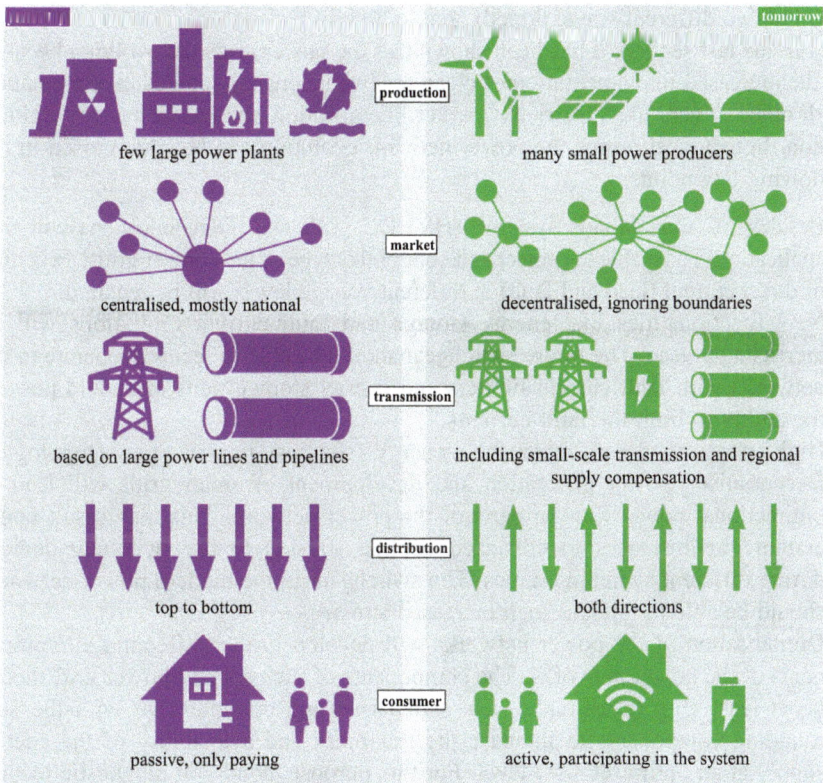

Figure 1.9   Modern vs. conventional power networks reproduced after [17]

impact of this driver for power network configurations has been an increasing number of cross-border interconnections between electricity grids.

Digitalisation is considered to have the potential to either help or hinder the fight against climate change [20]. Digitalisation has become an indispensable tool for achieving the objectives of a green economy, but technology itself is also responsible for a significant amount of pollution [21], so that digitalisation may reduce greenhouse gas emissions in medium or long term. The negative short-term impact is mainly related to the higher energy consumption and environmental footprint of the digital sector, i.e., information and communication technology sector, while optimisation leading to higher efficiency and lower energy consumption may lower emission intensities in the mid-term or long-term.

1.1.3   Expectations on modern HV switchgear

In this section, the question is addressed if there is any new expectation on high-voltage switchgear in the modern power networks, and if the answer is affirmative, what those expectations are.

The endeavour to significantly reduce the anthropogenic emissions of power grids will raise expectations on high-voltage switchgear of future modern power grids in two different ways, directly and indirectly.

In the last section, it has been shown that the power grids are evolving because of the increased implementation of clean renewable energy sources in combination with other drivers like electricity market liberalisation and power system digitalisation. In technical terms, the power network evolution can be summarised in the following few points:

- Besides AC transmission systems, (HV) DC grids and interconnections will find application. To be able to protect and control this type of new power grids, switching of direct current (load and fault) at different voltage levels will be required.
- New functionalities like energy storage and fault current limitations will be increasingly used. The future switchgear should be able to securely operate in the networks with fault current limiters and energy storage units or should possess the ability to limit the fault currents.
- High penetration of renewable energy sources implies new topologies. Decentralised power generation and development of smart grids will lead to bidirectional power flow in parts of the power network. This may result under certain circumstances to enhanced applied stresses to the switching devices during different switching duties. The switchgear in the modern power networks should be able to tolerate such increased stresses.
- Digitalisation of the power networks will develop further, affecting all components of the power networks. The components of such a smart power grid should be at least able to estimate their own status and communicate with the surrounding components to increase the reliability and availability of the energy supply at an optimised cost level. For this purpose, condition diagnostic techniques together with the processing and data communication capabilities will be integrated parts of the switchgear of future power grids.

- Electricity market liberalisation will result in a growing number of cross-border interconnections. This may result in larger short circuit current levels, which should be securely interrupted by the switchgear of future power grids unless fault current limitation possibilities are available.

These changes will affect the power equipment at all voltage levels used for transmission, distribution, and generation of electrical energy. Many of the desired functions related to the switchgear in modern power networks are not available in the conventional mature switching technologies. In Section 1.2.2, it will be detailed what these expectations mean for the switchgear in terms of the required functionalities.

In addition, following the decarbonisation thought of the entire electrical energy systems, it is required to minimise the GHG emission of power equipment used in transmission and distribution systems throughout their whole life cycle including their environmental footprint during the manufacturing, transport, commissioning, and operation. Estimation of the environmental footprint of any equipment, the so-called life cycle assessment, is a cumbersome task, as many different parameters indirectly affect this analysis. Staring with the manufacturing of switchgear, the minimisation of their environmental footprint means the use of environmentally benign materials and to optimise the resources, e.g., energy, necessary for processing the required materials. A simple rationale would be therefore to avoid materials with large environmental impact, e.g., gases with large global warming potential. This simple rationale was also followed by major political decision-makers like the United Nations and European Union. This is not favouring matured switching technologies to be used in modern power networks, as it will be discussed in Section 1.2.1. When it comes to emission related to the switchgear operation, it would be desired to prolong the lifetime, e.g., by appropriate implementation of condition assessment techniques to avoid faults in switchgear resulting in major failure in other power equipment and significant undelivered energy. This also falls under the category of new functionalities required for the switchgear and will be detailed in Section 1.2.2.

1.2   Challenges of conventional HV switchgear

High-voltage switchgears are based on different technologies depending on the voltage level, which have been matured over decades. They are switching arc based and able to interrupt alternating currents at their current zero-crossings. They consist of different parts, including interruption chamber (also called interrupter), driving (or operating) mechanism, and control unit. A switching arc is generated within the interruption chamber, and the actual switching functions, i.e., current interruption during switching operation, current carrying in a closed position as well as high voltage insulation in open position, will take place in the interruption chamber. Depending on the medium used in the interruption chamber, the switching arc can be a low-pressure (e.g., in vacuum circuit breakers) or a high-pressure (e.g., in gas circuit breakers) one. The physical mechanisms responsible for switching arc control are different in the two aforementioned switching arc types. The vacuum circuit breakers have been dominating the applications in the medium voltage range corresponding to distribution networks, and gas circuit breakers with SF_6 gas have been dominating applications at higher voltage levels. Furthermore, from the high-voltage insulation perspective, the switchgear can be air-insulated, where atmospheric air in combination with some solid insulations is employed for high-voltage insulation, or gas-insulated, where all components are placed inside an enclosure, and the high-voltage insulation is governed using an insulating gas (in many cases SF_6) at elevated pressures.

1.2.1   Adverse environmental impact

Vacuum switchgear is known as an environmentally friendly solution with small carbon footprint. SF_6 switchgear is, in turn, considered as very critical with respect to its environmental impact. SF_6 is a gas with superior dielectric and thermal characteristics, which makes it a brilliant candidate for high-voltage insulation and current interruption purposes but is, unfortunately, one of the gases with the highest global warming potential, as shown in Table 1.2. Its long atmospheric lifetime makes its impact on global warming even worse for longer time spans.

The electric power industry is the largest application area of SF_6, almost 80% of the annual production is used in power equipment [23]. Even though some studies claim that the adverse impact of SF_6 use in switchgear is balanced with the reduced amount of the resources used for their manufacturing and operation compared to the other possibilities if the total life cycle analysis is performed [24], but there are several indications showing that this is most likely not the case. As discussed in Section 1.1.1, the emission related to fluorinated gases has dramatically increased by almost 250% over the last 30 years. Another indication is the increased SF_6 atmospheric concentration [26] as shown in Figure 1.10, where it has more than tripled over the last 25 years. Even in comparison with the other greenhouse gases, this increase is noticeable: in comparison, N_2O atmospheric concentration has increased by 10% over the last 45 years [26]. These facts are the main drivers of many political entities to get away from SF_6 in applications where technical alternatives exist, reflected in the United Nations Climate change actions and F-gas regulations of EU as discussed in Section 1.1.1.

Table 1.2   Global warming potential and atmospheric lifetimes of some of greenhouse gases based on data provided in [22]

Greenhouse gas	Global warming potential (over 100 years)	Atmospheric lifetime (years)
Carbon dioxide (CO_2)	1	Hundreds
Methane (CH_4)	21	12
Nitrous oxide (N_2O)	310	114
Hydrofluorocarbons (HFCs)	140–11,700	1.4–270
Sulphur hexafluoride (SF_6)	23,900 (or 25,200 according to [25])	3,200

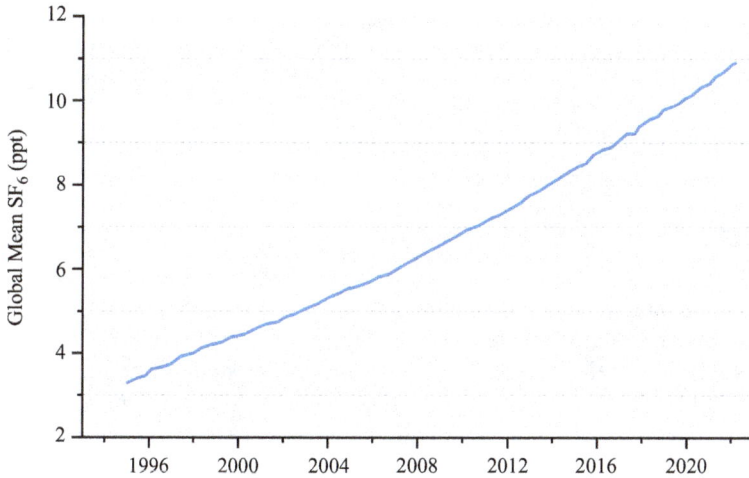

Figure 1.10 *SF₆ atmospheric concentration over the last 25 years, based on the information provided by NOAA global monitoring laboratory on April 27, 2022 [26]*

In a recent publication [27], the life cycle carbon footprint assessment of switchgear containing SF_6 is compared with other solutions, and clearly concluded that the replacement of SF_6 would significantly reduce the carbon footprint of the switchgear. It seems there is no way around the replacement of SF_6 for switching applications, but what the viable possibilities would be. This question is answered in detail in Chapters 3 and 4 of the present book, where switching technologies for high-voltage applications based on other gases rather than SF_6 (the so-called SF_6 – alternatives) and the possibility to employ the environmentally favourised vacuum switching technology to higher voltage applications will be thoroughly discussed.

1.2.2 New required functionalities

As stated in Section 1.1.3, decarbonisation of the electric energy generation forces new functionalities of power switching devices. Some of these new functionalities are not self-evident for conventional power switchgear. Direct current power transmission will gain importance, and to be able to control and protect such grids, HVDC switching devices will be required. The conventional power switchgears are only capable of interrupting alternating currents at their current zero crossings, but in the case of direct currents, no natural current zero is available, and therefore the conventional power switchgear cannot be used for HVDC current interruption. In addition, the transient response of a DC grid to short circuit faults is completely different compared to that of AC grids, resulting in fast-rising current waveforms with a rate of rise of currents of up to several kiloamperes per milliseconds. So, tens of milliseconds delay in the operation of circuit breakers would result in huge instantaneous current levels, which can hardly be controlled and interrupted. Thus, completely new concepts and technologies are required for HVDC switches (or circuit breakers). Requirements for such DC circuit breakers will be detailed in Chapter 2 of the present book. Chapter 5 of the present book thoroughly covers the developments in concepts and technologies for HVDC switching.

Another needed functionality in modern power networks would be the fault current limitation. As the arcing voltage of the conventional high-voltage switchgear is negligibly small compared to the rated voltage of the power network, initiation of a switching arc during their operation has almost no effect on the current to be interrupted, i.e., no inherent fault current limitation capability is available for high-voltage circuit breakers.

The change in grid configuration and high penetration of distributed energy sources will affect the stresses applied to the power switchgear during different switching duties. The open question in this context would be if and to what extent the conventional switching devices will be able to cope with the elevated stresses.

1.3   Outline of the book

A systematic approach has been taken in the following chapters of this book to explain the context of the switching in power networks and give extensive explanations on the recent development of the switchgear in response to make them more environmentally friendly and at the same time able to fulfil the high technical requirements on their performance.

Chapter 2 of this book explains the stresses a circuit breaker (CB) in high-voltage networks could be exposed to. The first five sections cover the stresses applied to HVAC CBs under different switching scenarios, which mostly are also of relevance for the new SF_6 free switchgear. In the following sections, the peculiarities related to switching to new applications, ranging from offshore applications to HVDC networks, are detailed. This chapter provides a comprehensive introduction to the requirements for circuit breakers for various applications.

Chapters 3 and 4 cover the most important concepts for SF_6-free switchgear, namely application of vacuum switching technology in high voltage, and gas circuit breaker design adapted based on SF_6 alternative gases.

Chapter 3 starts with a review of the vacuum switching technology which has been extensively employed in medium-voltage circuit breakers and sheds light on the main challenges related to upscaling the vacuum technology to higher voltage applications. The concepts for addressing the challenges and the relevant research are presented. It reveals countless and delicate details on the design of high-voltage vacuum circuit breakers from several perspectives.

Chapter 4 starts with a brief introduction to the concepts of current interruption in high-pressure gases followed by a detailed presentation of all relevant material properties of the alternative gases both from dielectric insulation point of view and from the current interruption point of view. It puts the focus on some natural gases (i.e., CO_2 and O_2) as well as the mixtures based on C4-FN (Fluoronitrile) and C5-FK (Fluoroketone) with different natural gas combinations. It extensively covers all the recent activities by different research groups and leading manufacturing companies on the development of gas circuit breakers with alternative gases. In the last sections, different mechanisms relevant to lifetime of high-voltage gas circuit breakers with SF_6 alternative gases are discussed in detail including gas decomposition (e.g., due to arcing or partial discharge) in switchgear and its impact on possible degradation of the dielectric and interruption capability of the switching equipment.

Chapter 5, in contrast to Chapters 3 and 4, is not related to the material constraints on switching equipment (e.g., SF_6 replacement) but addresses switching equipment for newly emerging types of stresses in modern power networks, namely HVDC networks. It discusses the basics of switching in HV DC networks and the proposed concepts. It gives insightful details on their performance and test requirements, by thoroughly discussing different aspects of various designs.

Many of the research fields related to high-voltage switching devices have not resulted so far in mature solutions but have the potential to improve the performance of the switching equipment, extend its lifetime, and widen its functionality. Chapter 6 tries to briefly cover some of the most relevant research fields related to high-voltage switching devices.

Acknowledgment

The author would like to thank Naghme Dorraki from the Leibniz Institute for Plasma Science and Technology (INP), Greifswald, Germany, for the help in the preparation of some of the figures.

References

[1] United Nations Environmental Programme, Ozone Action, https://www.unep.org/ozonaction/who-we-are/about-montreal-protocol, accessed April 26, 2023.

[2] United Nations Climate Change, https://unfccc.int/kyoto_protocol, accessed April 26, 2023.

[3] United Nations Climate Change, Paris Agreement, https://www.un.org/en/climatechange/paris-agreement, accessed April 26, 2023.

[4] P.R. Shukla, J. Skea, R. Slade, *et al.* (eds.), IPCC, 2022: summary for policymakers, In: *Climate Change 2022: Mitigation of Climate Change. Contribution of Working Group III to the Sixth Assessment Report of the Intergovernmental Panel on Climate Change.* Cambridge University Press, Cambridge and New York, NY. doi:10.1017/9781009157926.001.

[5] Proposal for a Regulation of European Parliament and of the Council on Fluorinated Greenhouse Gases, Amending Directive (EU) 2019/1937 and Repealing Regulation (EU) No. 517/2014, Strasbourg, April 2022.

[6] Regulation (EU) No. 517/2014: F-Gas Regulations, Regulation (EU) No. 517/2014 of the European Parliament and of the Council of 16 April 2014 on

Fluorinated Greenhouse Gases and Repealing Regulation (EC) No. 842/2006, May 2014.

[7] Regulation (EU) 2021/1119: European Climate Law, Regulation (EU) 2021/1119 of the European Parliament and of the Council of 30 June 2021 Establishing the Framework for Achieving Climate Neutrality and Amending Regulations (EC) No. 401/2009 and (EU) 2018/1999 ('European Climate Law'), July 2021.

[8] State of California Air Resources Board Final Regulation Order 2021 Proposed Amendments to the Regulation for Reducing Sulfur Hexafluoride Emissions from Gas Insulated Switchgear, https://ww3.arb.ca.gov/board/15day/sf6/fro.pdf, accessed July 20, 2022.

[9] United Nations Environmental Programme: Affordable and Clean Energy, https://www.unep.org/explore-topics/sustainable-development-goals/why-do-sustainable-development-goals-matter/goal-7), accessed July 20, 2022.

[10] T. Kober, H.-W, Schiffer, M. Densing, and E. Panos, Global energy perspectives to 2060 – WEC's world energy scenarios 2019, *Energy Strategy Reviews*, vol. 31, p. 100523, 2020, https://doi.org/10.1016/j.esr.2020.100523.

[11] A. Elahidoost and E. Tedeschi, Expansion of HVDC offshore grid: an overview of contributions, status, challenges and perspectives, NTNU Open, 2017.

[12] C. Kjaer, The North Sea offshore grid: a distant dream or soon reality?, A Presentation from European Wind Energy Association (EWEA), https://www.statkraft.com/globalassets/old-contains-the-old-folder-structure/documents/kjaer-29nov10_tcm9-12386.pdf, accessed April 26, 2022.

[13] G. Mazzanti, High voltage direct current transmission cables to help decarbonisation in Europe: recent achievements and issues, *High Voltage*, vol. 7, no. 4, pp. 633–644, 2022, doi: 10.1049/hve2.12222.

[14] National Grid Workgroup Report on Grid Codes: GC0062 – Fault Ride Through, https://www.nationalgrideso.com/codes/grid-code/modifications/gc0062-fault-ride-through, accessed April 26, 2022.

[15] A. Jalilian, M. Tarafdar Hagh, S. Abapour, and K. Muttaqi, C-link fault current limiter-based fault ride-through scheme for inverter-based distributed generations, *IET Renewable Power Generation*, vol. 9, no. 6, pp. 690–699, 2015.

[16] M.R. Barzegar-Bafrooie, J. Dehghani-Ashkezari, A. Akbari Foroud and H. Haes Alhelou, *Fault Current Limiters, Concepts and Applications*, Springer, 2022, ISBN: 978-981-16-6650-6.

[17] R.M. Aryblia, A. Bertram, A. Bolle, *et al.*, Facts and figures about renewables in Europe, In: *Energy Atlas*, Heinrich Böll Foundation, Berlin, Germany, Friends of the Earth Europe, Brussels, Belgium, European Renewable Energies Federation, Brussels, Belgium. Green European Foundation, Luxembourg, 2018, https://gef.eu/wp-content/uploads/2018/04/energyatlas2018_facts-and-figures-renewables-europe.pdf, accessed April 26, 2023.

[18] A. Marhold, The interplay between liberalization and decarbonization in the European internal energy market for electricity, In: *Energy Law and Economics*, Springer, 2018, ISBN: 978-3-319-74636-4.

[19] W. Zappa, M. Junginger, and M. van den Broek, Can liberalised electricity markets support decarbonised portfolios in line with the Paris Agreement? A case study of Central Western Europe, *Energy Policy*, vol. 149, p. 111987, 2021.

[20] G. Martin Quetglas and A. Ortega, Digitalisation with Decarbonization – Working Paper 8/2021 – Elcano Royal Institute, May 2021, https://media. realinstitutoelcano.org/wp-content/uploads/2022/02/wp8-2021-martin-ortega-digitalisation-with-decarbonisation.pdf, accessed April 26, 2022.

[21] J. Ye, Using Digitalization to Achieve Decarbonization Goals, A Research Publication of Center for Climate and Energy Solutions (C2ES), September 2021, https://www.c2es.org/document/using-digitalization-to-achieve-decarbonization-goals/, accessed April 26, 2022.

[22] S. Solomon, D. Qin, M. Manning, *et al.* (eds.), *IPCC, 2007: Climate Change 2007: The Physical Science Basis. Contribution of Working Group I to the Fourth Assessment Report of the Intergovernmental Panel on Climate Change, 2007*, Cambridge University Press, Cambridge and New York, NY.

[23] A.H. Powell, *Environmental Aspects of the Use of Sulphur Hexafluoride*, ERA Technology Ltd., ISBN: 0700807527, 2002.

[24] E. Preisegger, R. Durschner, W. Klotz, *et al.*, Life cycle assessment—electricity supply using SF_6 technology, In: *Proceedings of 2001 International Symposium on Electrical Insulating Materials (ISEIM 2001). 2001 Asian Conference on Electrical Insulating Diagnosis (ACEID 2001)*, pp. 371–374, 2001, doi:10.1109/ISEIM.2001.973676.

[25] Intergovernmental Panel on Climate Change (IPCC), Climate Change 2021, IPCC AR6 WGI, table 7.SM.7, 2021. https://www.ipcc.ch/report/ar6/wg1/downloads/report/IPCC_AR6_WGI_Full_Report.pdf.

[26] National Oceanic and Atmospheric Administration (NOAA), Global Monitoring Laboratory, Earth System Research Laboratories, SF_6 Combined Data, https://gml.noaa.gov/hats/combined/SF6.html, accessed April 26, 2022.

[27] P. Billen, B. Maes, M. Larrain, and J. Braet, Replacing SF_6 in electrical gas-insulated switchgear: technological alternatives and potential life cycle greenhouse gas savings in an EU-28 perspective. *Energies*, vol. 13, p. 1807, 2020, doi:10.3390/en13071807.

Chapter 2

Requirements for circuit breakers

René P.P. Smeets[1]

2.1 Introduction

Circuit breakers are the switching devices that are designed to interrupt every possible current that arises in the power system. Most prominent are the fault currents resulting from an unintended flow of current in a system. Mostly, this fault current arises as a short-circuit current when the current chooses a 'path of least resistance' created by a breakdown somewhere in the system or an unintentional galvanic connection, e.g., an earthing switch closing into an energized conductor or erroneously left in closed position at energization. Other examples of fault currents are unwanted flow of current due to a mis-synchronization of parts of the system, most commonly at locations where power generation is feeding a system.

In this sense, the circuit breaker is the hardware part of the power system protection and as such the last defence against major damage caused by the thermal and electro-dynamical impact of short-circuit currents.

Generic, idealized technical requirements for circuit breakers are the following:

- A very good conductor in a closed state. Circuit breakers are practically always in the closed state, conducting current. Since mechanical contacts in switching devices have resistance (actually the dominant resistance in the conducting chain), it is important to minimize ohmic (heat) losses, not in the last place because these losses also add up to the environmental impact of power systems, as long as the generation is not CO_2 emission-free.
- Shall be a perfect insulator in an open state. In open state, the switching device must be able to withstand all voltages and their transients, imposed across the switching gap between the contacts. When not, breakdown occurs that may cause (fault) current to resume again after an initially successful interruption or send electrical shockwaves into the system that may be harmful.
- Must be able to make the transition from insulating to conducting state. The closing operation may unintentionally occur under fault conditions resulting in the flow of very large currents through the closing circuit breaker. The associated stresses should not lead to a malfunction (e.g., contact welding) of the switching device.

[1]KEMA Labs, Arnhem, the Netherlands

- Must make the transition from conducting to insulating state fast enough. Operation and current interruption of a circuit breaker must be as fast as possible, to minimize the duration that the system is suffering from the fault.
- The change from conductor to insulator may not lead to unacceptable overvoltages. Interruption of current is basically a matter of energy re-arrangement in the system. Electric and magnetic energies are exchanged continuously, with magnetic energy having by far the biggest share. Therefore, interruption strategy is to interrupt when magnetic energy is zero: at the zero crossing of AC current (*'current zero'*). When the interruption occurs away from current zero, the remaining magnetic energy manifests itself as an overvoltage.
- Shall be able to interrupt a number of times. Current interruption in all switching devices is associated with an *'electric arc'*, that starts upon contact separation and shall extinguish at a current zero crossing. The electric arc is a high-temperature plasma that causes wear (melting, erosion, ablation) of internal parts of the arcing chamber in direct contact with the arc. Since circuit breakers are expected to interrupt a number of fault currents in their life, the wear of the essential breaker parts needs to be acceptable.
- Shall operate immediately when called upon. The duration of faults must be minimized, so as quick as possible after a trip command comes in from the protection relay, the breaker must act without appreciable delay.
- Shall fulfil extreme requirements regarding reliability. Since the circuit breaker is the only component to prevent major damage to the system, its reliability must be guaranteed, even after long idle time, under all relevant environmental conditions indoor, but also outdoor when designed for this.

In the last decade, a new requirement has been added:

- Low impact on environment. The past 50 years, the extreme greenhouse gas sulphur-hexafluoride (SF_6) has been the workhorse taking care of both high-voltage insulation as well as fault-current interruption, combining seamlessly these tasks in an almost ideal way. However, since the global warming potential of SF_6 is 25,200 and has an atmospheric lifetime of 3,200 years, it is considered by many as an unacceptable medium for the future.

This requirement is the leading theme of this book, more precisely how to accommodate the technical requirements with the environmental ones.

In this chapter, a summary is given of the electrical requirements that circuit breakers must fulfil, guided by the list above. Section 2.2 goes into the AC current interruption process. Section 2.3 lists a number of relevant standards. Depending on the location of the circuit breaker in the power system and the fault, a variety of fault current switching duties is defined, which are highlighted in Section 2.4.

Although circuit breakers have as their main tasks in the interruption of fault currents, in a number of cases, the interruption of (much smaller) load current needs attention from the user. These cases are switching of reactive loads: capacitive and inductive loads which are highlighted in Section 2.5.

In Section 2.7, a selection of old and new non-standard switching applications is highlighted.

Dielectric requirements, such as withstand capability of various overvoltage transients, are covered in Section 2.8.

Since a few years, SF_6-free high-voltage circuit breakers come to the market. Their fundamental principles of the current interruption process are dealt with in Chapters 3 and 4. Differences that may have an impact on requirements are highlighted in Section 2.9 of this chapter, both for SF_6-free gas circuit breakers as well as for SF_6-free high-voltage vacuum circuit breakers.

Section 2.10 describes the interruption process of fault currents in HVDC systems, together with its impact on the requirements for HVDC circuit breakers.

In order to create a common understanding for users as well as a level playing field for the manufacturing industry, a set of standards has been developed over the last century, e.g., compiled by IEC (International Electrotechnical Commission), IEEE (Institute of Electrical and Electronics Engineers) and various national standardization committees.

Literature on circuit breakers is extensive. The most important books published in the past 50 years are Refs. [1–10].

2.2 The AC interruption process

2.2.1 Circuit breaker operation

Two types of AC circuit breakers are generally distinguished:

- Gas circuit breakers. These devices use a gas plasma for the interruption of the current and cold gas for insulation, normally the same gas or gas mixture.
- Vacuum circuit breakers. These devices use the metal vapour plasma from their contacts as a medium to interrupt current and the high vacuum as insulation between the contacts. For insulation outside the sealed vacuum interruption chamber, a gas or a solid is used.

For current interruption, the arcing chamber, commonly called '*interrupter*', is the key component of circuit breakers. In its most simple concept, the interrupter of a high-voltage gas circuit breaker can be sketched as in Figure 2.1.

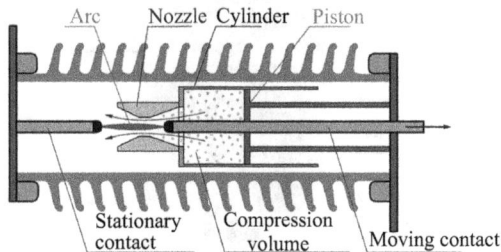

Figure 2.1   Principle lay-out of the interrupter of a gas circuit breaker

Gas circuit breakers always have two sets of contacts, each with a stationary (or fixed) and a moving contact:

- The *main contacts* for the conduction of the continuous current and voltage withstand in the open state.
- The *arcing contacts* able to withstand the electric arc.

In Figure 2.1, for simplicity, only the arcing contacts are shown. Upon activation of the breaker, first the main contacts separate without arcing, because the arcing contacts remain closed, and current is transferred to these contacts. The arc starts when the arcing contacts separate. In the meantime, mechanically, the gas inside the *compression volume* is compressed by a piston in a cylinder. The compressed gas is released, initiating an axial flow along the arc that removes the arc energy. The (insulating) *nozzle* guides the flow of gas for an optimum energy transfer, while under the influence of the arc, it releases gases (this process is called *ablation*) that augment the energy drain from the arc. The way how the compression is realized has led to many different arcing chamber designs. When the compression is purely mechanically from outside (as in Figure 2.1), this is called the *puffer principle*. In case there is additional (or mainly) compression by the arc heat, this concept is called *self-blast principle* (or *self-compression, auto-expansion, arc-assisted, thermal-assisted, auto-puffer, etc.*). Details can be found in Chapter 4. After successful cooling, the current is interrupted when the remaining arc plasma disappears quickly enough to withstand the rising voltage, see Section 2.2.4. After this recovery process voltage withstand is basically guaranteed by the main contacts.

Vacuum circuit breakers have a much simpler interrupter (Figure 2.2) having only a single set of contacts that must handle the continuous current conduction in the closed state, arcing in opening state and insulation in the open state.

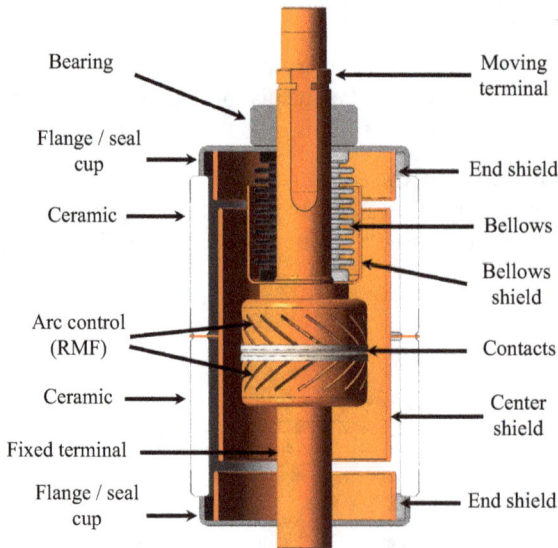

Figure 2.2   Principal lay-out of the interrupter of a vacuum circuit breaker (courtesy of Siemens AG).

The vacuum interrupter is a sealed ceramic bottle under high vacuum. The mobility of the moving contact is through a metal bellow. Internally a vapour shield collects the metal vapour released from the contacts by the arc in order to prevent the formation of a conducting layer on the inside of the ceramic enclosure. Behind the contact surfaces, *arc control devices* force current to flow in a certain geometry in order to create magnetic fields that control the high-current arc. Arc control is distinguished by the direction of the magnetic field it generates, axial or radial, see for details Chapter 3. Vacuum circuit breakers do not need any external means to remove energy from the arc, because the energy content of the arc in a metal vapour plasma is much smaller than that of an arc in a gaseous medium. After reaching current zero, the residual plasma expands immediately towards the vacuum background.

2.2.2   Impact of current

Short-circuit is the situation of an unintended low-impedance path in the system, that causes large current to flow. This situation is outlined in Figure 2.3, showing a three-phase short-circuit current starting at $t = 0$.

After the *relay time* (typically in the order of 10 ms), the system protection detects the abnormal current and sends a trip command to the circuit breakers, usually at both ends of the faulted overhead line. After receipt of the trip command, it takes the *opening time* to separate the contacts. This time is normally several tens of milliseconds, depending on the type of actuator, or drive, that makes the contacts separate.

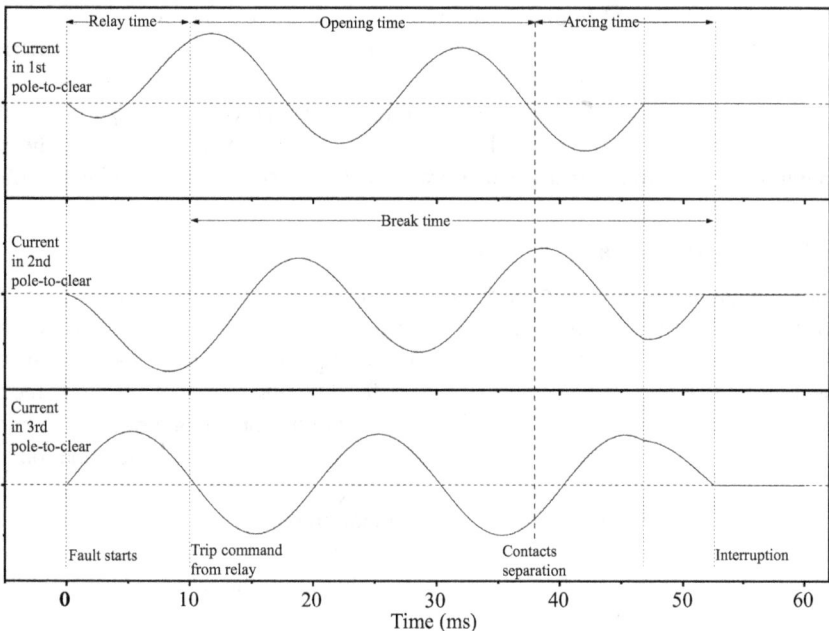

Figure 2.3   Onset of a short-circuit current and interruption by the circuit breaker

In Figure 2.3, at the time of contact separation ($t = 38$ ms), all three contacts of the circuit breaker open, and arcs start more or less simultaneously (within one-sixth of the power frequency period) in the three breaker poles. Depending on the type of circuit breaker and the circuit, it takes a certain *minimum arcing time* before a zero crossing of current actually leads to the extinction of the arc and interruption of the current. In Figure 2.3, the current in the upper phase is interrupted first and this breaker pole is therefore called *first pole-to-clear*. After the arcing time, the current in all three phases is interrupted. The time from trip command to the final interruption is called *break time*.

An important feature of the short-circuit current is its *asymmetry*. In Figure 2.3, it can be seen that in the upper phase, the power frequency AC current is shifted upwards. This is because an exponentially decaying DC component is super-imposed on the AC component. The characteristics of the DC component are:

- The initial value of the DC components at the moment the fault starts. This value depends on the moment in the power system voltage sinewave where the fault starts. Worst case is the start of short-circuit current at a zero crossing of the system voltage leading to the maximum asymmetry on one phase.
- DC time constant (τ) of the exponential decay. This parameter depends on the system configuration and the location of the circuit breaker. Its value varies widely from 10 to 300 ms [11].

Asymmetry of short-circuit current leads to higher short-circuit current peak value and longer arcing time in the circuit breaker and is therefore a major stress factor to be considered in design and its verification, see Section 2.4.1.1.

During the short-circuit current passage, the current, including its asymmetrical peak, passes through the closed contacts until contact separation. At contact separation, the arc starts due to the thermal melting of the final contact bridge [12] and it appears immediately, automatically, and inevitably. From that moment on, there is arcing with a duration from a few to a few tens of milliseconds in the arcing chamber.

2.2.3   *Impact of the switching arc*

In gas circuit breakers, during the arcing period, the pressure in the arcing chamber is not only increased by the heat of the arc but also by an active compression of the gas. For the breaker chamber, this transient interruption pressure means additional mechanical stresses. High pressure is needed for directing a flow of gas along the arc to remove its energy, creating favourable conditions for the interruption. Energy exchange mechanisms are through thermal conduction, convection, and radiation, see Section 4.1.

In vacuum circuit breakers, the arcing medium is the ionised metal vapour that is released from the contacts by the thermal energy of the arc [13,14]. The arc control device distributes the arc energy evenly across the contact area [10]. This limits the contact erosion, see Section 3.4.1.

Close to current zero, current becomes smaller, depending on its rate of change ($\text{d}i/\text{d}t$). As a result, the arc will become thinner and thinner and cools down. Upon reaching current zero, the arc plasma has not decayed completely, and thermal energy will remain in the switching gap for a short time, depending on fault current, extinction medium and circuit impressed voltage across the gap. The initial density, temperature, and temporal decay of the residual plasma are major parameters for the interruption.

In gas circuit breakers, due to the high background pressure, residual plasma decay is assisted by active flow of gas along the former arc channel. Therefore, a well-designed system of fluid dynamics shall ensure the quick removal of the residual plasma.

In vacuum circuit breakers, the residual plasma will diffuse towards the vacuum background in a matter of microseconds, ensuring a very fast recovery to the pre-arcing state.

In all switching devices, the arc is an inevitable means for the current interruption: on the one hand, its high pressure, density, and temperature greatly affect the internal parts of the circuit breaker chamber. On the other hand, its high electrical conductivity guides the current to zero, thereby releasing the magnetic energy from the system in a natural way.

Although current zero is the only opportunity for a switching device to interrupt a current, this does not imply that every current interruption is finally successful. The arc present between the contacts may have disappeared, but the hot remnants, e.g., ionised gases in gas circuit breakers and metal vapour in vacuum breakers, will reduce the dielectric breakdown strength, thus influencing the ability of the circuit breaker to withstand the *Transient Recovery Voltage* (TRV).

2.2.4   Transient recovery voltage

TRV is the voltage that appears across the gap immediately following current interruption, as a reaction of the network to the interruption of current. It is important to realize that this voltage is imposed by the external circuit in AC networks, in contrast to DC breakers, where the breaker itself has a strong effect on the recovery process, see Chapter 5. Only in a very short time interval around current zero, there is an interaction between the (residual) arc plasma and the circuit. *This arc-circuit interaction* only plays a role in the very short thermal interruption window of several microseconds [6].

Recovery of the switching gap is a race between the decay of the residual plasma on the one hand and the rising voltage on the other hand. If voltage rises too fast, and/or the plasma decays too slowly, *re-ignition* occurs followed by another loop of power–frequency current. Eventually, after several unsuccessful attempts, the device may not be capable to interrupt the current and will be damaged itself leading to a short-circuit.

Actually, the TRV arises as a result of the re-arrangement of electric energy at both sides of the open breaker after the magnetic energy has disappeared. Therefore, its origin and temporal development can be simply described in electrical terms with the help of the basic electric circuit diagram of Figure 2.4.

In this circuit, the current flowing through the circuit breaker is in general resistive-inductive (the impedances of the parallel RC paths are very high

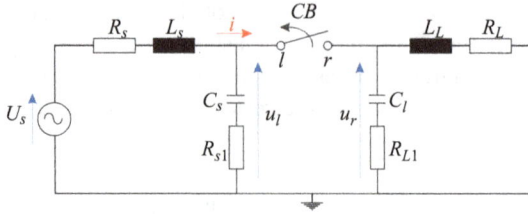

Figure 2.4   Basic electric circuit diagram for understanding TRV

compared with the stray series inductances since the capacitances are stray ones). In the case of short-circuit currents, the reactance is much larger than the series resistance, i.e., the short-circuit current is inductive. Therefore, current zero occurs practically simultaneous with the maximum of the AC power–frequency voltage. This has the important implication that the TRV is superimposed on the power–frequency voltage peak. The power–frequency voltage after decay of transients is called *recovery voltage*.

Two separate *RLC* circuit can be distinguished in Figure 2.4:

- One as a simplified equivalence of the source side, having inductance, capacitance and resistance L_S, C_S, R_S and R_{S1}, respectively and an ideal AC source U_S with angular power frequency ω. The reactance $X_S = \omega L_S$ is usually called the *short-circuit impedance*, since its reactance limits the short-circuit current to a maximum value of U_S/X_S through the circuit breaker (CB) in case of a short circuit on the right-hand side of the breaker. The source side circuit can represent a supply overhead line, cable, transformer, substation, etc.
- One as a simplified equivalence of the load side, having equivalent inductance, capacitance and resistance L_L, C_L, R_L and R_{L1}, respectively. The load side can represent a short-circuit immediately at the breaker terminal, a short-circuit on a line, a load or even a complete power system.

Once the current has been interrupted, the source- and load-side load will oscillate independently from each other. Due to damping, the oscillation will vanish after a short time, a fraction of a millisecond. The steady-state source-side voltage u_l will then be almost equal to the source voltage U_S, whereas the load-side voltage u_r will tend to zero. The voltage (u_{lr}), i.e., the TRV across the gap, is now the difference between the voltages at either side: $u_{lr} = u_l - u_r$.

In Figure 2.5, the TRV components at load- and source side are given for different load and source impedances ($\omega L_S = 0.5\,\omega L_L$). Both transients contribute to the TRV. As can be seen in this example, the initial rate-of-rise of TRV is determined by the TRV component of the highest frequency (here the source side), whereas its peak value is determined by the oscillation with the largest amplitude. Note that the steady-state voltage drop across load- and source-side impedances ($u_{0,S}$, $u_{0,L}$) determines the initial values of the oscillations and therefore their amplitudes.

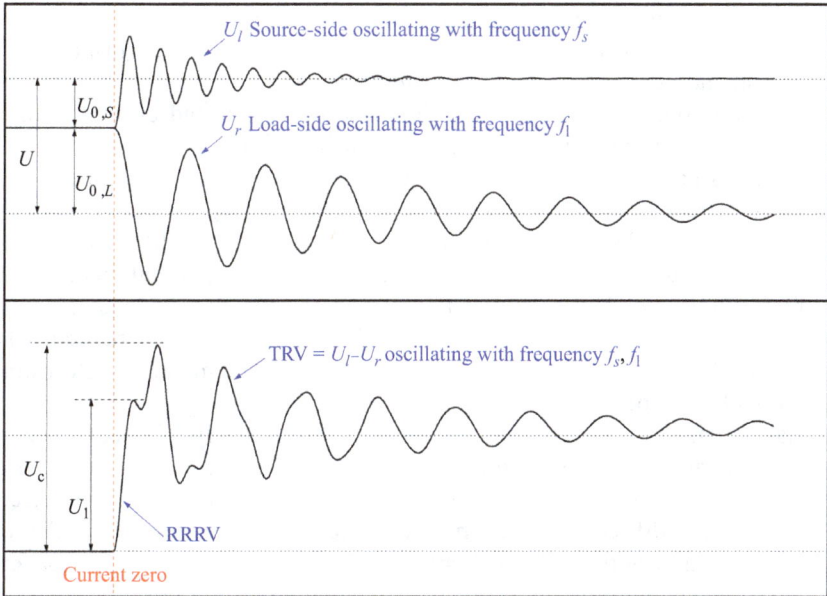

Figure 2.5   TRV (lower graph) as a superposition of source- and load-side voltages (upper graph)

Depending on the actual load, the characteristic of current interruption and TRV varies:

- In case there is zero load-side impedance ($R_L \to 0$ and $L_L \to 0$), there is maximum current through CB.
- In case there is an appreciable value of L_L, the current to be interrupted is mainly determined by the total impedance $\omega(L_S + L_L)$. Herein, the impedances of the parallel RC branches, and also the series resistances, are neglected. This, rather general case is the one depicted in Figure 2.5.
- In case $L_L \gg L_S$, the current is much smaller than the short-circuit current, and this reflects the interruption of an inductive load current, for more details, see Section 2.5.3.
- In case $L_L \to \infty$, the parallel impedance $1/(\omega C_L)$ is the current limiting impedance, and this reflects interruption of capacitive load, see Section 2.5.2.

The TRV calculation and evaluation method, using the lumped element method is described in detail in [15].

2.2.5   Critical interruption parameters

Circuit breakers are designed to interrupt fault current in a system. The critical parameter of the severity of interruption is the following:

- Magnitude of power frequency current to be interrupted; more specifically its derivative (di/dt) prior to current zero. This parameter is directly related to the residual plasma at current zero, and thus to the capability to recover to the insulating state fast enough. In vacuum circuit breakers, the magnitude of current and di/dt are separate parameters.

- Arc duration: the arc starts upon contact separation and current is interrupted at current zero, but not necessarily at the first appearing one. Breakers have a minimum arcing time, because the arc quenching system must be ready to interrupt, either by having built up sufficient gas blow pressure or by having reached a sufficient gap spacing. Minimum arcing time is not a fixed number, but depends on switching conditions like current, TRV, etc. When arcing time is below the minimum acing time, interruption at current zero will fail, see Figure 2.3. Breakers are required to interrupt after an arcing time that their contact system can handle from a thermal point of view.

- Rate-of-rise of recovery voltage (RRRV). Immediately after current zero, the system reacts to the energy re-arrangement due to current interruption by the TRV, while the residual plasma is cooling down (in gas circuit breakers) or diffuses (in vacuum circuit breakers). This period is called the 'thermal interruption window', since during this very short period (several microseconds), thermal processes dominate the interruption. This period is often described as a race between cooling down of the residual plasma (decline of its conductivity) and the rise of transient recovery across that plasma. Failure to interrupt during the thermal interruption window is called '*thermal re-ignition*'.

- Peak value of TRV. After the thermal interruption window, TRV rises further to its peak value, which is well above the peak value of the steady-state recovery AC voltage. As a result, the gap between the contacts, though no longer thermally stressed, experiences considerable dielectric stress. Therefore, the period between current zero and peak value of TRV is called the *dielectric interruption window*. For transmission systems, the TRV is normally composed of a local component (originating from the station) and a distant one (from line reflections), leading to a waveshape that has an initially fast rise to a local maximum (u_1 at time t_1) and a further slower increase towards the absolute peak value (u_c at time t_2). Figure 2.6 shows the envelope of such a so-called *four-parameter TRV* as standardized by IEC [16] for a full short-circuit test requirement of a circuit breaker having rated voltage (U_r) of 245 kV. Failure to interrupt (breakdown) due to dielectric weakness shorter than a quarter of the power–frequency cycle after an interruption is called '*dielectric re-ignition*'.

A schematic overview of the interruption process is drawn in Figure 2.7, with an enlargement of the interruption period in Figure 2.8. Both figures assume a purely inductive circuit and a single-frequency TRV. As such, this is the situation for the T100s (symmetrical 100% short circuit) fault current interruption requirement.

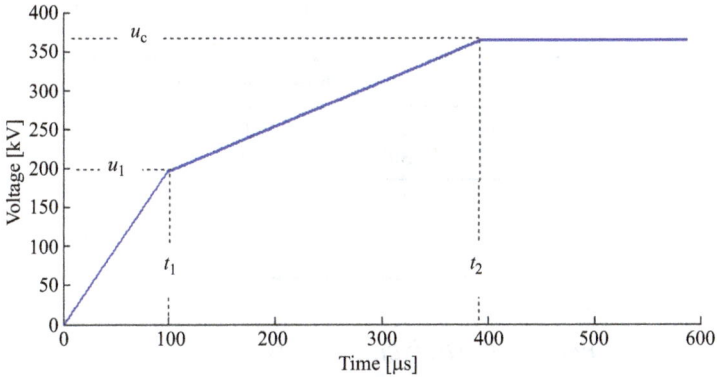

Figure 2.6   Four-parameter IEC envelope constructed from the parameters t_1, u_1, t_2, u_c for rated voltages 245 kV and 100% short-circuit current (test-duty T100)

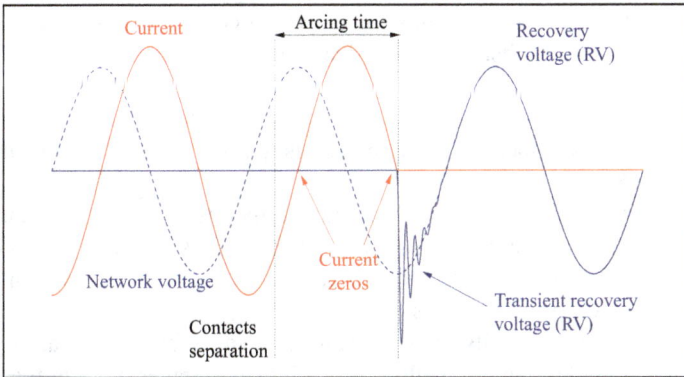

Figure 2.7   AC current interruption in an inductive circuit

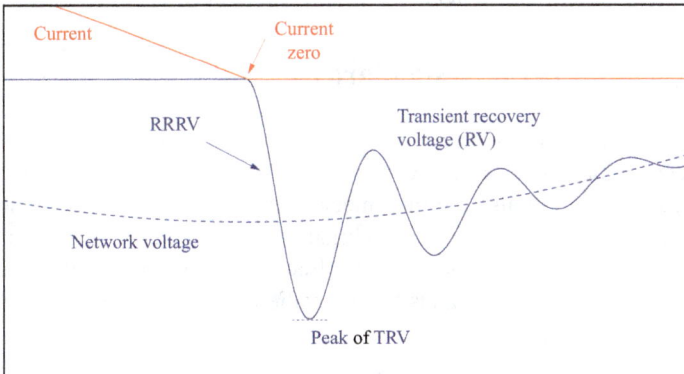

Figure 2.8   Current zero and TRV in an inductive AC circuit

Table 2.1   *Number of short-circuit faults per 100 km × year of overhead lines [17]*

Number of faults	Voltage class (kV)					
	<100	≥100 to 200	≥200 to 300	≥300 to 500	≥500 to 700	≥700
Average	11.5	5.1	3.1	2.0	2.0	1.7
90 percentile	17.3	8.3	4.8	3.3	4.2	not available

2.2.6   Occurrence of short-circuit events in service

Though quite dated, the most reliable source of information on the occurrence of short-circuit in service is the world-wide survey that CIGRE WG 13.08 conducted in the late 1990s [17]. Information is collected on the number of faults on transmission overhead lines (OH-lines). The survey covered 900,000 circuit-breaker-years* and 70,000 overhead-line-years in the voltage class of 63 kV and above.

Most short-circuits in a transmission network occur on the overhead-lines, i.e., more than 90% of all faults. Table 2.1 gives an overview by voltage level.

As a very general rule of thumb, based on the statistical distributions of number of faults per 100 km of an overhead line and the line lengths per voltage class, the average number of short-circuits is 1.7 per year (i.e., for all voltage classes).

The reported number of faults per transformer is 3–4 per 100 transformer-years. The number of faults per substation busbar(s) is 2–3 per 100 substation-years. The number of faults per cable circuit is very low in comparison with the number of faults on overhead lines [18].

Around 70% of the faults are single-phase-to-earth faults, 20% are two-phase faults, and 10% are three-phase faults. For the higher voltages, 90% of the cases are single-phase-to-earth faults. Approximately 80% of the faults disappear within one O–C (Open–Close) operation. Another 5% are cleared within an O–CO–C operation, and about 15% of all faults are permanent and need action on site.

2.3   AC circuit breaker standards

Circuit breakers, as all electrical equipment, have to fulfil an extensive list of requirements, which cover a very wide range of stresses to be expected in service situations. Electrical but also thermal, mechanical, and seismic stresses need to be dealt with in all kinds of electrical and climatical environments. The latter receives more and more attention because the more frequent occurrence of adverse weather conditions sets increasing demands to the resilience of power supply systems and thus of its components.

---

*Circuit-breaker years: the number of circuit breakers considered times the number of years in service.

Requirements can be most easily derived from the mandatory type test that equipment must pass for certification. For circuit breakers, these type tests are the following:

- Dielectric tests. These are for the demonstration that the device can withstand a number of prescribed overvoltage waveshapes (such as lightning, switching and AC) in a number of climatic conditions (wet, polluted environment, etc.).
- Measurement of resistance of the main circuit. This test is to verify the change of contact resistance by arcing and/or repetitive switching.
- Temperature-rise tests. This test is to verify that prescribed temperature limits are not exceeded by continuous passage of the rated continuous current.
- Short-time withstand current and peak withstand current tests. The short-time withstand requirement reflects the capability of the breaker to be able to withstand a specified duration of short-circuit current, whereas the peak withstand current test verifies the withstand of the maximum asymmetrical short-circuit current, both in the closed position.
- Additional tests on auxiliary and control systems. This is a set of requirements for the secondary, auxiliary and control systems of the circuit breaker.
- Mechanical operation tests at ambient temperature. Since the actuator of circuit breaker also have a key function, strict requirements are set to the correct mechanical functioning.
- Short-circuit current making and breaking tests. These are the main electrical requirements of circuit breakers. These are the main topic of this chapter.

Apart from the above set of mandatory tests, an extensive number of requirements are translated into tests depending on the specific application and (voltage) rating of circuit breakers. Those directly related to switching are: Capacitive current switching tests, inductive load switching, short-line fault- and critical current interruption tests, out-of-phase (OOP) making and breaking tests, single-phase and double-earth fault tests. Other ones related to endurance against internal and external influences are: Extended mechanical and electrical endurance tests, tightness, humidity and water ingress tests, low- and high-temperature tests, tests to prove operation under severe ice condition and electromagnetic compatibility. Related to the impact of environment are requirements regarding X-ray emission and radio interference voltage.

In general, a standard is a set of rules that establish uniform behaviour, criteria, methods, processes and/or practices with the aim to ensure the quality of specific equipment. For almost every requirement, international standards have been developed.

The following functions can be attributed to standards related to switchgear:

- Definition of the target for designers and developers of switchgear. Defining clear performance criteria for all relevant functions, standards greatly enhance product reliability and effectiveness, and enable easy exchange of apparatus in various locations in the system.

- Reference to network operators. Standards are designed to cover approximately 90% of the possible conditions occurring in service. System operators should compare the conditions, including the transients that can arise in their particular system with the standardized parameters in order to decide whether the equipment can handle the fault situation, or whether mitigating measures are needed. In some cases, even custom-made equipment needs to be ordered to meet specific requirements. Another risk mitigating measure is testing the equipment under non-standardized, specific duties, set-up in a test laboratory.
- Guidelines for testing institutions. Test circuits must be developed that provide adequate stresses.

The set of standards that is by far most used in the world is the series issued by the *International Electrotechnical Commission* (IEC) [19]. General, common, characteristics of switchgear, not related to switching performance, are standardized by IEC 62271-1 [20]. Regarding the typical tasks of circuit breakers, such as fault-current making and breaking, as well as capacitive-current switching, the relevant standard is IEC 62271-100 [16], on AC circuit breakers, which was revised in 2021. Moreover, under the heading IEC 62271, a large number of standards are developed to cover almost every conceivable switching device. In IEC 62271-306 [21], an in-depth guidance is given on the technical background of the above-mentioned standards. At present (2023), work is in progress on the development of technical specifications of HVDC switchgear. HVDC switchgear also comes in a large variety [22].

Other major standardizing bodies harmonize with IEC; currently, harmonization of IEC and the United States standards from IEEE/ANSI (Institute of Electrical and Electronics Engineers/American National Standards Institute) is well on its way.

The most important US standards are the following:

- IEEE C37.04-2018: 'IEEE Standard for Ratings and Requirements for AC High-Voltage Circuit Breakers with Rated Maximum Voltage Above 1000 V'.
- IEEE C37.06-2009: 'IEEE Standard for AC High-Voltage Circuit Breakers Rated on a Symmetrical Current Basis – Preferred Ratings and Related Required Capabilities for Voltages Above 1000 V'.
- IEEE C37.09-2018: 'IEEE Standard Test Procedures for AC High-Voltage Circuit Breakers with Rated Maximum Voltage Above 1000 V'.

Many national standards all over the world are based on or derived from the IEC standards. More and more influential are Chinese standards, since China started to play a leading role in ultra-large scale power transmission projects both in AC [23,24] and DC [25].

Standards have been developed extensively for AC switchgear. Sections 2.4–2.9 summarize the requirements for AC circuit breakers based on these standards.

Standards and technical specifications for DC switchgear having rated voltage of 100 kV and above, at the time of writing, are under development with IEC Technical Committee 17A.

Type-testing is the usual way of verifying the fulfilment of the standardized circuit breaker requirements. Passing such tests is not straightforward. A six-year survey of 1241 standardized IEC short-circuit test-series of high-voltage circuit breakers 72.5–800 kV showed that 32% does not pass the requirements for certification [26].

2.4 AC fault-current interruption

2.4.1 Switching duties of circuit breakers

Circuit breakers need to be versatile, and therefore need to be capable to interrupt a variety of fault currents, each having a different combination of the critical parameters listed in Section 2.2.5. In AC systems (in contrast to DC systems), the critical parameters are completely determined by the location of the circuit breaker in the electrical system and the type of fault it needs to handle.

As an illustration in Figure 2.9 (after [9]), shows a systematic drawing of a substation with several faults (in Section 2.4) and load switching (in Section 2.5), discussed in the following.

Apart from fault current interruption, circuit breakers may have to switch load current. In Section 2.5, it is argued that especially capacitive (unloaded cables, lines, and capacitor banks) and reactive (shunt reactors) load switching deserve attention. In that case, it is the release of energy that remains trapped in the load that may lead to issues for the system: capacitive loads by the sudden discharge of the trapped voltage at a 'restrike', inductive loads by a release of magnetic energy at 'current chopping'.

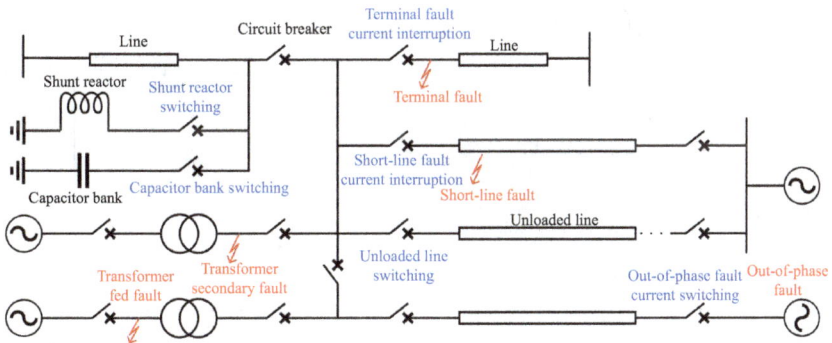

Figure 2.9   Schematic drawing of a station within coming supply and outgoing power lines, including several fault locations

2.4.1.1   Terminal fault-current interruption requirement

Terminal faults are faults that occur directly at the terminals of a circuit breaker. This implies that the impedance at the faulted side of the breaker is very low. Therefore, the terminal fault current is determined by the impedance of the source side of the fault current. This leads to the maximum possible fault current (the rated short-circuit breaking current I_{SC}). In specifying a circuit breaker, this current value shall be leading in the selection of the short-circuit current rating, normally 31.5 kA, 40 kA, 50 kA, 63 kA or sometimes even 80 kA.

It should be kept in mind that in AC systems, the inductance of the shorted system limits the fault current. Therefore, all fault currents are inductive so current will lag the voltage by 90°, leading to a maximum power frequency AC voltage at current zero.

In defining the requirements of circuit breakers, the asymmetry of short-circuit current shall be considered. In most situations, upon occurrence of a fault, the power frequency current AC component is tilted by a superposed exponentially decaying DC component, that decays with the DC time constant (τ) of the power system [27]. In the IEC [16] standard, the standard value is $\tau = 45$ ms. In worst-case, the asymmetrical current peak is higher by a factor 2.5 times the rated short-circuit breaking r.m.s. current in 50 Hz systems. The (temporary) tilting of the AC component has its impact on arc duration (increases with respect to the symmetrical case) and TRV (peak value is reduced with respect to the symmetrical case since current zero no longer coincides with the maximum of power frequency voltage). This is explained in Figure 2.10.

Since the impact of current asymmetry has an impact on critical parameters di/dt, arcing time and TRV, it is required to test circuit breakers under full

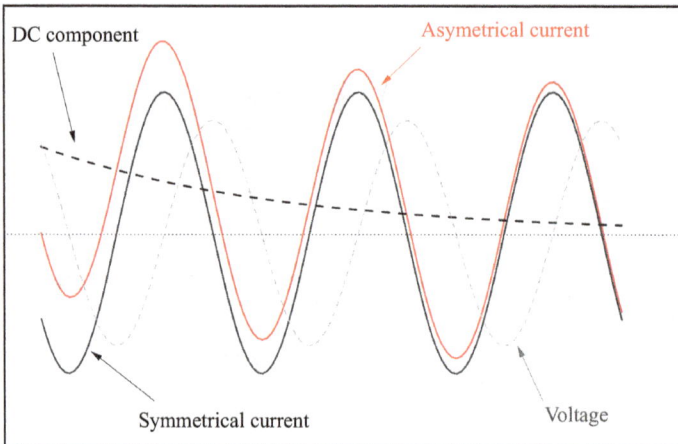

Figure 2.10   The DC component of a fault current leading to an asymmetrical peak, increased arc duration and shifting of current zero with respect to AC voltage peak.

asymmetrical and symmetrical conditions, both with 100% of the rated short-circuit breaking current. The associated test-duties are named T100a (asymmetrical) and T100s (symmetrical), respectively.

The TRV parameters associated with the interruption of very large current are moderately severe, compared with those from lower current. The reason is that when very large short-circuit currents arise, a large part of the system contributes to the fault current. Thus, also in the generation of TRV, a considerable stray capacitance of multiple cables/lines are involved. The large capacitance leads to a reduction of TRV oscillation frequency and considerable damping. Therefore, the T100 requirements need special attention because of the large current and somewhat less because of TRV parameters.

Circuit breakers must interrupt fault current up to the rated short-circuit breaking current, including any current below this level. For several reasons, interruption of current below the maximum may not be easier than at maximum current:

- Modern gas circuit breakers generate the necessary pressure to cool the arc and remove the residual plasma with the assistance of the arc itself. During the arcing phase, the thermal energy of the arc is used to increase the internal gas pressure, which is used later to extinguish the arc, the self-blast principle. Although this principle greatly reduces the need to generate pressure by mechanical means from outside (puffer action), it is required that also at lower arc current/energy, still sufficient pressure is available.
- Since at lower fault current, generally a smaller part of the station/system is involved in the supply of the fault current, the associated stray capacitance and damping is smaller, resulting in faster rising TRV having higher RRRV and peak values.

Therefore, circuit breakers are required to demonstrate an interruption capability of 10–60% of the maximum current in the following three test-duties:

- T10: Covering faults, where current is supplied by a single transformer with relatively high short-circuit impedance. The fault current is standardized to 10% of the rated short-circuit breaking current and is applied only symmetrical. The specified TRV is considered a single-frequency TRV created by the oscillation circuit of the transformer stray capacitance and inductance. This two-parameter TRV is characterized by a high value of RRRV.
- T30: Equivalent to T10, but with 30% of the rated short-circuit breaking current.
- T60: Covering faults, where fault current is supplied by transformer(s) and overhead line(s). The TRV has basically two components, a fast one originating from the substation side, and a slower one originating from the line side (see Figure 2.6). The fault current is standardized to 60% of the rated short-circuit breaking current and is applied only symmetrical.

A very important verification of a circuit breaker function is to ensure its capability to interrupt the complete arcing window. The arcing window is the range of possible arcing times from minimum arcing time (determined by the circuit breaker) and the maximum arcing time (determined by circuit and circuit breaker). During each test-duty, it is necessary to demonstrate this basic arcing window. The breaking capability at an arcing time just above minimum, at maximum and at the intermediate arcing time must be verified. The tests must be performed at the rated operating sequence.

A voltage condition check (VCC) is required after certain making and breaking short-circuit test-duties, e.g., after T100s, L90 (see Section 2.4.1.6) in order to verify the basic dielectric withstand capability of the open contacts even after intense arcing and associated breaker degradation. The test-voltage depends on the voltage rating of the circuit breaker but is lower than the basic AC, switching and lightning voltages [16].

2.4.1.2   Three-phase interruption requirements

Since the current interruption in AC system involves three phases, further requirements must be considered. Arcing times and TRV parameters depend on the earthing situation of the power system. In the standard requirements, two earthing situations are distinguished:

- Non-effectively earthed (unearthed) system. In this case, after the interruption of current in the first phase, the floating neutral will assume an AC voltage of 0.5 p.u., which adds up to the 1 p.u. AC recovery voltage. Thus, the resulting TRV across the first-pole-to-clear of the circuit breaker will oscillate around 1.5 p.u. AC voltage peak value, resulting in a multiplication factor (the '*first-pole-to-clear factor*') of $k_{pp} = 1.5$, with respect to the single phase case. The remaining two phases will clear at the same moment with a reduced current of $\frac{1}{2}I_{sc}\sqrt{3}$ and TRV (for each of the last clearing poles) will oscillate around a value of $\frac{1}{2}\sqrt{3}$ p.u. of AC voltage. In practice, this situation is prevalent in distribution systems.
- Effectively earthed system. In this case, the system's neutral is earthed through an impedance causing the system to have zero-sequence impedance. This is standard practice for transmission systems. The first-pole-to-clear-factor defined to cover this situation is $k_{pp} = 1.3$, implying a less severe TRV for the first-pole-to-clear than in the non-effectively earthed situation. However, arcing times and TRV peak values for the remaining poles are increased.

In Figure 2.11, the making and breaking, including its transients, are shown of a circuit breaker in an unearthed test set-up.

2.4.1.3   Transformer limited fault current interruption requirements

The requirements in test-duties T10–T100 apply for all circuit breakers. In addition, there is a special transformer-limited fault (TLF) duty specified in the IEC standard. This covers faults located downstream from the circuit breaker in radial networks with the current supplied by a transformer. The TLF duty (IEC 62271-100, Annex F [16]) is defined for systems of rated voltages below 100 kV and

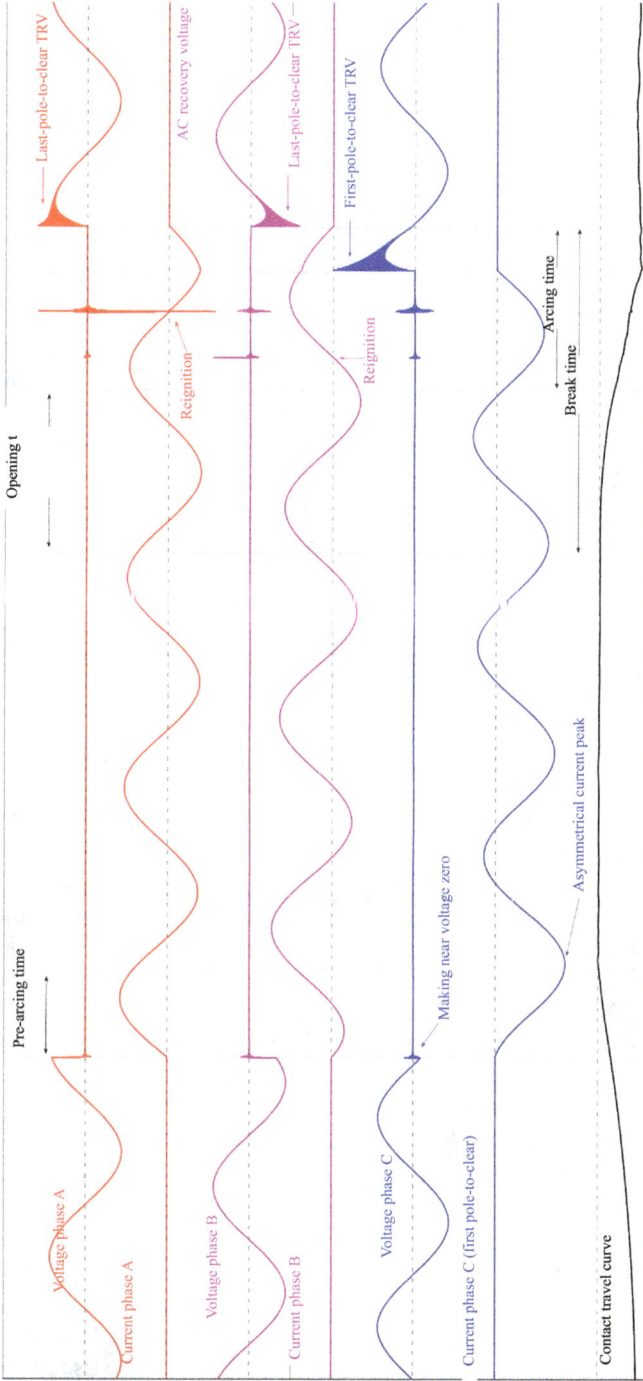

Figure 2.11 Complete make-break cycle from a short-circuit test of a T100a test-duty in an unearthed system set-up [7]

above 800 kV. The fault current is standardized as 30% of the rated short-circuit breaking current for voltages below 100 kV. Two topologies are distinguished (see Figure 2.9):

• Transformer-secondary fault (TSF): In this case, the circuit breaker is at the primary (HV or EHV) side of the transformer and the fault is at the secondary (MV or HV) side.
• Transformer-fed fault (TFF): Here, the circuit breaker is at the secondary (MV or HV) side of the transformer and the fault is located behind the circuit breaker.

The inherent RRRV is very high, several times that of the T10 duty. This is because the transformer, as an isolated network element, creates a single-frequency oscillation (at the load side in the transformer-secondary fault and at the source side in the transformer-fed fault) with a high-frequency due to its small stray capacitance. When a long cable connects the transformer to the circuit breaker (longer than a few tens of meters), the cable capacitance reduces the transformer natural frequency; as a result, the RRRV is considerably reduced.

2.4.1.4   Single-phase and double earth faults

The arcing time in this condition is longer than in three-phase conditions, and the combination of current and recovery voltages is not covered by the terminal-fault requirements. The following test duties are defined:

• Single-phase (Earth) fault (SEF): For applications in effectively earthed systems where circuit breakers must clear single phase faults. The tests imply a breaking current of I_{sc} and a recovery voltage of $U_r/\sqrt{3}$.
• Double earth fault (DEF): Covering the cases of earth faults in non-effectively earthed neutral systems. The faults are on two different phases, one of which occurs on the one side of the circuit breaker and the other on the opposite side. This situation of two simultaneous faults is more likely to occur than it might seem at first sight. A phase-to-earth fault in a non-effectively earthed neutral system leads to a voltage rise in the other phases, enhancing the probability of the second fault. Current to be interrupted is $\frac{1}{2}\sqrt{3}\,I_{sc}$ and the recovery voltage equal is to the rated voltage U_r.

2.4.1.5   Short-circuit making requirements

The performance of circuit breakers is often focused on the interruption, but also the energization of a faulted circuit by a closing, or making operation needs attention. During the closing operation with the contacts approaching each other, the withstand voltage of the gap decreases with time. At a certain distance, the breakdown voltage of the gap becomes equal to the voltage imposed by the circuit across the contacts. At that moment, the gap will break down, the so-called pre-strike. During the closing operation, an arc will be established lasting until the contacts touch mechanically. Due to this pre-arcing, the pressure rise in the interruption chamber of gas circuit breakers challenges the so-called close-and-latch capability. In such cases, contacts may slow down, especially when a single mechanism is used to drive all three poles. In vacuum circuit breakers, the combination of high temperature and high contact forces involved in closing may lead to welding of the ultra-clean contacts.

The two situations having contrasting requirements are:

- Closing at a low-voltage across the contacts (close to voltage zero): in this case, the duration of the pre-arc will be very short, but the resulting current will have a high asymmetry with an appreciable peak value.
- Closing at a high-voltage across the contacts (close to voltage peak): in this case, the pre-arc will last long and can input a large energy between the closing contacts, but the resulting current will be (close to) symmetrical.

Both situations are schematically outlined in Figure 2.12.

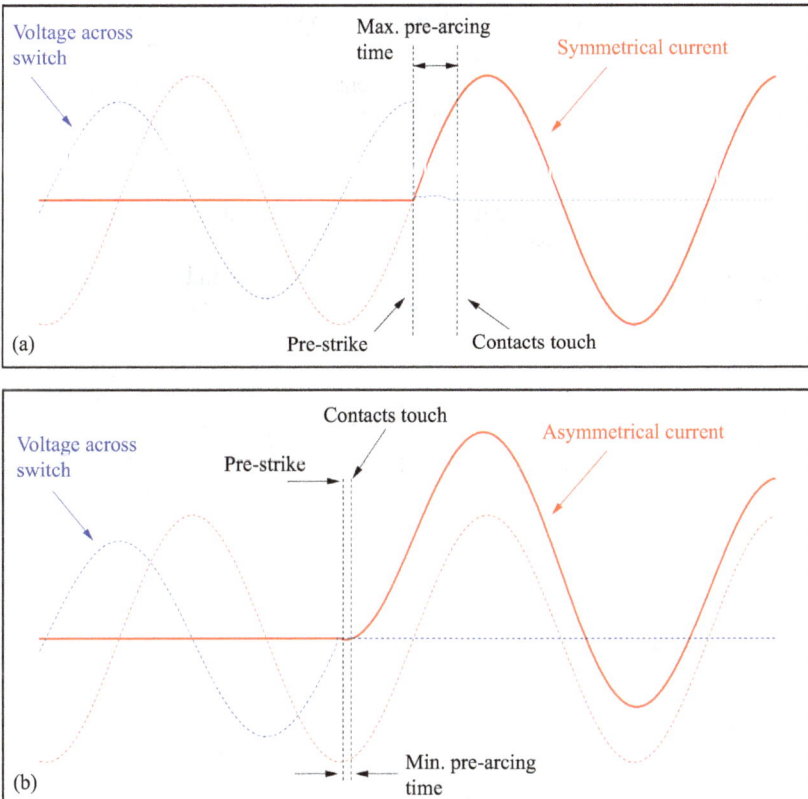

Figure 2.12   Pre-arc duration at making operations: (a) making operation with pre-strike near voltage maximum resulting in symmetrical current and maximum pre-arcing time; (b) making operation with pre-strike near voltage zero resulting in asymmetrical current and minimum pre-arcing time.

2.4.1.6   Short-line fault-current interruption

A short-line fault is a fault on an overhead line at a short distance (a few kilometres) from the substation. In the occurrence of such an event, travelling waves bounce back and forth between the open breaker and the fault location. In this case, the simplified *RLC* lumped-element approach from Section 2.2.4 no longer holds, and the faulted line section must be represented as a distributed element. This results in a triangular voltage pattern across the open contacts [6]. The 'frequency' of this pattern is inversely proportional to the (faulted) line length, the peak value is proportional to the line length. Thus, looking at the critical TRV parameters, with very short lines, the RRRV is very high, whereas the first peak remains limited. With longer lines, the RRRV is limited but the initial TRV peak is large. In this case, the physical parameters of the interruption medium determine where its sensitivity lies, high RRRV or high initial peak.

For SF_6, it was found that the most severe requirement comes with a line length that causes 90% of the rated short-circuit breaking current to flow, so only 10% current reduction is due to the additional line impedance. This is the reason why this duty is called 'short-line fault' current interruption. Its main critical feature is the very high initial RRRV whereas the first local peak value remains limited.

In this test-duty, fault current is standardized to 90%, 75% and 60% of the rated short-circuit breaking current, and the IEC-coded names for these test-duties are L90, L75 and L60, respectively.

The TRV is standardized separately for source and load (line) sides. Figure 2.13 shows the complete TRV (source- and line sides) of a high-voltage SF_6

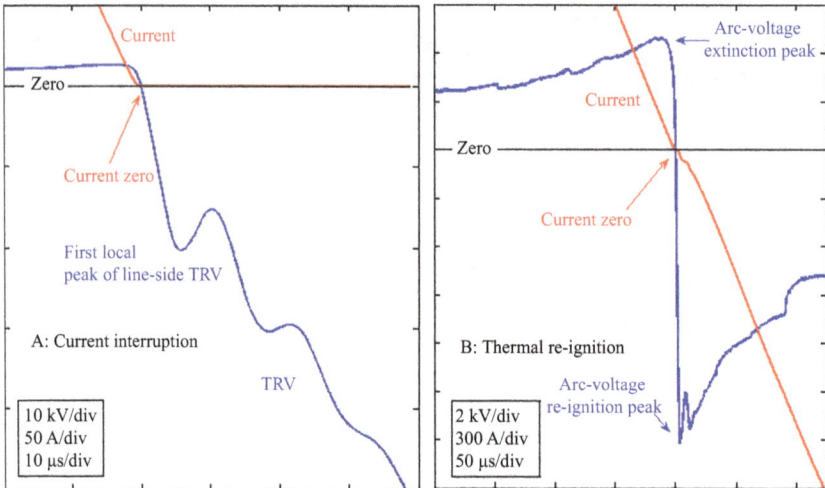

Figure 2.13   *Measurement of a short-line fault current interruption and TRV (left) as well as thermal re-ignition of a high-voltage SF₆ circuit breaker (right)*

circuit breaker in a test with successful current interruption (left) and thermal re-ignition (right). Note the arc voltage peak prior to current zero (*arc-voltage extinction peak*).

The source circuit is specified in the same way as for the terminal fault test-duties, i.e., with a relatively low RRRV. This is adequate because the SLF duties are intended to verify the thermal breaking capacity, which is relevant during the first few microseconds after current zero. Failure to pass this test is normally by a thermal re-ignition immediately following current zero.

From test failure statistics, the short-line fault-current interruption turns out to be the most challenging requirement for SF_6 circuit breakers. From 185 L90 test-series, 57% did not pass the standardized tests, for L75, the test-failure rate reported is 22% [28].

The short-line fault-current interruption requirement is relevant for circuit breaker intended to be installed in overhead line systems. For cable system breakers, there is no such requirement.

Because of the very fast recovery of vacuum circuit breakers, compared to gas circuit breakers, a short-line fault requirement has less relevance for vacuum circuit breakers.

2.4.1.7   OOP fault

OOP switching conditions occur at a coupling of two system parts having the same voltages but where the sources have different phase angles, partly or entirely 180° OOP. A difference in the phase-angle causes OOP currents across the connection, which must be interrupted by a circuit breaker at both ends of the connection.

In the following situations, the system or system parts must be separated:

1.   In the connection of a single generator/transformer unit, when the generator becomes instable, for example due to clearing of a nearby fault taking longer than a critical clearing time.
2.   In case of system stability problems due to reactive power unbalance, over-loading, load rejection or other major disturbances, that necessitates decoupling of system parts.

Note that, unlike the faults above, where usually an external arc is the consequence of the fault, this is not the case here. The OOP current is much smaller than the rated short-circuit breaking current.

In this situation, the load circuit, in Section 2.2.4, a passive circuit is an active one, regarding topology a mirror of the source circuit from Figure 2.4. Regarding the TRV, the specialty of this switching duty is the presence of active sources at both sides of the circuit breaker. Considering the fault-switching duties discussed before, in all cases, the load-side TRV component decays to zero. In the OOP situation, however, the TRV component at both sides of the breaker will decay to the power–frequency recovery voltage of their respective sources.

As a result, the OOP switching duty is characterized by a very high TRV peak with a moderate RRRV and a moderate current. The rated OOP breaking current I_d is standardized to 25% of the rated short-circuit breaking current. Test-duty OP1, with test current $0.3I_d$ and OP2, with test-current I_d, are defined.

Because the TRV of the OOP test duty shows the highest peak value of all switching duties, it is often used as a reference for other special switching conditions, such as clearing long-line faults or faults on series-compensated lines [29].

The OOP requirement applies only to breakers intended to be installed in systems that may have to deal with this issue.

2.5   AC load-current interruption

Interruption of load-current implies the interruption of low current (up to the rated continuous current). Since the current is far below the short-circuit breaking current value, the current is no longer a critical parameter. The low current will be easily interrupted by the circuit breaker, even after a short arc duration. Since (most) breakers are opened at a random moment in the current waveshape, there is a chance to have contact separation close to current zero. In other words, the arcing time is short, and the breaker can clear with very small minimum arcing time. Because of the ease of interruption, in such a case, TRV will be developed already before the contact gap has reached its maximum length and is thus prone to a breakdown. Although the breaker is usually not at risk, the breakdown and its consequences may endanger nearby equipment. Therefore, requirements are stipulated for circuit breakers regarding the interruption of (even very) low values of current.

The hazard of breakdown depends strongly not only on the arc duration but also on the nature and magnitude of TRV. Since TRV is always an oscillation superimposed on the momentary power frequency AC voltage at current zero, the value of that voltage at current zero is of prime importance.

Load switching can be categorized in the following situations.

2.5.1   Interruption of normal-load current

Normal load has a high power factor. Therefore, current and voltage are almost in phase. At current zero, the momentary power–frequency AC voltage is low, thus TRV is also very limited and plays an insignificant role. The situation is outlined in Figure 2.14.

Normal-load current interruption is not a standardized requirement for circuit breakers. The T10 terminal fault duty (with inductive current so 90 degrees phase shift of current and voltage) covers this situation well enough.

2.5.2   Interruption of capacitive-load current

Capacitive load switching can usually be encountered in the following situations.

2.5.2.1   Switching of local station components

Various station components draw capacitive current when unloaded: current transformers typically have 1 nF of stray capacitance, capacitive voltage transformers (CVT) have a few nano-farad, whereas busbars typically have 10–15 pF/m. The capacitive current through these components is very small, usually below 1 A.

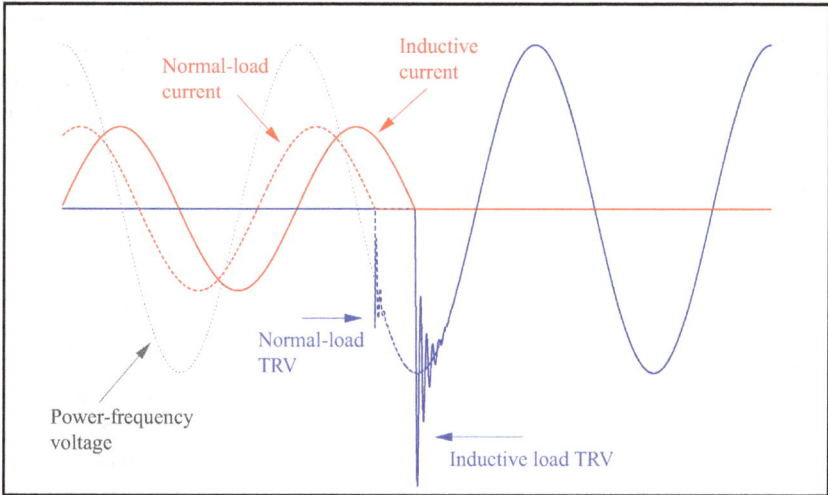

Figure 2.14   Impact of phase angle between current and voltage on TRV

The current of this value is small enough to be interrupted by disconnectors [30] Although normally designed to insulate sections of the system by opening without current, disconnectors can switch very small currents below 1 A. Air-break disconnectors can switch long busbars, sometimes even short sections of overhead line, whereas GIS disconnectors switch sections of GIS bus ducts [7,31]. GIS disconnector switching shall not lead to propagation of the switching arc to the enclosure and is often accompanied by very fast transient overvoltages [32], see also Section 2.8.1.2. Requirements for disconnectors are formulated in the relevant IEC standard [33], and the relevant requirement is called 'switching of bus-charging current'.

2.5.2.2   Switching of unloaded lines and cables

Almost all circuit breakers have the duty to de-energize unloaded lines or cables. These are capacitive loads by their inherent stray capacitance.

For overhead lines, the typical stray capacitance varies in the order of 10 nF/km per phase for single-conductor overhead lines [34]. The current to be interrupted is in the range of several amperes at medium voltage to several hundreds of amperes at high voltage, which corresponds to a line length of a few tens to hundreds of kilometres.

At interruption of capacitive loads, the normal load is already switched off by a breaker at the remote end of the line, but due to the stray capacitance of the transmission system, a small capacitive current is still flowing in the system, to be interrupted by the station breaker. The current to be interrupted depends on the voltage level and the length of the line or cable, and possibly on some station components.

Due to the relatively high capacitance of cables as compared with overhead lines, the current to be interrupted by cable-connected breakers is significantly higher, although the length of cable connections is usually shorter. The capacitance of cables has a wide variation in value depending very much on the type, design, and voltage level. No-load cable currents can be up to several hundreds of amperes.

The switching of unloaded transmission systems is a relatively rare event, and it is especially the interruption operation that must be considered.

2.5.2.3   Switching of capacitor banks

Capacitor banks are normally used for the control of reactive power for voltage regulation in AC systems and are normally switched on and off daily as the load pattern varies. Thus, the circuit breaker's switching performance of capacitor banks must be considered statistically, considering a very large number of switching operations in the breaker's lifetime.

Unlike the lines and cables, having distributed and small stray capacitance, capacitor banks have a large and concentrated capacitance. Therefore, capacitor banks generally draw much more current than unloaded cables or lines – in practical cases up to several hundreds to a few thousands of amperes.

Regarding the energization, by switching on of capacitor banks, the concentrated capacitance causes another onerous phenomenon for circuit breakers: the '*inrush current*', a very high transient current, drawn by the capacitor bank upon energization because of the very low surge impedance, see Section 2.5.2.5.

2.5.2.4   Requirements regarding de-energization of capacitive loads

At interruption of a capacitive current, the power frequency AC voltage is lagging by 90° and is at maximum at current zero. The disconnected load will remain charged at a constant (DC) value at 1 p.u. At the other terminal of the circuit breaker, the power–frequency AC voltage continues to vary, resulting in a power–frequency voltage offset by 1 p.u. across the interruption gap. This recovery voltage is often referred to as *1-cosine* waveshape and has a maximum value of 2 p.u. This voltage is no longer called a TVR, but recovery voltage, because it is in a steady state.

Looking at critical parameters, current and RRRV are not an issue, but arc duration and recovery voltage peak. Arc duration can be very short, and peak value of TRV can be large, i.e., 2 p.u. in a single phase, 2.5 in three-phase. In case the gap has not reached sufficient distance to withstand the 1-cosine recovery voltage, a breakdown will follow, usually close to the maximum voltage that comes half a power–frequency period after current zero. This breakdown is called '*re-strike*', and it is distinguished from '*re-ignition*' by its 'late' (more than a quarter power frequency period after interruption) occurrence. Generally, restrike is considered a highly unwanted phenomenon. Even more harmful are repeated restrikes, since they can lead to *voltage escalation*, an increase of voltage by a process of repeated charge accumulation of the load capacitance after successive restrikes [9].

In Figure 2.15, this process is depicted in a test environment.

Apart from damage to nearby equipment by switching overvoltages, restrikes can damage critical parts of gas circuit breakers. By design, the breakdown of the contact gap shall take between the arcing contacts. A failure mode, observed from

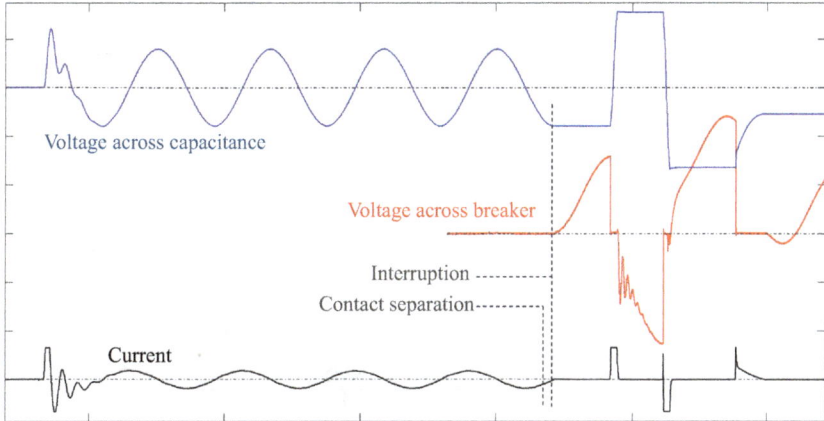

Figure 2.15   Multiple re-strike and voltage escalation in a test of a 72.5 kV vacuum circuit breaker

time to time, is a discharge path from one arcing contact to one main contact, thereby punching the nozzle. Proper dielectric coordination of the contact sets should prevent this to happen.

A third type of breakdown, apart from re-ignition and restrike, encountered (almost) exclusively in vacuum circuit breakers, is the '*non-sustained disruptive discharge*' (NSDD), an unexpected breakdown up to several hundreds of milliseconds after interruption. After the breakdown the contact gap recovers its insulation in a matter of microseconds [7,35]. Although this type of breakdown is not only observed in capacitive current interruption, its occurrence is most prominent at capacitive switching because (in non-earthed systems) it leads to a shift of the 1-cos recovery voltage in the other phase.

All three types of breakdown, including the recovery voltage shift, can be observed in the test-example of Figure 2.16.

The standardized requirements regarding the interruption of capacitive current focus towards minimising the risk of occurrence of restrike and provide confidence that the circuit breaker will restrike only in exceptional cases. Re-ignition and NSDD are considered a natural part of the interruption process. From type-test statistics, it is reported that NSDD shows up mostly (up to 80% of all tests) at the high medium voltage end of vacuum circuit breakers and when switching off capacitor banks. In single-phase tests of 72.5 kV vacuum breakers [36], an occurrence of 3% is reported [35].

Through extensive test programs circuit breakers can be attributed a class of restrike performance [16]: C1: low probability of restrike and C2: very low probability of restrike. In IEEE, also a class C0 (a probability of restrike up to one restrike per operation) is defined [37]. Capacitive switching tests consist of a number of tests, of which one quarter is at minimum arcing time condition, i.e., at the most hazardous situation of restrike occurrence. The total number of tests can

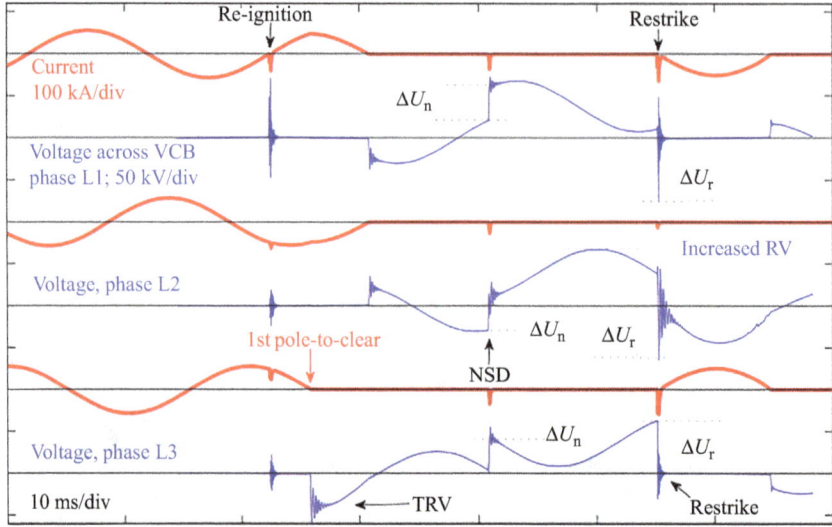

Figure 2.16   Three-phase capacitive current interruption with re-ignition, NSDD (leading to recovery voltage shift ΔU_n, ΔU_r) and restrike [7]

be up to 120 close-open operations and depends on the intended application (line/cable or bank switching), single- or three-phase and capacitive switching class assignment of the circuit breaker. Breakers having the lowest restrike probability class C2 need to demonstrate this after pre-conditioning with short-circuit current.

Apart from the capacitive load, also the supply circuit plays a role. When for example a capacitor bank is installed in a weak part of the system, the local short-circuit current is reduced because of a relatively high source side impedance (e.g., a long line of low capacity). This impedance creates an initial TRV ('*voltage jump*') that might already initiate a re-ignition before the recovery voltage can reach a higher value. In this way, the voltage jump excludes restrike, and thus has a beneficial effect. The opposite is true for capacitive loads fed by strong systems. In that case, the voltage jump is very small and cannot prevent restrike. Both conditions are covered by the requirements in the standard.

Anyhow, in any application, the occurrence of restrike must be considered. Even very low probability of restrike class circuit breakers installed for frequent switching applications will restrike. The standard allows at maximum one restrike for the C2 class breakers (having one restrike in the first test-series but zero in a repeated on) and three for the C1 class.

A common way to avoid restrikes is to apply controlled switching [38,39]. Controlled switching, see Section 6.1.8, prevents opening of the switching device at short arcing time.

2.5.2.5   Requirements of energization of capacitive loads
Regarding energization of capacitive loads, it is the switching-in of (shunt) capacitor banks that deserve the most attention. Since the surge impedance of capacitor

banks is far smaller than that of cables and lines, capacitor-bank inrush current can be very high. Management and its consequences for the breaker and the system are of considerable concern to users of switchgear [40].

The onerous action of the inrush current is that it starts to flow at pre-strike when the contacts do not yet touch galvanically. This means arcing exists between the closing contacts, and because the inrush current can be of high frequency (several kHz) and may reach considerable peak values (up to a few tens of kilo-amperes) even during short pre-arcing times the thermal energy released between the closing contacts can be considerable.

Practically, two switching situations need to be distinguished:

- Energization of a single, stand-alone capacitor bank: single-bank energization. In this topology, the inrush current path is through the source circuit, limiting frequency and magnitude of inrush current. Since the inrush current flows through the supply system impedance, the busbar to which the bank is connected will experience a voltage transient, affecting power quality. However, the impact on the circuit breaker is modest.
- Energization of a capacitor bank while one or more parallel banks are already in operation: back-to-back energization. This configuration allows inrush current to pass through the very low impedance of one or more parallel capacitors. The frequency (standardized at 4.25 kHz) and peak value (20 kA$_{pk}$) are considerable. Therefore, this switching configuration is very severe. This switching requirement is a special challenge for vacuum circuit breakers. Since vacuum circuit breakers do not have separate main and arcing contacts, pre-arcing is on the same contacts that have to withstand the voltage in the open position. Due to the pre-arcing, the pre-arcing energy causes local contact melting, and when the contacts touch each other, local welding may occur. The opening mechanism should be designed to mechanically break this weld, but remnants of the weld may still cause local surface irregularities and sharp protrusions that act as electric-field enhancing sites. If these irregularities are not sufficiently removed by arcing during the opening of the contacts, they may impair the dielectric strength of the contact gap. A common failure mode observed at testing is welding of the contacts after closing and stuck contacts. Also, the observed frequent occurrence of NSDD during back-to-back testing [36] may be explained by an impaired dielectric integrity due to pre-arcing and subsequent welding.

2.5.3 Interruption of inductive-load current

2.5.3.1 Re-ignitions

Inductive-load switching is a requirement where very small to modest inductive current (typically few A to several hundred of A) shall be interrupted, typically from unloaded transformers, starting motors and shunt reactors. Similar to capacitive current interruption, the small current can lead to interruption after (very) short arcing time, when the contact gap is not yet long enough to withstand the TRV, and re-ignition will follow.

TRV is of high-frequency (1–10 kHz, depending on rated voltage) since it is basically determined by the load inductance and its stray capacitance only. Because of the high-frequency TRV, several too many trials to recover fail, depending on breaker technology,[†] resulting in a multitude of re-ignitions: '*multiple re-ignition*'. Unlike a restrike in capacitive circuits, the energy delivered by the inductive re-ignition discharge is relatively low, being the discharge of the stray capacitance only. Upon re-ignition, a high-frequency re-ignition current will flow briefly, and the gap may or may not recover from the event.

Re-ignition, even multiple re-ignition is considered unavoidable, and they are allowed by the standard [41]. Testing is aimed to quantify the re-ignition behaviour. The parameters of the interruption process, more specifically the re-ignition pattern, can be used to predict overvoltages in practical situations [42].

Another feature of small current interruption is the fact that the current will be interrupted slightly before natural current zero. This is called '*current chopping*' and is due to the very strong gas blast that is needed for fault-current interruption. Since at tripping, the breaker is not aware of the magnitude of the current it needs to interrupt, its action is always prepared for the maximum current. This causes the current to be chopped at a level of a few amperes (chopped current) before reaching power–frequency current zero. In vacuum breakers, the same phenomenon is observed, in this case caused by the inability of the arc to exist stable at very small current. Chopping current of vacuum circuit breakers is of similar level as that of gas circuit breakers.

In inductive-load switching, the magnetic energy stored in the inductive load, even at chopped current of a few amperes, increases the magnitude of the subsequent TRV, see Figure 2.17, and enhances the probability of re-ignition.

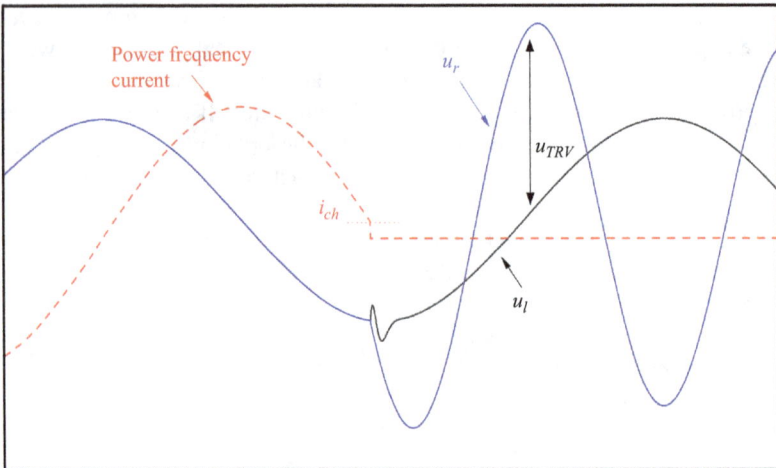

Figure 2.17   TVR applied to the breaker in the case of current chopping

[†]Usually, vacuum circuit breakers show much more re-ignition attempts during interruption with small arcing times than SF_6 circuit breakers, see also Section 2.9.3.3 and Figure 2.36.

In the relevant IEC standard IEC 62271-110 [41], two inductive switching situations are standardized for verification through testing:

1. Motor-current switching.
2. Shunt-reactor current switching.

Switching of unloaded transformers (interruption of magnetizing current) is no longer considered in the standard, because it is not possible to correctly simulate this switching duty using linear components used in a test laboratory. Tests using a laboratory transformer are not representative of other transformers.

2.5.3.2   Motor-current switching requirements

Motors act as an inductive load during the start or when stalled. The requirement of the circuit breaker is to switch motors in this situation without unacceptable overvoltages. Because the inductive switching process is an inter-action between the breaker and the circuit, overvoltage data from tests in a laboratory cannot be transferred directly to service practice. Testing and record-ing its waveshape parameters then serve to characterize its behaviour with respect to switching overvoltages, re-ignitions and current chopping. The waveshape characteristics may serve as a basis for estimates of the breaker performance in other motor circuits.

However, to predict with some accuracy the overvoltages to be expected in the field based on the tests is a major challenge in the motor switching case because of the extremely high frequencies that are involved in the (multiple) re-ignition pro-cesses. The relevant phenomena are entirely governed by parasitic network elements. Especially with vacuum breakers, that have an excellent capability to interrupt re-ignition current, multiple re-ignition is frequently observed. This has led to a strict definition (of component values as well as of topology) in the standard of motor substitute test circuits regarding their high-frequency behavior [41].

It is essential to carry out these tests as three-phase tests, to verify the absence of '*virtual current chopping*': current chopping with very large chopped current in one phase, induced by high-frequency re-ignition current in a neighbouring phase. Virtual current chopping is known to cause very large overvoltages. In laboratories, values of up to 10 p.u. have been observed [43]. Also, multiple re-ignition tran-sients can reach a very high value. Their repetitive nature, together with their very high rate-of-change (du/dt) upon breakdown, can potentially harm equipment with windings, such as transformers, when unprotected.

In the IEC standard [41], 80 tests are prescribed across four different test-duties. The difference in the test-duties is in the current (100 and 300 A) and in the source-side capacitance of the test circuit.

From the test experience, the following conclusions can be drawn regarding motor switching:

• Current chopping overvoltages can be neglected.
• Multiple re-ignitions are a normal phenomenon that can lead to high overvoltages.

- Virtual current chopping (in one or two phases) occurs regularly in testing, but it is observed in circuits with small supply side capacitance only. Overvoltages can be tremendous. Higher load current produces higher overvoltages.

2.5.3.3   Shunt-reactor current switching requirements

Shunt-reactor switching is by far the most common practice of inductive-load switching. Shunt reactors are installed for overhead-line capacitance compensation and switched depending on the momentary load of the system. This results in very frequent switching operations (like capacitor-bank switching), and thus merely from the number of required switching operations, this duty needs special attention, not only electrically but also mechanically.

Accompanying shunt-reactor switching, multiple re-ignition occurs very frequently and may be onerous for gas circuit breakers. This is the reason why shunt-reactor switching is sometimes termed 'a circuit breaker's nightmare' – and not the least because it is a daily switching operation [44]. In a world-wide reliability survey, shunt-reactor switching is found to be the service operation having by far the highest number of major failures (2.5 major failures per 100 circuit breaker years) [45].

In Figure 2.18, a typical shunt-reactor switching process is shown, including multiple re-ignitions at the first current zero and successful interruption at the

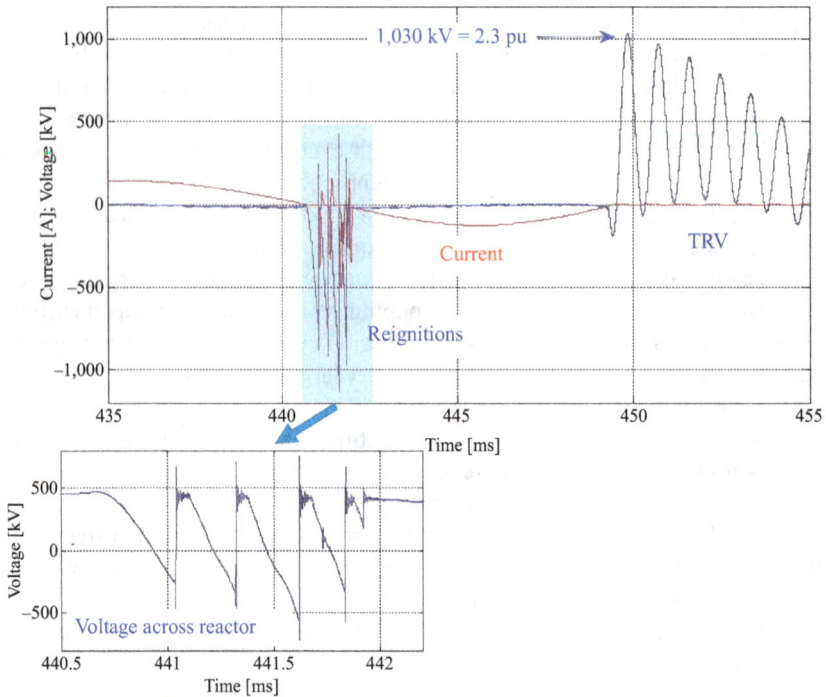

Figure 2.18   Switching process during interruption test of 100 A shunt reactor current per IEC 62271-110. Lower: magnification of voltage across reactor during multiple re-ignitions at the first arcing current zero [43].

second. Note that the voltage across the reactor jumps from −700 kV to +700 kV in a time much shorter than a microsecond.

Extremely high switching overvoltages with values above 3 p.u. have been observed after multiple re-ignition.

Contrary to common belief, it is the multiple re-ignition process, rather than the current chopping itself that makes reactor switching potentially hazardous to both circuit breaker parts and external equipment.

Critical gas circuit breaker chamber parts, like the nozzles, can be damaged by the very high and high-frequency re-ignition current that excites shock waves in the interruption chamber [46]. In addition, the internal voltage coordination in the interruption chamber may change because of the extremely high di/dt associated with the re-ignition current, causing the re-ignition to take place between other parts of the contact system than the arcing contacts. Punctures of the nozzle have been observed from time to time [47], as well as discharge traces on and damage to the main contacts.

The influence of shunt-reactor switching on the circuit breaker condition has been the subject of several studies. Practical application guidelines are provided in the relevant IEC [42] and IEEE application guides [48].

Due to the breakdown processes in switching gaps that take place in tens of nanoseconds, the du/dt values are very high, causing extremely steep-fronted waves. This poses a threat to the neighbouring equipment in substations – notably transformers. The voltage across the windings in transformers is equally distributed across each turn as long as the frequency content of the voltage remains below approx. 1 MHz. However, as soon as voltage surges become so steep that higher frequency components become dominant, the voltage distribution inside the winding becomes non-homogeneous. Where the voltage wave enters the winding, there will be disproportional stresses across the terminal turns. Although HV transformers are electrically re-enforced at their terminals, stresses due to steep-fronted waves, initiated by re-ignitions, may be a challenge. Any capacitance in the system (such as a bushing capacitance in metal-enclosed switchgear and transformers) mitigates these stresses.

In addition, there are electromagnetic compatibility (EMC)-related concerns due to the high penetration potential of steep-fronted waves in secondary equipment for control, measurement, diagnostics, etc.

Multiple re-ignitions in vacuum circuit breakers are unlikely to damage the contacts, but due to 'vacuum's' capability of interrupting a current of very high frequency, the conducting period of the re-ignition current is very short, making the number of re-ignitions significantly higher than in SF_6 circuit breakers.

Since the shunt reactor can be treated as a lumped circuit element having a stray capacitance, the equivalent load circuit in testing can be simplified to a single-frequency circuit, with an oscillation frequency standardized by IEC 62271-110 [41] to values between 6.8 kHz at the rated voltage of 72.5 kV and 1.5 kHz at 800 kV. In a CIGRE Technical Brochure 50 [49] and IEEE C37.015 [48], practical values of practical shunt-reactor frequencies are listed.

Traditionally, in medium-voltage applications, dry-type and oil-filled shunt-reactors are applied. Electrically, one major difference is their stray capacitance.

Dry-type shunt reactors have very small stray capacitance that results in notably higher oscillation (and TRV) frequencies. The IEC standard on shunt-reactor switching is based on these values and covers the application range well.

Air-core shunt reactors are more and more applied directly connected to the HV system [50]. Air-core HV reactors in this voltage class may have a stray capacitance of no more than 300–500 pF. Using an estimate of 500 pF, against 2,000–4,000 pF of the traditional, oil-insulated HV dead-tank shunt-reactors, the air-core shunt-reactor frequencies can be expected to reach well beyond the standardized range, for which 1,750 pF as stray capacitance is standardized. When directly connected to a breaker, air-core shunt reactor switching may not be covered by the IEC standard.

In Figure 2.19, standardized and practical shunt-reactor TRV frequencies are plotted against rated voltage.

The aim of shunt-reactor switching tests is not to impose a limit on overvoltages or on the number of re-ignitions, but to verify that re-ignitions occur at the first arcing current zero only. This implies that the breaker should recover fast enough to have a minimum arcing time below half a power frequency period, to ascertain that the arc cannot re-ignite at the second or later power–frequency current zero [41].

Two test circuits are defined to verify the requirements: the higher-current circuit (at 315 A for rated voltages > 72.5 kV) creates the highest-frequency TRV causing the higher probability of re-ignition, whereas the lower-current circuit (at 100 A) has the highest inductance with a higher energy stored at current chopping and causing the highest TRV peak.

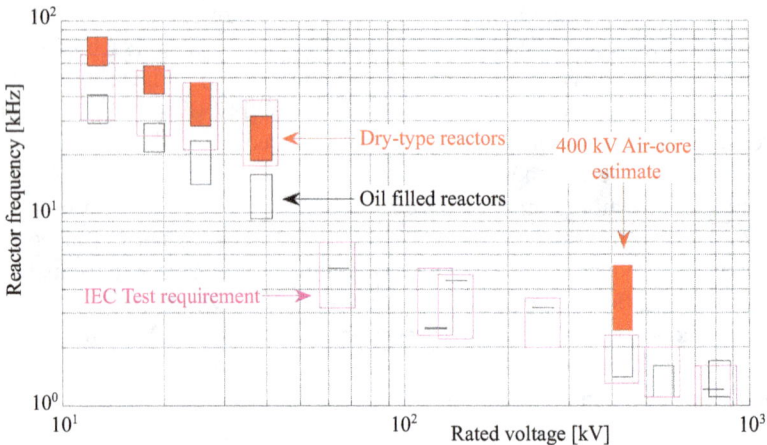

Figure 2.19   Natural frequencies of shunt reactors (oil filled and dry-type) and TRV frequencies of IEC 62271-110 test-circuits compared with EHV air-core shunt reactor frequency estimate [43]

Concluding, the small (reactive) current is rather a disadvantage than an advantage, and interruption of lower current might be onerous for the system. It is recommended to perform shunt-reactor switching tests at the actual service current when this current is lower than the standardized value.

The criteria to pass a test are:

- The circuit breaker shall consistently interrupt the current with (one or more) re-ignitions at one current zero crossing only.
- Re-ignitions shall always occur between the arcing contacts only.

Generally, re-ignition can be avoided by utilizing controlled switching, i.e., the strategy to realize contact separation outside the 're-ignition window', when dielectrically sufficient gap-spacing is always guaranteed at the current interruption. Defining the re-ignition-free window is one of the aims of testing.

2.6   Summary table of AC current interruption duties

Table 2.2 shows the main features of the various current interruption requirements for AC circuit breakers, as presented before.

2.7   Non-standard applications of AC circuit breakers

2.7.1   Offshore application

2.7.1.1   General considerations

The number of offshore wind farms has increased considerably in recent years. Depending on their distance to the onshore station, either a DC or an AC submarine cable connection is used for power transmission. This section is on the application of AC cables and the switching issues that might be related to the special character of this AC transmission system, its main feature being the cable length that exceeds

Table 2.2   Overview of current interruption requirements of circuit breakers

Requirement	Keyword	Current	TRV steepness	TRV peak	Reference
Fault-current interruption					
Terminal fault	Asymmetrical current	Very large	Moderate–high	Moderate–high	2.4.1.1
Short-line fault	Travelling voltage wave	Large	Very high	Moderate	2.4.1.6
OOP fault	Two sources	Moderate	Moderate	Very high	2.4.1.7
Load-current interruption					
Capacitive load current	Restrike	Small	Very low	Moderate	2.5.2
Inductive load current	Re-ignition	Very small	High	(Very) High	2.5.3

in many projects the typical onshore cable length. For the transfer of energy to offshore oilrigs, extremely long AC cables (up to 250 km) are projected that need special power quality control methods.

Generally, the reliability of offshore operating switchgear must be very high, given the complications of maintenance, repair, and replacement. At the same time, footprint and dimensions must be minimized because of the expensive floorspace of platforms or the possible installation of wind turbines directly. Moreover, adverse weather conditions, salty environment, vibrations, etc. are new influential factors.

2.7.1.2 Switching of long AC cables

Long high-voltage cables generate sizeable quantities of reactive power that can limit the active power transmission capability and produce overvoltages. In order to avoid these undesirable effects, it is quite common to compensate for the reactive power generated by cables with shunt reactors at both ends of the link. The energization and de-energization of long compensated AC cables represent a certain challenge in case the reactive compensation exceeds 50%.

During the de-energization of the link, capacitive current must be interrupted. Because of length (and voltage class), this current can be much higher than the (highest) standardized value of 500 A. High capacitive current, as such, is not an issue for the circuit breaker. Higher current can even be advantageous because the longer associated minimum arcing times reduce the risk of restrike (see Section 2.5.2.2).

When energizing shunt reactor and cable simultaneously, the current is a superposition of the cable current and the reactor current. The latter might, when the moment of energization is near a voltage zero, have a significant DC component. Its DC time constant is usually large because the shunt reactor bank can be of considerable reactive power. As a result, the current through the circuit breaker may fail to have current zero for a considerable time, and current zero is delayed in a '*missing zero*' scenario until the DC component has decayed. If, for example, at a single-phase-to-earth cable fault, the breaker is required to interrupt, the healthy phases maintain the missing zero and their interruption will fail, leading to the destruction of the circuit breaker [51]. Such a scenario is shown in Figure 2.20. Another 'missing zero' scenario occurs at an earth fault of the station busbar under energized conditions, where the faulted phase of the shunt reactor current may miss zero [52].

The generation of transient overvoltages because of the excitation of low-frequency oscillations generated by cable capacitance and shunt reactor inductance is also studied [54].

These phenomena, and the countermeasures to avoid them, are studied in detail for very long AC submarine cable projects, such as the links Mallorca – Ibiza (132 km) [55], Malta – Sicily (118 km) [56], Crete – Peloponnese [57] and metropolitan Tokyo [52] and other major projects [58].

Possible countermeasures are the following:

- Controlled closing at voltage peak, thus avoiding high DC component in the shunt reactor current. This requires single-pole operation capability and needs dedicated controllers.

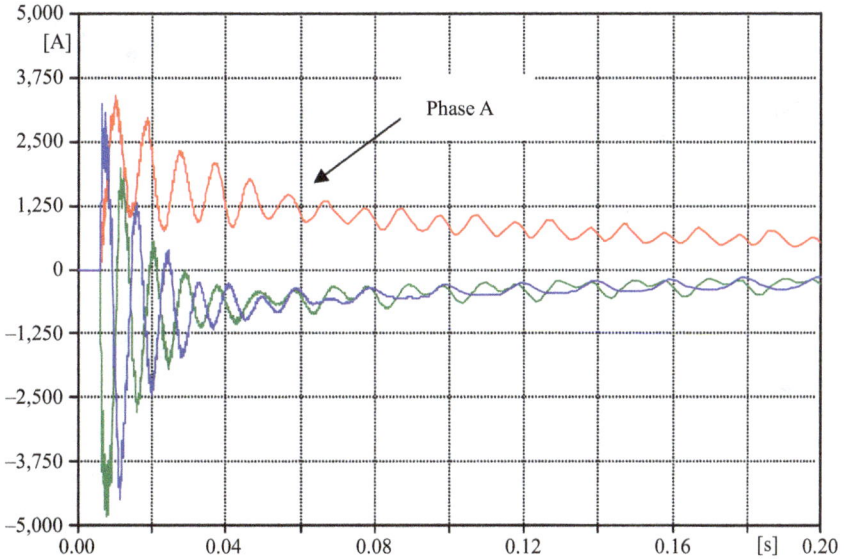

Figure 2.20   Missing zero phenomenon [53]

- Sequential switching of shunt reactor and load, which allows load switching after decay of shunt-reactor transient. This would mean that additional circuit breakers have to be installed at the shunt reactor.
- Closing resistor, to damp the DC component. This means more mechanical complexity, whereas circuit breakers equipped with closing resistors are not available below a rated voltage of 362 kV.
- Delay (in the protection strategy) of the tripping of the circuit breaker until the DC component has decayed sufficiently.

2.7.2   Missing current zero in transmission systems

For transmission circuit breakers close to generating stations, the occurrence of missing zeros requires certain generator operating modes depending on excitation (under-excited, over-excited), load (full- or no-load) and fault type, usually three-phase (simultaneous or non-simultaneous) [59]. A study on the application of 550 kV circuit breakers at Itaipu substation in Brazil assumed a theoretically highest DC component resulting from a two-phase line-to-line fault at zero line-to-line voltage that develops into a three-phase fault at zero voltage to earth on the third phase [60]. Subsequent short-circuit testing demonstrated the capability of the circuit breaker to force current zero crossings due to the high arc voltage of the arc. Such circuit breakers have designs to increase the arc voltage.

In Figure 2.21, an example from a test shows how the arc voltage of a SF_6 circuit breaker advances current zero significantly.

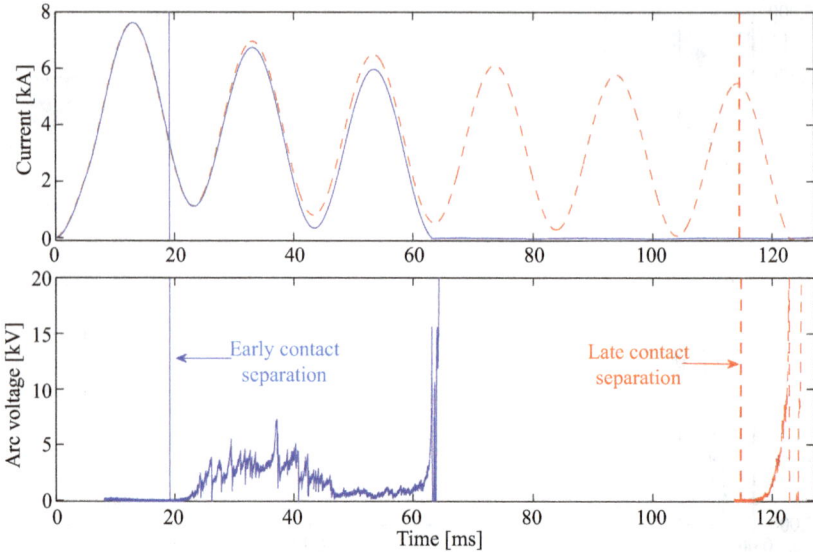

Figure 2.21   Interruption of a 245 kV SF$_6$ circuit breaker in a missing zero test-situation with early contact separation (drawn traces) and late contact separation (dashed traces). Upper traces: current; lower traces: voltage across the breaker [7].

A further study on the series compensated 735 kV system also showed that for certain circuit breaker locations and sequence of tripping, missing zeros could occur [61]. Later testing demonstrated the positive influence of high arc voltages.

All in all, given the multiple contingencies required for missing zeros, such actual events are conceivable but not probable for transmission circuit breakers.

2.7.3   Switchgear in mobile substations

Increased dependence on electricity requires the preparedness for catastrophic events like cyberattacks, floods, fires, explosions and last but not least pandemics. Floods may arise from sea (tsunami, broken sea-defence), rising rivers or heavy rainfall. Submersible equipment is deemed to be required for underground solutions, where flooding is a risk to be mitigated [62]. Automatic, self-power-restoring switchgear can to a certain extend mitigate outages due to dysfunctional MV overhead lines. Critical restoration strategies are being developed to withstand natural events, physical- and cyber-attacks, while advanced data analytics, Artificial Intelligence (AI) and Machine Learning (ML) will provide early warnings to detect and prevent direct attacks.

The availability of mobile HV substations and emergency restoration lines that can be rapidly deployed after an unintended significant outage greatly mitigates the risk of social unrest and huge compensation claims. Moreover, the deployment for planned outage of certain key grid components/stations makes the availability of mobile stations attractive by increasing the operational flexibility of the system [63].

Operational differences with conventional stations, reflecting additional requirements are the following:

- Mobile stations are expected to be used infrequently and for a short time, remaining mostly in idle storage mode. The impact and conditioning of this operation mode must be investigated since manufacturers often advice storage times much shorter than operational times. For example, long storage of GIS can lead to contamination/oozing of grease.
- The transportation, multiple assembly, cable (un)winding, multiple use of cable termination, bushing (dis)mounting cause additional stresses. Nevertheless, the must-run capability must be verified, e.g., by mobile (dielectric) test facilities on site.
- Necessary weight/size reduction (400 kV transformer in a three single-phase configuration, multiple filling, reduced lightning impulse withstand requirements [63]) for the sake of transportability must not compromise reliability and stress withstand capability.
- Equipment must be plug-and-play, fool-proof, maintenance-free, fit-for-purpose and modularly designed to allow assembly also by less-experienced personnel.
- Mechanical and electrical endurance needs to be ascertained, given the unusual and intermittent stresses and the expected impact on availability. Due to its operation mode, the probability of terminal faults in mobile substations is higher than in permanent substations.
- Since it might be expected that cyberattacks are a reason to activate a mobile station, tele-control and external data access must be extremely well shielded or even made non-existing.
- Functional testing at regular intervals must verify the continued availability of the key (standard) components operated in a non-standard manner.

2.7.4 Generator circuit breakers

2.7.4.1 Requirements

Generator circuit breakers are medium voltage circuit breakers applied to interrupt fault currents between a generator and the system it supplies. In a power plant, they are located between the generator and the step-up transformer. When no generator circuit breaker is applied, a circuit breaker at the high-voltage side of the step-up transformer can be applied. The advantage of this solution is the less complicated high-current connection ('*generator busduct*') between the generator and the transformer. The advantage of having a generator circuit breaker is the possibility to connect the auxiliary plant to the medium voltage side of the (permanently energized) step-up transformer. This is schematically outlined in Figure 2.22.

The electrical and mechanical performance requirements of generator circuit breakers are very different from standard MV distribution switchgears. The only standard available worldwide that covers specifically the requirements for generator circuit breakers is IEC/IEEE 62271-7-013 [64]. Apart from the ratings and

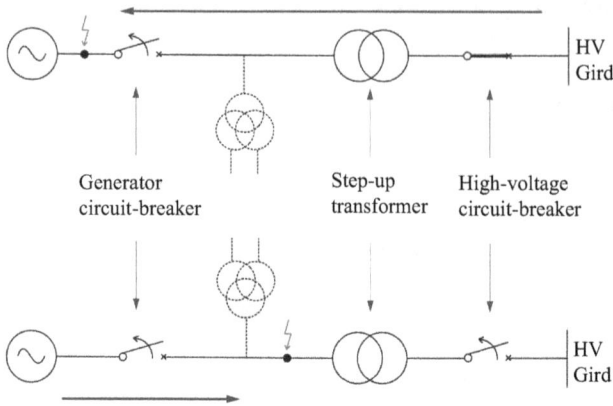

Figure 2.22   Topology of generator circuit breaker and faults

other relevant characteristics, this standard contains guidelines for the type-testing of generator circuit breakers.

The main points of distinction between generator circuit breakers and general-purpose circuit breakers as far as the interrupting duties concerned are:

(a) *Load-current switching.*
Load currents for large generation units can be as high as 30–50 kA, often this makes forced cooling necessary. Following the interruption of the load current, as with any circuit breaker the two circuits at both sides of the generator circuit breakers oscillate independently, creating a TRV that is the difference of the oscillating source and load component.

(b) *System-source fault current interruption.*
In this fault case, a major part of the fault current is supplied through the step-up transformer by the system upstream. The total current in this scenario has the highest value of all other fault situations, because of the low short-circuit reactance of the transformer and the HV system compared to the (sub-) transient reactances of the generator. Unlike the situation for general-purpose HV circuit breakers, the maximum TRV stress on generator circuit breakers appears after the maximum short-circuit current stress. A very high RRRV value originates from the small, distributed, capacitances of the step-up and the auxiliary transformers. In Figure 2.22 (upper part), the circuit topology of this fault is outlined in a basic circuit.

For smaller generators, the capacitance of a connecting cable can reduce the RRRV considerably, depending on its length [65]. For the same reason, in many designs of SF_6 generator breakers, TRV mitigating capacitors are built inside the circuit breaker enclosure.

(c) *Generator-source fault current interruption.*
In this case, a major part of the fault current is supplied by the generator (see Figure 2.22, lower part) causing a DC component that can be higher than the

symmetrical short-circuit current, resulting in very high asymmetry, possibly with delayed current zeros [66]. The AC component, having a constant amplitude in faults in substations and on overhead lines, in a generator-source fault has a decreasing amplitude. This results from the specific transient behaviour of the generator [67].

This transient behaviour of generators is shown in Figure 2.23. In the lower part, the unique characteristic of this type of fault is highlighted: missing- or delayed current zeros. The first current zero can be delayed for several cycles. This implies that during this period, the generator circuit breaker is not able to interrupt the fault current.

The circuit breaker arc and the fault arc(s) reduce the circuit DC time constant because the effective arc resistances add to the circuit resistance [68]. Current zero may therefore be advanced with respect to the situation without arc(s). The arc voltage of the fault arc contributes to this reduction. Values of around 10 V/cm are measured for fault currents up to 70 kA [68]. For advancing current zero crossing, high arc voltage (as for air-blast arcs and to a lesser extent for SF_6 arcs) can be advantageous, although a high arc voltage also implies a higher thermal stress for the interruption chamber. Vacuum circuit breakers have a very low arc voltage but have been shown to be effective as

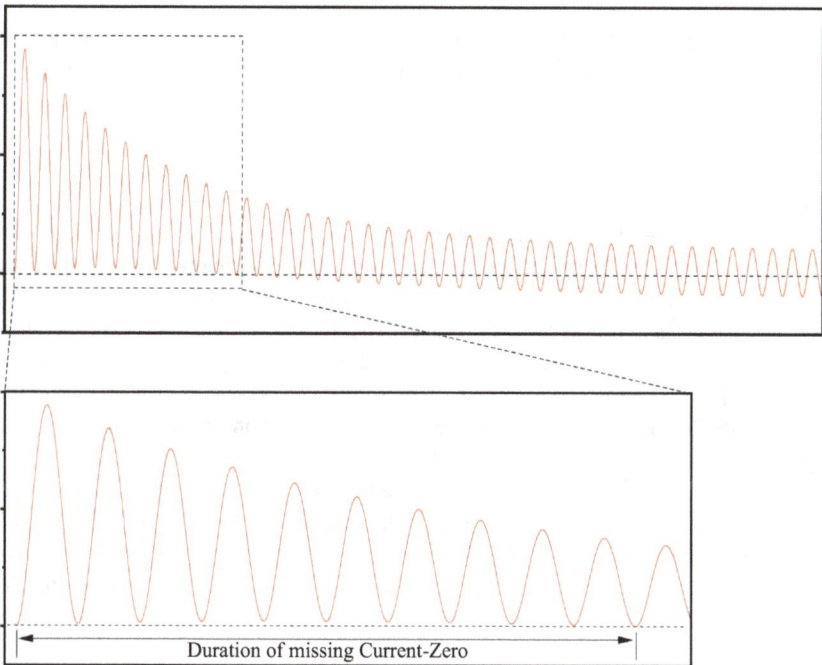

Duration of missing Current-Zero

Figure 2.23   Generator-source fault current with magnification of the period of 'missing zeros'

well [69]. Vacuum generator circuit breakers with significant generator-source fault clearing capability have been tested successfully [70].

DC time constant, inherent to this situation, is standardized as 133 ms [64]. The value of the generator reactance limits the short-circuit current to a value below the system-source-fault case. The same applies to the associated TRV stress. Dominated by the relatively large inherent capacitance of the generator, the TRV rate-of-rise (RRRV) is lower than the value of the system-source fault RRRV and standardized at a maximum of about half that value.

(d) *OOP fault current interruption.*

An *OOP condition* occurs when a making operation of the generator circuit breaker is performed at the instant when there is no synchronism between the voltage phasors of the generator at one side of the circuit breaker and the external grid at the other side, similar to HV circuit breakers in that switching duty, see Section 2.4.1.7. Another example is when a generator running OOP as a result of system instability and the generator circuit breaker has to be tripped [29].

The severity of this interruption depends on the OOP angle δ. Since the generator is at risk for values of δ larger than 90°, the protection relay trips around $\delta = 90°$. The standardized OOP TRV values are based on a 90° OOP angle at rated voltage. However, for smaller generator units, a large OOP angle can occur [71].

In the case of an OOP angle of $\delta = 90°$, the current is about 50% of the fault current supplied by the system. On the voltage side, the generator circuit breaker experiences a TRV with a RRRV roughly of the same order as in the system-source fault, but with a peak value nearly two times higher.

The OOP current specified in the IEC/IEEE standard [64] is half the system-source-fault current.

In Figure 2.24, TRV wave shapes based on the IEC/IEEE standard [64] are shown for various interruption duties of a generator circuit breaker, rated voltage 12 kV, connected to a generator having a rated power between 200 and 400 MVA. For comparison, standardized TRV waveshape for the 100% short-circuit current interruption duty for a general-purpose 12 kV circuit breaker, in accordance with IEC 62271-100 [16], is added. It is apparent that, apart from current stresses, the TRV requirements for generator circuit breakers are far more severe than the TRV requirements for general-purpose circuit breakers.

For lower generator-power ratings, even below 100 MVA, the RRRV of the system-source-fault TRV is still 2.6 times higher than the T10 duty standardized for general-purpose circuit breakers [72].

In general, the generator neutral is unearthed. Therefore, the first pole-to-clear faces a power–frequency recovery voltage of 1.5 times the phase-to-earth voltage of the system.

Since the generator circuit breaker is located directly at the output of the power plant, the reliability of generator circuit breakers must be extremely high [73].

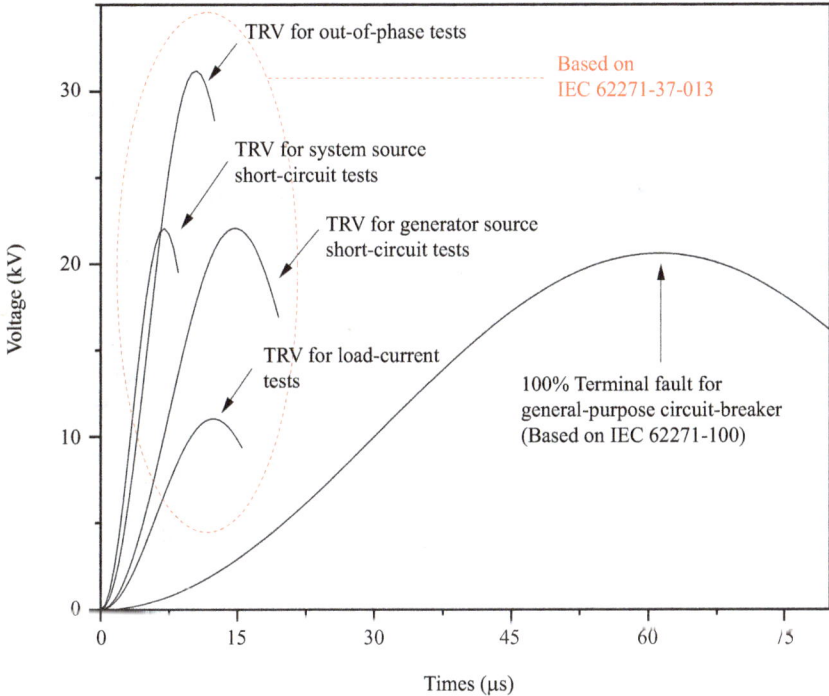

Figure 2.24   Standardized TRV shapes of various generator faults compared with TRV of a 100% fault for a 12 kV general-purpose circuit breaker

2.7.4.2   Technology of generator circuit breakers

Current interruption in generator circuit breakers was at first done with compressed air in extinction chambers assisted by a high-pressure air blast. Air as arc-extinguishing medium has a rather long time constant to recover to its non-conducting state. Often, an opening resistor is connected in parallel to the extinction chamber to reduce drastically the RRRV in order to facilitate the interruption. The disadvantage of an opening resistor is that a second interruption chamber is required to interrupt the current through the resistor.

Generator circuit breakers using SF_6 as the arc extinction medium came on the market in the 1980s. Exploiting the thermal energy of the arc combined with a puffer action, high breaking capacity is realized with relatively low operating energy, through self-blast technology, see Section 2.2.1. Reduction of the RRRV by capacitors connected in parallel to the interruption chamber (*surge capacitor*, see Figure 2.25, Part 9) has a positive influence on current interruption.

Recently, powerful SF_6-free vacuum generator circuit breakers became available. The largest designs use multiple vacuum interrupters in parallel per phase in order to deal with the very high continuous current [70]. In vacuum generator circuit breakers, RRRV mitigating capacitors can be omitted.

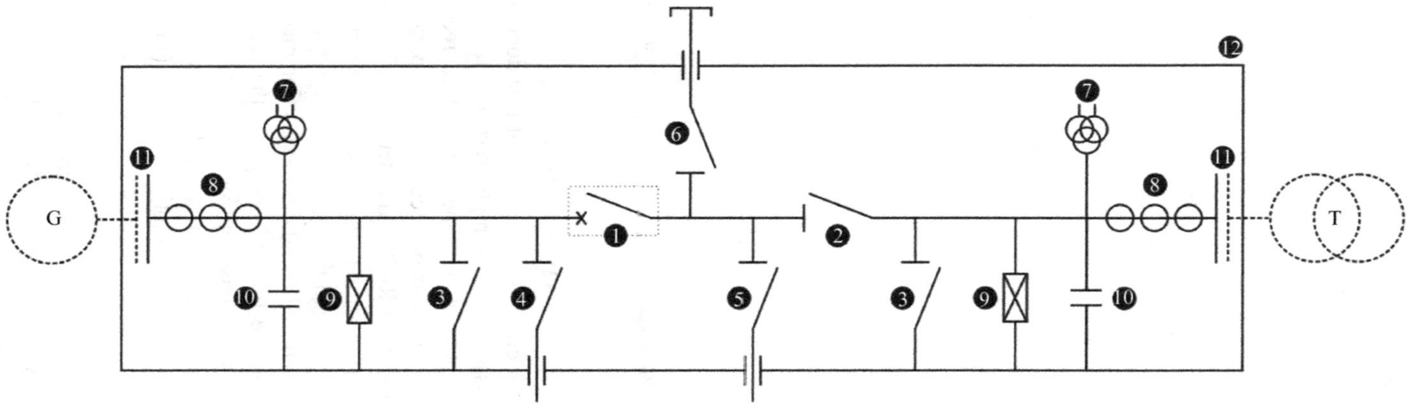

Figure 2.25 Generator circuit breaker system. 1: Circuit breaker; 2: disconnector; 3: earthing switches; 4: starting switch; 5: starting switch (back-to-back); 6: short-circuiting switch/braking switch; 7: voltage transformers; 8: current transformers; 9: surge arrester; 10: surge capacitors; 11: terminals; 12: enclosure.

In most cases, the available short-circuit power of test laboratories is insufficient to perform generator circuit breaker test in a direct circuit, and synthetic test methods have to be applied [7]. It is recommended to apply synthetic test circuits that support constant AC recovery voltage instead of AC recovery voltage with a decaying amplitude as in standard synthetic tests [74]. Applying constant-amplitude AC recovery voltage is similar to the service situation and is essential to demonstrate the absence of late restrikes.

In addition to the obvious requirement that a generator circuit breaker must be able to carry the full load current of the generator and ensure a sufficient insulation level at all times, it must also be capable of performing the following functions:

- Synchronise the generator with the main system.
- Disconnect the generator from the main system.
- Make, continuously carry, and interrupt load currents up to the full load current of the generator.
- Make, short-time carry, and interrupt the system-source short-circuit currents and generator-source short-circuit currents.
- Make, carry, and interrupt currents under OOP conditions.

All the associated items of switchgear can also be integrated into the generator circuit breaker enclosure as an option besides their separate installation. Such items comprise a series disconnector, earthing switches, a short-circuiting switch, current transformers, voltage transformers, protective capacitors, and surge arresters.

Depending on the type of power plant, additional items, such as starting switches in gas-turbine and hydro power plants and braking switches in hydro power plants, can also be mounted in the generator circuit breaker enclosure, see Figure 2.25 [75].

2.8   Dielectric requirements of AC circuit breakers

2.8.1   Classification

Dielectric requirements are designed to assure that electrical equipment can withstand all voltages, including overvoltage transients (also called overvoltages) that occur in the power system. A distinction can be made based on their origin:

- From external sources: Most common are transients caused by lightning strokes that have their origin in atmospheric disturbances.
- From the power system itself: These are mostly generated by changes in the configuration of the network, most often by switching operations, but also during and after the occurrence of faults. Power system transients are the subject of many studies, literature is collected in [76–79].

Voltages and overvoltages are classified in IEC 60071-1 [80] by their waveshape and duration as summarized in the following.

2.8.1.1 Low-frequency voltages and overvoltages

These overvoltages are characterized by their constant r.m.s. value, their low frequency, their relatively modest amplitude, and their long duration:

- Continuous voltages with a power frequency of 50 or 60 Hz and duration of at least 1 hour.
- Temporary overvoltages (TOV) are power-frequency and harmonic overvoltages of relatively long duration, from 20 ms to 1 h. They may be undamped or weakly damped. Frequencies in practice are in the range from 10 to 500 Hz. The most common temporary overvoltages occur on the healthy phases of a system during phase-to-earth faults. Some other well-known events leading to the generation of temporary overvoltages are load rejection, Ferranti rise [79], resonance, and ferro-resonance [81,82]. Temporary overvoltages may lead to overstressing of surge arresters, magnetic saturation of transformers, and shunt reactors. Temporary overvoltages are considerably lower in magnitude than transient overvoltages.

2.8.1.2 Transient overvoltages

These overvoltages have a duration of a few milliseconds or less and are usually highly damped:

- Switching overvoltages or slow-front overvoltages (SFO) are usually caused by switching operations as well as by faults occurring on the system. Circuit breakers do not generate SFOs directly; they just initiate them by changing the topology of a circuit. If present, re-ignition, restrike, inrush current, current chopping, multiple re-ignition, and NSDD (Non-Sustained Disruptive Discharge) phenomena can be also responsible for the generation of SFOs. In practice, the SFOs have time-to-peak between 20 μs and 5 ms and tail duration of less than 20 ms.
- Lightning overvoltages or fast-front overvoltages (FFO) can appear in a substation either due to a lightning stroke directly to the substation or a strike to the transmission line feeding the substation [83]. Alternatively, flash-overs, back-flash-overs, re-ignitions, and restrikes at a short distance may cause the FFOs. Typical FFOs have time to peak between 100 ns and 20 μs and a tail duration of 300 μs.
- Very-fast transient overvoltages (VFTO) belong to the highest frequency range, 30 kHz up to 100 MHz, with time to peak between 3 ns and 100 ns and duration shorter than 3 ms [32]. They are mainly produced by gas-insulated switchgear, especially by switching with disconnectors, see Section 2.5.2.1. Disconnector switching is a slow process and is associated by a multitude of breakdowns between the contacts and instant recovery. The resulting transients may have a direct impact on primary equipment, transformers, or on secondary control systems due to their strong electromagnetic interference potential by their high-frequency content. The impact of VFTOs on equipment can be divided into internal and external VFTOs [84,85].

o Internal VFTOs: Travelling waves that are initiated between the inner conductor and the enclosure, putting high stresses on the internal insulation.
o External VFTOs: At discontinuities of a GIS enclosure, such as sealings, connectors, windows and bushings, the electromagnetic wave is partially transmitted to the outside potentially resulting in:
   ▪ Transient electromagnetic fields (TEMF) causing stresses and electromagnetic interference (EMI) in connected primary equipment like transformers and instrument transformers.
   ▪ Travelling waves on overhead lines causing stresses in the connected equipment (transformers, instrument transformers).
   ▪ Transient enclosure voltages (TEV) causing stresses and electromagnetic interference in secondary equipment. According to IEC 62271-1 [20], switching a small portion of (unloaded) GIS is the most severe situation with regard to the generation of electromagnetic disturbances in the secondary system [86].

For applications with rated voltage above and including 800 kV, the maximum VFTO in GIS systems can reach the insulation level of lightning impulse withstand voltage in certain cases. VFTO can become the limiting dielectric parameter that defines the dimensions at rated voltage levels above 800 kV [32].

Practical guidelines on various aspects of insulation coordination can be found in IEC 60071-2 [87].

2.8.2   Requirements for AC circuit breakers

Withstand voltage is one of the earliest determined rated parameters of circuit breakers. Minimum dielectric characteristics of circuit breakers, prescribed by the standards, should ensure their reliable operation in power systems under the conditions of the various overvoltages described in Section 2.8.1.

The rated insulation level of a circuit breaker refers to the standard atmospheric air conditions and shall be selected from the values given in Tables 1–4 from IEC 62271-1 [20]. The values include the altitude correction to a maximum altitude of 1,000 m, and they apply to both indoor and outdoor circuit breakers. The rated insulation level is specified by the rated phase-to-earth lightning impulse withstand voltage.

For transient overvoltage withstand capability, not only the magnitude but also their duration is of importance. This is expressed in the voltage–time (V–t) characteristic of media and is directly related to the physical mechanism that leads to breakdown. For SF_6 gas, the breakdown voltage is relatively independent of duration, making the higher and faster lightning test requirement most severe, whereas for air the lower but longer lasting switching impulse becomes more onerous at longer gaps.

Up to and including a rated voltage of 245 kV (Tables 1 and 2 [20]), this value is indicated as 'common value' because the same value applies to phase-to-earth, between phases and across the open switching device. The withstand voltage value 'across the isolating distance' is valid only for the switching devices where the clearance between open contacts is designed to meet the functional requirements specified for disconnectors. Circuit breakers normally do not have a disconnecting function; this is a duty of disconnectors. When the circuit breaker is not in use, it is normally insulated by a disconnector and may be stressed only for a short period. There is no switching impulse requirement for rated voltages ≤ 245 kV.

Above a rated voltage of 245 kV (Tables 3 and 4 [20]), also switching impulse withstand capability is required. Switching overvoltages are the primary dimensioning parameter for air-clearances in EHV and UHV systems. The reason is that, while clearances for lightning impulse withstand voltages increase linearly with gap distance, those for switching impulse withstand levels tend to saturate with increasing gap distances. Thus, for large air gaps (roughly above 1 m), breakdown voltage for switching impulse is lower than for lightning impulse. Switching-impulse breakdown involves both streamer and leader development with the latter being the main driver, whereas lightning-impulse breakdown involves streamers only [88]. Optimal leader development occurs at front times in the range of 100–400 μs, leading to a selection of a front time of 250 μs for standardization purposes. Detailed descriptions of electric discharges and breakdown in air can be found in [89,90].

For the verification of the dielectric requirements, the following waveforms have been defined [91]:

- One-minute AC voltage: The standardized voltage shape for testing is in the range 48–52 Hz with a duration of 1 min.
- Switching impulse: Standardized switching impulses (SI) having time to peak of 250 μs and decay time to half-value of 2.5 ms.
- Lightning impulse: Standard lightning impulses (LI) having a front time of 1.2 μs and decay time to half-value of 50 μs, commonly referred to as 1.2/50 impulse.

In addition, IEEE includes a chopped wave impulse requirement for circuit breakers. This is a lightning impulse voltage during which (after the peak) a discharge causes a rapid collapse to zero. Its duration is 2 μs with peak voltage 1.29 times the full wave lightning impulse peak [92]. The earlier, additional 3 μs chopped-wave duration requirement (with 1.15 times the lightning impulse) is abandoned because it was found that longer duration is not more critical to SF_6 and vacuum [4]. This requirement is in recognition of the fact that the voltage at the terminals of a surge arrester has a characteristically flat top, but at some distance from the arrester, the voltage is somewhat higher.

There exists no defined waveshape for studying the impact of VFTOs. The test requirement is designed to verify that arcing between the disconnector contacts does not lead to breakdown between live parts and enclosure. Nevertheless, for their most common occurrence, disconnecting a busbar section (bus-charging switching duty), the test-circuit shall be designed such as to produce a voltage with a risetime to 1.4 p.u. within 500 ns [33].

Figure 2.26 shows a comparison of the relevant phase-to-earth insulation levels[‡] for various rated voltages. The insulation levels refer to:

- Amplitude of the short-duration withstand voltage.

[‡]These rated insulation levels are specified by IEC 62271-1 (2017), Tables 1.1–1.4. The value of the short-duration withstand voltages $U_{d,p}$ have been calculated from the specified r.m.s. rated insulation levels U_d, i.e. $U_{d,p} = U_d \sqrt{2}$.

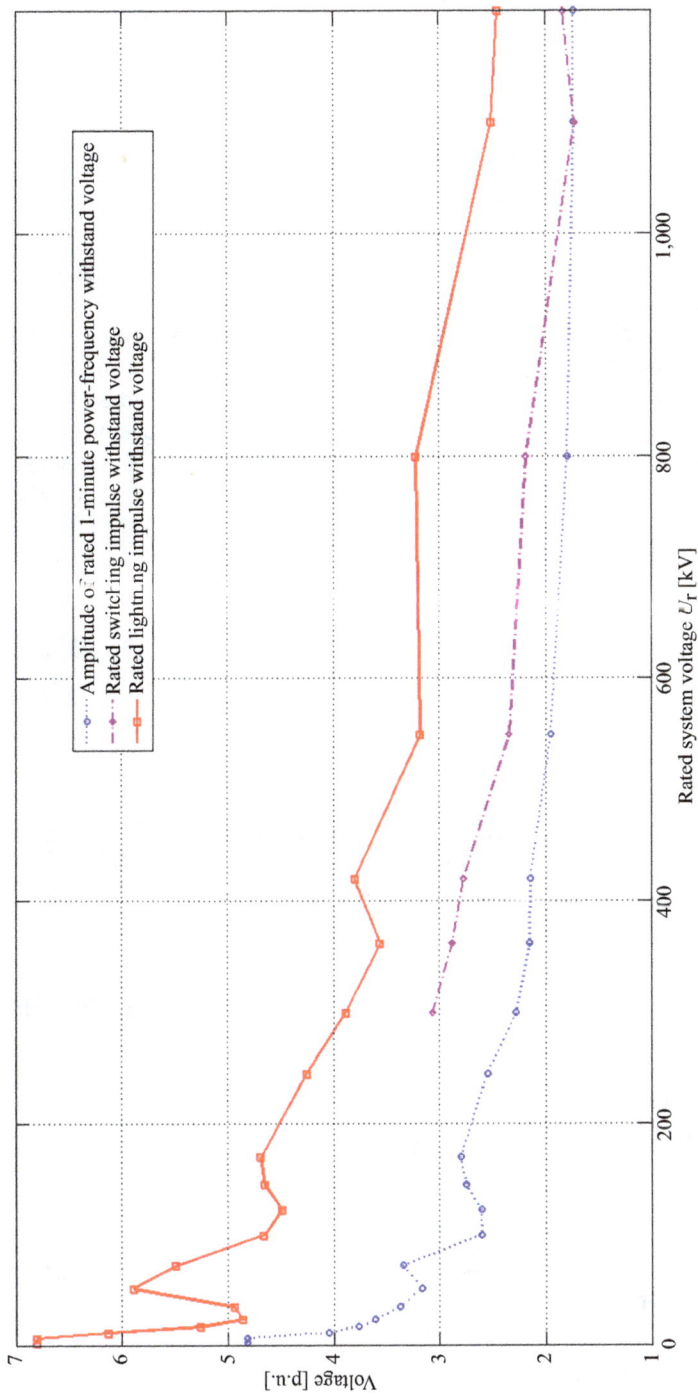

Figure 2.26 Rated short-duration power-frequency-, switching impulse- and lightning impulse withstand voltages vs. rated system voltage

- Peak of switching-impulse voltages.
- Peak of lightning-impulse voltages.

  In each case, 1 p.u. $= U_r \sqrt{2}/\sqrt{3}$.

2.9   SF$_6$-free HVAC circuit breakers

2.9.1   Introduction

In electricity generation, huge steps and investments are ongoing to meet the internationally established CO_2 reduction targets. Not surprisingly, similar efforts are being undertaken in the electricity delivery industry. At present, the biggest share (>95%) of the CO_2 equivalent from T&D grids are transmission losses. While the energy generation mix will eventually tend to zero carbon footprint, the remaining CO_2 equivalent will be due to SF$_6$, stored in T&D equipment. The past decades, it is recognized that SF$_6$ is the most potent man-made greenhouse gas, having a CO_2 equivalent (Global Warming Potential, GWP) of 25,200 [93] and an atmospheric lifetime of 3,200 years.

At the time of writing, alternative gases to SF$_6$ in high-voltage equipment are based on:

- Natural-origin gases (GWP < 1) such as air, or mixtures of its components (N$_2$, O$_2$ and CO$_2$).
- A mixture of natural-origin gases and a fraction of fluoronitriles[§] (C4-FN) or fluoroketones[||] (C5-FK). The C4-FN mixtures have GWP in the range 300 750, and are mostly applied in transmission equipment, whereas C5-FK mixtures have GWP < 1 and are found in distribution equipment. These mixtures are commonly called 'F-gases', as classified by the F-gas regulation of the European Union [94]. However, distinction must be made by their GWP: C4-FN (in pure form) has GWP = 2,750, whereas C5-FK has GWP = 0.29, actually not a greenhouse gas.

These gas mixtures can be applied either for insulation only, but also for insulation and switching. Details of physical properties can be found in Sections 4.2 and 4.3. An overview can be found in [95].

[§]2,3,3,3-Tetrafluoro-2-(trifluoromethyl) propane nitrile, also described as $(CF_3)_2CFCN$ or C_4F_7N is a fluoronitrile. For easier naming, reference and identification, it is also named C4-FN (FN: fluoronitrile). Its commercial name is Novec[TM] 4710 and it is a component of the gas, marketed as 'g^{3}'. It is a synthetic fluid that has in its pure form a GWP of 2750 [92] and an atmospheric lifetime of 34.5 years.

[||]1,1,1,3,4,4,4-Heptafluoro-3-(trifluoromethyl)-2-butanone, also described as $CF_3C(O)CF(CF_3)_2$ or $C_5F_{10}O$ is a fluoroketone. For easier naming, reference, and identification, it is also named C5-FK (FK: fluoroketone) or C5-PFK (PFK: perfluoroketone). Its commercial name is Novec[TM] 5110 and it is a component of the gas marketed as 'AirPlus'. It is a synthetic fluid and has in pure form a GWP < 1 [92].

2.9.1.1 SF₆-free mixtures for insulation only

SF$_6$-free gases are in use for the insulation of non-switching GIS compartments, gas-insulated lines (GIL) [96,97], gas-insulated busbars (GIB) but also of instrument transformers (IT) [98,99]. At the time of writing, research is ongoing on the possibility of retro-fill, i.e., the direct replacement of SF$_6$ with an SF$_6$-free gas in transmission system parts like GIS compartments that do not contain switchgear [100]. Due attention must be paid to the material compatibility of SF$_6$-free CO$_2$-based mixtures, since permeation of CO$_2$ through SF$_6$-grade gaskets and seals is much larger than for example of N$_2$-based mixtures [101]. This would end up in a higher leakage rate, compromising the equipment lifetime. Also, pressure withstand might become an issue, since non-SF$_6$ gases usually have to be operated at higher pressure than SF$_6$ in order to maintain similar equipment dimensions and footprint [102].

Another function is the insulation of vacuum interrupters in SF$_6$-free vacuum switchgear. In this case, switching is with (one or more) vacuum interrupters whereas the hermetically sealed interrupters are externally insulated with natural-origin gas, mostly pressurized air. This technology is already applied the past 50 years in medium voltage vacuum circuit breakers, where the use of air and solid insulation material is very common. Since the 1990s of past century vacuum switching with gas insulation is extended to high voltage having SF$_6$ as external insulation [103]. At present, there is a strong development towards insulation with compressed technical (dry) air, but also solid insulation is emerging in the lower transmission level, e.g., in reclosers [104].

2.9.1.2 SF₆-free mixtures for switching and insulation

When it comes to switching, high-voltage SF$_6$-free switchgear can be categorized into the following technologies, each with a different class of medium for arc interruption:

- Switching in natural-origin gas and insulation with natural-origin gas. This technology in principle is very old, since air-blast circuit breakers led to a breakthrough in ultra-high-voltage systems in the 1960s of the last century [9]. As a new development, mostly a mixture of CO$_2$ and O$_2$ is used.
- Switching in C4-FN or C5-FK mixtures for switching and insulation. In high-voltage switchgear, at the time of writing, only C4-FN mixtures are applied. C5-FK mixtures have been abandoned for high-voltage application in 2021, although they are still in use in distribution switchgear for insulation and/or switching. Every manufacturer has its own composition of this mixture which can even depend on the application, e.g., the minimum operation temperature. In CIGRE Technical Brochure 871, an inventory is presented of the state of products and projects [105]. At the time of writing, it can be concluded that SF$_6$-free C4-FN- and vacuum circuit breaker technology are available up to a rated voltage of 170 kV and are applied in various projects worldwide. Announcements have been made of 420 kV SF$_6$-free circuit breakers in double-break technology [106].

Table 2.3 gives an overview of SF_6-free equipment available, operating in pilots or in full service, or announced.

In Figure 2.27, the increasing development of arc extinction media over the years is outlined. This gives the interruption power of a single interruption chamber

Table 2.3   Summary table of SF_6-free gases for switching and insulation (applied and announced)

Switchgear function		GWP of gas mixture	Highest voltage (kV)	Technology DT: dead tank; LT: live tank; LBS: load-break switch
Insulation	Interruption			
		SF_6		
SF_6		25,200	1,200	GIS / DT / LT
SF_6	Vacuum		204	GIS
		Natural-origin gases		
Technical air	Vacuum	0	170	GIS / DT / LT / LBS
O_2 / CO_2		< 1	145	LT
Technical air	–	0	420	GIB
		Mixtures with fluoronitriles (C4-FN) and natural-origin gases		
C4-FN/air	Vacuum	> 300	38	Circuit breaker
C4-FN/CO_2	Vacuum		170	GIS
C4-FN/air			38	LBS
C4-FN/O_2/CO_2		300–750	170	GIS/LT/DT
			420	GIS (announced)
C4-FN/CO_2			170	GIS
C4-FN/CO_2	–		1,000	GIB / GIL
C4-FN/O_2/ N$_2$	–		420	GIL / GIB
		Mixtures with fluoroketones (C5-FK) and natural-origin gases		
C5-FK/air	Vacuum	< 1	40.5	Circuit breaker
C5-FK/air			40.5	LBS
C5-FK/O_2/CO_2			170	GIS
C5-FK/O_2/ N$_2$	–		420	GIB/GIL

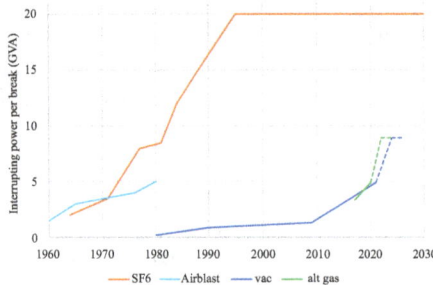

Figure 2.27   Single-interruption chamber interruption power of media for current interruption over time

(as a product of rated line-to-earth voltage and rated short-circuit breaking current) vs. time.

Experimental ('hybrid') low-volume SF_6 designs with SF_6 and vacuum interrupters in series have been reported as well [107,108]. The idea is to use the very fast recovery of a vacuum interrupter to withstand the initial TRV (such as appears in short-line fault interruption), whereas an SF_6 interrupter with a reduced amount of SF_6 should withstand the peak value of the TRV.

2.9.2 Gas circuit breakers

2.9.2.1 Differences with SF_6 breakers that impact performance

The gas mixtures that are applied in SF_6-free gas circuit breakers have physical properties that differ from SF_6, which might have an impact on operational and switching requirements. It is the task of the switchgear designer to compensate, by design, for the consequences of these different properties. The most relevant ones are listed below.

2.9.2.2 Low-temperature performance requirements

The gas in circuit breakers that have an insulating and/or interruption performance has to remain at sufficient pressure during the operational life of the device. This implies that losses of the gas by leakage and permeation (or by consumption) must remain limited, but also condensation must be kept below a limit.

This can be understood with the aid of Figure 2.28, showing the operational trajectory of the SF_6 gas in an SF_6 circuit breaker in the phase diagram of SF_6. This diagram relates the pressure, temperature, and phase (liquid, gas) of a fluid, in this case SF_6. The key curve is the *saturation curve*, the separator between liquid and gas (or vapour) state. In this diagram, *isochores* are drawn, straight lines of constant gas density, actually the gas mass since the volume of the breaker does not change. Nominal pressure (pressure at 20°C) is taken as 8 bar in this example, point A. Starting from this nominal condition, gas pressure loss can occur by two processes:

1. Loss of gas pressure through leakage and permeation, this process is acceptable until the *minimum pressure for insulation and/or switching* (at room temperature) at point D is reached, here at 7 bar. With more loss of gas, the

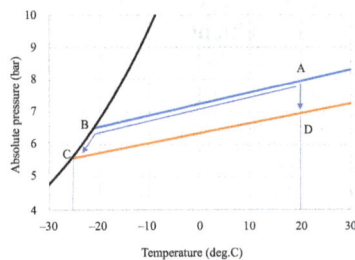

Figure 2.28 Operation limits of an SF_6 circuit breaker in the phase diagram

breaker will not fulfil its function. A large part of type tests must be performed at this condition, since also in this minimum condition, the operation must be guaranteed. Because gas is physically lost, point D is on an isochore of lower gas density.

2. Loss of gas pressure by cooling. This occurs when ambient temperature drops, in the state diagram, the transition to a lower temperature along an isochore, in this example from point A to point B. Point B is on the saturation curve, which means that at further cooling, part of the gas will condensate, quickly leading to lower pressure. This is acceptable until point C is reached, the *minimum ambient temperature*, here −25°C. Further loss of gas pressure by cooling or leakage will lead to an unacceptable situation.

Points A and B are on the isochore corresponding to SF_6 density 53.4 kg/m³, points C and D are on the 46.0 kg/m³ isochore. For a 0.4 m³ volume (live tank) breaker, transition from points A to D would mean a loss of approx. 3 kg of SF_6. For an SF_6-free breaker with a C4-FN/O_2/CO_2 (3.5/13/83.5%) mixture, isochore A–B would correspond to a gas density of 16.7 kg/m³, 3.3 times less mass than in the case of SF_6.

Breakers are designed for the minimum ambient temperature, one of the key parameters for the user. IEC defines ranges of temperature; the lowest limit is −50°C for extremely cold climates [20].

In the case of a gas mixture, as for SF_6-free gas circuit breakers, the phase diagram becomes very complicated, and various phases of the components may coexist. Details can be found in [105] and in Chapter 4.

From thermodynamic point of view, the major issue of the fluorinated additives C4-FN and C5-FK is their (very) high boiling temperatures: −4.7°C for C4-FN and 26.9°C for C5-FK (at standard pressure), whereas for SF_6, the boiling temperature is −64°C. This makes C4-FN and C5-FK inappropriate for use in a pure form. Since boiling temperature drops with a reduction of pressure (see Figure 2.28), the reduction of partial pressure of the fluorinated additives is commonly used as a standard solution for application in equipment. By adding a limited fraction of the additive (5–10%), also the partial pressure reduces in proportion and condensation is brought within an acceptable temperature region. As a general rule, the lower the required ambient temperature, the lower the fraction of C4-FN/C5-FK must be. However, too small fractions of additives might comprise the dielectric performance. The choice of the proper fraction of C4-FN and C5-FK is thus directly related to the minimum ambient temperature.

For high-voltage applications, the use of C5-FK is abandoned because at typical high-voltage switchgear pressures (> 6 bar), the partial pressure of C5-FK should be too low to remain effective at lower temperatures. For medium voltage application, allowing much lower filling pressures (< 2 bar) because of less stringent dielectrical requirements, the use of reasonable fractions (10–15%) of C5-FK is still possible for low temperatures.

The remainder of the gas is a carrier gas, which is usually CO_2 as the majority component and (mostly) O_2 as the third component. Natural-origin gases have very

Pure substance vapor pressures P_{vo}

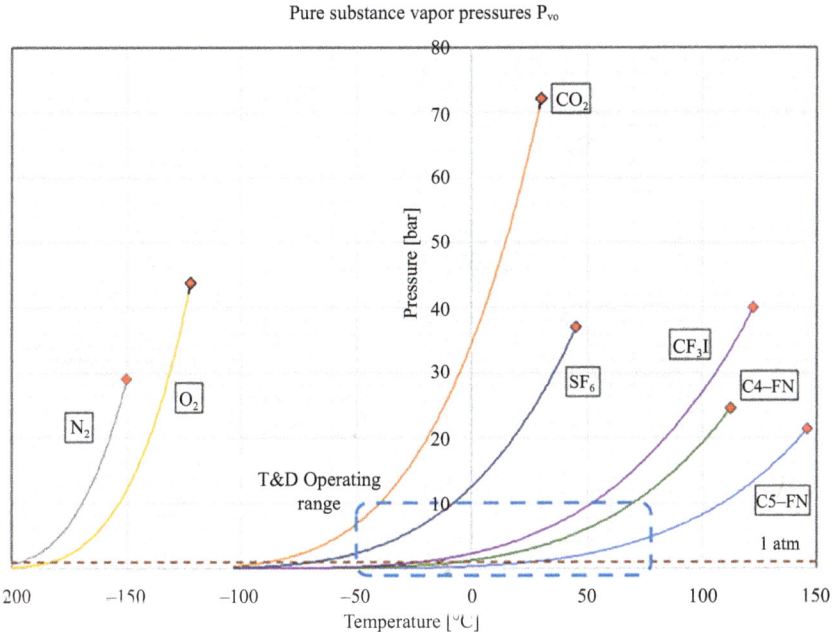

Figure 2.29   Saturation curves of various relevant pure gases in the phase diagram from 104, reprinted with permission from GIGRE, @2022

low boiling points (CO_2: $-78.5°C$, N_2: $-195.8°C$, O_2: $-183.0°C$) so they will not condensate in any ambient temperature.

Figure 2.29 shows the saturation curves of the various pure gases discussed.

For very and extremely cold climates, gas circuit breakers exclusively using natural-origin gases (CO_2/O_2) are reported, using a high gas pressure to compensate for the absence of C4-FN additive that improves dielectric performance [109].

The pressure of the SF_6-free gas mixtures must be high, actually higher than SF_6 because of dielectrical requirements.

2.9.2.3   Insulation performance requirements

Withstand against dielectrical stresses is a key requirement of any high-voltage equipment, see Section 2.8. Independent experimental research was carried out in a round-robin, multi-laboratory approach to the breakdown voltage of gas mixtures with C4-FN and C5-FK additives, compared to SF_6 [102]. For this, under the same experimental conditions (6 bar gas pressure, sphere-to-plane electrode topology with a 15 mm gap), the pure SF_6 gas was compared with the mixtures C4-FN/O_2/CO_2: 5.0/5.0/90% and C5-FK/O_2/CO_2 5.5/11/83.5%. Figure 2.30 shows the summarized results combining data from over ten laboratories participating in the study.

From these results, it can be recognized that at the same pressure, SF_6 has the superior performance. In order to reach similar insulation distances in SF_6-free

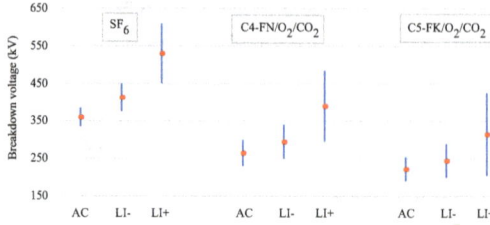

Figure 2.30 Breakdown voltage ranges of SF_6, C4-FN and C5-FK mixtures for AC, negative and positive lightning impulse voltages [102]

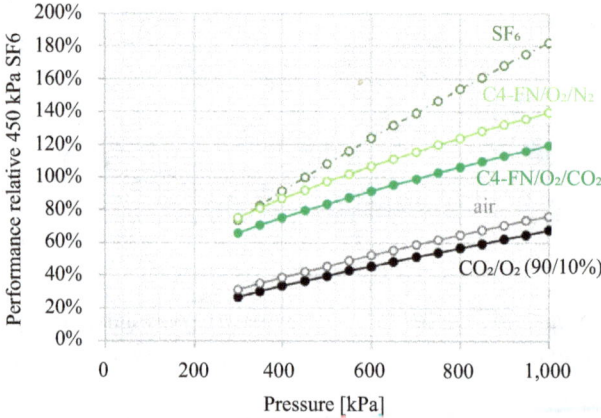

Figure 2.31 Breakdown voltages ranges of various mixtures under typical GIS conditions from [100]. The C4-FN mixtures are defined to have a condensation temperature of $-30°C$, reprinted with permission from CIGRE, @2022.

equipment, preserving similar footprint and dimensions as with SF_6 insulation, the pressure must be increased in the order of 2–3 bar beyond the usual SF_6 insulated equipment gas pressure of 4–6 bar absolute. This can be understood from Figure 2.31 which schematically depicts pressure dependence of breakdown voltage of SF_6 and various mixtures [110].

As a pure gas, C4-FN has an insulation performance of approx. 2.2 times better than of SF_6 [111]. Nevertheless, it is observed that using only a minority fraction of C4-FN, C5-FK, the insulation strength of SF_6 can be approached. This can be explained by the synergy effect, i.e., a non-linear increase of the dielectric strength with the additive ratio, as is known in SF_6/N_2 mixtures [112]. Due to this synergy effect, already a 2% C4-FN fraction in CO_2 results in approximately half the dielectric strength of SF_6, which is roughly 150% of the dielectric strength of pure CO_2 [102]. At a mixture ratio of 20% C4-FN in CO_2, the mixture has the same dielectric strength as SF_6 in the tested arrangement [102]. Figure 2.32 shows the

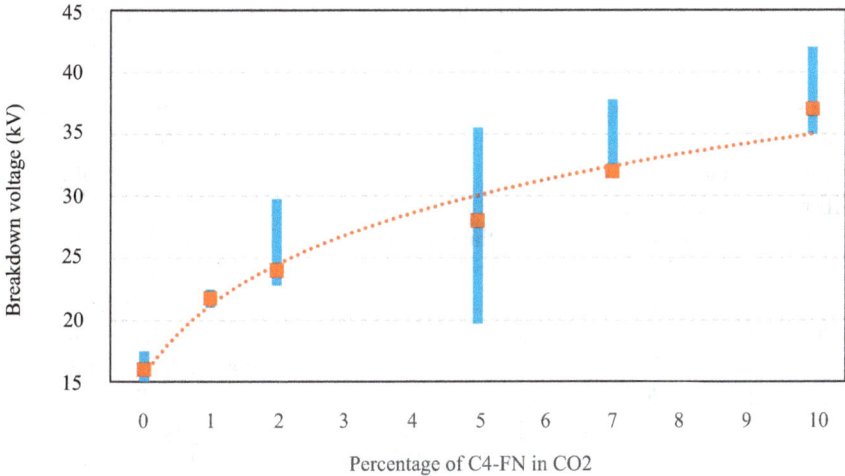

Figure 2.32 *AC breakdown voltage ranges of various fractions of C4-FN in CO₂
at 1 bar total pressure, results from three laboratories combined.
Dots are the 50 percentiles of the research-leading laboratory [102].*

measured impact of a low C4-FN percentage in CO_2 on the AC breakdown voltage of the gas mixture at 1 bar. For this arrangement, the breakdown voltage of SF_6 is in the range 45–48 kV.

Thus, the addition of C4-FN (or C5-FK) enhances the dielectric performance of its carrier gas CO_2 significantly.

2.9.2.4 Lifetime requirements regarding gas composition

A new phenomenon, not present in SF_6, is the consumption of the effective fluorinated additive by arcing. Under the high temperature of the arc, SF_6 gas dissociates and recombines again. However, the fluorinated additives C4-FN and C5-FK when exposed to arcing also dissociate but do not recombine. Therefore, these compounds are consumed, at a rate of 0.2–0.4 mol/MJ of arc energy [105]. Natural-origin gases or gas components are not consumed by arcing. As a result, over switchgear lifetime, the fraction of the fluorinated additive reduces, which has an impact on the dielectrical performance, as illustrated in Figure 2.32.

The volume of the circuit breaker containing fluorinated gas mixtures in direct contact with the arc is of importance as well. In smaller volume breakers (e.g., live tank breakers), the end of the effective gas life will be reached earlier than in large volume breakers (e.g., GIS, dead tank designs), simply because there is more gas available. For a gas circuit breaker, any test duty that includes a voltage condition check after severe arcing should be performed in the design with the smallest gas volume intended for service. This helps to ensure maximum decomposition prior to the voltage condition check.

A strong requirement is that during the estimated lifetime of the circuit breaker, possibly including replacement or top-up, provisions must be taken not to 'run out' of fluorinated additives. This new wear mechanism, which adds to the existing wear mechanism of contacts and nozzle, has been investigated, claiming that critical 'traditional' wear of contacts and nozzle is faster than the consumption of the fluorinated mixture additives [105]. Thus, it is argued that the end of gas life will not be a new lifetime-limiting factor.

Due attention shall be paid to the capability of performing the basic switching duties after extended arcing has taken place. In type-testing, the extended electrical endurance qualification (E2) has been defined [113]. It imposes electrical wear followed by a comprehensive acceptance phase including high RRRV, short-line fault, capacitive switching, and voltage condition check on a single test object. The basic electrical endurance class E1 requires only the voltage condition check as performance verification [16]. Therefore, in consideration of gas consumption in SF_6-free gas circuit breakers, more proof of the capability of basic breaking performance might be needed.

An additional new factor is the presence of more species in a gas mixture. Whereas the state of SF_6 is characterized by its pressure only, in the mixtures the same pressure can be realized with a variety of partial pressures of its constituents and even a change (e.g., by permanent decomposition as argued above) during lifetime. Therefore, the term '*rated pressure for insulation and interruption*' is no longer a unique classifier and shall be replaced with '*rated composition and pressure*'. Similarly, '*minimum pressure*' shall be replaced with '*type test composition and pressure*' [105]. The word 'minimum' is no longer applicable, since the quantity of gas present is not 'minimum' but shall have sufficient reserve to pass through the whole test-duty series starting from the 'type test composition and pressure'. This alternative terminology considers the possibility of gas mixtures having both variable pressure and composition over the lifetime of the equipment.

Long-term studies on the use of fluorinated compounds are small in number. In one, crystalline decomposition products (types of amids and ligands), through a chemical reaction between C4-FN and moisture, were detected [114].

2.9.2.5   Interruption performance requirements

For successful current interruption in gas circuit breakers, the gas flow that enables energy removal from the arc plasma and that ultimately 'sweeps' the residual plasma out of the switching gap is of key importance. The gas flow is driven by transient pressure, realized by mechanical compression supported by the thermal energy of the arc, see Section 2.2. Sufficient pressure shall be maintained near current zero, after any possible arc duration, which is normally in the range 10–30 ms. In SF_6, its low sonic velocity (or speed of sound, 135 m/s at 20°C and 1 bar) enables a relatively slow flow over time, which also covers the longer arcing times. The SF_6-free gas mixtures, consisting mostly of CO_2, have a higher sonic velocity (267 m/s at 20°C and 1 bar) leading to a different pressure pattern, outlined in Figure 2.33 [115]. Pressure in CO_2 rises earlier, reaches a higher peak, and drops faster.

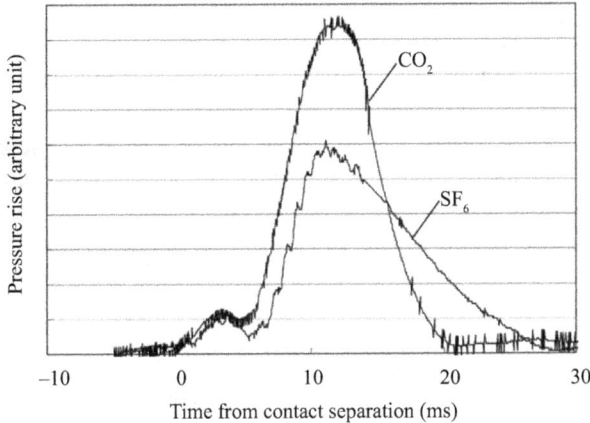

Figure 2.33   Transient pressure measurement in SF$_6$ and CO$_2$ interruption [115], reprinted with permission from CIGRE, @2022

It is a requirement that the pressure drop occurs after the longest possible arcing time, as to ensure that sufficient pressure can sustain the necessary gas flow at current zero. Through design, the speed of gas outflow from the compression volume shall be reduced. This can be realized by modifying the shape of the nozzle, flow channels and by increasing the compression volume of the arcing chamber [109,116].

In the SF$_6$-free mixtures, soot (carbon particles resulting from the incomplete combustion of hydrocarbons) may be generated under the influence of the arc. When deposited on insulating surfaces, this can affect the dielectric withstand capability. The addition of a sufficient fraction of O$_2$ is often used to reduce or avoid soot formation [117]. Moreover, the addition of O$_2$ is reported to reduce contact- and nozzle degradation by arcing [118]. In testing, the voltage condition check after severe arcing can be applied to verify that dielectric strength has not been reduced below an acceptable level. Oxygen-free HV circuit breaker developments having only C4-FN and CO$_2$ as a mixture are also reported [116,119].

Short-line fault current interruption, as one of the most severe interruption duties, has drawn considerable attention. Generally, it is concluded that the presence of CO$_2$ is the main factor to make this duty successful in SF$_6$-free gases. However, suitable design modifications need to accommodate for the physical differences, see also Section 4.3. As a positive factor, the transient pressure of a CO$_2$-based mixture is higher than SF$_6$ and is possibly augmented by O$_2$ addition. As a negative point, it is reported that arc conductance very shortly before current zero decays significantly slower than in SF$_6$ [120,123]. As a net result of the influential factors, it can be understood that short-line fault interruption requirement is a very severe one for SF$_6$-free gases (as it is for SF$_6$) and in general, these gases do not reach the performance of SF$_6$ [101]. It is reported that arc voltage extinction peak needs to increase by more than 30% to upgrade high-current interruption capability [121]. Research suggests that short-line faults occurring at longer distances from the circuit breaker (e.g., L75, L60) might be more onerous for the SF_6-free gases, however, there is no evidence that these gases cannot cope with this [105,122].

A requirement for capacitive current switching is the absence of restrike. This sets strong requirements for the dielectric withstand capability several milliseconds after current zero, see Section 2.5.2. Because of the lower dielectric withstand capability, for SF_6-free gas circuit breaker, it may be necessary to increase the contact separation velocity, or even use a double-motion contact system, where both contact sets move apart, thus increasing the relative contact separation velocity [109].

2.9.2.6 Miscellaneous requirements

Conduction of continuous current without exceeding standardized temperature limits is an important requirement. The thermal properties of SF_6-free gases lead to inferior performance with respect to SF_6 in this aspect. This can be explained mainly by the density difference between SF_6 and CO_2, technical air. Thus, with respect to convective heat transfer, SF_6 gas performance remains much higher than that of CO_2 and technical air. Temperature rise is reported to be around 20% higher than in SF_6 [123], though pressure increase and additional O_2 can compensate this partly [117]. Also here, careful design can bridge the gap [124].

The rate at which 'fresh' gas can be replenished to the interruption region to replace the C4-FN arc-depleted mixture is a critical parameter for switching duties that require multiple opening operations in a short time, like auto-reclosure.

Pressures, both steady state and transient are higher for SF_6-free gases. This has its consequences for pressure withstand and overpressure safety valve settings. In some cases of sealed-for-life (medium voltage), switchgear transportation regulations may restrict factory filling pressure [125]. It might be necessary to adapt the mechanical actuator that drives the kinematical chain of the circuit breaker.

One important loss mechanism for gases is leakage and permeation. Leakage is the penetration of gas through a very narrow channel (scratch, hair). Permeation is the emission of gas due to diffusion through material of various components. In SF_6 equipment, EPDM rubber (Ethylene Propylene Diene Monomer) material is used for sealings. The permeation of natural-origin gases, including CO_2, through this material is substantially higher than that of SF_6. This might lead to a change of the mixture composition and pressure. As an alternative, butyl rubber (IIR/XIIR) is used, although the permeation of CO_2 is still factors higher than the permeation of SF_6 through EPDM rubber [105], so adequate measures are required to manage material loss through this mechanism. EPDM rubber, on the one hand, has shown good permeation performance towards non-SF_6 mixtures with N_2 as a carrier gas. On the other hand, N_2-based SF_6-alternatives are not suitable as arc interruption media.

As a consequence, in re-filling 'passive' equipment, N_2-based mixtures are preferred because of their compatibility with existing SF_6-grade EPDM rubber seals [101]. Refilling of circuit breakers with any of the alternative mixtures is not possible since design has to be adapted to the different properties of the mixtures. For non-switching gas-insulated applications, there are possibilities [100].

2.9.2.7 Gas handling

Several conclusions are formulated in CIGRE Technical Brochure 802 [126], on the aspect of SF_6-alternatives gas handling and management:

- Time-based maintenance strategy for gas-related maintenance works is easily adoptable to high-voltage gas-insulated equipment with non-SF_6 gases and gas mixtures.
- Compared to SF_6, if gas mixtures are applied to HV equipment, the mixing ratio is one new, additional gas-related indicator to be measured or monitored to evaluate the need for maintenance actions.
- When handling SF_6-free gas mixtures, the main difference to SF_6 is that additional efforts for maintaining the correct gas mixing ratio during filling and reclaiming works are needed. The liquefaction behaviour of single gas components has to be considered which may affect the required gas handling time during maintenance work.
- For gas mixtures, it is necessary to evaluate case by case the most economical way of providing big volumes of gas by mixing on-site or by using pre-mixed containers.
- Recycling of gas mixtures is technically feasible (e.g., by separation of gas components) but not yet established for most non-SF_6 gases and gas mixtures on the market.
- Today, no standards similar to IEC 60376 or IEC 60480 exist for non-SF_6 gases and gas mixtures. Procedures for handling of decomposed non-SF_6 gases and gas mixtures are specified by each equipment manufacturer.
- End users should discuss case by case with the equipment manufacturer whether different assessment criteria apply when equipment failures occur.
- Methods for life extension, like enhanced maintenance or refurbishment, are generally adoptable to equipment with non-SF_6 gases or gas mixtures. However, retrofit or replacement of existing SF_6 equipment is not feasible with non-SF_6 gases or gas mixtures available on the market, mainly due to different design requirements.
- When end of life has been reached, the planning of the dismantling of the equipment and the disposal (including material recycling) can be done according to the best practice experiences with SF_6 equipment.

2.9.3 *Vacuum circuit breakers*

2.9.3.1 Introduction

In this section, aspects of high-voltage vacuum circuit breakers are discussed, as far as they differ from medium voltage vacuum circuit breakers, see also Chapter 3. High-voltage circuit vacuum circuit breakers became available in the mid-nineties of the last century. From that time on, a relatively large installed base was realized in Japan, mostly at 72.5–84 kV level, predominantly with SF_6 insulation of the vacuum interrupter(s) [103].

For vacuum switchgear technology, the major technology change associated with SF_6 elimination is not in the switching technology (this remains the current interruption in vacuum) but in the insulation of the vacuum interrupter, which is shifting from SF_6 to technical air. Since high-voltage vacuum circuit breaker technology is a natural extrapolation from the well-known medium-voltage vacuum circuit breaker technology, the technology change, and its impact, is not as large as in the case of gas circuit breakers, where the switching medium must be replaced. At the time of writing, SF_6-free vacuum circuit breakers are available up to 204 kV in double break technology and 145 kV in single break technology. An overview (2022) is presented in CIGRE Technical Brochure 871 [105].

Since SF_6-free (high-voltage) vacuum circuit breakers in practice do not need fluorinated gases, their GWP is close to one, and they are marketed as 'F-gas free'. This very low GWP, however, is not per se specific to HV vacuum technology, since also C4-FN-based mixtures could be used for external insulation.

2.9.3.2   Insulation performance requirements

Insulation of vacuum circuit breakers consists of two aspects, the internal insulation between the switching contacts and other parts inside the vacuum interrupter, and the external insulation across the outside of the vacuum interrupter.

- External insulation is the insulation between the end caps (see Figure 2.2) of the vacuum interrupter and between interrupter and all other circuit breaker parts. The interrupter is embedded either in a solid, or in a gas. Whereas solid insulation has been demonstrated to be effective in vacuum switchgear products up to 72.5 kV, atmospheric air can only be applied up to a much lower voltage level. In present-day high-voltage vacuum circuit breaker technology, the external insulation is with pressurized technical air. From Figure 2.31, it can be recognized that air is an inferior insulator compared to fluorinated gas mixtures at the same pressure, so that air pressure must be raised to a level in the order of 8–10 bar, depending on the voltage level. The high pressure of the insulating gas puts high mechanical stresses to the metal parts of the vacuum interrupter enclosure (at 0 bar inside) that might lead to deformation. Also, leakage and permeation into the high-vacuum interrupter should be avoided during lifetime. The main challenge regarding dielectric testing for the external air insulation is the chopped wave test (see Section 2.8.2). Although the pressure of the insulating air is high, condensation is not an issue: it will not take place above approx. $-50°C$.
- Internal insulation is the insulation inside the vacuum interrupter, between the contacts, but also between the contacts and the vapour shield, see Figure 2.2. Vacuum insulation is entirely different from gaseous insulation, since the former is mainly determined by the contact surface, whereas the latter involves the volume between the contacts. Another clear feature of vacuum insulation is the fact that the relationship between gap length and breakdown levels off for the higher voltages, while in gases, there is a proportionality between gap length and breakdown voltage. This is outlined in Figure 2.34.

As a consequence, the design of a high-voltage vacuum interrupter is not a matter of proportional scaling of gap length, as in gas circuit breakers [127]. The application of more than one shorter vacuum gaps in series, instead of one longer gap is a common way out to deal with the non-linearity as outlined in Figure 2.34 and to reach high voltage ratings [128]. This is demonstrated in Figure 2.35 where the curves are interpolations of experimental results [129].

Vacuum circuit breakers operating with a single gap (per pole) are reported up to a voltage of 170 kV [130,131]. Since the breakdown voltage of the vacuum gap

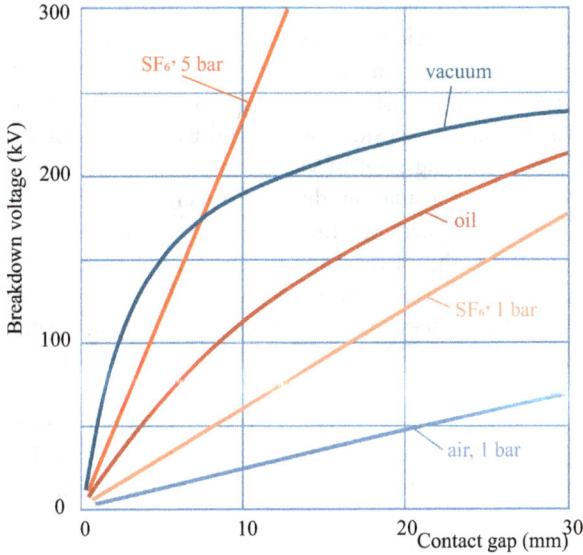

Figure 2.34   Breakdown voltage vs. gap length of various insulating media

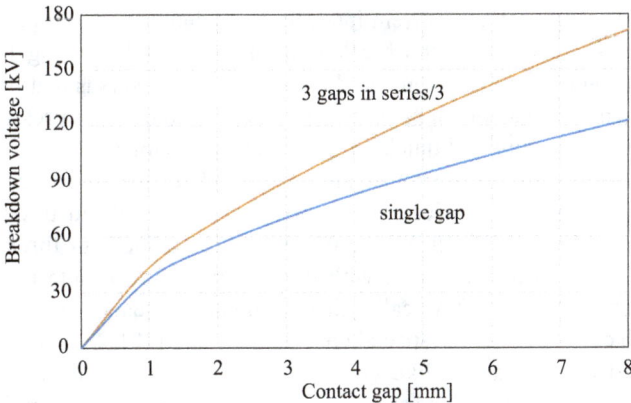

Figure 2.35   Breakdown voltage of a single vacuum gap and of three vacuum gaps in series with the same total contact gap. For the upper curve, each gap has one-third of the gap length of the single gap in the lower curve.

is basically determined by the surface condition, a proper selection of contact material is of key importance, which is nowadays an alloy of copper and chromium. The detailed breakdown mechanism in vacuum is still not completely resolved, although there is an agreement that micro-particles as well as field electron emission current play a role [10]. Micro-particles, loosely adherent to the contact surface, might get detached, charged, and accelerated in the electric field. On impact on the opposing contact, they release material that could initiate breakdown. Field electron emission results from micro-protrusions that cause high electric field magnification because of their sharp geometry. The extreme electric field pulls electrons directly out of the cold metal (by the quantum mechanical 'tunnel-effect') which leads to a net field emission current that might assist in breakdown. Both micro-particles and micro-protrusions are natural by-products of surface deformation by arcing. Since the single set of vacuum interrupter contacts has to perform arc interruption as well as insulation, this dual function ensures that the switching history has a direct impact on the dielectric condition of the switching gap. Therefore, conditioning of the contact surface, either reducing or improving dielectric performance is an often-discussed and multi-faceted topic [10]. As an example, prior to dielectric testing, in the case of vacuum interrupters, it is allowed to apply up to 25 preliminary voltage pulses up to the rated withstand voltage. Its rationale is to positively condition the contacts for a successful dielectric test series [20]. Specific to insulation of (switching) vacuum gaps is the non-sustained disruptive discharge (see Section 2.5.2.4), a late (up to 1 s after interruption) and unexpected short-lasting (microseconds) loss of dielectric strength.

2.9.3.3   Interruption performance requirements

The interruption performance on high-voltage vacuum circuit breakers is not different from the medium-voltage ones. Typically, the recovery is very fast, without the need of external cooling of the arc plasma. Arcing times are shorter than those of gas circuit breakers. Fault current interruption is demonstrated up to the highest currents, e.g., in generator circuit breakers even with very fast rising TRV.

An advantage of vacuum compared to gas circuit breakers is that once vacuum breakers internally break down unexpectedly, there is no need for pressure build-up of a medium to interrupt and quick recovery can be assumed.

All vacuum circuit breakers need arc control to keep the arc in a condition that guarantees interruption, see Section 3.2. Arc control is realized through either a radial magnetic field (to bring the arc in a rotational motion) or through an axial magnetic field (to diffuse the arc), with the intention to distribute the arc energy evenly across the surface. Physically, these magnetic fields are generated by passing the high current through a special contact geometry [10]. Axial magnetic fields are generated through a coil-like current path below both contact surfaces. At longer contact distances, which are necessary for high-voltage vacuum interrupters, it is a requirement to maintain sufficient axial field strength in the arcing zone to control the arc up to the rated short-circuit breaking current. A comprehensive overview of the specific arcing features, including challenges regarding high-current arc control in HV vacuum interrupters, is provided in [132].

Regarding load current interruption, there are two high-voltage switching duties that deserve attention, capacitor-bank current interruption (see Section 2.5.2.3) and shunt-reactor switching (see Section 2.5.3.3).

- Capacitor-bank switching. Capacitive switching is influenced by the inherently wide variation of vacuum in breakdown statistics, which becomes significant for frequently switching capacitor banks. Test-statistics show that capacitor-bank switching, especially in the case of large inrush currents, at higher voltages is associated with an increasing occurrence of late breakdown [36]. At energization of capacitor banks, a high-current high-frequency inrush current occurs at pre-strike which might lead to welding of the closing contacts. After breaking of the weld at the subsequent opening operation, sharply protruding remnants of the weld might promote local high electric fields and decrease withstand. In a recent study of 72.5 kV-class vacuum interrupters, of 475 tests under inrush conditions, in 3% of the switching operations, late restrikes were observed [35]. A special design of HV vacuum interrupters may sometimes be advisable for switching of capacitor banks [103]. Also, controlled switching or damping reactors can avoid high inrush current.
- Shunt reactor switching. In shunt-reactor (inductive-load) switching, the number of re-ignitions (not the probability of re-ignition) is usually much larger than with SF_6 circuit breakers, see Figure 2.36. This is because 'vacuum'

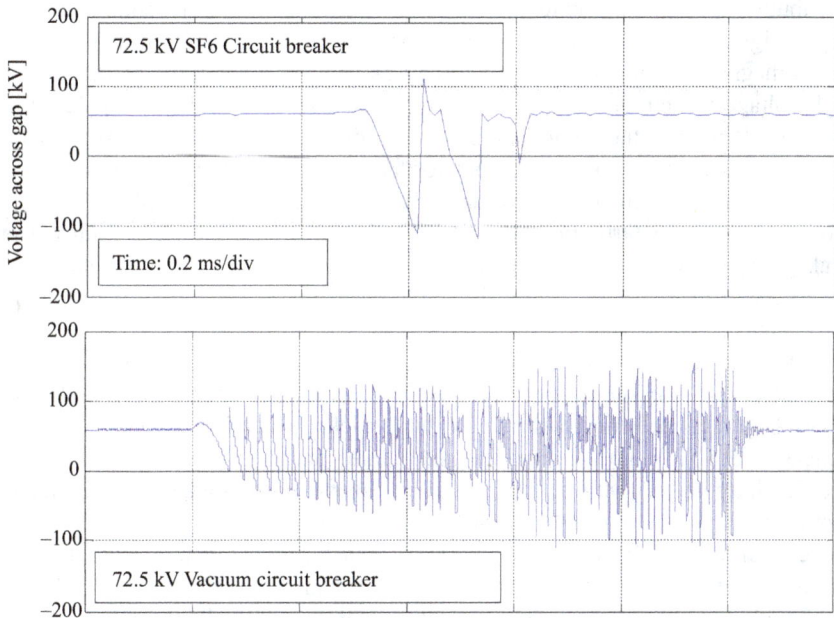

Figure 2.36   Comparison of multiple re-ignitions at shunt reactor switching with a 72.5 kV SF_6 and a vacuum circuit breaker [43]. Vertical axis is voltage across the breaker gap in [kV].

has an excellent capability to interrupt the high-frequency re-ignition currents that flow after every breakdown. Protective measures when switching small HV reactors, especially when directly connected to the circuit breaker, are sometimes recommended. Investigations show that RC filter circuits are an appropriate measure to operate shunt reactors in a secure manner with high-voltage vacuum circuit breakers. It can be expected that multiple re-ignitions can be avoided as long as the parameters of the filter are tuned to the specific system transient behavior [133].

Alternatively, designs that are optimized for shunt-reactor switching may be used. The chopping current level of vacuum circuit breakers does not differ essentially from those of SF_6 circuit breakers and does not depend on voltage level [134].

2.9.3.4   Vacuum circuit breakers with multiple vacuum interrupters

Recently, high-voltage vacuum circuit breakers have been studied and designed that consists of a relatively large number of vacuum interrupters designed for medium voltage. Application of multiple interrupters in series increases the rated voltage, whereas the addition of parallel path(s) enables larger continuous current ratings, such as applied in certain designs of vacuum generator circuit breakers [70].

An advantage of the application of multiple smaller series interrupters is that the relatively low contact separation forces and contact system mass of smaller distribution vacuum interrupters (up to rated voltage of 40.5 kV) – compared to the larger high-voltage interrupters – can more effectively benefit from the use of electromagnetic repulsion coil technology as actuator, a spin-off from ultra-fast high-voltage DC circuit breaker technology, see also Section 2.10.4.1 [135]. With such developments, the combination of vacuum interrupters with ultra-fast actuators opens a new range of applications for fault current limitation, since switching, including the creation of current zero, can become extremely fast, by realizing 'non-zero current interruption' even before the (asymmetrical) first fault current peak [136]. Only vacuum interrupters are suitable for such technology since they have very short minimum arcing time and are ready to interrupt almost immediately after contact separation, in contrast to gas circuit breakers.

Developments have been reported on a fast 363 kV/63 kA/5,000 A air-insulated vacuum circuit breaker having six vacuum interrupters (each 40.5 kV/31.5 kA/ 2,500 A ratings) in series with two of such stacks in parallel per phase having a dimension of 10.3 × 3.6 × 6.4 m [137]. For a three-phase design, this would mean 36 vacuum interrupters (with one fast actuator controlling two adjacent interrupters). Similar multiple-interrupter-based switching devices have been under study for a (concept of a) fast 550 kV/80 kA/5,000 A six-break vacuum circuit breaker, fast fault current limiter as well as for fast series capacitor bank bypass switches [138].

In all cases where more than one interrupter is used, voltage shall be equally shared across every interrupter. This applies for power frequency – as well as for transient voltages. Voltage grading is usually carried out with capacitors for AC application and a series combination of capacitors and resistors for DC.

Requirements of such a voltage grading system are very severe since it must guarantee the correct voltage across every interrupter. Tests shall be performed using the complete stack of series interrupters since also their contact gaps and the contact travel curves shall be similar within a small margin. This sets high requirements for the synchronicity of the actuators, see also Section 2.10.4.1.

After high-current interruption in 'vacuum', the residual metal vapour plasma will conduct a '*post-arc current*' (under the influence of the TVR) that is relatively high, compared to post-arc current in SF_6 [7]. When multiple interrupters are arranged in series, there may be a difference in post-arc current in every interrupter, even when current is interrupted at the same instant. The post-arc current interacts with the grading capacitance and may distort a theoretical even voltage distribution [139]. Understanding the processes coming along with the interruption process lays the basis for the successful realization of high-voltage vacuum switchgear based on (series-connected) vacuum interrupter units [140].

2.9.3.5 Miscellaneous requirements

A few differences exist between high-voltage vacuum and gas circuit breakers regarding operation:

(a) *Vacuum quality assessment.* A key requirement for vacuum interrupters is to provide a high vacuum $(10^{-1}–10^{-5}$ Pa) over a long time, which is usually required as at least 30 years. This ensures a high breakdown strength and half-cycle interruption of currents throughout the lifetime of a circuit breaker. Vacuum interrupters are sealed units, that must remain under this high-vacuum condition during their entire lifetime. It is not straightforward to obtain information on the vacuum state inside. High-voltage gas circuit breakers have pressure gauges to detect pressure (not gas composition) changes that can be used for continuous pressure/density monitoring. In the case of vacuum circuit breakers, by application of high voltage from specialized portable diagnostic units, an impression can be gained on the voltage withstand capability. Measurement of the vacuum state can be performed on separate interrupters, taken out of the circuit breaker. No on-line method is generally available [10], though one product is reported [141]. In service, it is not common to have any vacuum-quality measurement or detection system for indication of loss of vacuum, because a vacuum-quality monitoring system is in general less reliable than the vacuum interrupter sealing technology itself. This means that adding these vacuum-measurement or loss-of-vacuum detection systems may reduce the reliability of the vacuum circuit breaker.

(b) *Continuous current conduction.* Vacuum circuit breakers have butt contacts. This implies that the contact resistance is relatively high, and normally increases by high-current arcing, roughly in the order of 20% after a short-circuit make- and break test-duty. This limits the rated normal current that can be handled within the temperature-rise limits. In order to reduce the contact resistance and to counteract the 'popping' caused by 'blow-off' electromagnetic forces that try to open the contact under high-current conditions, additional contact closing force must be applied by a set of springs that are energized by the mechanism during the closing operation [142]. A design challenge is to drain the heat, generated at the contact interface, which cannot take place by convection, such as in gas interrupters. Instead, the heat has to be conducted by the supporting contact stems to the external environment. This limits the rated normal current, especially for high-voltage applications where the contact supporting stems are relatively long because the requirements of the external dielectric strength need longer ceramic enclosures [143]. Radiators are sometimes provided to the contact stems in order to increase the rated continuous current, which might especially be relevant to high-voltage transmission systems that require higher continuous current than distribution systems.

(c) *X-ray emission.* X-ray emission is generated in high-voltage vacuum devices because electrons, accelerated by the electric field in the gap collide with the metal target contact. In this process, electromagnetic radiation is generated, the energy of which is determined by the voltage across the gap and the intensity of which is determined by the electron current. The biological effect of X-ray radiation on human tissue is expressed as equivalent radiation dose. Its SI unit is the sievert [Sv]. 1 Sv = 1 J/kg, the dose rate is expressed in sievert per hour [Sv/h]. At high-voltage vacuum circuit breaker application the X-ray emission is considered, though the situation of open contacts is rare for circuit breakers. The emission depends on the voltage level, design and on the contact surface roughness. The IEC standard sets limits of 5 μSv/h at the rated voltage and 150 μSv/h at rated power frequency withstand voltage (such as in a test setup) [20]. Measurement and evaluation show that a typical X-ray dose rate for an (open) 145 kV vacuum circuit breaker at its rated voltage is in the range of few tenths of a micro-sievert per hour and can reach close to 100 μSv/h when a voltage of 275 kV r.m.s. is applied across the same gap [144].

2.10 HVDC circuit breakers

2.10.1 Introduction

The interruption process in DC systems is completely different from that in AC systems, and so are the requirements of DC circuit breakers. DC circuit breakers are relatively new components although they have a long research history [145]. The concept of multi-terminal and meshed DC systems is under study in various parts of the world [146,147], mainly for the transmission of renewable energy, and is realized in China in a few projects [148–150]. In order to achieve selective protection of these grids, the importance of DC circuit breakers has gained momentum and now draws major research and product development attention. At present, several products and prototypes have been tested and installed in HVDC networks [151].

International standards do not yet exist.¶ At the time of writing, IEC Technical Committee 17 is working on a standard covering general requirements of HVDC switchgear for 100 kV and above (including dielectric requirements) whereas IEC Technical Committee 17A is compiling technical specifications of various types of HVDC switchgear, including HVDC circuit breakers.

¶A Chinese national standard exists: Chinese draft standard GB/T 38328-2019: 'Common specifications of high-voltage direct current circuit breakers for high-voltage direct current transmission using voltage sourced converters' (2019, in Chinese).

2.10.2 The DC interruption process

In the following sections, information is provided on the interruption process and the requirements of the breaker and its components are collected.

For understanding current interruption in DC systems, the circuit in Figure 2.37 serves as a guidance:

Its circuit equation is (i is current, U_s the system voltage, u_{cb} is the voltage across the circuit breaker, L the inductance):

$$U_s = L\frac{di}{dt} + u_{cb} \rightarrow \frac{di}{dt} = \frac{1}{L}(U_s - u_{cb}) \tag{2.1}$$

This implies that a reduction or suppression of current ($di/dt < 0$) can only be achieved by creating a voltage across the circuit breaker that exceeds the system voltage: $u_{cb} > U_s$. This voltage opposes and counteracts the supply voltage and is therefore called '*counter voltage*'.

All DC breakers are based on the above principle: generation of counter voltage that exceeds the system voltage for a sufficiently long duration. During the presence of the counter voltage, the fault current is suppressed to zero within the '*fault current suppression time*'. It is important not only to generate enough counter voltage but also to develop it very quick (within few milliseconds) during the '*fault neutralization time*', and maintain it sufficiently long, until the fault current is suppressed to zero. Counter voltage is (the first) part of the transient voltage across the circuit breaker called '*Transient Interruption Voltage*' (TIV). After fault current suppression, a residual current switching device then interrupts the residual current, which can be small enough to be interrupted with a disconnecting switch.

The schematized interruption sequence of a HVDC interruption process and its nomenclature is shown in Figure 2.38.

DC fault current is eventually limited by the system's resistance (R) (not shown in Figure 2.37). In the simplified concept the 'steady-state' fault current (I_{ss}) is: $I_{ss} = U_s/R$.

Many different technologies have been proposed to generate and to maintain counter voltage [150]. In the 'conventional' way, the arc voltage between separating contacts can be used as counter voltage, as long as technical means are available that raise (and maintain) arc voltage above the system voltage [7]. Obviously, this is not an option for high-voltage DC systems.

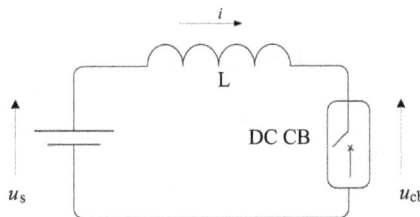

Figure 2.37 Fundamental circuit for DC current interruption

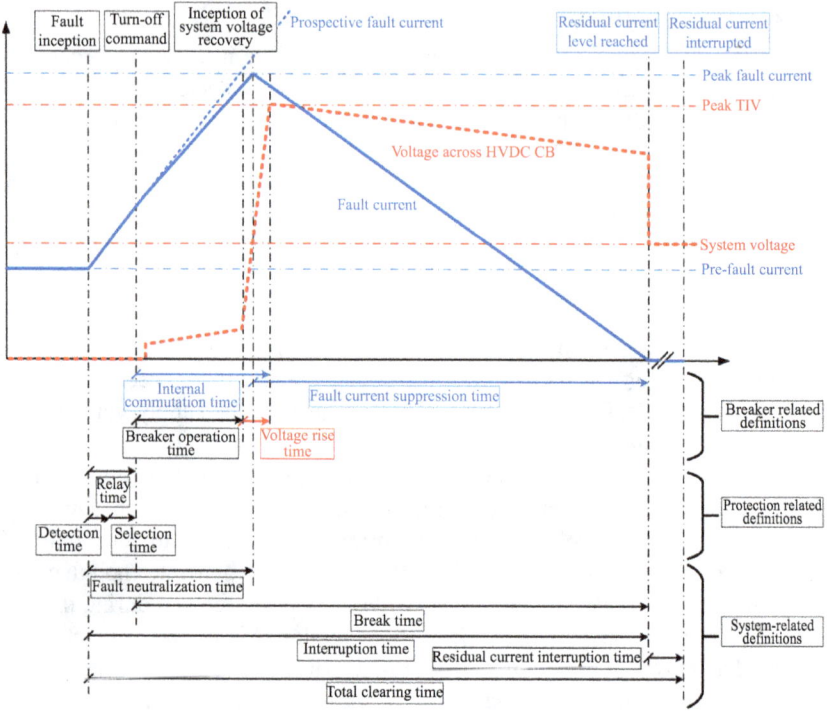

Figure 2.38   DC interruption process and relevant nomenclature

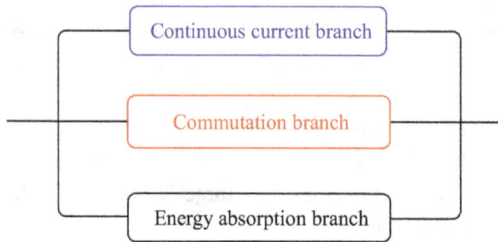

Figure 2.39   Generic topology of DC circuit breakers

In fact, unlike the AC circuit breaker, the HVDC circuit breaker is no longer just a mechanical contact system, but rather a system of components arranged in multiple parallel current branches to which current is successively commutated in a controlled manner to achieve DC interruption. HVDC circuit breakers consist of (at least) three parallel branches, see Figure 2.39:

• Continuous current branch: this is a low-resistance branch for the conduction of the continuous current with minimum conduction losses.

- Commutation branch: this branch creates a high impedance and thus generates an adequate counter voltage when current is forced to commutate into this branch.
- Energy absorption branch: this branch limits and maintains the counter voltage while absorbing the magnetic energy trapped in the HVDC system during the suppression in the fault current suppression period.

The sequence of events that becomes effective upon activation of a HVDC circuit breaker is as outlined below, see also Chapter 5:

1. Local current interruption in the continuous current branch. The fault current is quickly interrupted on its rising edge in this branch. This is achieved by power-electronic switches or by mechanical switchgear with an auxiliary circuit (active current injection circuit) or by a combination of both.
2. Counter voltage generation in the commutation branch. After the continuous current branch is blocked for current passage, current is forced to transfer into the commutation branch. This branch then generates TIV, either immediately upon commutation (active current injection technology) or slightly later after a large stack of series power-electronic switches has interrupted the commutated current (hybrid technology).
3. Energy dissipation in the absorption branch. The TIV rises until the protection level of a metal-oxide surge arrester (MOSA) bank in the third parallel branch is reached. From that moment on, current starts to flow through this branch. Because MOSA protection voltage (U_{MOSA}) is higher than the system voltage ($U_{MOSA} > U_s$), now the current through the MOSA will steadily decline to zero with a negative di/dt. When the current is suppressed to near-zero after the fault current suppression time (t_{FS}), the very small residual current can be interrupted by another switch (residual current switching device) and system voltage appears across the open DC circuit breaker.

The system voltage starts to recover as soon as TIV has been fully developed, limiting the impact of the fault on the system basically to the fault neutralization time (t_{FN}), see Figure 2.38. After this time, the fault (current) is not yet removed from the system, but its impact is compensated by the MOSA temporarily acting as the system voltage source.

HVDC breakers basically differ in the way local current interruption in the continuous current branch is achieved, for example, whether a mechanical switching element is used for interruption or semiconductors do this job. Details on the technology of HVDC circuit breakers are discussed in Chapter 5. Requirements on the key components of the branches are discussed in Section 2.10.4.

In every design, mechanical switchgear is present to interrupt and/or to isolate. This can be vacuum or SF_6 switchgear, each having their '*breaker operation time*', but the one common key requirement is that it must be very fast acting (<10 ms of contact separation) and therefore differs from AC switchgear that never achieves contact separation on the first rising edge of the fault current.

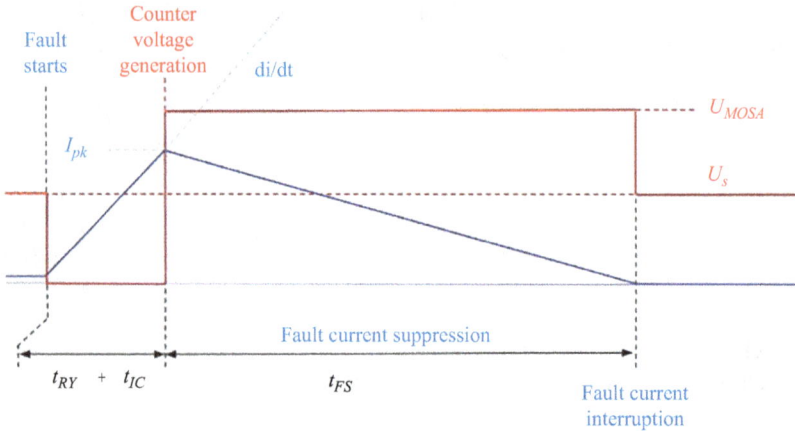

Figure 2.40   Voltage (at supply side of the breaker) and fault current through breaker in the simplified DC interruption process

To formulate the behaviour of a HVDC circuit breaker quantitatively, a simplified mathematical description is provided below using the schematized interruption process in Figure 2.40.

Since fault-current interruption involves the disciplines of circuit breaker design, protection system design, and power system studies, the following input parameters are necessary and need to be defined initially:

System parameter:
• System voltage: U_s
Protection parameter:
• Relay time: t_{RY}
Circuit breaker parameters:
• Peak fault current (interruption capability) I_{pk}.
• Internal current commutation time t_{IC}.
• Protection level and TIV developed and maintained by MOSA: U_{MOSA}.

The above parameters are interdependent. For example, a DC series current limiting reactor (in series with the circuit breaker) is designed depending on the following parameters (in addition to the system voltage):

1. The achievable fault neutralization time considering the worst-case relay time of the protection system and internal commutation time of the HVDC circuit breaker.
2. The maximum current withstand capability of the system components considering the total current breaking time. The peak fault current must not exceed this value.
3. The maximum current interruption capability of the HVDC circuit breaker. The circuit breaker must be able to interrupt any short-circuit current occurring

in the system. Thus, the maximum current interruption capability must be coordinated with the maximum allowable short-circuit current in the system with a sufficient margin.

Given the above parameters, the maximum rate-of-rise of fault current that the HVDC circuit breaker is able to handle, the minimum equivalent total inductance of the system, the fault current suppression time (t_{FS}) and the energy absorbed (E) can be derived using the following basic set of equations described below:

$$\frac{di}{dt} = \frac{I_{pk}}{t_{IC} + t_{RY}} = \frac{I_{pk}}{t_{FN}} \tag{2.2}$$

Thus, to ensure the third condition, the inductance (L) of the system has a value of at least:

$$L = \frac{U_s}{di/dt} \tag{2.3}$$

The fault-current suppression time (t_{FS}) can be estimated as:

$$\frac{I_{pk}}{t_{FS}} = \frac{1}{L}(U_{MOSA} - U_s) \rightarrow t_{FS} = \frac{I_{pk}L}{U_{MOSA} - U_o} \tag{2.4}$$

The energy to be absorbed by the HVDC circuit breaker is:

$$E = \frac{1}{2}U_{MOSA}I_{pk}t_{FS} = \frac{1}{2}\frac{U_{MOSA}I_{pk}^2 L}{U_{MOSA} - U_s} = \frac{1}{2}LI_{pk}^2\frac{1}{1 - \frac{U_s}{U_{MOSA}}} > \frac{1}{2}LI_{pk}^2 \tag{2.5}$$

In order to get an impression of practical values of parameters, the data of a 500 kV HVDC circuit breaker are used as shown in [152], see Table 2.4.

The actual inductance of the system is higher than the value shown in the Table 2.4. Hence, the peak fault current is also lower than value indicated in the

Table 2.4 Parameters from the simplified model of a 535 kV HVDC circuit breaker

Name	Symbol	Value	Unit
	Input parameters		
System voltage	U_s	535	kV
Peak fault current	I_{pk}	25	kA
Internal commutation time	t_{IC}	3	ms
Relay time	t_{RY}	3	ms
MOSA voltage	U_{MOSA}	800	kV
	Derived parameters		
Fault current rate of rise	di/dt	4.2	kA/ms
Minimum system inductance	L	130	mH
Fault current suppression time	t_{FS}	12	ms
Energy supplied to MOSA	E	125	MJ
Stored energy	$\frac{1}{2}LI_{pk}^2$	40.7	MJ

Table 2.4 – showing sufficient margin between the maximum breaking capability of the breaker and the peak fault current.

Note that the energy absorption requirement (E) shown in the Table 2.4 refers to a single breaking operation. In cases where faults are persistent, like a secondary arc appearing in overhead line systems after the main fault is cleared, two or more opening operations may be required, basically doubling the required energy dissipation, since the cool-down time of the large MOSA bank is much longer than the time between successive opening operations (in AC applications usually 300 ms).

2.10.3   Requirements for DC circuit breakers

Compared to AC fault-current interruption, DC fault current interruption is challenging because of the following reasons:

- There is no natural current zero in DC systems. This implies that there is no instant where the inherent magnetic energy ($\frac{1}{2}Li^2$) in the system is zero. For AC current interruption, current zero provides the opportunity to interrupt when there is no magnetic energy in the system. Thus, the AC circuit breaker does not need to absorb magnetic energy in the system whereas the DC circuit breaker must have a provision to absorb many megajoules of the energy in the faulted part of the system. For example, an HVDC circuit breaker needs to absorb over 10 MJ of energy when interrupting 15 kA of fault current flowing in a 100 km long overhead line. Using a mechanical analogy, this is equivalent to absorbing the kinetic energy in a few milliseconds of a 30-ton truck running at 100 km/h by abruptly stopping its motion.
- Fault current in HVDC systems rises rapidly to a peak value limited only by the resistance in the current path, whereas in AC systems the peak value is limited mainly by the inductance of the conductors. This implies that, in HVDC systems, very high fault currents can emerge in a matter of milliseconds. Therefore, HVDC breakers need to act fast, around ten times faster than AC breakers to clear a fault current on its rising edge before reaching its peak value. This requirement can only be met by ultra-fast electro-magnetical systems that actuate mechanical switches in the continuous current branch. With rate-of-rise of fault current (di/dt) in the range of a few to several kA/ms [153], breaker operation time may not exceed 8–10 ms in order to handle technically feasible values of peak fault current (I_{pk}). Though this current is much lower than rated short-circuit breaking currents of AC breakers (up to 63–80 kA), the challenge in DC current interruption is in realizing a short breaker operation time to limit undesirable consequences for system and converter. In addition, a rapid response also calls for the need of very fast DC fault protection.
- To interrupt fault current, HVDC circuit breakers need to quickly generate and sustain counter voltage exceeding the system voltage, whereas there is no such need for an AC current interruption. Rather an AC circuit breaker needs to sustain a system-generated TRV imposed by the system.

In this sense, an AC circuit breaker is a passive device, the stresses in terms of fault current to-be-interrupted and TRV come from the system, and are not influenced by the breaker, whereas the HVDC circuit breaker is an active device that, by design, sets its breaking current and TIV.

In defining requirements, the critical stages of current interruption process and the actual stresses that need to be reproduced (for complete stress) must be identified. To this aim, six critical stages of DC fault interruption are defined:

1. Rise of fault current (high di/dt) → breaker needs to act very fast.
2. Local current interruption in continuous current branch.
3. Internal commutation → initiation of counter voltage generation (du/dt).
4. Limitation and maintenance of TIV → fault current suppression.
5. Energy absorption.
6. System recovery voltage withstand.

In Table 2.5, these stages are schematically outlined, and the critical parameters are identified.

The main motivation for including the stages 4–6 in interruption requirement and tests is the following:

• Mechanical switches are key and novel components of all HVDC circuit breakers. In some applications, they switch high current, in some they isolate high voltage, or both. Short-time dynamic (stage 4) overvoltage withstand and

Table 2.5 Stages of interruption and critical parameters

Stage of interruption	Breaker action	Components mainly stressed	Critical parameter
1 Rise of fault current	Acts and trips very fast	Electro-magnetical actuator(s)continuous current branch	$t_{IC}di/dt$
2 Local current interruption	Blocks the continuous current branch	Interrupting device (vacuum interrupter, power electronics)	I_{pk}
3 Internal commutation	Generates counter voltage disconnector insulates	Active current injection circuit breaker in commutation branch	U_{MOSA}
4 Limitation and maintenance of TIV	Withstands short duration overvoltage ($\approx$1.5 p.u.) fault current suppression	Switching gap of mechanical switch breaker in commutation branch	$U_{MOSA}t_{FS}$
5 Energy absorption	Absorbs energy	MOSA	E
6 System voltage recovery	Withstands long-duration dielectric stress	Switching gap of mechanical switch (residual current switching device)	U_s

long-duration static dielectric withstand (stage 6) shall be an essential part of a verification program.

• Surge arresters in HVDC circuit breakers are used in a different application than for overvoltage protection (the usual application). The unusual amount of energy to be absorbed requires a large number of parallel arrester columns and an equal current sharing between the non-linear zinc-oxide (ZnO) elements [154].

In order to harmonize the verification of stress withstand of HVDC circuit breakers during fault current interruption tests, within a large European demonstration project [155], agreement among participating manufacturers and a test-authority was reached on a set of test-duties to which their breakers would be exposed. Table 2.6 shows these test-requirements.

Table 2.6   HVDC circuit breaker fault current interruption test requirements as agreed in the PROMOTioN project

Test-duty	Current	Breaking test	Number of tests
TC10+	10% of rated continuous current	Tests in positive current direction	2
TC10−	10% of rated continuous current	Tests in negative current direction	2
TC100+	100% of rated continuous current	Tests in positive current direction	2
TC100−	100% of rated continuous current	Tests in negative current direction	2
TF100+	100% of peak fault current	Test at specified energy absorption[a], positive current direction	2
TF100−	100% of peak fault current	Test at specified energy absorption[a], negative current direction	2
TDT+	To be estimated[b]	Test at rated fault current suppression time[c], positive current direction	2
TDT-	To be estimated[b]	Test at rated fault current suppression time[c], negative current direction	2
		In all tests, a constant DC voltage U_s (considering 10–15% overvoltage) will be supplied during 300 ms after main current interruption. All tests are single opening operations	

[a]Specified energy absorption based on specified value of energy absorption (MJ) of the test-object delivered to the laboratory, which for EHV breakers may be much smaller than planned in a project.
[b]Value to be estimated with (2.2)–(2.4) based on U_s, U_{MOSA}, t_{IC}, I_{pk}, as would be present in service.
[c]This test-duty is to verify dynamic dielectric withstand of the mechanical switching gap during the maximum duration of the current suppression time as would occur in service condition

In many cases, especially for testing EHV DC circuit breakers, it may not be possible to apply full-rated energy stress due to either a limitation of a test laboratory or for some practical reasons the test breaker is supplied with reduced energy capability. In such a case, an alternative test duty which can replicate the magnitude and duration of TIV is defined. The test-duty named 'TDT' in Table 2.6 is introduced to fill this gap. This duty is intended to demonstrate the TIV withstand capability of the circuit breaker during the full duration of TIV that would occur under full rated energy (e.g., as specified in a specific project).

The test program of Table 2.6 is taken as a guideline in the testing of three HVDC circuit breakers rated 80 kV 16 kA (*'VSC assisted resonant current technology'*), 160–200 kV 16 kA (*'active current injection technology'*) and 350 kV 20 kA (*'hybrid technology'*) [156].

A novel test method that uses existing installations at a high-power AC test laboratory is developed [157] following an analysis of correct stresses to subcomponents in the HVDC circuit breaker [158]. To supply the complete stress from Table 2.6 in 'one shot', the test circuit is composed of four parts: the power source (short-circuit generators running at 16 ⅔ Hz), over-current protection, a DC voltage source for post current-suppression dielectric stress and an arcing time prolongation circuit described in detail in [159].

2.10.4 Component stresses and requirements

All designs of HVDC circuit breakers have components applied in a non-conventional way, or they include new types of components. In order to reduce the risk of failures of these subcomponents in the application of HVDC circuit breakers, there is a need to analyse the new stresses, typical for HVDC circuit breakers, that these components face. Table 2.7 lists the 'standard' components applied in a 'non-standard' way.

2.10.4.1 Mechanical switching devices

Every practical HVDC circuit breaker is equipped with one or more mechanical switching devices. Their function is to enable low losses in continuous operation and to alleviate (when in open state) dielectric stresses on the power-electronic components. In HVDC circuit breakers, every mechanical switching device has to achieve contact separation very fast, which is realized by electromagnetic repulsion drives. Such drives are electronically controlled, which implies a certain susceptibility to EM interference from transients of the primary sources (arcing, reignition, fast switching, high di/dt-current, high du/dt, etc.). In most designs, a (considerable) number of mechanical switching devices are put in series. This implies that power to the individual drives cannot be supplied through galvanic connections. Usually, transformers that have enough insulation capability are used. For example, several isolation transformers are stacked in series to achieve adequate insulation from the earth for 500 kV HVDC circuit breakers.

High-speed drives and their isolated power supply are not used in such a way before in power equipment and service experience is very limited or non-existing.

Table 2.7 HVDC circuit breaker components facing 'non-standard' stresses and potential issue(s)

	Subcomponent	Characteristic in standard operation	Non-standard stress in HVDC circuit breaker application	Potential issue(s)
Mechanical switching devices	(Multiple) actuator(s) for mechanical switching device	Speed 1 – few m/s	• Ultra-high speed • High impact forces • Control electronics on board	• Mechanical reliability • Compatibility with equipment attached • Electromagnetic sensitivity • Synchronicity
	Power supply to actuator	At earth potential	At high potential	Non-galvanic power supply
	Vacuum interrupters for interruption	Power frequency AC current interruption	• HF current interruption at high di/dt • Recovery at small gap length against very high du/dt • Arc stays at same polarity on the contacts	• Interruption of high-frequency current • Very high du/dt recovery • Uneven contact conditioning
	Vacuum interrupters for insulation	AC voltage insulation	DC voltage applied after interruption	DC voltage withstand capability
	High-speed making switches (vacuum, power electronics, triggered gap)	Capacitor bank inrush current making	Injection current making above highest IEC standardized value	• Contact welding by pre-strike arc • Mechanical synchronicity
	Multiple vacuum breakers in series	Usually single break	Several to many interrupters in series	• Voltage grading for transients and DC • Redundancy • Mechanical synchronicity
	SF_6 gap(s) for insulation	Very low opening speed in GIS AC application	Ultra-high contact separation speed	• Shall not switch current • Dynamic DC voltage withstand capability

semi-conductor switching devices	Power electronics in continuous current branch	Conduct continuously and switch only occasionally	• Extreme mechanical consistency over time • Thermal stability • Unequal thermal distribution
	Power electronics in commutation branch	Never/hardly conduct and switch only occasionally	• Reliability after long idle time
MOSA	MOSA consisting of multiple columns	Significant energy absorption (similar application in series capacitor tank protection)	• Thermal overload, runaway • Current sharing between columns
	MOSA columns	Always under (AC) voltage Occasionally stressed by voltage	• Conditioning • Stability of U-I characteristic • DC stability

Due attention needs to be paid to the verification of the mechanical endurance of the total kinematic chain.

In addition, the proper overall voltage withstand of one stack consisting of a larger number of smaller interrupters needs a well-synchronized contact separation as well as a built-in redundancy to overcome the functional loss of one or more individual interrupters. Differences of ± 0.1 ms in switch opening times are reported in a stack of 10 fast disconnecting vacuum switches which achieves 1,000 kV isolation in 2 ms [160].

The special high-speed drives cause huge impact forces on the contact systems of vacuum or SF_6 interrupters. Care must be taken when applying standard AC vacuum interrupters in combination with fast drives. Especially the bellows of standard spring-actuated vacuum interrupters are not designed to withstand a certain number of high-impact force opening operations.

The series combination of several/many interrupters has its challenges, not only mechanically but also electrically. Sharing (grading) of voltage needs to be considered seriously, not only regarding DC voltage but also during transients. For AC applications, niche products like capacitor bank switches consisting of up to 9 series vacuum interrupters do exist but have a poor service record [103]. In general, for AC, commercial high-voltage vacuum circuit breakers are developed with as few as possible series interrupters, not more than two at present.

The use of a limited number of HV vacuum breakers is an advantage compared to a much higher number of MV vacuum breakers, although it is a challenge to realize a high enough opening speed with the heavier HV vacuum interrupter contact systems.

Vacuum is a very good 'medium' regarding interruption of high-frequency current and very fast recovery of the gap against steep rising recovery voltage. Nevertheless, the application of vacuum interruption in active current injection type of HVDC circuit breakers may approach performance limits.

Mechanical gaps (vacuum/SF_6) break down electrically when they are not able to withstand the voltage. Most critical is the fault current suppression phase, where the overvoltage is around 1.5 p.u. whereas at the same time, the gaps are recovering from interruption and/or switching. After fault current suppression, there is a much longer exposure to the recovering system voltage and its (slow) transients, until the residual current switch takes over that voltage stress. Breakdown of (a) vacuum gap(s) is less critical than of an SF_6 gap, since a vacuum is able to restore its insulating state very fast, while SF_6 generally cannot. Upon breakdown, in addition to the damage of its contacts, the impact to SF_6 switchgear is that power-electronic switches in the continuous current branch will be damaged which results in a complete malfunctioning.

In HVDC circuit breakers, SF_6 disconnectors need to open with very low current and with very low voltage in order to avoid arcing. Once the current (at contact separation) exceeds a certain threshold, the arc will persist during a time, depending on the voltage across the commutation branch. The current at disconnector contact separation is determined by the leakage current through the snubbers/grading elements of the series (power-electronic) switches in the continuous current branch (up to a few amperes). The voltage against which the disconnector is opening is determined by the on-state voltage of the power electronics in the commutation branch (main breaker), which can be several hundreds of volts to a few kilovolts. The latter could be sufficient to keep current conduction in the disconnector if opened before the current is fully commutated. Therefore, the design of (ultra-)fast disconnectors is extremely critical. Opening of the disconnector needs to be synchronized carefully, after current transients in the continuous current branch have decayed sufficiently.

Once the breaker in the commutation branch has interrupted the fault current, soon the full TIV appears across the switching gap which must have been sufficiently open to isolate. Breakdown of this gap would lead to dielectric overload of the load commutation switch and a free burning arc in the SF_6 disconnector. Therefore, the dielectric coordination of this disconnector allows only a very small variation in its opening time over service life. Requirements for mechanical stability are more severe than for controlled (capacitor bank) SF_6 circuit breakers for HVAC applications [161].

In many cases, standard AC vacuum circuit breakers are used as ultra-fast disconnector. Such devices are not optimized for DC voltage withstand, so proper verification is necessary. Moreover, standard vacuum breakers contain arc control devices (axial or radial magnetic field arc control devices). They add unnecessary weight to the interrupters and additional limits to the actuator, given the fact that currents that DC vacuum breakers have to deal with are far below those in AC systems. In addition, the standard contact material, a CuCr alloy, may not be the optimum choice for DC conduction and insulation. From dedicated studies, it was found that different designs of commercially available 36–38 kV AC vacuum circuit breakers acted very different regarding the interruption of HF counter current and TIV withstand capability [157].

In the active current injection schemes using capacitor discharge as the counter current source, an ultra-fast switching device must be used to start the discharge. This may be a mechanical switch (vacuum-making switch, triggered spark gap) or a semiconductor stack (IGCTs, thyristors). In both applications, the very large di/dt and peak current need to be evaluated as a non-standard stress. In the case of vacuum-making switch, provisions must be made to avoid contact welding, originating from the pre-strike arc that is comparable to the back-to-back capacitor bank-making function in AC application (see Section 2.5.2.5). For EHV (extra high voltage) applications, several of these switches need to be connected in series. Synchronous operation of these making switches is essential to avoid premature current injection as a result of pre-strike. In the case of semiconductor-making switches, di/dt and short-time thermal and dynamic stresses can be extreme and far from standard. In addition, during normal operation, the making devices in the open position are subjected to continuous DC voltage stress.

2.10.4.2 Power-electronic switches

The application of large stacks of power-electronic switches is not new and ample experience exists in AC/DC converters.

In the commutation switch that needs to commutate the current into the commutation branch, large current is ramped down to zero with a very high di/dt. This switch consists of a limited number of power-electronic switches. The proper choice of the number of elements is critical from thermal point of view and continuous cooling is necessary.

Commutation switches are mostly made from the state-of-the-art IGBTs (*Insulated-Gate Bipolar Transistor*) slabs which consist of IGBTs and diodes at different sections of the semiconductor package. These packages have IGBTs and diodes to be able to conduct current in reverse direction. Since the load commutation switch is conducting DC current continuously in normal operation, only the IGBT parts of the semiconductor volume are heated due to on-state loss. The results of this unequal heating must get sufficient attention as there is not much operational experiences. During functional tests of HVDC circuit breakers which focus on the proof of concept, steady-state effects like unequal heating cannot be observed. The latest developments of IGBTs, known as BIGTs (*Bi-mode Insulated Gate Transistor*), which uses the same power-electronics volume for both IGBT and diode are not susceptible to this situation [162].

In specification, one might make sure what the thermal limit in the case of a full through short-circuit current (short-time current) is. Power electronics are sensitive, and contrary to converters, HVDC breakers may have no protection from power electronics (like blocking and bypassing current into freewheeling diodes) in the case of an unexpected short-time current exceeding the rated breaking capability.

By activating only a limited number of cells in the main breaker power-electronics stack, certain designs of hybrid HVDC circuit breakers can operate in fault current limiting mode [163].

In hybrid HVDC breakers, the commutation switch is conducting current continuously (the breaker in the commutation branch may be as well, but at very low current) without switching. This is a different operation than in 'normal' operation, like in converters, where power-electronic switches are switching continuously. In such a way, semiconductors conduct during on-state while diodes might conduct during off-state. This results in a lower thermal gradient across the package volume. Alternatively, the breaker in the commutation branch is idle, or conducting small current while under 'standard converter operation', its semiconductors are operating continuously. It is not known how the 'one time only' activation to switch off all power-electronic switches might have an impact on a possible conditioning of the semiconductor junction and/or package.

2.10.4.3   MOSA energy absorbers

A large volume of MOSA is needed for absorbing the energy from the faulted system and maintaining the counter voltage. Many columns are needed in parallel to cope with large energy absorption. This means the individual zinc-oxide (ZnO) varistor discs composing each column needs to be carefully selected to have an equal current flowing through the column. Given the high non-linearity of the u–i characteristic, equal stacking and paralleling can change its characteristic unfavourably [164,165]. Therefore, very careful selection of individual varistor disks and matching of the parallel columns is essential. A small mismatch in conduction voltage would lead to a large current difference. This, in turn, would heat the columns.

The total mass of ZnO material in HVDC circuit breakers can be over one thousand kilograms, which implies that cooling down (after interruption of a significant fault current) is very time consuming, in the order of several hours after rated fault current interruption. The consequence of this is that when a reclosure and re-open function is required, the design should be able to absorb at least double the energy that is associated with a single interruption (and proportionally more counter voltage creation, etc. when more than two reclosures are expected). Moreover, the other functions of the breaker (local current zero creation) should be accommodated for quickly repeated operation.

Multi-reclosure of HVDC breakers is required in overhead line (OHL) systems. OHL arcing faults most often disappear after a reclosure and a subsequent opening (O–CO sequence) of the breaker. HVDC circuit breakers for the Zhangbei (overhead line) project in China have been specified to deal with total energy absorption exceeding 150 MJ [166]. When the actual short circuit, carrying the large fault current is removed, in many cases, a low-current secondary arc to earth persists, which is fed through the stray impedances of the transmission system. Reclosure should then be delayed until the secondary arc ceases, mostly by natural reasons, like wind or by thermal elongation. CIGRE studies have indicated that in HVAC overhead line systems, after a single open-close action, the fault is removed in the vast majority of the cases [17]. Further study needs to reveal the persistence of secondary arcs in HVDC OHL systems.

In cable systems, reclosure does not seem to be a suitable action since faults in cable systems are normally destructive and need repair.

References

[1]   Ragaller, K., *Current Interruption in High-Voltage Networks*, Plenum Press (1977), ISBN 0-306-40007-3.

[2]   Browne, T.E., Jr., *Circuit Interruption*, Marcel Dekker, New York, NY (1984), Chapter 3, ISBN 0-824771-77-X.

[3]   Nakanishi, K. (ed.), *Switching Phenomena in High-Voltage Circuit-Breakers*, Marcel Dekker, New York, NY (1991), ISBN 0-8247-8543-6.

[4]   Garzon, R.D., *High-Voltage Circuit-Breakers – Design and Application*, 2nd ed., Marcel Dekker, New York, NY (2002), ISBN 0-8247-0799-0.

[5]   Kapetanović, M., 'High voltage circuit-breakers', ETF – Faculty of Electrotechnical Engineering, Sarajevo (2011), ISBN 978-9958-629-39-6.

[6]   Ryan, H.M., 'Gas-filled interrupters – fundamentals', In: Jones, G.R., Seeger, M., and Spencer, J.W. (eds.), High-Voltage Engineering and Testing, 3rd ed., IET, Stevenage, UK (2013), ISBN 978-1-84919-263-7.

[7]   Smeets, R.P.P, van der Sluis, L., Kapetanovic, M., Peelo, D.F., and Janssen, A.L.J., *Switching in Electrical Transmission and Distribution Systems*, 1st ed., John Wiley & Sons (2015), ISBN 978-118-38135-9. Chinese edition, John Wiley, China Machine Press (2019) (in Chinese), ISBN 978-7-62086-0.

[8] Niayesh, K. and Runde, M., *Power Switching Components. Theory, Applications and Future Trends*, Springer (2017), ISBN 978-3-319-84656-9.

[9] Ito, H. (ed.), *Switching Equipment*, 1st edn., CIGRE Green Book, Springer Int. Publ. Co. (2019), ISBN 978-3-319-72537-6.

[10] Slade, P.G., *The Vacuum Interrupter. Theory, Design, and Application*, 2nd ed., CRC Press (2021), ISBN 978-0-367-27505.

[11] Smeets, R.P.P. and Lathouwers, A.G.A., 'Economy motivated increase of DC time constants in power systems and consequences for fault current interruption', In: IEEE PES Summer Meeting (2001), 0-7803-7173-9/01.

[12] Slade, P.G., *Electrical Contacts*, Marcel Dekker, Inc. (1999), ISBN 0-8247-1934-4.

[13] Lafferty, J.M. (ed.), *Vacuum Arcs, Theory and Application*, John Wiley & Sons, New York (1980), ISBN 0-471-06506-4.

[14] Anders, A., *Cathodic Arcs. From Fractal Spot to Energetic Condensation*, Springer (2008), ISBN 978-0-387-79107-4.

[15] Peelo, D.F., *Current Interruption Transients Calculation*, John Wiley & Sons (2014), ISBN 978-1-118-70719-7.

[16] IEC 62271-100, 'High-voltage Switchgear and Controlgear – Part 100: Alternating-Current Circuit-Breakers', Ed. 3.1 (2021).

[17] CIGRE Working Group 13.08, 'Life management of Circuit-Breakers', *CIGRE Technical Brochure* 165 (2000).

[18] CIGRE WG B1.57: 'Update of service experience of HV underground and submarine cable systems', *CIGRE Technical Brochure* 815 (2020).

[19] IEC website: www.iec.ch

[20] IEC 62271-1, 'High-Voltage Switchgear and Controlgear – Part 1 Common Specifications', Ed. 2.0 (2017).

[21] IEC 62271-306, 'High-Voltage Switchgear and Controlgear – Part 306: Guide to IEC 62271-100, IEC 62271-1 and Other IEC Standards Related to Alternating Current Circuit-Breakers', Ed. 1.1 (2018).

[22] CIGRE Joined Working Group A3/B4.34, 'Technical Requirements and Specifications of State-of-the-Art HVDC Switching Equipment', *CIGRE Technical Brochure* 683 (2017).

[23] CIGRE Working Group A3.22, 'Field Experience and Technical Specifications of Substation Equipment up to 1 200 kV', *CIGRE Technical Brochure* 362 (2008).

[24] CIGRE Working Group A3.22, 'Background of Technical Specifications for Substation Equipment Exceeding 800 kV AC', *CIGRE Technical Brochure* 456 (2010).

[25] CIGRE Joint Working Group B4/A3.80, 'Design, Test and Application of HVDC circuit Breaker', *CIGRE Technical Brochure* 873 (2022).

[26] Smeets, R.P.P., Verhoeven, S.A.M., 'Reliability of key T&D equipment: Test laboratory- and field experience', *CIGRE Electra*, No. 323, (2022).

[27] CIGRE Working Group 13.04: 'Specified Time Constants for Testing Asymmetric Current Capability of Switchgear', *Electra* 173, pp. 19–31 (1997).

[28] Smeets, R.P.P., Baum, B., Nijman, R., Petropoulos, D., and Ohtaka, T., 'High-Voltage Circuit Breaker Test Statistics 2011-2016 and Test Analysis Tools', *CIGRE Conference*, report A3-102 (2018).

[29] CIGRE Joint Working Group A3/B5/C4.37: 'System Conditions for and Probability of Out-of-Phase. Background, Recommendations, Developments of Instable Power Systems", *CIGRE Technical Brochure* 716 (2018).

[30] Peelo, D.F., 'Current interruption using high-voltage air-break disconnectors', Ph.D. Thesis Eindhoven University (2004), ISBN 90-386-1533-7, http://alexandria.tue.nl/extra2/200410772.pdf.

[31] Smeets, R.P.P., Linden, van der, W.A., Achterkamp, M., de Meulemeester, E.M., and Damstra, G.C., 'Disconnector switching in GIS: three-phase testing and – phenomena', *IEEE Trans. Pow. Del.*, vol. 15, no. 1, pp. 122–127 (2000).

[32] CIGRE Working Group D1.03, 'Very Fast Transient Overvoltages (VFTO) in Gas-Insulated UHV Substations', *CIGRE Technical Brochure* 519 (2012).

[33] IEC 62271-102, 'High-Voltage Switchgear and Controlgear – Part 102: Alternating Current Disconnectors and Earthing Switches', Ed. 2.0 (2018).

[34] CIGRE Working Group 13.04, 'Line-Charging Current Switching of HV Lines – Stresses and Testing, Part 1 and 2', *CIGRE Technical Brochure* 47 (1996).

[35] Surges, B., 'Late breakdown behaviour of 72.5 kV vacuum interrupters during capacitive switching with a synthetic test method', Ph.D. Thesis Technical University of Darmstadt (2022).

[36] Smeets, R.P.P., Wiggers, R., Bannink H., Kuivenhoven, S., Chakraborty S., and Sandolache, G., The impact of switching capacitor banks with very high inrush current on switchgear, In: *CIGRE Conference*, Paper A3-201 (2012).

[37] IEEE C37.09-2018, 'IEEE Standard Test Procedures for AC High-Voltage Circuit Breakers with Rated Maximum Voltage Above 1000 V' (2018).

[38] CIGRE Working Group 13.07: 'Controlled Switching of HVAC Circuit-Breakers – Guide for Application Lines, Reactors, Capacitors, Transformers (1st Part)', Electra No. 183, pp. 43-73, February 1999; (2nd Part), Electra No. 185 pp. 37-57 (1999).

[39] CIGRE Working Group A3.07, 'Controlled Switching of HV AC Circuit-Breakers – Part 1: Benefits and Economic Aspects', *CIGRE Technical Brochure* 262 (2004); Part 2: 'Guidance for further Applications including Unloaded Transformer Switching, Load and Fault Interruption and Circuit-Breaker Uprating', *CIGRE Technical Brochure* 263 (2004); Part 3: 'Planning, Specification and Testing of Controlled Switching Systems', *CIGRE Technical Brochure* 264 (2004).

[40] CIGRE Working Group A3.38, 'Shunt capacitor switching in distribution and transmission systems', *CIGRE Technical Brochure* 817 (2020).

[41] IEC Standard 62271-110, 'High-Voltage and Controlgear – Part 110: Inductive Load Switching', Ed. 4.0 (2017).

[42] IEC Technical Report 62271-306, 'High-Voltage Switchgear and Controlgear – Part 306: Guide to IEC 62271-100, *IEC 62271-1 and Other IEC Standards Related to Alternating Current Circuit-Breakers'* (2018).

[43] Smeets, R.P.P. and te Paske, L.H., Recent standardization developments and test-experiences in switching inductive load current, In: *3rd International Conference on Electric Power Equipment* – Switching Technology, Busan (2015).

[44] Peelo, D.F., Avent, B.L., Drakos, J.E., Giudici, B.C., and Irvine, J.R., 'Shunt reactor switching tests in BC Hydro's 500 kV System', *IEE Proc.*, vol. 135, Pt. C, no. 5, pp. 420–434 (1988).

[45] CIGRE Working Group A3.06, 'Final Report of the 2004 – 2007 International Enquiry on Reliability of High Voltage Equipment. Part 2 – Reliability of High Voltage SF$_6$ Circuit Breakers', *CIGRE Technical Brochure* 510 (2012).

[46] Tanae, H., Matsuzaka, E., Nishida, I., Matori, I., Tsukushi, M., and Hirasawa, K., 'High-frequency reignition current and its influence on electrical durability of circuit-breakers associated with shunt-reactor current switching', *IEEE Trans. Pow. Del.*, vol. 19, no. 3, pp. 1105–1111 (2004).

[47] Bachiller, J.A., Cavero, E., Salamanca, F., and Rodriguez, 'The operation of shunt reactors in the Spanish 400 kV Network – Study of the suitability of different circuit-breakers and possible solutions to observed problems', In: *CIGRE Conference*, Paper 23–106 (1994).

[48] IEEE Switchgear Committee, 'IEEE Guide for the Application of Shunt Reactor Switching', IEEE Std C37.15 (2017).

[49] CIGRE Working Group 13.02, 'Interruption of Small Inductive Currents', CIGRE Technical Brochure 50 (1995).

[50] Papp, K., Sharp, M.R., and Peelo, D.F., 'High Voltage Dry-Type Air-Core Shunt Reactors', In: *CIGRE Conference*, paper A3.101 (2014).

[51] CIGRE Working Group C4.502, 'Power system technical performance issues related to the application of long HVAC cables,' *CIGRE Technical Brochure* 556 (2013).

[52] Tsukao, S., Yoshida, I., Kobayashi, T., Kosakada, M., and Sugiyama, H., 'Considerations for equipment of underground cable networks', In: *Int. Conf. on Cond. Mon. Diagn. and Maintenance*, Paper 4-2 (2019).

[53] Ohno, T., 'Dynamic study on the 400 kV 60 km Kyndbyværket – Asnæsværket Line', Ph.D. Thesis, Aalborg University, 2012.

[54] Zheng, Z., Sun, Q., Yang J., Chen, S., Wang, F., and Zhong, L., 'Investigation on overvoltage caused by vacuum circuit breaker switching off shunt reactor in offshore wind farms', *High Voltage*, vol. 7, no. 5, pp. 936–949 (2022). https://doi.org/10.1049/hve2.12231

[55] Hernández, V.J., Álvarez, G., Burgos, A., García, B., Molina, G., and del Solo, D., 'Switching of long compensated cables. Transients and switching strategies applied in 132kV AC Mallorca-Ibiza submarine link', In: *CIGRE Conference*, Paper A3-308 (2018).

[56] Palone, F., Rebolini, M., Lauria, S., Maccioni, M., Shembari, M., and Vassallo, J., 'Switching transients on very long HV AC cable lines: simulations and measurements on the 230 kV Malta-Sicily Interconnector', In: *Cigre Conference*, Paper C4-210 (2016).

[57] Spathis, D., Boutsika, T., Prousalidis, J., Tsirekis, K., Kabouris, J., and Georgopoulos, A., 'Zero missing effect transient analysis on the 150 kV AC interconnection between Crete and Peloponnese', In: *2018 IEEE International Conference on High Voltage Engineering and Application (ICHVE)*, pp. 1–4 (2018) doi: 10.1109/ICHVE.2018.8642102.

[58] Khalilnezhad, H., Popov, M., van der Sluis, L., Bos, J.A., de Jong, J., and Ametani, A., 'Countermeasures of zero-missing phenomenon in (E)HV cable systems', *IEEE Trans. Pow. Del.*, vol. 33, no. 4, pp. 1657–1667 (2018) doi: 10.1109/TPWRD.2017.2729883.

[59] IEC 62271-306, 'High-Voltage Switchgear and Controlgear – Part 306: Guide to IEC 62271-100, IEC 62271-1, and IEC Standards Related to Alternating Current Circuit-Breakers' (2012).

[60] Kulicke, B. and Schramm, H-H., 'Clearance of short-circuits with delayed current zeros in the Itaipu 550 kV substation', *IEEE Trans. Pow. Appl. Syst.*, vol. PAS-99, pp, 1406–1414 (1980).

[61] Bui-Van, Q., Khodabakhchian, B., Landry, M., Mahseredjian, J., and Mainville, J., 'Performance of series-compensated line circuit breakers under delayed current-zero conditions', *IEEE Trans. Pow. Del.*, vol. 12, pp. 227–233 (1997).

[62] Kerr, B., Ache, J., Linn, S., and Uzelac, N., 'Smart switchgear for extreme installation environments', In: *CIRED Conference*, Paper 1100 (2019).

[63] Osborne, M., Waldron, M., Jarman,P., Zhang, R., and Gardner, R., 'The development of a 400 kV mobile substation bay for flexible transmission services', In: *CIGRE Conference*, Paper B3–109 (2018).

[64] IEC/IEEE 62271-37-013: 'High-voltage switchgear and controlgear. Alternating-current generator circuit-breakers' (2021).

[65] Dufournet, D. and Willieme, J.M., 'Generator circuit-breakers: SF$_6$ breaking chamber – Interruption of current with non-zero passage – influence of cable connection on TRV of system fed faults', In: *CIGRE Conference*, Paper 13–101 (2002).

[66] Lim, L.S. and Smith, I.R., 'Turbogenerator short circuits with delayed current zeros', *Proc. IEE*, vol. 124, no. 12, pp. 1163–1169 (1977).

[67] Boldea, I., *The Electric Generators Handbook*, CRC Press (2005), ISBN 7949314810808.

[68] Kohyama, H., Kamei, K., Yoshida, D., Ito, H., and Wilson, H., 'Study of interrupting duties of delayed zero crossing current in generator main circuit', In: CIGRE A3/B2 Colloqiuum, Paper 421, Auckland (2013).

[69] Venna, K.R., Urbanek, H., and Lechner, A., 'Short circuit analysis of a doubly fed induction generator and their impact on generator circuit breakers', In: *CIGRE Conference*, Paper 10805 (2022).

[70] Leufkens, P., Nayar, R., and Venna, K.R., 'Vacuum generator circuit breaker as a reliable SF6 alternative with reduced life cycle costs for power plants up to 400 MW', In: *CIGRE Conference*, Paper A3–302 (2018).

[71] Jansen, A.L.J., Smeets, R.P.P., Linden, van der, W.A., and Riet van, M.J.M., 'Distributed generation in relation to phase opposition and short-circuits', In:

10th Int. Symp. on Short-Circuit Currents in Pow. Syst., Lodz (Oct. 28–29, 2002).

[72] Dufournet, D. and Montillet, G., 'Transient recovery voltages requirements for system source fault interrupting by small generator circuit-breakers', *IEEE Trans. Pow. Del.*, vol. 17, no. 2, pp. 472–478 (2002).

[73] Palazzo, M., Braun, D., Cavaliere, G., *et al.*, 'Reliability analysis of generator circuit-breakers', In: *CIGRE Conference*, Paper A3–206 (2012).

[74] Smeets, R.P.P., te Paske, L.H., Kuivenhoven, S., *et al.*, 'Interruption phenomena and testing of very large SF_6 generator circuit-breakers', In: *CIGRE Conference*, Paper A3–307 (2014).

[75] Smeets, R.P.P., Barts, H.D., and Zehnder, L., 'Extreme stresses of generator circuit-breakers', In: *CIGRE Conference*, Paper A3–306 (2006).

[76] Slamecka, E. and Waterscheck, W., 'Schaltvorgänge in Hoch- und Niederspannungsnetzen', Siemens Aktiengesellschaft Erlangen, ISBN 3 8009 1106 X (1972).

[77] Rüdenberg, R., *Elektrische Schaltvorgänge*, 5th ed., Springer Verlag (1974), ISBN 978-3642503344.

[78] Greenwood, A., *Electrical Transients in Power Systems*, John Wiley and Sons Ltd. (1991), ISBN 0-471-62058-0.

[79] Sluis, van der, L., *Transients in Power Systems*, John Wiley and Sons (2001), ISBN 0-471-48639-6, Ltd.

[80] IEC 60071-1, 'Insulation co-ordination – Part 1: Definitions, principles and rules', Ed. 9.0 (2019).

[81] Iravani, M.R., Chaudhary, A.K.S., Giesbrecht, W.J., *et al.*, 'Modeling and analysis guidelines for slow transients – Part III: the study of ferror-esonance', *IEEE Trans. Pow. Del.*, vol. 15, no.1, pp. 255–265 (2000).

[82] CIGRE Working Group C4.307, 'Resonance and Ferroresonance in Power Networks', *CIGRE Technical Brochure* 569 (2014).

[83] CIGRE Working Group C4.407, 'Lightning parameters for engineering applications', *CIGRE Technical Brochure* 549 (2013).

[84] CIGRE Working Group 33/13.09, 'Monograph on GIS Very Fast Transients', *CIGRE Technical Brochure* 35 (2005).

[85] CIGRE Working Group A3.22, 'Background of Technical Specifications for Substation Equipment Exceeding 800 kV AC', *CIGRE Technical Brochure* 456 (2011).

[86] CIGRE Working Group C4.208, 'EMC within Power Plants and Substations', *CIGRE Technical Brochure* 535 (2013).

[87] IEC 60071-2, 'Insulation Co-ordination – Part 2: Application Guidelines', Ed. 4.0 (2018).

[88] Kuffel, E. and Zaengl, W.S., *High-Voltage Engineering. Fundamentals*, Pergamon Press (1984), ISBN 0-08-024212-X.

[89] Allen, N.L., 'Mechanism of air breakdown', In: Haddad, M., and Warne D. (eds.), Chapter 1 of *Advances in High-Voltage Engineering*, IEE Press (2004), ISBN 0 85296 158 8.

[90] Cooray, V., 'Mechanism of electrical discharges', In: Chapter 3 of *The Lightning Flash*, IEE Press (2003), ISBN 0 85296 780 2.

[91] IEC 60060-1, 'High-Voltage Test Techniques – Part 1: General Definitions and Test Requirements', Ed. 3.0 (2010).

[92] IEEE C37.04-2018, 'IEEE Standard for Ratings and Requirements for AC High-Voltage Circuit Breakers with Rated Maximum Voltage Above 1000 V' (2018).

[93] Intergovernmental Panel on Climate Change (IPCC), 'Climate Change 2021', IPCC AR6 WGI, tabe 7.SM.7, https://www.ipcc.ch/report/ar6/wg1/downloads/report/IPCC_AR6_WGI_Full_Report.pdf, (2021).

[94] Proposal for a Regulation of the European Parliament and of the Council on Fluorinated Greenhouse Gases, amending Directive (EU) 2019/1937 and Replacing (EU) No. 517/2014 (2022).

[95] Franck, C.M., Chachereau, A., and Pachin, J., 'SF$_6$-free gas-insulated switchgear: current status and future trends', *IEEE Elec. Insul. Mag.*, vol. 37, no. 1 (2021).

[96] Gao, K., Yan, X., Li, G., Wang, W., and He, J., Research of UHV Gas-insulated Transmission Line (GIL) with perfluoronitrile (C4F7N) gas, In: *CIGRE Conference*, Paper 1118 (2022).

[97] Holaus, W., Schueller, M., and Schneider, M., 'Pressurized air insulated cables: a novel, compact GIL design for 12 kV–420 kV: design, simulation, and test results', In: *CIGRE Conference*, Paper 10657 (2022).

[98] Giovanelli, L., Prucker, U., and Ferdjallah Kherkhachi, E., 'Instrument transformers and bushings using alternative and eco-friendly high voltage insulation systems', In: *CIGRE Conference*, Paper 10657 (2022).

[99] Christen, R., Lüscher, R., Szymczak, J., Juge, P., Dupouy, M., and Saint-Marc, J., 'Optimized LPIT (Low Power Instrument Transformer) applications in GIS using SF$_6$ and climate-friendly insulating gas g3', In: *CIGRE Conference*, Paper 10659 (2022).

[100] Loizou, L., Chen, L., Liu, Q., Waldron, M., Wilson, G., and Owens, J., 'Application of SF$_6$ alternatives for retro-filling existing equipment', In: *CIGRE Conference*, Paper 10657 (2022).

[101] Gatzsche, M., Straumann, U., Stoller, P., *et al.*, 'Moving towards carbon-neutral high-voltage switchgear by combining eco-efficient technologies', In: *CIGRE Conference*, Paper 10656 (2022).

[102] CIGRE Working Group D1.67, 'Electric performance of new non-SF$_6$ gases and gas mixtures for gas insulated systems', *CIGRE Technical Brochure* 849 (2021).

[103] CIGRE Working Group A3.27: 'The Impact of the Application of Vacuum Switchgear at Transmission Voltages', *CIGRE Technical Brochure* 589 (2014).

[104] See for example https://www.gwelectric.com/products/distribution-reclosers-and-overhead-switches/introducing-viper-hv-recloser/

[105] CIGRE Working Group A3.41, 'Current interruption in SF$_6$-free switchgear', *CIGRE Technical Brochure* 871 (2022).

[106]   Gregoire, C., Rognard, Q., Berteloot, T., *et al.*, 'Switchgear scalability demonstration using environment friendly C4-FN / O_2 / CO_2 gas mixture in 420 kV GIS substations', In: *CIGRE Conference*, Paper 10858 (2022).

[107]   Smeets, R.P.P., Kertész, V., Dufournet, D., Penache, D., and Schlaug, M., 'Interaction of a vacuum arc in a hybrid circuit-breaker during high-current interruption', *IEEE Trans. Plasma Sci.*, vol. 35, no. 4, pp. 933–938 (2007).

[108]   Cheng, X., Liao, M., Duan, X., and Zou, J., 'Study of breaking characteristics of high-voltage hybrid circuit-breaker', In: *XXIV International Symposium on Discharges and Electrical Insulation in Vacuum*, Braunschweig (2010).

[109]   Stengard, P., Stoller, P., Buffoni-Scheel, S., *et al.*, 'SF_6-alternative 145 kV live-tank circuit breaker', In: *CIGRE Conference*, Paper 10507 (2022).

[110]   Chachereau, A., Hösl, A., and Franck, C.M., 'Electrical insulation properties of the perfluoronitrile C4F7N', *J. Phys. D: Appl. Phys.*, vol. 51, no. 49, pp. 495201 and 335204 (2018).

[111]   Li., X., Zhao, H., and Murphy, A.B., 'SF_6-alternative gases for application in gas-insulated switchgear', *J. Phys. D: Appl. Phys.*, vol. 153001 (2018).

[112]   CIGRE Task Force D1.30.10, 'N2/SF6 Mixtures for Gas Insulated Systems', CIGRE Technical Brochure 230 (2004).

[113]   IEC Technical Report 310 Ed.2.0, 'High-Voltage Switchgear and Controlgear – Part 310: Electrical Endurance Testing for Circuit-Breakers Above a Rated Voltage of 52 kV' (2008).

[114]   Juhre, K., Haupt, H., Kessler, F., and Goll, F., 'Investigations on the long-term performance of Fluoronitrile-containing gas mixtures in gas-insulated systems', In: *CIGRE Conference*, Paper 11114 (2022).

[115]   Uchii, T., Hoshina, Y., Kawano, H., Suzuki, K., Nakamoto, T., and Toyoda, M., 'Fundamental research on SF_6-free gas insulated switchgear adopting CO2 gas and its mixtures', In: *Proceedings of the International Symposium on Eco Topia Science* (2007).

[116]   Park, J., Ha, M., Seo, K., Kim, H., and Lee J., 'Experimental and numerical analysis of the interruption capability of SF_6-free 245kV 63kA GCB', In: *CIGRE Conference*, Paper 10317 (2022).

[117]   Majima, A., Uchii, T., Yasuoka, T., Inoue, T., and Schiffbauer, D., 'Properties of CO_2/O_2 gas mixtures as an alternative medium for gas circuit breakers', In: *Proceedings 22nd International Conference on Gas Discharges and their Applications*, pp. 367–370 (2018).

[118]   Uchii, T., Yoshida, D., Tsukao, S., Taketa, K., and Tsuboi, K., 'Recent development of SF_6 alternative switchgear using natural-origin gases in Japan', In: *CIGRE Conference*, Paper 10643 (2022).

[119]   Kim, J.C., Kim, H.C., Tanasic, Z., Ye, X., and Mantilla, J., 'Experience in the development of a 170kV/50kA/60Hz HVCB using a $CO_2+C_4F_7N$ mixture', In: *CIGRE Conference*, Paper 966 (2022).

[120]   Lee, W.-Y., Jun, J.-U., Oh, H.-S., *et al.*, 'Comparison of the interrupting capability of gas circuit breaker according to SF_6, g3, and CO_2/O_2 mixture', *Energies*, vol. 13, p. 6388 (2020), doi:10.3390/en13236388

[121]   Ozil, J., Biquez, F., Ficheux, A., *et al.*, 'Return of experience of the SF_6-free solution by the use of Fluoronitrile gas mixture and progress on coverage of full range of transmission equipment', In: *CIGRE Conference*, Paper A3-117 (2020).

[122]   IEEE C37-100.7, 'Guide for the Evaluation of Performance Characteristics of Non-Sulfur Hexafluoride Insulation and Arc Quenching Media for Switchgear Rated above 1000 V' (2023).

[123]   Schiffbauer D., Majima, A., Uchii, T., *et al.*, 'High voltage F-gas free switchgear applying CO_2/O_2 sequestration with a variable pressure scheme', In: *CIGRE-IEC2019 Conference on EHV and UHV (AC&DC)*, (2019).

[124]   Hermosillo, V., Leguizamon-Cabra, D., Catal, M., Darles, L., Gregoire, C., and Rodriguez, J.A., 'Comparative continuous and overload current performance of high voltage switchgear with SF_6 and alternative gases', In: *CIGRE Conference*, Paper 10126 (2022).

[125]   Laso, A., Tefferi, M., Glomb, S., Göppel, M., Uzelac, N., and Smeets, R., 'Design considerations for implementing SF_6 alternatives for distribution switchgear applications with focus on toxicity and load break performance', In: *CIGRE Conference*, Paper 136 (2022).

[126]   CIGRE Working Group B3.45, 'Application of non-SF_6 gases or gas-mixtures in medium and high voltage gas-insulated switchgear', *CIGRE Technical Brochure* 802 (2020).

[127]   Kosse, S., Heinz, T., Giere, S., Teichmann, J., and Helbig, D., 'First CO_2-neutral 145 kV and up to 63 kA dead tank circuit breakers based on vacuum switching and clean air insulation technology', In: *CIGRE Conference*, Paper A3–106 (2020).

[128]   Nicolic, P.G., Goebels, T., Teichmann, J., Weisker, J., and Huth, R., 'Basic aspects of switching with series-connected vacuum interrupter units in high-voltage metal-enclosed and live tank arrangements', In: *CIGRE Conference*, Paper A3–112 (2020).

[129]   Liao, M., Zou, J., Douan, J., Fan, X., and Sun, H., 'Dielectric strength and statistical property of single and triple-break vacuum interrupters in series', In: *Proceedings – 22nd International Symposium on Discharges and Electrical Insulation in Vacuum*, pp. 157–160 (2006).

[130]   Kim, K., Heo, S., Choi, B., *et al.*, 'First 170 kV / 50 kA GIS with clean air and vacuum interrupter technology as a climate-neutral alternative to SF_6', In: *CIGRE Conference*, Paper A3–301 (2020).

[131]   Gronbach, P., Butter, K., Sarkar, A., *et al.*, 'Experience with F-gas-free high voltage equipment for on- and offshore applications', In: *CIGRE Conference*, Paper 11069 (2022).

[132]   Liu, Z., Wang J., Geng, Y., and Wang, Z., *Switching Arc Phenomena in Transmission Voltage Level Vacuum Circuit Breakers*, Springer (2021), ISBN 978-981-16-1397-5.

[133]   Trunk, K., Lawall, A., Taylor, E., *et al.*, 'Small inductive current switching with high-voltage vacuum circuit breakers', In: *29th International Symposium on Discharges and Electrical Insulation in Vacuum* (2021).

[134]   Tokoyoda, S., Takeda, T., Kamei, K., Yoshida, D., and Ito, H., 'Interruption behaviours with 84/72 kV VCB and GCB', In: *2nd International Symposium on Discharges and Electrical Insulation in Vacuum* – Switching Technology, Paper 2-A1-P-1, Matsue, Japan (2013).

[135]   Wang, J., Guan, C., Zhang, B., Yao, X., Liu, Z., and Geng, Y., 'Study on arc evolution behaviour of breaking short circuit current for a fast vacuum circuit breaker', *B&H Elec. Eng.*, vol. 15, Special Issue, pp. 27–25 (2021), doi: 10.2478/bhee-2021-0003.

[136]   Ängquist, L., Nee, S., Modeer, T., Heuvelmans, M., and Norrga, S., 'VARC DC circuit breaker – a versatile concept for non-zero current interruption', In: *CIGRE Conference*, Paper A3–103 (2020).

[137]   Yao, X, Guan, C., Wang, C., *et al.*, 'Design and verification of an ultra-high voltage multiple-break fast vacuum circuit breaker', *IEEE Trans. Pow. Del.*, vol. 37, no. 5, pp. 3436–3446 (2022).

[138]   Yao, X., Wang, J., Ai, S., Liu, Z., Geng, Y., and Hao, Z., 'Vacuum switching technology for future of power systems', *Engineering*, vol. 13, pp. 164–177 (2022).

[139]   Goebels, T., Nikolic, P., Cernat, R., Weisker, J., and Giere, S., 'Investigation of the switching behaviour, voltage distribution and post-arc current of series-connected vacuum interrupter units for live tank and dead tank circuit breakers $\geq$ 420 kV', In: *CIGRE Conference*, Paper 11068 (2022).

[140]   Huang, D., Shu, S., and Ruan, J., 'Transient recovery voltage distribution ratio and voltage sharing measure of double- and triple-break vacuum circuit breakers', *IEEE Trans. Comp. Pack. Manuf. Technol.*, vol. 6, no. 4, pp. 545–552 (2016).

[141]   Schellekens, H., Preve, C., Ferraro, V., and Bonnard, F., 'Continuous vacuum monitoring for air-insulated vacuum circuit breakers', In: *23rd International Conference on Electrical Distribution*, Paper 0128 (2015).

[142]   Slade, P., *Electrical Contacts. Principles and Applications*, 2nd ed., CRC Press (2017), ISNB 978 11380 77102.

[143]   Liu, S., Wang, Z., Li, H., *et al.*, 'Temperature-rise performance of 126 kV single-break vacuum circuit breakers with 4 types of AMF contacts', In: *28th International Symposium on Discharges and Electrical Insulation in Vacuum* (2018).

[144]   Giere, S., Heinz, T., Lawall, A., *et al.*, 'X-radiation emission of high-voltage vacuum interrupters: dose rate control under testing and operating conditions', In: *28th International Symposium on Discharges and Electrical Insulation in Vacuum* (2018).

[145]   Franck, C.M., 'HVDC circuit breakers: a review identifying future research needs', *IEEE Trans. Pow. Del.*, vol. 26vol. 2, pp. 998–1007 (2011).

[146] van Hertem, D., Gomis-Bellmunt, O., and Liang, J., *HVDC Grids for Offshore and Supergrid of the Future*, 1st ed. (2016), John Wiley & Sons, ISBN 978-1-118-85915-5.

[147] Antoine, O., Henneux, P., Karoui, K., *et al.* 'Towards a deployment plan for a future European offshore grid: development of topologies', In: *CIGRE Conference*, Paper B4–131 (2020).

[148] Li, X., Yuan, Z., Fu, J., Wang, Y., and Liu, T., 'Nanao multi-terminal VSC-HVDC project for integrating large-scale wind generation', In: *IEEE PES General Meeting* (2014).

[149] Zhou, H., Li, H., Xie, R., *et al.*, 'Research of DC circuit breaker applied on Zhoushan multi-terminal VSC-HVDC project', In: *IEEE Asia Pacific PES Conference* (2016).

[150] Tang, G., He, Z., Pang, H., *et al.*, 'Characteristics of system and parameter design on key equipment for Zhangbei DC grid', In: *CIGRE Conference*, Paper B4–121 (2018).

[151] Smeets, R.P.P. and Belda, N.A. 'High-voltage direct current fault current interruption: a technology review', *High Voltage,* vol. 6, pp. 171–192 (2021), https://doi.org/10.1049/hve2.12063

[152] Tang, G., 'Development of 500kV modular cascaded hybrid HVDC breaker for DC grid application', In: *CIGRE Conference*, Paper A3–105 (2018).

[153] Belda N.A., Plet, C.A., and Smeets, R.P.P., 'Analysis of faults in multi-terminal HVDC grid for definition of test requirements of HVDC circuit breakers', *IEEE Trans. Pow. Del.*, vol. 33, no. 1, pp. 403–411 (2018).

[154] Liu, S., Popov, M., Belda, N.A., Smeets, R.P.P., and Liu, Z, 'Thermal FEM analysis of surge arresters during HVDC current interruption validated by experiments', *IEEE Trans. Pow. Del.*, vol. 37, no. 3, pp. 1412–1422 (2022).

[155] The EU Horizons 2020 project PROMOTioN ('Progress on Meshed HVDC Offshore Transmission Networks') seeks to develop meshed offshore HVDC grids on the basis of cost-effective and reliable technological innovation, https://www.promotion-offshore.net/

[156] Belda, N.A., Smeets, R.P.P., Ito, H., *et al.*, 'Recent HVDC circuit breaker development and testing', In: *CIGRE Conference*, Paper 10545 (2022).

[157] Belda, N.A. and Smeets, R.P.P., 'Test circuits for HVDC circuit breakers', *IEEE Trans. Pow. Del.*, vol. 32, no. 1, pp. 285–293 (2017).

[158] Belda, N.A., Smeets, R.P.P., and Nijman, R.M., 'Experimental investigation of stresses on main components of HVDC circuit breakers', *IEEE Trans. Pow. Del.*, vol. 35, no.6, pp. 2762–2771 (2020).

[159] Belda, N.A., Plet, C.A., and Smeets, R.P.P., 'Full-power test of HVDC circuit-breakers with AC short-circuit generators operated at low power frequency', *IEEE Trans. Pow. Del.*, vol. 54, no. 5, pp. 1843–1852 (2019).

[160] Wei, L., Contribution to panel session HVDC circuit breakers at 5th Int. Conf. on Electr. Pow. Equipment – Switching Techn., Kitakyushu, Japan (2019).

[161] IEC TR 62271-302: 'High-voltage switchgear and controlgear – Part 302: Alternating current circuit-breakers with intentionally non-simultaneous pole operation' (2010).

[162] Rahimo, M., Storasta, L., Dugal, F., *et al.*, 'The Bimode Insulated Transistor (BIGT), an ideal power semiconductor for power electronics based DC breaker applications', In: *CIGRE Conference*, Paper B4–302 (2014).

[163] Jovcic, D., Jamshidi Far, A., and Hassanpoor, A., 'Control methods for fault current limiting using hybrid HVDC breakers', In: *CIGRE Conference*, Paper A–104 (2018).

[164] Hock, P., Belda, N.A., Hinrichsen V., and Smeets, R.P.P., 'Investigations on metal-oxide surge arresters for HVDC circuit breaker applications', In: *INMR World Conference*, Tucson, USA (2019).

[165] Le Roux, R., Roche, S., and Belda, N.A., 'Utilization of metal oxide surge arresters in HVDC circuit breakers and similar application', In: *International Conference on Condition Monitoring, Diagnosis and Maintenance*, Bucharest, pp. 4–8 (2019).

[166] 张北柔性直流电网示范工程 500 kV 机械式直流断路器试验方案 (Zhangbei Flexible DC Grid Demonstration Project, 500 kV mechanical DC circuit breaker test proposal), State Grid Economic and Technical Research Institute Co. (2018) (in Chinese).

Chapter 3

Environmentally friendly high-voltage AC switching technology: vacuum circuit breakers

Erik D. Taylor

3.1 Why vacuum?

3.1.1 What existed before vacuum

Early circuit breakers relied on a medium to provide the dielectric insulation between the open contacts and to reduce the energy and external effects of arcing. Oil-based circuit breakers began with simple knife switches placed in large containers filled with oil. These later evolved into more sophisticated minimum-oil breaker designs that dramatically reduced the amount of oil in the breaker while concurrently improving interruption performance with the motion of cool oil into the arcing region. Although oil provides excellent dielectric insulation, the arcs in oil could generate gas, which could eject oil out of the breaker during short-circuit operations. While later designs reduced or eliminated the ejection of oil, issues of expansion/contraction of the oil with temperature, disposal of the oil at the end of the breaker life, and flammability remained. Although these issues were not automatically fatal and could possibly be solved, the focus of circuit-breaker development shifted to other interruption mediums.

The simplest interruption medium was to use what was free and always available, namely atmospheric air. This option worked well for low-voltage (LV) applications and is still the dominant medium at this voltage level. Air-based circuit breakers were developed and used in the field for medium-voltage applications. However, the large amounts of arc energy made the short-circuit operation of these devices unpleasant, to put it mildly. Research started focusing on electronegative gases to improve the dielectric performance and reduce the arcing energy. Sulfur hexafluoride (SF_6) came into focus, offering many electronegative fluorine atoms in a stable, easy-to-handle gas. SF_6 offered better dielectric performance than air or nitrogen [1], together with the ability to make sealed interrupter units with no external arcing. This allowed SF_6 to become the primary interruption medium in high voltage (HV), along with some limited penetration into the medium voltage (MV) market.

The primary disadvantage of SF_6 is its global warming potential, which identified it as a potent greenhouse gas. The electrical power industry was only one of many users of SF_6, some of whom actively emitted SF_6 into the atmosphere. For example, some main sources of SF_6 emission in Germany include the disposal of sound-insulating windows filled with SF_6, the casting of aluminum, magnesium foundries, and the filling of car tires during the 1990s [2]. Increasing regulations and outright bans stopped most of these applications leaving the electrical power systems as one of the few remaining users. Although the emission of SF_6 from circuit-breakers is limited largely to leaks and accidental emission and the emission rate was predicted to drop to negligible levels [3], the increasing burdens of legal documentation and upcoming bans on SF_6 usage are pushing HV circuit-breakers to alternative interruption mediums such as vacuum.

A note on units: the units for voltage and current can be confusing. Current and voltage oscillating at a power frequency f_r are root mean squared (RMS) unless specifically noted. This includes the rated system voltage U_r, short circuit and load currents, power frequency withstand voltages U_d, etc. All other currents and voltages not oscillating at f_r, such as the current in pressure measurements on a vacuum interrupters (VI), field emission, lightning impulse voltage U_p, the peak of the transient recovery voltage (TRV) u_c, etc. are in standard units. Also, all units derived from current and/or voltage are in standard units, such as electric field stress, transferred charge, I^2t, magnetic field, rate of rise of recovery voltage (RRRV), etc. Sadly, these two separate conventions are standard in electrical engineering and physics.

3.1.2   Why is there interest in using vacuum

3.1.2.1   Environmental advantages

Since environmental issues are driving the switch from SF_6, any analysis of why a vacuum would be a replacement option must first focus on this point. VI have no specific environmental problems in their manufacture, usage, and disposal. The materials and manufacturing methods for VI are widely used in other products and present no special environmental burden.

VI are made of alumina ceramic; stainless steel; copper; contacts typically made of Cu–Cr, Ag–WC, or Cu–W [4]; and a plastic bearing to guide the moving terminal motion. During the lifetime and usage of VI, there are no emission products or other problematic outputs. Finally, the disposal of the VI at the end of its use is straightforward. The interrupters can be disposed of as standard mixed waste without any hazardous components. With some effort, the interrupters could even be separated into high-value copper for recycling and mixed material for disposal. This contrasts with the need to remove and recycle SF_6 from old interrupters and the problematic disposal of fluorine-based arc by-products remaining inside the interrupters [5–7].

3.1.2.2   Vacuum versus gas

Vacuum behaves in some fundamentally different ways from gases. Figure 3.1 shows a range of gas pressures and some examples of where the pressures occur [8]. A key value is the transition pressure of $p_{trans} \sim 10^{-2}$ hPa. Pressure above

pressure in hPa

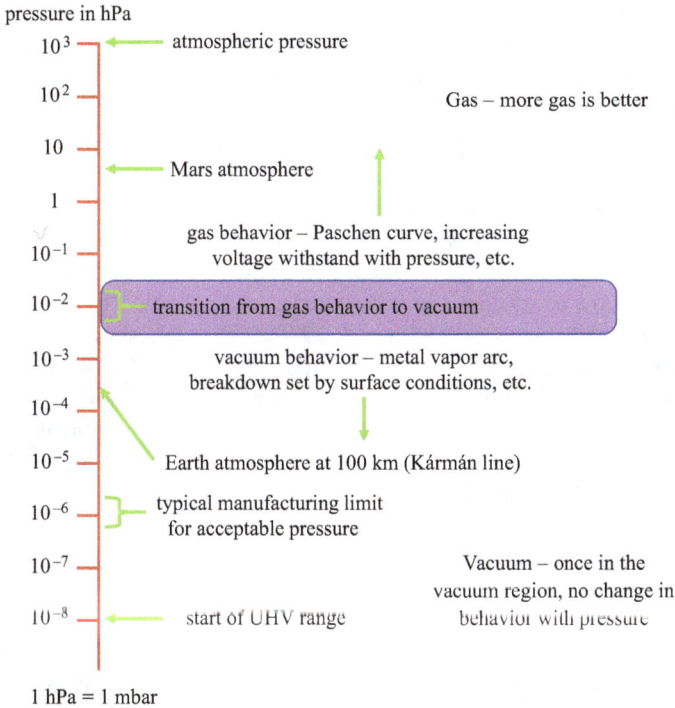

Figure 3.1   Plots the pressures for the vacuum and gas behavior regions including the transition between the two regions, along with some practical examples of the pressure values.

p_{trans} are in the gas region. In general, for the gas region more gas/higher pressure is better. Both current interruption and dielectric performance increase with the pressure. Below p_{trans} is the vacuum region. Here the performance is independent of the pressure for both interruption [9] and dielectric withstand [10]. The performance in the vacuum region is dictated by the contact material, surface conditions, arc behavior, etc. The background gas has no significant influence on the performance.

3.1.2.3   VI history especially regarding HV

There is a long history of using a vacuum as an interruption medium in electrical power systems. The first demonstration of switching at 41.5 kV and 926 A using handmade metal/glass seals and active pumping occurred in 1926 [11]. Unfortunately, commercial vacuum sealing techniques lagged far behind these results. Only after rapid advancements in vacuum processing and vacuum furnaces during and shortly after World War II [12] did VI production become feasible [13]. Early experiments noted a very HV hold-off at small contact gaps of a few millimeters, which was then extrapolated to extremely HVs at some tens of millimeters. The current interruption was limited to a few kiloamps with disk contacts [14], which is consistent with the need for magnetic field to achieve higher interruption currents [15]. The combination of very HV and modest interruption current led to early applications as load break switches for HV systems [16].

The field advanced rapidly, to the point that two "historical" review articles were published in 1963 [13,14]. One of these reviews reported on the successful interruption of a 43 kA short-circuit current at 15.5 kV with a 67 mm diameter spiral contact [14]. Other tests on the same interrupter demonstrated interruption at 18 kV and 27 kA, with a RRRV of 6 kV/μs. Despite this auspicious start, it took until the late 1980s before VI achieved a major fraction of the MV market. Nevertheless, some companies already offered a range of MV VI, including designs for arc furnace switching, as early as 1961 [17]. HV applications expanded from load break switches in the late 1950s [16] to circuit breakers for 145 kV/40 kA [18] and 132 kV/3,500 MVA [19]. The only country with significant deployment of VI for HV was Japan, which installed around 8,300 vacuum circuit breakers between 1970 and 2010; nevertheless, this remained a small fraction of the total installed base of HV circuit breakers [20]. Although the application of vacuum in HV was in niche applications, it is vital to understand that vacuum in HV is not new and has a history of over half a century. Table 3.1 collects a sample of papers describing short-circuit testing and/or field applications of VI in HV before the turn of the century.

Vacuum developed into the principal interruption medium for MV. Applications exist in LV where enclosed arcing and/or long electrical life are required, however, the additional cost compared to simple contacts in air limits the application of vacuum. In HV, the earlier establishment of SF_6-based breakers combined with their perfectly adequate performance in the field reduced the interest in new interruption mediums, especially for economic reasons [24]. It is only with increasing legal pressure, such as in the European Union (EU) [25], that vacuum is now an active option for HV.

The key feature for HV VI applications is the basis of the MV experience. HV applications do present active research and development challenges. Nevertheless,

Table 3.1   List of technical publications in the 20th century along with the application for VI in HV equipment

Reference	Equipment year	Indoor	Outdoor	Load-break switch	Circuit breaker	Used in field
[16]	1961		X	X		X
[19]	1968		X		X	X
[21]	1979		X		X	X
[22]	1980	X			X	
[23]	1982		X		X	X
[18]	1991	X			X	X

Indoor or Outdoor indicates the setting for the application and used in the field indicates that the equipment was used on a power system.

MV design and background are the foundation of the HV VI. This is very beneficial, as the vacuum is not an exotic, new technology, but rather one that can call on over a half-century of development and application work. For HV applications specifically, the vacuum has been used in the field in both outdoor and indoor applications. These applications included short-circuit and load break switching, as well as line and cable switching. Many of the early applications used multiple VI in series. New developments in HV VI include the extensive use of single breaks for $U_r \leq 170$ kV, completely SF_6-free external insulation around the VI [23], and shunt reactor switching [26].

3.2 MV experience

3.2.1 VI basics

VI are the principal interruption technology for MV. Fortunately, this extensive experience in MV can be extended to both HV and LV applications. Figure 3.2 shows an example of the range of VI. Although there are some differences to account for, which will be discussed in this chapter, a good understanding of the MV performance provides the fundamental basis for applications at higher and lower voltages. Reference [27] provides an excellent, detailed compilation of work on VI. It includes background on the science behind vacuum arcs and vacuum breakdown, as well as extensive background on the testing and applications of VI particularly at MV. This section will focus on the key behaviors and performance needed to understand how to apply VI to other voltage ranges.

Figure 3.3 shows a typical VI. Two contacts are contained in a vacuum housing, with one moving and one fixed contact. The circuit breaker mechanism connects to the moving terminal and controls the opening and closing of the VI. The contacts handle the arcing during current interruption, withstand HV when the

Figure 3.2 An example of a VI portfolio for MV (courtesy of Siemens AG).

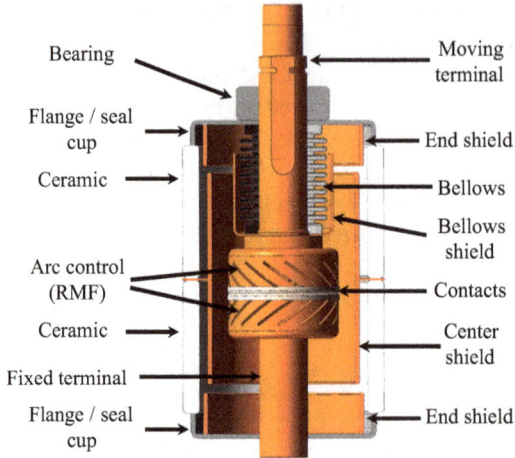

Figure 3.3   Cross-section view of a typical VI, with the labeling of key parts (courtesy of Siemens AG).

contacts are open after the interruption, and pass the continuous current when the contacts are closed.

Virtually all VI applications above a few kiloamps use arc control by the magnetic field generated from the short-circuit current [15]. The contacts and the arc control system are then joined to terminals, which provide the connection to the circuit breaker and the current path. A metallic bellows on the moving terminal allow the contacts to open and close. Alumina ceramics form the housing of the interrupter and provide external dielectric insulation when the contacts are open. Metallic shields protect the ceramic from metal deposits during arcing and dielectrically protect the triple points where the metal parts, ceramics, and vacuum come together.

Thorough cleaning of the parts is a vital part of making the parts vacuum compatible [28]. Vacuum joints between parts are sealed with Ag–Cu-based brazing material. All modern VI designs are sealed in vacuum furnaces which remove the gas, outgas the components, and melt the braze material at approximately 800°C. Detailed information on vacuum furnaces and VI brazing processes is given in [29]. Temperatures above 450–500°C help remove water on the surfaces of the parts and temperatures above 550–600°C help remove hydrogen [27, Figure 3.83], particularly from stainless steel parts [30].

3.2.2   Production steps on sealed VI

Production steps on the brazed interrupters universally include voltage conditioning and internal pressure measurement. Voltage conditioning is a well-established step for improving and stabilizing the internal breakdown behavior of all vacuum gaps. Even vacuum gaps used in linear accelerators, using custom contacts with as perfect as possible surface conditions, depend on voltage conditioning to maximize and stabilize the breakdown performance. The voltage conditioning process begins with a ramp of the voltage with time. Breakdowns will start at voltages well below the goal of the design. If many breakdowns occur, then the voltage ramp will generally stop at this voltage and wait for the breakdown activity to stop before continuing. Once the ramp reaches the end value, the voltage will hold at this magnitude for some length of time. The exact details of the contact gap, voltage application, voltage ramp, reaction to breakdowns, and holding voltage vary based on the empirical experience of the manufacturer. The amount of breakdown activity during the conditioning and/or the time required for the conditioning is affected by the cleaning processes and surface treatment of the contacts [31]. Nevertheless, voltage conditioning is always necessary to reach a high and stable breakdown voltage.

3.2.3 Internal pressure measurement

3.2.3.1 General principles

One of the last production steps is to measure the internal pressure of the individual VI. The interrupter itself can function as a cold cathode pressure gauge when placed in an external DC magnetic field with a DC voltage applied to the VI, Figure 3.4 [32]. Electrons in the VI are confined by the magnetic field to small cyclotron orbits orthogonal to the magnetic field. The voltage places an electric field on these electrons. The component of the electric field is orthogonal to the magnetic field creates an $\vec{E} \times \vec{B}$ drift of the oscillating electrons. When the magnetic field is along the axis of the VI, and the electrical field extends radially from the contacts and terminals to the center shield, then the electrons will drift azimuthally. The electrons can collide with the neutral gas in the VI to produce ions,

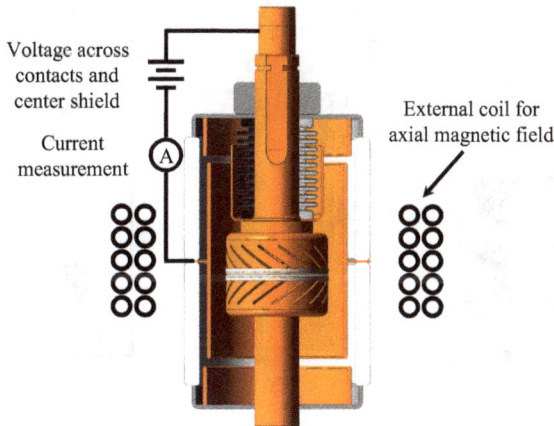

Figure 3.4 *VI in a magnetron pressure measurement. The coils generate a magnetic field along the axis of the VI, and voltage is applied between the center shield and the contacts (inverted magnetron shown in the figure).*

which then move towards the negatively charged electrode. This is detectable as a current, and the magnitude of the current is a monotonic increasing function of the pressure. The relation between the magnetic field strength and the voltage across the VI is discussed in [33].

Although a DC magnetic field and DC voltage are applied to the VI, the current signal from the VI is not a constant value. Cold cathode gauges are well known to pump the vacuum chamber during their operation [34], and this behavior appears in VI [32]. The discharge starts after the voltage application, and very rapidly rises to a peak value. This current peak is the value used to calculate the pressure inside the VI. The current then exponentially decays, with a decay time ranging from a few milliseconds [27, Figure 3.85] to some hundreds of milliseconds [32]. These decay times are consistent with values measured for cold cathode gauges in vacuum systems [35]. The discharge requires an initial electron to begin the current flow, normally provided by field emission or stray ionization by a cosmic ray or similar. A certain time is required before the discharge starts [36, Section 5.2].

$$t \cdot p = 5 \times 10^{-7} - 5 \times 10^{-6} \text{ hPa} \cdot \text{s} \tag{3.1}$$

where t is the time and p is the measured pressure. For $p = 10^{-8}$ hPa, the discharge can take $\sim 1 - 10$ min to start, for example.

3.2.3.2   Cold cathode gauge (magnetron) arrangements

When the magnetic field is applied along the axis of the VI, there are three possible ways to apply the voltage [37], Figure 3.5. The first voltage arrangement is an inverted magnetron discharge [38,39], where a positive voltage is applied to a contact, and a

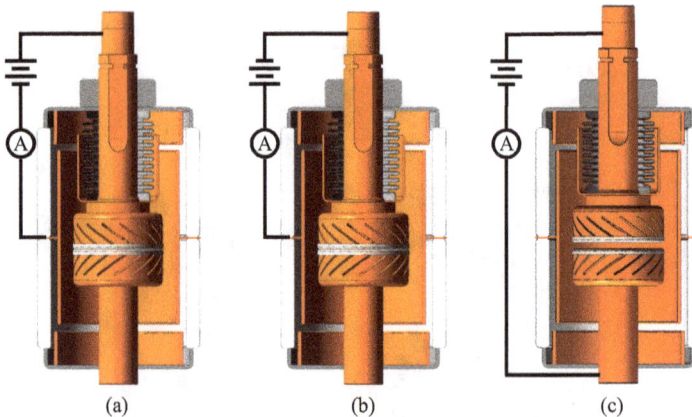

Figure 3.5   *Examples of the three types of magnetron discharges: (a) inverted magnetron with positive voltage on contacts and negative voltage on shield; (b) magnetron with negative voltage on contacts and a positive voltage on shield; and (c) voltage between the contacts resulting in inverted magnetron and magnetron arrangements in series.*

negative voltage to the shield around the contact. This arrangement generally provides the largest current signal for a given pressure. A second arrangement, sometimes referred to as a magnetron discharge [37,38], flips the polarity of the voltage, with the cathode inside of the anode cylinder. The current magnitude for a given pressure is generally smaller. Finally, the voltage can be applied between the contacts; this method is generally only used when the center shield is not accessible. The largest magnitude electrical field between the contacts is parallel to the magnetic field, leading to zero $\vec{E} \times \vec{B}$ force. The electric field between the contacts and center shield does allow an inverted magnetron discharge between the positive contact and the center shield, and a magnetron discharge between the negative contact and the shield, giving a complex path for the current [33]. The current between the two discharges flows through the center shield, with the magnitude limited by the magnetron discharge [37].

The terminology for cold cathode gauges is confusing, particularly when related to VI. Cold cathode gauge is the general term from vacuum science [36], while most people in VI manufacture and research would refer to a magnetron pressure measurement, regardless of exactly which of the three methods is used. The first arrangement is referred to as an inverted magnetron both with VI and general vacuum science. For the second arrangement, although [37,38] refer to this arrangement as a magnetron, this does not necessarily agree with other usage where the magnetron discharge is a more complex arrangement [39]. The term Penning discharge is also used, although in vacuum science this tends to refer to an arrangement with an anode between two cathode plates [36,38]. Finally, older papers sometimes refer to a Redhead gauge. I will use the general term magnetron to refer to any type of cold cathode gauge, to be consistent with common usage in VI research and production, and the terms inverted magnetron discharge, magnetron discharge, and between the contacts when referring to a specific arrangement.

All three arrangements produce a roughly linear relation between the current drawn across the vacuum gap to the pressure inside the VI. The relation is:

$$I = k \cdot p^{\alpha} \tag{3.2}$$

where I is the current drawn, p is the pressure inside the VI, k is a proportionality constant, and $\alpha = 1$ in [40] and more generally is $\alpha = 1.0 - 1.15$ for pressures $\geq 10^{-8}$ hPa [41]. Below 10^{-8} hPa, the value of k is no longer constant [42]. The value of k is a function of the gas being detected, as well as the specific VI design [37,40], and the voltage and magnetic field [40]. Modeling of the discharge between the contacts gave $k = 2A/hPa$ [33]. Experiments varying the voltage between the contacts from 2 kV to 6 kV produced a range $k \sim 1 - 5.5A/hPa$, and varying the magnetic field from 40 to 80 mT produced a complex variation in the proportional constant [40]. The voltage and magnetic field values can also affect the linearity of the measurement, especially when the pressure $\leq 10^{-6}$ hPa [41]. In addition, the details of the internal design can significantly affect the magnitude of the measured current. In [43], modest changes in the contact design led to two orders of magnitude changes in the measured current. Figure 3.6 plots data from [32] from measurements on a VI where $k = 8A/hPa$.

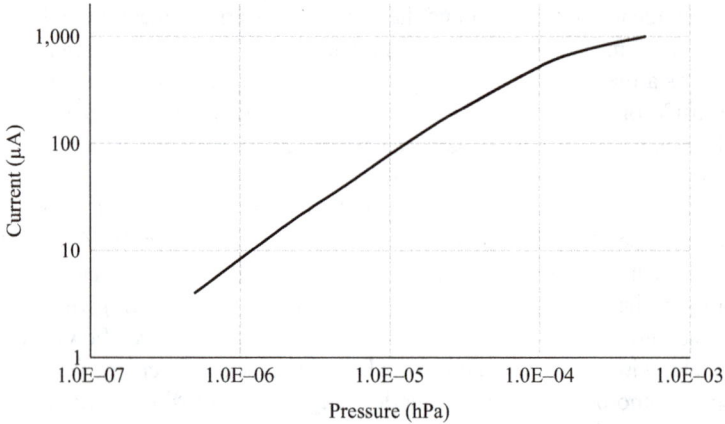

Figure 3.6   *An example of a VI calibration curve connecting the output current to the actual pressure in the VI from [32]*

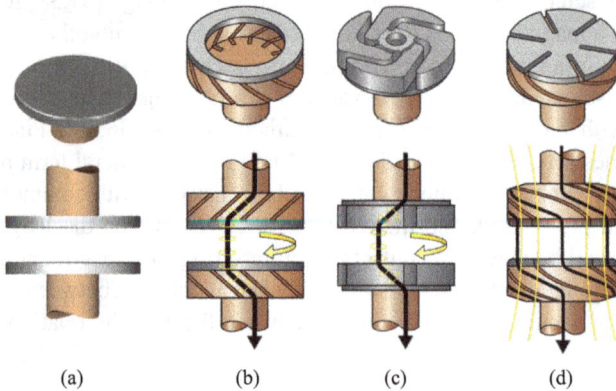

Figure 3.7   *Typical examples of the (a) disk contact without arc control for low currents; (b) RMF (also called TMF) contact with a coil to drive arc rotation; (c) RMF contact using the contact itself to drive arc rotation; (d) AMF contact which drives arc diffusion (courtesy of Siemens AG)*

3.2.4   VI contact design

Applications other than load break switches and contactors, where the short-circuit currents are below ~5 to 6 kA, require some method of arc control to prevent thermal overloading of the contacts during the arcing. All commercial designs use one of two basic techniques, both of which depend on the magnetic field generated by the short-circuit current [15], Figure 3.7. The first design generates a magnetic field radially along/transverse to the contact surface and is therefore referred to as a radial magnetic field (RMF) or transverse magnetic field (TMF) design. The radial field when combined with the current flowing through the arc produces a $\vec{J} \times \vec{B}$

force on the arc, causing it to rotate around the contact surface. The arc is constricted, but the arc energy is spread over large portions of the contact surface through the motion. The second design creates a magnetic field along the axis between the contacts and is referred to as an axial magnetic field (AMF). The AMF contacts force the arc to remain diffuse even at very high short-circuit currents.

Both designs are widely used in medium-voltage applications. Although various works attempt to compare the relative advantages and disadvantages of the two designs [4,44,45], ultimately the background and experience of the individual manufacturer plays the main role in which design is used. There is a tendency for AMF use in generator circuit breakers [46] and other applications at high short-currents greater than 40–63 kA [47] as well as in HV [48,49]. However, even in these cases, there are examples of successful applications with RMF designs for both generator circuit breakers [50] and HV applications [20,23]. Although the arc behavior is significantly different (see later sections), for the practical user the decision of which arc control technique to use is best left to the VI manufacturer based on the specific requirements.

3.2.5 VI dimensions

3.2.5.1 VI diameter

The general dimensions of VI are set by the key performance parameters. Although the exact dimensions vary depending on the design choices of the manufacturer, the general trends of the dimensions are straightforward to understand. The first dimension is the diameter. The contact diameter is the critical driver of the overall diameter. The interruption current increases with the contact diameter as a power law, with an exponent ranging from 1 to 2 [44,51,52]. Figure 3.8 compares the AMF data from [51]

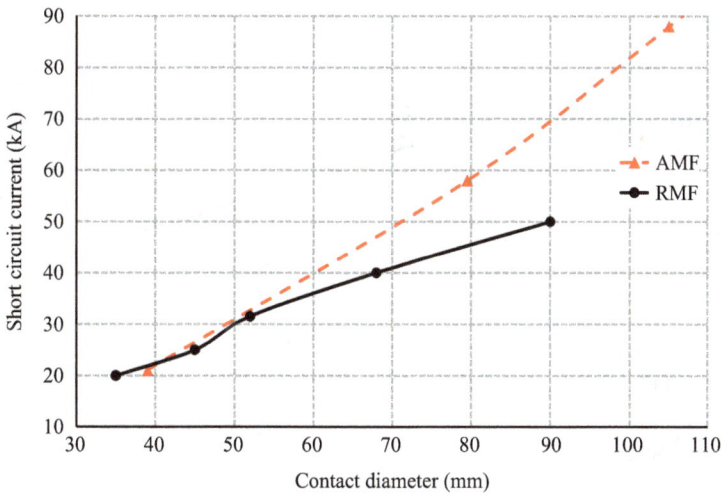

Figure 3.8 *An example of a symmetric short-circuit current that can be interrupted at $U_r = 12$ kV as a function of the contact diameter for AMF [51] and RMF [52] contact designs*

and the RMF data from [52] at $U_r = 12$ kV. The exact value depends on the details of the contact material composition, internal VI layout, rated system voltage, and contact motion/contact gap. Nevertheless, the general trend holds.

The contact diameter is then added the spacing to the center shield, the space for the center shield and its connections, and finally the thickness of the ceramic housing. The spacing to the center shield is set by the dielectric requirements and the distance to minimize or control the interaction between the center shield and the arc [53]. Higher system voltages require larger spacing to the center to maintain acceptable electric field stress. In addition, higher system voltages also require larger contact gaps (discussed later in this section), which especially for RMF designs increase the opportunity for interaction with the center shield. The space for the center shield and its connections, and the thickness of the ceramic tend to be constant with only a slowly increasing trend with increasing diameter. These values then combine to give the final diameter of the VI, with the contact diameter and spacing to the shield being the primary variables.

3.2.5.2   VI length

The second dimension is the length, in particular the length of the ceramic insulators. The ceramics provide external dielectric insulation when the contacts are open. Higher system voltages have higher lightning impulse U_p and power frequency withstand voltage U_d requirements [54], which increase the necessary ceramic length. In addition, the ceramic length itself also affects how much voltage it can withstand, with longer ceramics holding off proportionality lower voltages, Figure 3.9 [55]. The ceramic length can be reduced through improvements in the

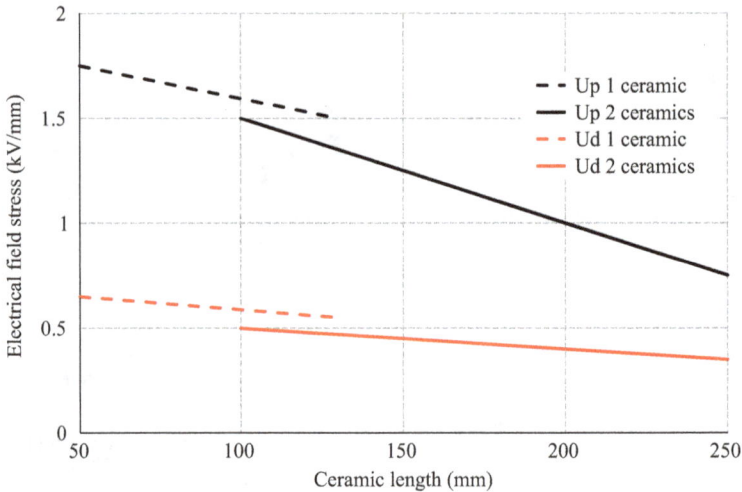

Figure 3.9   Electric field stress resulting in an external breakdown in air as a function of the ceramic length for U_d and U_p voltages for single and two ceramic cases [55]

external insulation around the VI. This can include insulating gases such as SF_6 [56], high-pressure gas, or solid insulation [57]. The balance between the rated voltage and the external insulation sets the ceramic length. The overall VI length is a function of the ceramic length, plus any additional space to accommodate the contact gap and bellows length. Both increase with the rated voltage, specifically with U_p.

3.2.5.3   VI contact gap

The contact gap is primarily a function of U_p. Both RMF and AMF contact designs have a minimum contact gap below which the arc control does not function. This gap is $\sim$3 to 4 mm for RMF and AMF designs [27]. However, the contact gap required for U_p in MV already exceeds these values. The TRV during the interruption process also sets a limit on the minimum contact gap, particularly at short arcing times of a few milliseconds. However, the gap required for interruption is smaller than the ultimate gap required to withstand U_p. This created the opportunities such as [58] for HV VI, where the contact gap moves at different speeds to an ultimate open gap. The vacuum breakdown process is mostly independent of the frequency of the applied voltage [59]. Therefore, the highest voltage across the open gap, namely U_p, determines the full contact gap. Computer modeling is a common tool to calculate the electric field stress and compare it to limiting values, such as given in [60].

3.2.6   What vacuum has to offer

3.2.6.1   Enclosed arcing

The vacuum required some advantages over existing interruption medium to justify the time and expense to develop new manufacturing methods and certify the resulting products. A key basic advantage over oil and air-based interrupters is the enclosed arcing. During the interruption process in a vacuum, the arc is only in contact with the VI contacts and occasionally with the shield surrounding the contacts. It does not interact with explosive, corrosive, or toxic materials in the surrounding environment, nor are explosive, corrosive, or toxic materials generated by the arcing. This contrasts strongly with arcs in the air, where the energy in the arc plasma can create undesired material and effects. The pressure generated by the vacuum arc is completely confined to the VI. Again, in contrast, interrupters using air eject hot gas, leading to the term "dragon's breath" for MV air-magnetic breakers. The pressure from the arc can eject oil out of the interrupter into the surroundings for some simple oil-based designs. The only external effects of short-circuit current interruptions in VI are some light emissions through the ceramics (the ceramics have very low transparency but are nevertheless transparent) and very minor warming of the interrupter housing during short-circuit testing.

3.2.6.2   Long electrical life

VI offer a very long application life even after particularly hard switching duties. Four factors determine the VI end of life at load and short-circuit currents [61]:

contact erosion, loss of the contact shape/arc control, loss of dielectric strength, and excessive increase in the contact resistance. The vacuum arc is in fact a metal vapor arc, burning in the plasma generated from the metal surface. Nevertheless, the amount of material removed from the metal surface is very small. In addition, most of the material lost from the cathode to generate the arc plasma is re-deposited on the anode contact. As the polarity randomly varies over operations, most of the vaporized metal simply moves back and forth between the contacts. Only material that leaves the contact gap, or molten material that flows off the arc-facing surface contributes to contact erosion. Erosion rates are often given in units of µg/C, which are optimal for comparing different materials. For VI, since one material is in focus and since the primary use is to calculate the loss of contact thickness from switching, a volume erosion rate is more useful. The volume erosion rate for Cu–Cr contact material (typical material for circuit breaker applications) is 2.3×10^{-6} cm^3/C for switching at load currents up to 4,000 A [61], increasing with short-circuit current to 1.5×10^{-4} cm^3/C for maximum short-circuit currents on RMF contacts [62]. Both these rates are extremely low and result in contact erosion that is only a concern for tap changer [63,64] and arc furnace switching [65,66] applications, where the switches are operated several times a day at currents of some thousands of amperes. For comparison, the erosion rates of vacuum can be a factor of 12 smaller compared to contacts in the air [67] and a factor of 27 smaller compared to contacts in oil [63].

The loss of contact shape and/or the arc control through contact melting is only a concern for short-circuit currents [68,69] or load-current switching where the number of operations greatly exceeds that of a standard circuit breaker, such as arc furnace switching. Switching of load currents up to 4,000 A and out to 30,000 operations do not result in enough contact melting to affect the arc control techniques, Figure 3.10 [61]. Most VI can pass the short-circuit electrical life test from the International Electrotechnical Commission (IEC) standard for circuit breakers, Class E2 [70]. Numerous VI can further pass the electrical life type tests for 20, 30,

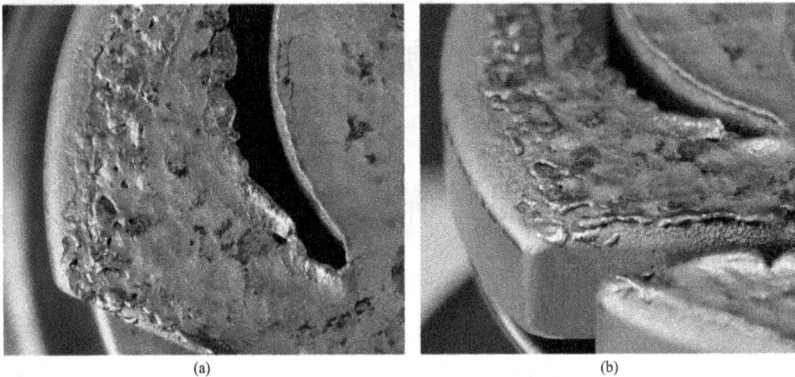

(a)   (b)

Figure 3.10   (a) Overhead and (b) angled photographs of Cu–Cr contacts after 30,000 switching operations at 2,500 A [61]

or 50 operations at the full short-circuit current [71, Section 6.112.2, Table 30]. These types of tests greatly exceed the practical requirements. Data on short-circuits in the field was compiled in [72], and summarized in [73, Section 2.4]. The average short-circuit current is less than 20% of the rated value. Therefore, virtually no circuit breakers come near the end of life due to the short-circuit current and loss of contact shape.

Loss of dielectric strength can occur through the deposit of metal vapor onto the ceramic insulators, thereby creating a breakdown path when the VI contacts are open [74]. Maintaining at least 80% of U_d after electrical life testing is the standard condition check to verify the continued adequate performance of the circuit breaker [70]. Passing this condition check during electrical life testing at continuous current levels is straightforward for a wide range of VI designs [61]. For short-circuit electrical life, passing the condition check for IEC Class E2 [70] tests and GB Class E2 list 4 [71] tests at 20–30 short-circuit operations is very straightforward. Maintaining the dielectric strength to 50 short-circuit operations and even beyond is possible as well with reasonable VI designs [75]. Related to this point is the deposit of contact material onto the shields protecting the ceramics. The deposit on the shield can become thick enough that the mechanical strain in the layer causes it to peel away from the base material. The loose metal layer can then interfere with the proper operation of the VI. This is generally not a concern for circuit breaker or recloser VI but should be looked at for frequent switching of high-load currents in compact VI designs.

The last point is the increase in contact resistance. All mechanical switches have some increase in contact resistance due to mechanical operations and arcing. Therefore, the mere existence of a change in resistance is unimportant; what is important is the magnitude of the change compared to something. Figure 3.11 plots the increase in contact resistance after short-circuit testing and after electrical life tests at currents of 2,500 A with 30,000 operations for a breaker rated at 31.5 kA short-circuit current and 3,150 A with 18,500 operations for a breaker rated at 40 kA. The data is from [61,76,77] as well as some new data from the author. These values are compared to the average value of the largest resistance difference between the three phases of individual switchgear panels in the new state. The chart also plots the 25% and 75% values of the distribution. The bulk of the resistance increases are well below the average variation between phases of the same switchgear unit, and furthermore, all the points are well within the distribution of the switchgear values. Therefore, the increase in the resistance from arcing in the VI is indistinguishable from the random variation from a large number of con-nections inside of switchgear. In addition, these resistance increases are for the extreme cases of several full short-current circuit operations or full electrical life at maximum continuous current. As mentioned earlier, circuit breakers in the field do not anywhere near such extreme duties.

One advantage of the vacuum in MV applications is the long electrical life, both at load currents [61] and at short-circuit currents [62,78]. This advantage extends to HV applications as well. HV VI has passed the Class E2 electrical endurance at the short-circuit current [79] and demonstrated contact melting and

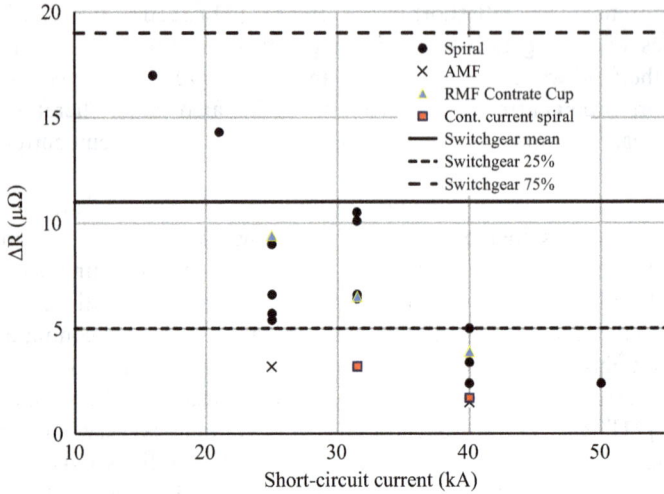

Figure 3.11   *Increase in the resistance of individual VI after a short circuit or*
>10,000 load current switching operations as a function of the rated
short-circuit current [61,76,77, plus new data], compared to the
distribution of the difference between phases on individual MV
switchgear panels

damage completely in line with the expectations from MV [48]. HV VI has also
passed 30 operations at 145 kV/40 kA [80]. The larger spacing between the contacts and the center shield could potentially leave more of the ceramic exposed to
metal vapor deposit from arcing, and hence limit the dielectric performance.
However, newer models of the vapor interaction and deposition onto the ceramic in
realistic designs [74] can be combined with dielectric modeling to optimize the
contact and shield designs. Various VI designs demonstrated the ability to maintain
dielectric strength out to 100 full short-circuit operations [75,78] and to the end of
the mechanical switching life of 10,000 operations at 1,250 A [78] and 30,000
operations at 2,500 A [61].

3.2.6.3   High short-circuit current interruption

VI offer very good and scalable short-circuit current performance. As mentioned
earlier, the initial VI designs focused on HV load break applications. The development of the AMF and RMF contact designs opened the door to very high short-circuit
currents. AMF designs have successfully performed generator circuit breaker applications at 100 kA [46] and interrupted symmetric currents up to 12 kV/200 kA [47].
In addition, AMF designs were used in DC switches for magnetic fusion experiments
with hundreds of operations at 50–70 kA DC currents [81–85]. RMF designs have
been tested out to 15 kV/63 kA for generator circuit breaker applications [50].

So far, no fundamental limit in the short-circuit current had been demonstrated
for either arc control technique. Practical limitations due to the size and mass of the

contacts and terminals to withstand forces induced from the current path, nearby phases, and the force necessary to prevent contact welding are expected to be more of an issue than any ultimate limit in the arc control method.

3.2.6.4   RRRV and d*i*/d*t*

The interruption process at current zero involves several competing processes [86]. As the current goes to zero, the rate of change of the current as a function of time di/dt can affect the ability of arcing contacts to interrupt. After the current zero, the RRRV also affects the interrupter performance. For SF_6, both are limitations on the interruption performance. While SF_6-based interrupters can handle the RRRV and di/dt values specified for basic short-circuit test duty [70], they can struggle with higher values and sometimes require additional circuity to modify the system response. In contrast, a vacuum can interrupt currents with an RRRV up to and beyond 10 kV/µs, Figure 3.12, as well as $di/dt \leq 450$ A/µs [87]. This covers applications for generator circuit breakers [88] as well as the high RRRV in short-line faults [89]. The performance limit for RRRV has not been clearly reached, while the performance limit that starts to appear at di/dt > 450 A/µs not an issue for any practical AC power system. Other interruption mediums such as SF_6 and air magnetic can require the addition of external capacitors to reduce the RRRV when it exceeds the values for standard circuit breakers [90], such as for generator breakers [88]. Documented examples exist where exceeding the rated RRRV values from the type tests led to field failures in air magnetic breakers [91].

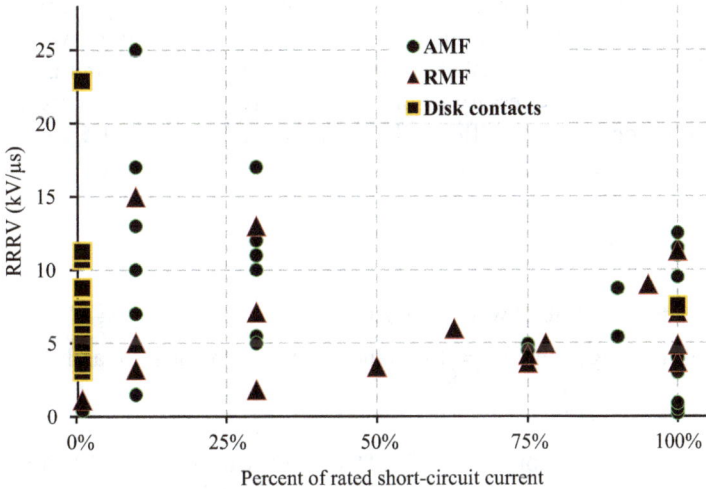

Figure 3.12   Experimental data on the RRRV of the TRV for successful current interruptions as a function of the rated short-circuit current [87]. No performance limit or boundary due to the RRRV has currently been identified.

While research often focuses on the interruption performance at the full-rated short-circuit current, it is important to remember that circuit breakers must interrupt currents from a few amps all the way to the maximum-rated current. Some other interruption methods use techniques that depend on the pressure generated by the arc to push cool gas or oil into the arcing region. When the currents are low, this process is not necessary. When the currents are around the rated short-circuit current, the arc generates plenty of pressure to drive this process. However, mid-range currents can struggle when the pressure from the arc is not enough to drive the required amount of cooling gas. This effect is referred to as a critical current and is one of the reasons for tests at multiple current levels in type testing [70]. The vacuum does not suffer from this behavior; no critical current issues have been identified. While tests at a range of currents are included in the standards to cover multiple interruption techniques, these tests are generally skipped in development tests of VI that instead focus on the maximum interruption current [92,93].

3.3 VI short-circuit type testing

3.3.1 Specific short-circuit tests

Chapter 2 provides an excellent general overview of type testing; the focus of this section details specific to short-circuit testing of VI. The testing standards include tests to ensure the operation of circuit breakers in a variety of different possible failure conditions. One recent addition is tests for effectively earthed neutral systems, also referred to as $k_{pp} = 1.3$, where k_{pp} is the first-pole-to-clear factor used to calculate the peak of the TRV u_c [70]. This arrangement is common for system voltages $U_r > 123$ kV, but rare in MV where non-effectively earthed neutral ($k_{pp} = 1.5$) is the standard arrangement [94]. Nevertheless, tests were added to ensure that MV and HV breakers could cover both conditions and are now performed even when unnecessary or of limited utility. Although the first and third-pole-to-clear conditions are effectively the same for both $k_{pp} = 1.5$ and $k_{pp} = 1.3$, the TRV peak u_c of the second-pole-to-clear is

$$u_c = k_{af} \cdot \frac{\sqrt{3}}{2} \cdot PU \tag{3.3}$$

for $k_{pp} = 1.5$ [95, Table 2] where the per-unit voltage is $PU = \sqrt{\frac{2}{3}} \cdot U_r$ (peak of the line-to-earth voltage) and $k_{af} = 1.4$ is the amplitude factor [73], and

$$u_c = 1.27 \cdot k_{af} \cdot PU \tag{3.4}$$

for $k_{pp} = 1.3$ [70, Section 7.104.2.3]. This gives a factor of 1.47 increase in the TRV peak in the second-pole-to-clear for the $k_{pp} = 1.3$ case versus the $k_{pp} = 1.5$ case. An extensive series of tests on VI that previously passed the $k_{pp} = 1.5$ case also passed over 20 three-phase asymmetric current tests at the full short-circuit current for the $k_{pp} = 1.3$ case, Table 3.2 [96]. These results suggest that a well-designed VI with good results on $k_{pp} = 1.5$ testing would also pass the $k_{pp} = 1.3$

Table 3.2   *Summary of tests on AMF and RMF designs previously tested and used in the field for $k_{pp} = 1.5$ applications to verify the ability to also interrupt in $k_{pp} = 1.3$ systems [96]*

40.5 kV/31.5 kA AMF		
kpp	**Test**	**Sequence**
1.5	T100s	O–CO–CO
1.5	T100a	3× O
1.3	T100s	O–CO–CO
1.3	T100a	27× O
1.3	T100s	O–CO
1.3	T100s	2× CO
17.5 kV/31.5 kA RMF		
Kpp	**Test**	**Sequence**
1.5	T100s	3× O
1.5	T100s	O–CO–CO
1.5	T100a	4× O
1.3	T100s	O
1.3	T100s	O–CO–CO
1.3	T100a	21x O

Notes: O is contact opening operations, and CO is closing operation followed immediately by an opening operation. T100s is the full symmetric current test, and T100a is the asymmetric current test, both according to [70].

case without concern, making the re-testing for $k_{pp} - 1.3$ of vacuum circuit breakers previously tested according to IEC 62271-100 Ed. 2.2 or earlier is generally unnecessary.

The testing standards also specify a variety of single-phase tests [70]. These tests include single-phase substitute tests for the $k_{pp} = 1.3$ case [97], as well as tests for specific single-phase and double-earth fault conditions [98]. In addition, there is also a test for out-of-phase breaking conditions, where the circuit breaker opens on a fault when the source and the load voltages are not yet synchronized. The short-circuit current is limited to 25% of the rated value, but u_c is quite high. The test is referred to as OP2 [70].

3.3.2   Comparison of short-circuit test duties

Table 3.3 compares the u_c formulas and values, arcing times, and current levels for the various single- and three-phase tests [70,95,99]. The u_c values are given in *PU* values. The arcing time Δt is

$$\Delta t = t_{arc} - t_{min} \tag{3.5}$$

Table 3.3 Summary of main short-circuit test duties including the TRV peak voltage in PU $(1 PU = \sqrt{2} U_r / \sqrt{3})$ and arcing time Δt (time added to t_{min} minimum arcing time to calculate the required arcing time for the test) with the location of the definition in the standard

Standard	k_{pp}	Test duty	Phase	U_c (PU)	$\Delta t/T$	I (%)	Source (voltage)	Source (time)
IEC6227-100	1.5	T100s	1st	2.10	0.167	100%	7.105.5.1a and Table 16	
IEC6227-100	1.5	T100s	2nd	1.21	0.417	100%	Cigre 305, Table 2	
IEC6227-100	1.5	T100s	3rd	1.21	0.417	100%	Cigre 305, Table 2	
IEC6227-100	1.5	T60	1st	2.25	0.167	60%	7.105.5.1a and Table 16	
IEC6227-100	1.5	T60	2nd	1.30	0.417	60%	Cigre 305, Table 2	
IEC6227-100	1.5	T60	3rd	1.30	0.417	60%	Cigre 305, Table 2	
IEC6227-100	1.5	T30	1st	2.40	0.167	30%	7.105.5.1a and Table 16	
IEC6227-100	1.5	T30	2nd	1.39	0.417	30%	Cigre 305, Table 2	
IEC6227-100	1.5	T30	3rd	1.39	0.417	30%	Cigre 305, Table 2	
IEC6227-100	1.5	T10	1st	2.55	0.167	10%	7.105.5.1a and Table 16	
IEC6227-100	1.5	T10	2nd	1.47	0.417	10%	Cigre 305, Table 2	
IEC6227-100	1.5	T10	3rd	1.47	0.417	10%	Cigre 305, Table 2	
IEC6227-100	1.3	T100s	1st	1.82	0.167	100%	7.105.5.1a and Table 16	
IEC6227-100	1.3	T100s	2nd	1.78	0.381	100%	Cigre 305, Table 2 for X_0/X_1=3.25	7.104.2.3 + 1ms
IEC6227-100	1.3	T100s	3rd	1.40	0.500	100%	Cigre 305, Table 2 for X_0/X_1=3.25	7.104.2.3 + 1ms
IEC6227-100	1.0	SPF		1.40	0.350	100%	7.108.2.2, Table 28	7.108.2.3
IEC6227-100	1.3	SPF	2nd	1.76	0.331	100%	7.104.2.3, Table 12	7.104.2.3
IEC6227-100	1.3	SPF	3rd	1.40	0.450	100%	7.104.2.3, Table 13	7.104.2.3
IEC6227-100		DEF		2.42	0.350	87%	7.108.2.2, Table 28	7.108.2.3
IEC6227-100	1.3	SPF asym	2nd	1.76	0.405	100%	7.104.2.3	7.104.2.3, Table 10/11
IEC6227-100	1.3	OP2		3.13	0.450	25%	Table 25	7.104.3.2
IEEE C37.09	1.5	T100s 1Ph		1.40	0.375	100%	4.8.4.5	4.8.4.5
IEEE C37.09	1.5	T100a 1Ph		1.40	0.440	100%	4.8.4.5	4.8.2.3.5b t_{arc2}
IEC6227-100	1.3	T100a	2nd	1.76	0.455	100%	Cigre 305, Table 2	7.104.2.3, Table 10/11
IEC6227-100	1.5	Substitute	Med	2.10	0.183		7.105.5.1a	7.104.3.2 t_{med}
IEC6227-100	1.5	substitute	Max	2.10	0.367		7.105.5.1a	7.104.3.2 t_{max}

Note: T is the cycle length (20 ms for 50 Hz and 16.67 ms for 60 Hz). IEC 62271-100 is [70], IEEE C37.09 is [99], and Cigre 305 is [95].

where t_{arc} is the arcing time in the test and t_{min} is minimum arcing time observed during the three-phase symmetric current tests (referred to as t_{a100s} in [70]). This allows for a simple comparison of the arcing time between the tests. For asymmetric current tests, t_{arc} was calculated using $\tau = 45$ ms and a minimum clearing time of $27.0 < t \leq 47.5$ ms for 50 Hz and $\tau = 45$ ms and a minimum clearing of $22.5 < t \leq 39.5$ ms for 60 Hz in [70, Tables 10 and 11]. Reference [70, Table 10 and 11] are identical to [99, Table 2 and 3]. Δt values are divided by the cycle length T for comparison (20 ms for 50 Hz and 16.67 ms for 60 Hz). Finally, the current used in the test is given as a percent of the rated short-circuit current. Some of these tests are for conditions where the full short-circuit current would not appear, but other issues such as longer arcing times, higher u_c values, or different RRRV and/or di/dt necessitate a separate test. Unfortunately, the information in Table 3.3 is scattered in different sections in [70,99], so the sources for u_c and arcing times are given in the table.

Figure 3.13 plots the u_c versus the percentage of the rated short-circuit current for the various tests, sorted by the $\Delta t/T$ values. The voltages are compared to the peak of the average U_d for $U_r = 12 - 36$ kV. This value represents an upper limit on u_c, above which there is a risk of dielectric breakdown in the system (not necessarily in the VI or even the circuit breaker). The upper envelope of values is the OP2 test at 25% and the double earth fault (DEF) test at 87% of the rated short-circuit current. Tests at $\Delta t/T = 0.167$ are for the first pole-to-clear, and the $\Delta t/T > 0.167$ tests have lower u_c values since they simulate the 2nd or 3rd phases to clear or correspond to line-to-ground fault conditions.

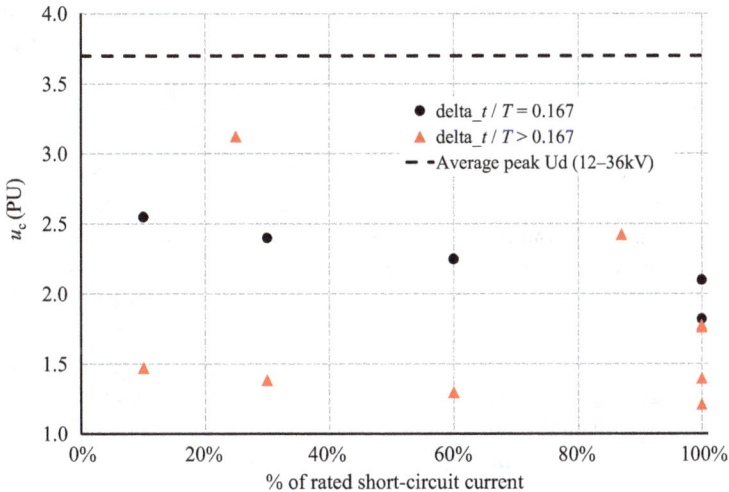

Figure 3.13   Data from Table 3.3 plotted as peak TRV voltage u_c in PU as a function of the percent of the rated short-circuit current. First pole clear cases at $\Delta t/T = 0.167$ and the later pole clears at longer times. u_c compared to the average peak of U_d for $U_r = 12 - 36$ kV in [70].

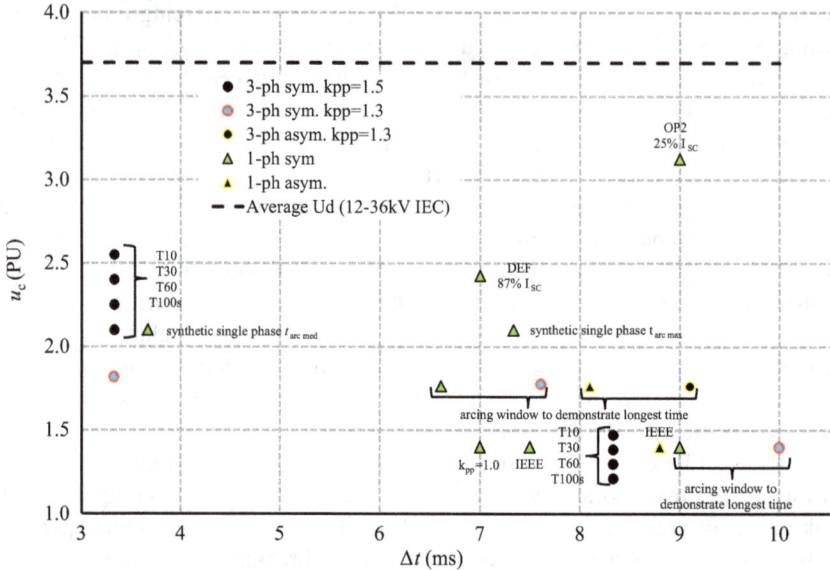

Figure 3.14   Data from Table 3.3 plotted as u_c in PU versus t for 50 Hz. 3-ph are three-phase short-circuit tests, and 1-ph are single-phase tests. Compared to the average peak of U_d for $U_r = 12 - 36$ kV in [70]. Nomenclature for tests follows Table 3.3 and the listed standards. Times for three-phase tests correspond to the longest possible times, and single-phase and synthetic test times are the minimum required values. Arc windows note the range between the minimum time for single-phase substitute tests and the corresponding three-phase tests.

Figure 3.14 plots the same data as Figure 3.13, but plots u_c versus Δt for 50 Hz. u_c from 1.82 to 2.55 per unit (PU) appear at $\Delta t = 3.3ms$ for the first-phase-to-clear, with u_c of 1.4 and 1.76 PU for single phase tests with $\Delta t > 6.5ms$. Again, the highest u_c are for the DEF and OP2 tests. The Δt values for three-phase tests correspond to the maximum possible arcing time. The Δt values for single-phase tests are the minimum arcing time required for a valid test. Tests with arcing time $t_{arc} < t_{min} + \Delta t$ are not valid. For longer times, arcing times between $t_{min} + \Delta t$ and $t_{min} + \Delta t + 1ms$ fall in the arc window are required to demonstrate interruption at the longest possible arcing time for the condition being tested. Arcing times longer than $t_{min} + \Delta t + 1ms$ are not meaningful and can overburden the test object. However, they are sometimes used for the convenience of the test lab to overcome scatter in the breaker opening time and/or the test lab timing to ensure that the minimum required arcing time is achieved.

3.3.3   Applications beyond the standards

The ability to meet conditions with higher TRV values and longer arcing times leaves the vacuum well-positioned to deal with new switching conditions that can arise. The OP2 test is particularly useful to extend applications of the VI when the current level is 25% of the rated short-circuit current or lower. The Δt for OP2 test matches the single-phase test for the third-pole-to-clear for $k_{pp} = 1.3$, and exceeds all other single- and three-phase Δt values. For example, the OP2 test on a VI rated for (U_r/I_{SC}) 12 kV/25 kA would exceed the u_c and Δt requirements for a DEF test for a rating of 15.5 kV/7 kA ($25kA \cdot 25\%/87\%$). Similarly, the OP2 test also exceeds the T100s tests for the first- and last-pole-to-clear for 17.5 kV/6.25 kA. A similar analysis can be applied to the DEF test for I_{SC} values up to 87% of the rating. Most special conditions do not occur at the full-rated short-circuit current, which involves a fault of all three phases directly at the breaker and is not generally influenced by details of the circuit load. Therefore, the OP2 and DEF tests together with the general insensitivity to RRRV and di/dt can allow coverage of some special requirements without further testing.

3.3.4   Synthetic type testing of VI

3.3.4.1   Comparison to direct testing

Type testing of vacuum circuit breakers for MV applications has almost exclusively been performed by direct testing. Direct testing refers to short-circuit tests where the same source provides both the short-circuit current and the recovery voltage after the interruption. If the breaker fails to clear at a CZ or later during the recovery voltage, the source also supplies the next half-cycle of current and recovery voltage at the next CZ where the breaker can again attempt to interrupt. The source is most commonly an MV generator behind a transformer to adjust the voltage level. In a few rarer cases the source can be the power grid itself, again with a transformer to adjust the voltage level. Passive components are often necessary to modify the TRV values to meet the detailed requirements of the standard. The main reason for this relatively happy situation is that generators exist that operate in MV, making the test setup reasonably straightforward.

In HV, the situation is different. While very limited options for direct tests at $U_r = 72.5$ kV exist, in general, HV testing requires synthetic testing [100]. The key feature of synthetic testing is that the current source and recovery voltage source are separated [101]. The current source can be a capacitor bank combined with the appropriate inductance supplying a power frequency current or can even be a generator. The main feature is the current source operates at voltages well below the tested U_r value. This increases the options for supplying the current and considerably reduces the size and cost of the current generation. For the current injection method of synthetic testing [100, Section 4.2.1], the TRV and recovery voltage are supplied by a separate capacitor bank that supplies a small current at a high frequency but at a HV. This current is superimposed on the main power frequency current. The tested breaker then interrupts the high-frequency current with the correct TRV and di/dt at the CZ. If the VI fails to interrupt, then only the high-frequency current will flow for one more half-cycle of high-frequency current as the power frequency current source has been switched out of the circuit. The disadvantage of this method is that the interrupter has only one chance to interrupt, whereas in a direct test, the interrupter would have another chance at the next CZ.

When arcing times longer than one half-cycle of power frequency current are required, then a re-ignition circuit is required. From Figure 3.14, when for power frequency $f_r = 50$ Hz common MV minimum arcing times of $t_{min} = 1 - 2$ ms could put a few single-phase tests to arcing times longer than one half-cycle of current $1/(2 \cdot f_r)$.

$$t_{arc} = t_{min} + \Delta t > 1/(2 \cdot f_r) \tag{3.6}$$

Longer minimum arcing times of $t_{min} > 3$ms, which can appear in HV VI designs [48], would push more single-phase tests beyond $1/(2 \cdot f_r)$, including the single-phase synthetic tests from [100]. Small t_{min} with values like MV are possible in HV but require very fast opening speeds of 2.5–5.5 m/s [102–104]. The re-ignition circuit injects a high-frequency current with a very high di/dt value. The goal is that the di/dt exceeds what the interrupter can interrupt and allows another half-cycle of current to flow. While this method is well established for SF_6 interrupters [105], it is harder to implement for vacuum given the very high di/dt values that can be interrupted [87]. Nevertheless, re-ignition circuits exist that can extend the arcing time of VI with good reliability.

This discussion focuses on single-phase testing using the current injection method. Other methods exist, as do three-phase synthetic methods. The primary challenge of synthetic testing is the time coordination of multiple sources and control of the applied currents and voltages. Despite these challenges, more synthetic testing of VI will be necessary for the future in both MV and HV. In HV, the lack of any feasible alternative will continue. In MV, applications of VI to generator circuit breaker applications [106] often require synthetic testing due to the lack of feasible alternatives at large short-circuit current ratings. Furthermore, the switch to SF_6-free MV GIS (gas-insulated switchgear) will require large amounts of testing to re-structure the portfolios of several manufacturers. Direct testing options will be very hard-pressed to manage the testing volume, especially given the recent concentration of test lab ownership [107,108].

3.3.4.2   Performing synthetic tests and Class E2

The basis of synthetic-type testing is described in [100]. It states in Section 1:

"This part of IEC 62271 mainly applies to AC circuit-breakers within the scope of IEC 62271-100. It provides the general rules for testing AC circuit-breakers, for making and breaking capacities over the range of test duties described in 7.102 to 7.111 of IEC 62271-100:2021, by synthetic methods. It has been proven that synthetic testing is an economical and technically correct way to test HV AC circuit-breakers according to the requirements of IEC 62271-100 and that it is equivalent to direct testing."

The use of single-phase substitution tests to overcome the difficulty and expense of three-phase tests is already established for $k_{pp} = 1.3$ tests [97]. The statement from [100] does not explicitly cover the electrical life test described in [70, Section 7.112], specifying the Class E2 test for $U_r \le 52$kV. This test is particularly important for applications in MV. The procedure outlined in [70, Table 34] can be performed on one phase of the breaker following the synthetic test procedures for each step. Reference [70, Table 34] can be combined with [100, Table 7] to make Table 3.4.

Figure 3.15 compares the transferred charge and I^2t value as a function of t_{min} for the direct test and the equivalent synthetic test. The values for the direct test are averaged over the three phases, including the behavior of the first and last poles to clear for $k_{pp} = 1.5$. The values were calculated for 1 kA, and could be scaled to other values by multiplying the charge by I_{SC} and the I^2t by I_{SC}^2 in the correct units. The synthetic test values are calculated at t_{min} and $t_{min} + 1$ms, to give the region

Table 3.4 *Possible procedure to adopt Class E2 list 3 test from [70] to synthetic testing using the terminology from [100]*

Test duty	Synthetic test method	Operating sequence
T10, T30, and T60	1	Os -t - (Cd)Os -t' - (Cd)Os
T100s	5	Csv_{sym}
		Os
		Od -t- CdOs-t' - CdOs

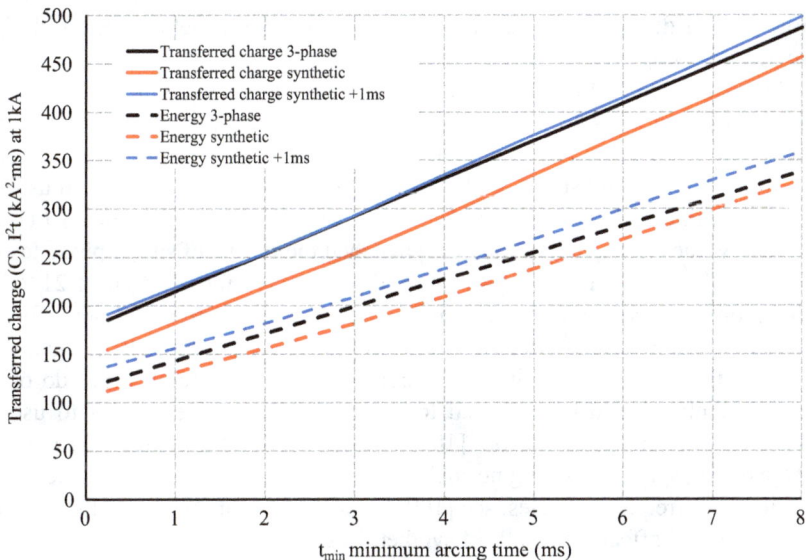

Figure 3.15 *Comparison of the total transferred charge and I^2t averaged over all three phases with distributed opening times during Class E2 list 3 testing following [70] and the procedure in Table 3.4. The comparison is as a function of the minimum arcing time t_{min} calculated at 1 kA. The synthetic tests are compared for the times of $t_{min} + \Delta t$ and at $t_{min} + \Delta t + 1$ ms.*

that tests would fall in. The I^2t for direct tests falls between the synthetic calculations. The transferred charge is also between the values for synthetic tests. The
results of direct and synthetic Class E2 tests were compared in [109], with very
good agreement between the test methods. Reference [100] does not explicitly
cover short-circuit tests from standards beyond [70], such as generator circuit
breakers [88] and reclosers [110]. Synthetic tests are discussed and allowed within
[88]. Although the recloser standard is currently limited to $U_r \leq 38$ kV, new
applications at higher voltages [111] should prompt a review of when synthetic test
methods are feasible for recloser applications. Synthetic tests can also be used to
verify the short-circuit electrical life of VI [80].

3.3.4.3   Differences compared to direct tests

The arcing times in synthetic tests are set by t_{min}, the same as the various single-
phase tests. This increases the importance of, or at least the visibility of, the contact
opening speed in setting t_{min} [102–104,112]. The t_{min} value for vacuum also has a
distribution. For MV applications most of the variation is within ± 1 ms, which is
within the practical timing adjustment range for test labs. The higher u_c in HV,
especially with more moderate contact opening speeds, could widen the distribution of t_{min}. Widening of the t_{min} distribution can generally be tolerated for short-
circuit test duties but could become problematic for electrical life tests. Increasing
t_{min} by $2 - 3$ ms, for example, produces a significant increase in the transferred
charge and I^2t during Class E2 testing, Figure 3.15. Nevertheless, certain designs
have achieved $t_{min} = 4$ ms for $U_r = 145$ kV [48].

Also, as noted in Figure 3.14, the TRV peak for the medium and maximum
arcing times are $u_c = 2.1$ PU, instead of $u_c = 1.21$ PU for the last poles to clear in
direct three-phase tests. This is done for practical reasons to reduce the number of
test setups required and speed up the testing. Formally the u_c could be adjusted to
the correct values. All the tests in [98,112,113] were performed at $u_c = 2.1$ PU, and
specifically, the results of the study [98] matched the results of single-phase tests at
other u_c values. In addition, [96] showed that increasing u_c from 1.21 PU to
1.78 PU on the second pole-to-clear in asymmetric current tests did not influence
the results. These results suggest that testing at $u_c = 2.1$ PU will not generally
overstress the VI. If failures in the medium or maximum arcing times do occur,
then the results should be reviewed to see if splitting the test series to use the
correct TRV would be applicable [100, Section 7.104.3.5]. Finally, the recovery
voltage during synthetic testing normally has a DC value with a gradual decay over
several power frequency cycles, see [100, Figure 2]. As for the u_c results, this does
not significantly affect the results of synthetic testing of VI. This stress is higher, so
the user should be aware of this difference if testing very near the performance
boundary of the VI.

Certain interruption behavior of vacuum in three-phase testing is not reproduced in single-phase testing, whether direct or synthetic. Some designs of gas
interrupters depend on the motion of cool gas into the arcing region for interruption, which severely limited the ability to interrupt outside this period. The vacuum
does not have this limitation and can successfully interrupt on a later CZ. In a

T100s test at 50 Hz for example, the breaker only must live through ~3.3 ms of arcing before another chance to interrupt occurs. For single-phase direct tests, the breaker must live through 10 ms of arcing by comparison, and synthetic tests are not able to tell if the VI would interrupt at the next CZ. In three-phase tests for a $k_{pp} = 1.5$ system, the last two poles-to-clear the interrupt in series. If one of the VI fails to hold the voltage, the second VI can still interrupt if it holds off the entire recovery voltage. This behavior has been observed in testing successful designs. Single-phase and synthetic testing could not observe this behavior and instead, the test could fail. For three-phase tests on $k_{pp} = 1.3$, a failure on the second pole-to-clear leads to an interruption condition with lower stress, which can then be interrupted [96]. Single-phase and synthetic tests would again not produce this behavior. These behaviors need to be considered when testing with single-phase substitution tests and/or synthetic tests.

3.3.4.4 Development tests and practical experience

Variations in synthetic and single-phase testing are the basis for virtually all VI development [92,93,114,115]. Each VI manufacturer has their own specific procedure; however, the common element is a focus on using long arcing times similar to the half-cycle length and varying the interrupted current to identify a performance boundary. Many of the tests specified in [70] such as T10, T30, T60, and short-line fault [89] are not the limiting factor for VI. The RRRV and di/dt values are also of limited importance in power frequency switching [87]. The experience in [96] implies that the peak of the TRV u_c is not the most critical factor, although it could play a role when the value is high in DEF and OP2 tests, Figure 3.14. All this together suggests the current, or various similar measures like the transferred charge or I^2t is an important short-circuit performance limitation.

It is important to understand that no VI operates independently; all VI are part of a circuit breaker, contactor, or load break switch. Therefore, type tests always apply to the circuit breaker and not to the individual VI type. There is a long history of VI manufacturers supplying VI to OEM (original equipment manufacturers), and the OEM successfully integrating the VI into their specific breaker design [116]. Nevertheless, even when the VI itself had an extensive pedigree of successful development and type tests, the OEM must separately perform type tests on their breaker.

Reference [20] summarizes the key experience in testing and field usage of VI in HV up to the year 2014. Table 3.5 compiles type tests described in the technical literature with clear statements of the tests performed from 2014 on. The table shows

Table 3.5   Recent technical publications on type tests of HV circuit breakers with VI

U_r (kV)	In series	T100s	T100a	DEF	OP2	Electrical life	Line/cable
72.5/84	Single VI	[118]	[118]		[118]		[118]
120–170	Single VI	[48,79,80]	[48,79,80]	[79]	[79]	[79,80]	[48,79,80]
>170	≥ 3	[119,120]	[119,120]				

Note: Terminology for the test duties is the same as used in [70].

the present focus on single VI solutions for the $U_r = 126 - 170$ kV range, covering a broad range of requirements. Work at higher voltages is also progressing, while the use of several $U_r = 40.5$ kV VI in series has passed the main interruption tests. Reference [117] provides an update to [20] regarding commercial products.

3.4   Application problems to be solved for HV

3.4.1   Contact design

3.4.1.1   AMF strength

Currently, the AMF contact is the primary contact design for magnetic arc control in HV VI. The key challenge in HV VI design is the long contact gaps of ≥ 20 mm required for the dielectric performance. The basic problem this poses for magnetic arc control is the reduction in the magnetic field strength as the gap increases. The AMF coils can be treated as two perfect current loops of current, Figure 3.16, with a current of I/n where I is the peak current and n is the number of coil arms, diameter D and separated by a gap d. The magnetic field strength along the axis of the coils in the middle of the gap is $|B_z|$.

$$|B_z| = \frac{\mu_0 \cdot \frac{I}{n} \cdot \left(\frac{D}{2}\right)^2}{\left(\left(\frac{D}{2}\right)^2 + \left(\frac{d}{2}\right)^2\right)^{3/2}} \tag{3.7}$$

Equation (3.7) illustrates how the magnetic field strength will decay with increasing contact gap, and hence becomes an issue as the lightning withstand voltage U_p requirements increase. The I/n dependence comes from the typical design of AMF coils, where each arm carries only a portion of the current. For

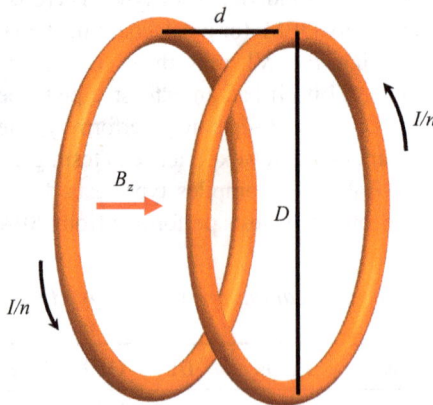

Figure 3.16   Two coils carrying current I/n in the same direction of diameter D separated by gap d. The simplest model of the magnetic field from AMF contacts with n arms.

example, in a typical three-arm AMF design each arm carries one-third of the total current approximately one-third of the way around the contact. Therefore, the three-arm coil is roughly equivalent to a single loop of one-third of the total current, hence $I/3$ where $n = 3$.

Several experiments indicate the importance of the magnetic field strength in the performance of AMF contacts. Some early work used an AMF generated by an external current allowing the magnitude of the field to be varied for different currents between the arcing contacts [121]. High-speed movies of the arc appearance identified a critical current as a function of the peak current between the contacts. If the applied field was higher than this critical field, then there was no anode activity for contact gaps up to 18 mm. If the applied field was below the critical value, then anode spots and/or anode jets appeared. Some even earlier work observed a U-shaped behavior of the arc voltage between the contacts as a function of the applied magnetic field for Cu and Mo contacts [122]. The arc voltage first dropped sharply with the application of a few tens of milli-tesla of AMF, reached a minimum, and then gradually increased with the applied field. Later work extended these results to Cu and Cu–Cr contacts at different contact diameters and gaps and observed the same trend, Figure 3.17 [123]. A simple theoretical model connected the change in the arc voltage to a change in the experimental arc behavior [124,125]. Finally, arc visualization experiments with AMF contacts with a large contact gap of 40 mm observed the transition from a constricted arc to a fully diffuse arc with increasing magnetic field strength [126]. A three-dimensional model of the vacuum arc reproduced the same transition with magnetic field strength, where the transition to a constricted arc appeared as the loss of convergence in the model.

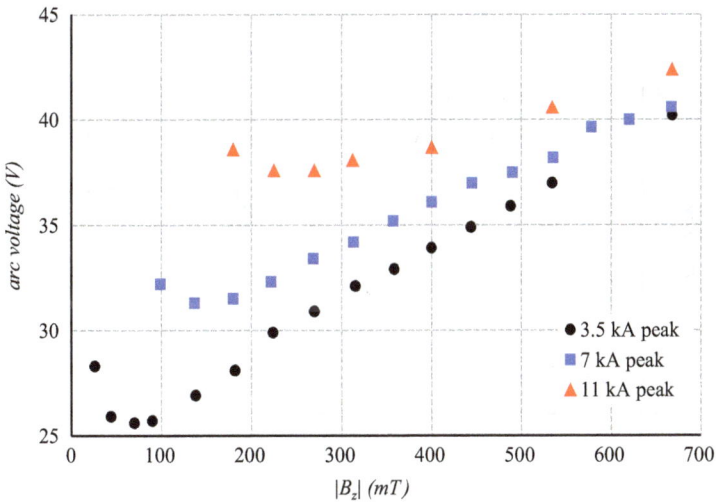

Figure 3.17   Effect of the AMF strength on the arc voltage for different current levels [123]. Cu–Cr30 contacts with a diameter of D = 30 mm and a contact gap of d = 12 mm.

3.4.1.2   Arc mode transition and performance limit

Some key recent work studied the dependence of the transition from a diffuse arc to a high-current anode mode as a function of the current, AMF strength, contact diameter, and contact gap [49]. The high-current anode mode was identified as the peak current level where a bright region of visible light emissions appeared on the anode surface. The visual light transition from a diffuse arc to some type of constricted arc and/or anode activity occurs over a region of a few kiloamps, Figure 3.18 [127], and therefore is useable as an experimental criterion. Figure 3.19 plots the threshold current of the transition from a diffuse arc to an anode mode I_{th} as a function of the AMF strength, contact diameter, and contact gap [49].

The key features are that the threshold current is a strongly increasing function of the AMF strength and the contact diameter. This agrees with empirical experience, where an increase in the AMF strength or an increase in the contact diameter can improve the current interruption performance of a VI. The threshold current increases as the contact gap decreases, but this change is considerably smaller in magnitude than the corresponding change with diameter and field strength. While this is encouraging for the use of AMF designs with large contact gaps, it does need to be remembered that the results plotted in Figure 3.19 are for changes in the gap for fixed AMF strength. In a practical VI design where the AMF coils are behind the contact, a larger contact gap would also reduce the AMF field strength, in line with (3.7). This would in turn reduce the threshold current and could require efforts to increase the AMF strength. A key result from [49] is that there is a smooth transition in the behavior from the contact gaps typical of MV applications ($d = 6$–20 mm) to those gaps typical of HV applications ($d = 25$–60 mm). While the

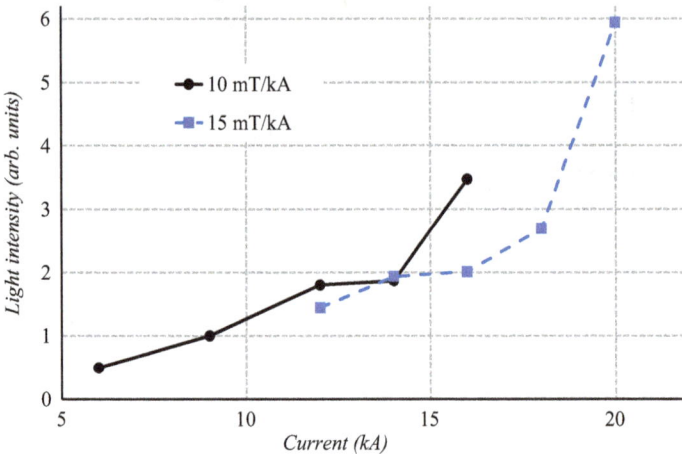

Figure 3.18   Effect of AMF strength and short-circuit current on the plasma visible light emission. Data integrated parallel to the contact surface, at the middle of the contact gap, at the time of peak light emission for D = 38 mm Cu–Cr contacts [127].

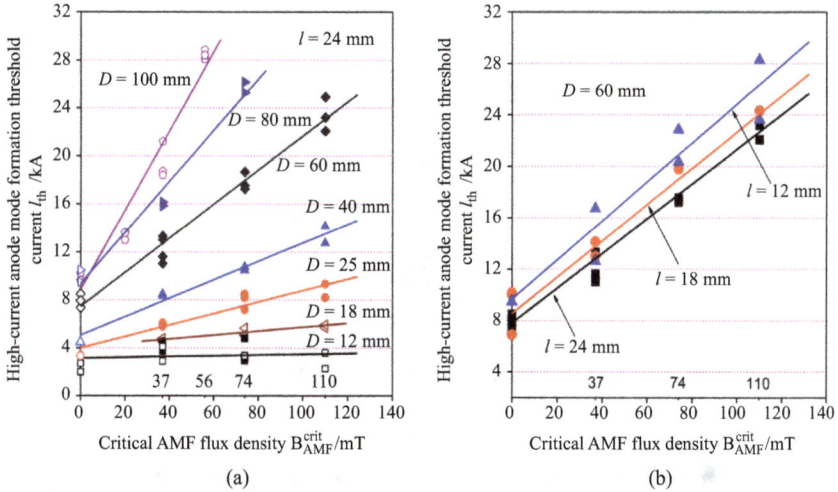

Figure 3.19   *Peak threshold current I_th for the transition to an active anode mode on Cu–Cr25 contacts as a function of the AMF strength for (a) different contact diameters D at contact gap l = 24 mm and (b) different contact gap l for contact diameter D = 60 mm. Note that the current is in kA peak values. Reproduced from [49] with permission.*

larger contact gap will slightly reduce the threshold current and make the required AMF strength more challenging to achieve, the vacuum arc behavior remains the same for the MV and HV ranges. Therefore, current interrupt performance in HV can be extended from and built upon the extensive experience in MV.

These experiments observed the transition from a presumably good state (diffuse arc, passive anode, minimum/low arc voltage) to a presumably bad state (constricted arc, light emissions from anode, higher arc voltage, anode melting). Although this is reasonable, the exact correlation between the onset of a high-current anode mode and the failure to interrupt remains open. Observations of VI contacts even a single short-circuit operation at full short-circuit current show that the anode is not passive during the arcing [128], and some amount of contact melting and constricted arcing occur before the VI fails, Figure 3.20. The data in Figure 3.19 implies a linear equation relating the critical magnetic field for the arc mode transition $\left|B_{AMF}^{crit}\right|$ of [49, Eq. (1.3)].

$$\left|B_{AMF}^{crit}\right| = k_{AMF}\left(I - I_0\right) \tag{3.8}$$

where I is the peak current, I_0 is the offset current, and k_{AMF} is a proportionality constant. This equation agrees with earlier work in [121]. k_{AMF} is a function of the contact diameter D and only weakly dependent on the contact gap d [49, Eq. (1.7)]

$$k_{AMF}(D) = \frac{4956}{(D/mm)^{1.6}} \frac{mT}{kA} \tag{3.9}$$

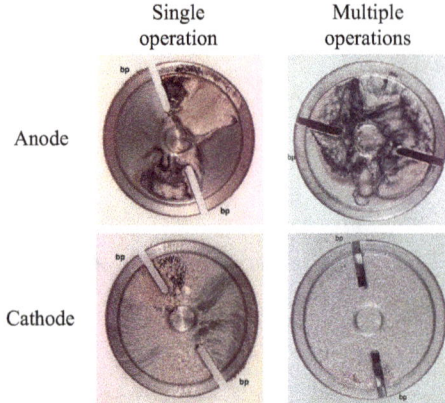

Figure 3.20   Cu–Cr25 AMF contacts of D = 60 mm after one and multiple short circuits breaking operations at a constant polarity [128]. Single operation at 25 kA, multiple operations consisted of 15 operations at 7–27.5 kA.

The experimental results in [49] can be compared to the interruption test results described in [129]. In [129], interruption tests on a three-arm AMF design with a contact diameter of 62 mm and a contact gap of 10 mm and arcing times of ~8.5 ms were performed. In addition, the AMF strength was calculated, with a peak value of 275 mT for 30 kA. This design saw a transition in the TRV breakdown voltage at 35 kA, where higher current saw a sharp decrease in the breakdown voltage. This corresponds to a peak current of 50 kA peak and an AMF strength of 321 mT at peak current. Using (3.9) requires an I_0. The I_0 value from Figure 3.19 can be taken as 9 kA peak, and the slope of the dependence between the critical field and the threshold current is roughly constant for contact gaps of 12–24 mm. Therefore, we can take the k_{AMF} value for 60 mm diameter contacts from (3.9) of 7.1 mT/kA as sufficiently accurate for this initial comparison. The critical magnetic field for a threshold current of 50 kA peak is 272 mT. The AMF strength in the VI is larger than the critical field, however, the critical magnetic field is similar in magnitude to the point where VI stops interrupting. This suggests that the threshold current and critical field can provide an initial guide to selecting the necessary AMF required for a particular interruption current.

3.4.1.3   Heat flux to contacts and CZ behavior

While the arc mode transition is one key limitation of the current interruption performance, a second key limit is the maximum heat flux to the contacts, particularly the anode. Even when the arc is diffuse at the peak current, the current density in the central portion of the anode is roughly double the current density on the outer half of the contact diameter, Figure 3.21. This concentration can lead to anode melting, particularly in the center of the contact, and eventual failure to interrupt. Reference [130] modeled the peak heat flux to the center of the anode at

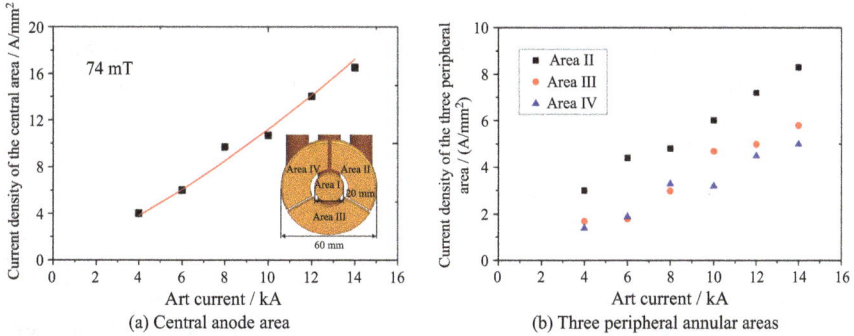

Figure 3.21   *Current density measured for a diffuse arc mode on an AMF contact D = 60 mm as a function of the peak current in four regions: (a) the central region of the contact and (b) three outer portions of the contact. Note the difference in vertical scale between (a) and (b). Reproduced from [49] with permission.*

the transition to a high-current anode mode on 60 mm diameter contacts of Cu–Cr. Varying the contact gap and AMF strength resulted in a narrow variation of the peak heat flux to the anode of 7.3×10^8 W/m^2 to 11.3×10^8 W/m^2. Ions contributed 23% of heat flux, with the remainder from the electrons. The narrow range for the peak heat flux suggests that it might be a more general parameter for the transition to an anode mode than the threshold current, which varied from $I_{th} = 18.4 - 39.6$ kA peak for the same conditions. The exact limiting value for the peak heat flux to the anode for current interruption performance remains open. Further modeling work in the same reference looked at the time evolution of the peak temperature of the anode as a function of the input energy. The melting point of copper contacts was reached at 5×10^8 W/m^2 in the model. Above this peak heat flux, the time for the anode to cool below the copper melting point increased significantly. Two factors contributed to this. First, the exponential decay of the temperature simply took longer as the peak temperature increased. Second, and more importantly, the heat of solidification took longer to conduct away as the heat flux increased. This had a large effect on the total cooling time.

While the previous discussion focused on the arc behavior, particularly around the peak current, it is very important to remember that the ability to interrupt is finally determined by the situation at and after the current zero (CZ). The arc behavior around the peak current can influence the conditions at the CZ but is not directly the source of failure and requires detailed effort to make a connection [131]. Experiments rapidly applied a high DC voltage after the extinction of a diffuse arc of 1.5 kA after 9 ms of arcing with a 10 mm contact gap on Cu and Cu–Cr contacts [49]. The voltage the gap could withstand before breakdown increased from ≤ 15 kV to ≥ 60 kV in ~4 μs. This can be compared to practical applications by looking at the rate that the applied voltage increases with time after the CZ. The RRRV is denoted as u_c/t_3 [70,87]. During an initial period after the CZ, the voltage rises at a much slower rate before assuming the RRRV value. This time delay is referred to as t_d and ranges from 1.34 to 17.1 µs for rated voltages U_r from 3.6 to 40.5 kV [70]. The magnitude of t_d could reduce or eliminate the effect of an initial period of lower breakdown voltage discussed above and help us understand the extremely high RRRV that a vacuum can withstand.

3.4.1.4   TRV breakdown process

When a circuit breaker fails to interrupt, the TRV level where the breakdown occurs is often used to characterize the failure. If the failure occurs at roughly $\leq 25\%\ u_c$, then the failure is labeled thermal and associated with thermal over-loading of contacts, often with large-scale contact melting. The standard solution to prevent the failures would be to increase the contact diameter, reduce the current level, or reduce the arcing time. In other cases, if the failure occurs at roughly $\geq 50\%\ u_c$, the failure is characterized as dielectric and thereby connected to the electrical field stress in the contact gap after CZ. Here the standard solution would be to increase the contact gap at CZ or use a lower u_c value.

These characterizations work well for SF_6, but the applicability to vacuum is less clear. Figure 3.22 plots the breakdown voltage on failures to interrupt normalized to the rated u_c value as a function of the percent of the limiting value of the transferred charge Q [113]. Three different contact designs are plotted. There is a sharp transition at $Q/Q_{limit} \sim 100\%$, from a very few to many breakdowns. The transition from almost no breakdowns to greater than 50% occurs over a transition

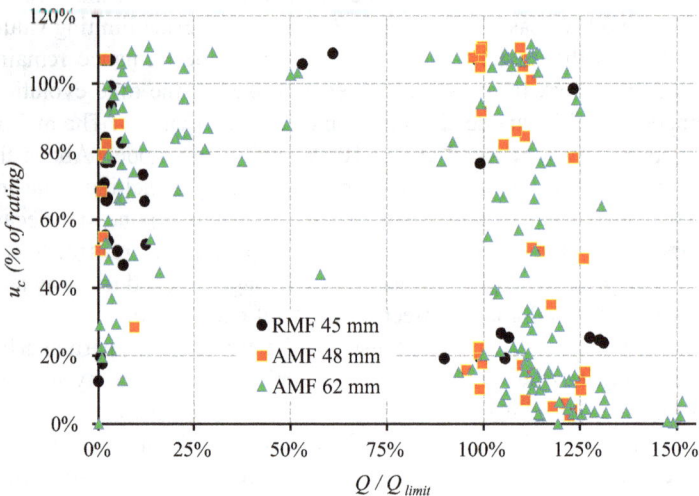

Figure 3.22   Breakdown voltage during current interruption tests on Cu–Cr normalized to u_c as a function of the transferred charge during the arcing, normalized to the transition between high rates of passing to high rates of failing Q_{limit}. Data from one spiral RMF contact D = 48 mm and two AMF contacts of D = 48 mm and D = 62 mm [113].

region of 25 A.s [132], or over 3 kA [93]. However, the voltage where these breakdowns occur varies significantly, all the way from slightly above the rated u_c value to below 25% of u_c. The TRV breakdown data in [129] also suggests a wide dispersion of the breakdown voltages during the interruption.

While the transition from interrupting to not interrupting is very sharp as a function of the current and/or the transferred charge during arcing, how the contact gap breaks down after the CZ does not necessarily follow the simple model from SF_6. More research is needed in this area. Also, care should be taken when interpreting the breakdown voltages of a small number of failures, since the potentially wide dispersion of the breakdown voltages could lead to misleading conclusions.

3.4.1.5   AMF contact designs

Two main AMF contact designs exist. The first directs the current flow in an azimuthal direction underneath the contacts, Figure 3.16, effectively creating two current loops. The number of coil arms directing the current corresponds to n in (3.7). Therefore, the smaller the number of arms, the greater the AMF strength. The disadvantage of a smaller number of arms is an increase in the resistance of VI, as well as more complex structures to support the coil [129]. Most designs range between two and six arms.

The second main design uses magnetic material such as steel or iron to redirect the azimuthal magnetic field from the current flow in the terminals into the contact gap. The most common design uses sheets of magnetic material in a U-shape underneath one contact, with a U rotated by 180° underneath the opposite contact, and are referred to as horseshoe contacts [133]. The U-shapes force some of the azimuthal magnetic fields into an axial field between the ends of the U. Such designs have been used in MV applications for many years. For HV applications, the main disadvantage is that the AMF strength has a step decrease in magnitude when the contact gap exceeds a certain value. Some researchers have studied the behavior of horseshoe contacts with large gaps [134], however, it would seem to be difficult to apply these designs to contact gaps larger than ~40 mm or would require unusually large contacts (120 mm diameter for arcing at 40 kA), as described in the reference.

3.4.1.6   RMF contact designs

RMF contact designs have also been used for HV VI [20,23] for limited applications. The primary advantage of the RMF contact design is the lower resistance of the entire VI, and this contact design is widely used in MV [135]. Since the current does not need to flow through the coil structure, the total resistance of the VI is reduced. For HV VI, this is an important issue, since the overall length of the VI for external dielectric performance has the unfortunate side-effect of increasing the VI resistance. The main limitation of the application of RMF contacts to HV VI is the instability of the constricted arc with long contact gaps [44]. Contact gaps beyond ~15 mm run the risk of plasma jets from contact towards the center shield, which reduces the current interruption performance of the VI. This compares poorly with the stable performance of AMF designs with contact gap and magnetic field strength for HV applications. Very little research is focused on improving the control of constricted arcs with large contact gaps. While improvements in the design of RMF contacts are conceivable, the present lack of effort limits the application of RMF designs for HV VI to specific applications with limited short-circuit current and critical continuous current I_r requirements.

3.4.2   Dielectric design

3.4.2.1   Distribution of breakdown voltage

The breakdown in a vacuum follows a distribution, with an increasing chance of a breakdown with increasing voltage. Although voltage breakdown in SF_6 also follows a distribution [136], the distributions for breakdown in gas tend to have a sharper transition from low to high probability of breakdown than in vacuum. Therefore, although breakdown voltages are quoted for vacuum gaps, it is important to know how the test was performed and how many voltage applications were used to understand where on the distribution the tests were performed. The Weibull distribution provides a good characterization of breakdown in a vacuum [27,137].

$$F(U) = 1 - e^{-(U/\lambda)^k} \tag{3.10}$$

where $F(U)$ is the cumulative probability of a breakdown by a peak voltage U, λ is the scale parameter where the breakdown probability reaches 63.2%, and k is the shape parameter that sets how the distribution increases with voltage. Practical experiments measuring the breakdown distribution often use the voltage U_{50} to characterize and compare distributions, where

$$F(U_{50}) = 50\% \tag{3.11}$$

Figure 3.23 plots the U_{50} values as a function of the contact gap from a variety of different experiments [137]. For comparison, the function for the breakdown voltage versus the contact gap compiled from many earlier experiments is also plotted [27]. This function is as

$$U_B(d) = \begin{cases} 97 \, kV \cdot \dfrac{d}{mm} & \text{if } d < 0.3 \text{ mm} \\ 58 \, kV \cdot \left(\dfrac{d}{mm}\right)^{0.58} & \text{if } 0.3 \text{ mm} < d < 23 \text{ mm} \\ 123 \, kV \cdot \left(\dfrac{d}{mm}\right)^{0.34} & \text{if } d \geq 23 \text{ mm} \end{cases} \tag{3.12}$$

Using the formula from [27] at a 40 mm, contact gap gives a voltage of

$$U_B(40\text{mm}) = 430\text{kV} \tag{3.13}$$

Assuming $U_{50}(d) = U_B(d)$, as supported by the data in Figure 3.23, a Weibull distribution (3.10) using the average shape parameter for Cu–Cr contacts with floating shields from [137] of $k = 14$ and a scale factor of $\lambda = 441$ kV peak gives

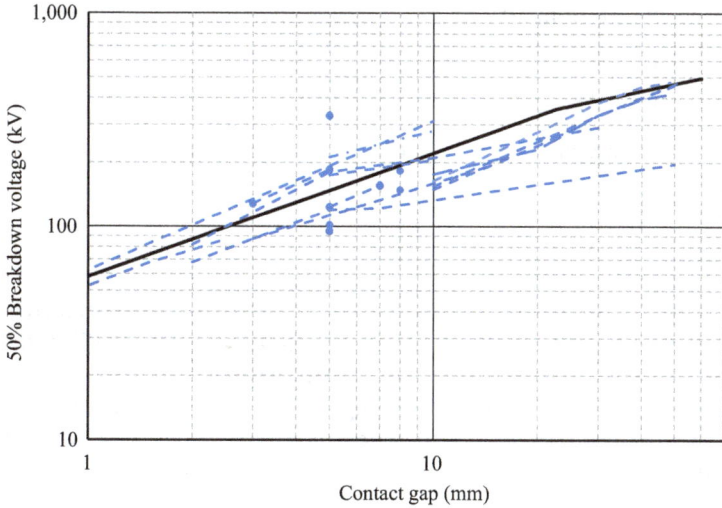

Figure 3.23   U_50 breakdown voltage as a function of contact gap. The black line is (3.12), and the blue dashed lines and points are from the references and data in [137].

the desired $U_{50}(40 \text{ mm})$ value. The Weibull distribution can then be used to calculate the chance of breakdown at the lightning impulse withstand voltage U_p for 325 kV, the required value for $U_r = 72.5$ kV [54]

$$F(325 \text{ kV}) = 1 - e^{-(325 \text{ kV}/441 \text{ kV})^{14}} = 1.4\% \tag{3.14}$$

The type tests in [54] allow a maximum of two breakdowns over 15 tests for each pole and each polarity, giving a maximum failure rate of $p_{BIL} = 2/15 = 13.3\%$, which for this distribution occurs at $U_{BIL} = 384$ kV. This short analysis demonstrates the general feasibility to use 40 mm contact gaps for applications at $U_r = 72.5$ kV.

3.4.2.2   Comparison of breakdown in linear accelerators

Most experiments on breakdown rates in VI focus on the range from 1% to 100%, with a key focus on p_{BIL}. To help confirm the application of the Weibull distributions calculated in this breakdown range to values $\ll 1\%$, we can compare the results on VI to the experience from linear accelerators [137]. The breakdown voltage distribution is a key parameter for these devices. Reference [138] compiled data from several different linear accelerator experiments and found that the breakdown rate (BDR) followed a distribution

$$BDR = E_a^{\gamma} \tag{3.15}$$

where E_a is the magnitude of the electric field on the accelerator part, and γ is an empirical factor used to fit the data, with a value of $\gamma = 30$. The BDR in these experiments ranges from 10^{-1} to 10^{-7}. This leads to the idea that (3.15) could be expressed as

$$F(U) = (U/\lambda)^{\gamma} \tag{3.16}$$

as $U/\lambda \rightarrow 0$. Comparing the ratio of (3.10) and (3.16) gives

$$\lim_{U/\lambda \to 0} \frac{1 - e^{-(U/\lambda)^k}}{(U/\lambda)^{\gamma}} = \frac{1 - \left(1 - (U/\lambda)^k\right)}{(U/\lambda)^{\gamma}} = 1 \tag{3.17}$$

when $k = \gamma$. Therefore if γ is equivalent to the Weibull shape parameter k, then the two empirical datasets converge. The value of $k = 14$ from [137] is lower than $\gamma = 30$ from [138], however, they are close enough in magnitude given the differences between the experiments to suggest the same fundamental behavior. Equation (3.16) also matches (3.10) up to $\sim p_{BIL}$. The values for k and γ observed in experiments lead to very steep distributions as $U/\lambda \rightarrow 0$. For example, applying the values used in (3.14) to the line-to-earth voltage across the open contacts leads to

$$F\left(\frac{72.5\text{kV}}{\sqrt{3}}\right) = 1 - e^{-(59.2\text{kV}/441\text{kV})^{14}} = 6.2 \times 10^{-13} \tag{3.18}$$

which is equivalent to roughly one breakdown every 500 years on a 50 Hz voltage, with two chances to break down per cycle (one on the positive and one on the negative).

While the frequency of the applied voltage can affect the breakdown process in gases, in vacuum the breakdown process is largely independent of the frequency. Earlier experiments noted at most a weak dependence of the breakdown voltage on the frequency [59]. Linear accelerator experiments regularly apply the results of DC voltage pulses to radio frequency (RF) applications [139]. Although it is wise to confirm the results at the exact frequency of interest, in general, vacuum breakdown results are applicable to a wide range of frequencies.

3.4.2.3   Influence of electric field stress on breakdown

One key finding to understand the breakdown of metal contacts in vacuum is the dependence on the local electric field stress on the contact. For the classic arrangement of two parallel disk contacts separated by contact gap d (Figure 3.24), the local electric field stress $|E|$ is

$$|E| = \frac{\beta_m \beta_g U}{d} \tag{3.19}$$

where d is the contact gap, U is the applied voltage, β_m and β_g are enhancement factors due to the microscopic surface and the specific geometry, respectively [27]. β_m depends on the contact material, machining processes, and cleaning processes. Once these are fixed, β_m should remain a constant value. Both $|E|$ and β_m are very

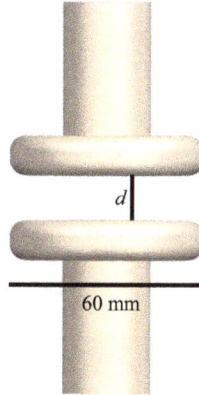

Figure 3.24   Example of contacts in vacuum for calculating the macroscopic electric field stress and β_g

difficult to measure. Assuming that β_m is a constant value over the contacts of interest allows one to look at the geometric electrical field stress

$$|E_g| = \frac{\beta_g\, U}{d} \qquad\qquad (3.20)$$

The values for $|E_g|$ are straightforward to calculate using two or three-dimensional electrostatic field modeling programs. Therefore, (3.12) and (3.20) can be combined to give the geometric electric field stress at a ~50% breakdown rate

$$|E_g(d)| = \frac{U_B(d)}{d} \qquad\qquad (3.21)$$

The values are plotted in Figure 3.25. These values provide a rough guideline for the possible electric field stress on contacts in a vacuum. Considerable work is occurring to both reduce the β_m values through material choice, improved machining, and improved cleaning as well as work to increase the value of $|E_g(d)|$ such as through improved design and conditioning. This work is described later in this chapter. Therefore, Figure 3.25 is only a rough guide to the current state of the art but is useful for evaluating the progress being made and where a current design lies in comparison.

When the ratio of the contact diameter D over the contact gap d is large, the geometric electric field stress is $|E_g| \sim U/d$ in the center of the contact and has a slight enhancement at the contact edge. As D/d decreases, $|E_g|$ increases on the contact edge as well as the side of the contact due to the increase in β_g, and for AMF designs, on the side of the AMF coil, Figure 3.26 [27,49]. This behavior is relevant for HV designs, where the contact gap can become a significant fraction of the contact diameter. However, practical experience in HV VI shows that break-downs on the contact surface are critical in determining the breakdown voltage,

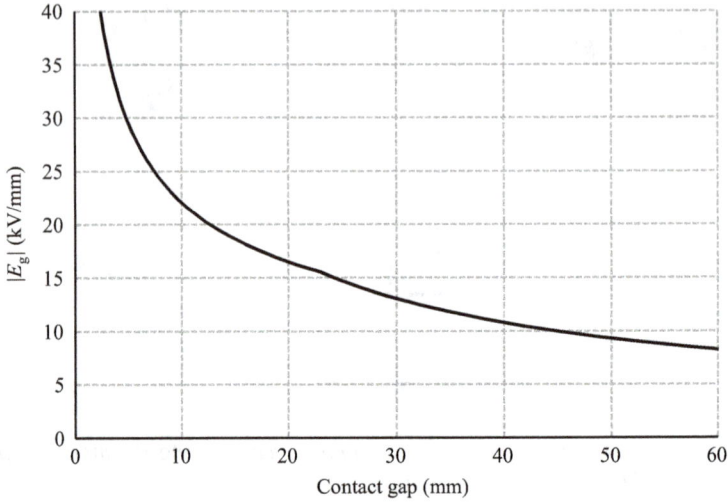

Figure 3.25   *Macroscopic electric field stress at U_50, calculated by dividing (3.12) by the contact gap d and plotting as a function of the contact gap*

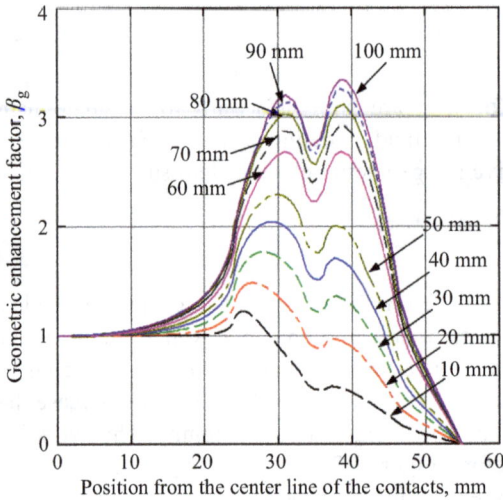

Figure 3.26 β_g *for D = 60 mm contacts at different contact gaps. β_g is plotted along the contact surface and then along the side and back of the contact, see Figure 3.23. Reproduced from [49] with permission.*

despite a potential higher $|E_g|$ on other parts of the contact/coil structure. Although a lower $|E_g|$ is always better, the electric field stress alone does not set the ultimate breakdown voltage. New models and ideas are described in Section 3.6.3 that seek to explain this behavior.

3.4.2.4   VI design and arrangements to improve breakdown behavior

The challenges in the breakdown voltage in HV drove many attempted improvements beyond the MV experience [140]. One example is an arrangement of multiple floating shields inside the VI [141]. The shields are designed to minimize the electric field stress when a partial breakdown of one of the floating shields occurs. This allows for a more compact design, where partial breakdowns between the contacts and one of the floating shields do not increase the electric field stress enough to drive a complete voltage breakdown.

Another focus area with a similar goal is putting two or more VI in a series. Here the idea is to take advantage of the higher $|E_g|$ at smaller contact gaps by having multiple small gaps instead of one large gap. Any breakdown must also occur in both (or all) gaps, further giving an additional improvement. Early work demonstrated successful current interruption with two VI in series [142], including cases where one VI held the entire TRV after a breakdown in the other VI [143]. External grading capacitors often must be added to such arrangements to drive an even voltage division between the VI, and tests demonstrated successful interruption with such capacitors in place [142,144].

Further work studies the breakdown behavior of two [145] and three [146] interrupters in series and the effects of grading capacitors on the voltage split. Because the addition of grading capacitors is expensive and a potential source of failure, further work used different VI designs in series to improve the voltage division [147]. Other work looked at using designs where the entire unit (VI, pole, and operating mechanism) operated at a floating potential [119,148]. This allows for compact designs and ease of placing the units in series, with increased risks from uneven voltage division over the units and potential issues with the timing of the mechanical motion of the separate units. Successful tests with four VI in series with grading capacitors passed interruption tests at 252 kV/40 kA [119]. Finally, rather than using external capacitors some authors looked at the option of asymmetric VI design for specific use in series applications with some initial modeling [149].

3.4.2.5   Influence of test equipment and chopped wave tests

When breakdowns occur during U_p testing, the design of the test set itself can affect the subsequent results. During breakdowns, a current is discharged into the VI, and the current is a function of the test set design [150]. Figure 3.27 plots data from [150] on the effect of test set parameters on the transferred charge during breakdowns. The transferred charge over the voltage peak Q/U_p increases linearly with C_S/n, the capacitance of the test set where C_s is the capacitance of each stage and n is the number of stages. Some test sets fall far below the expected Q/U_p value. This seems to be connected to the ratio of the test set capacitance C_S/n to the measurement capacitance C_m, which affects the load of the test set. Higher Q/U_p can damage or de-condition the VI on breakdowns. Therefore, these values are important to review if unexpected de-conditioning behavior occurs during testing, especially when using test sets that were not previously used with VI.

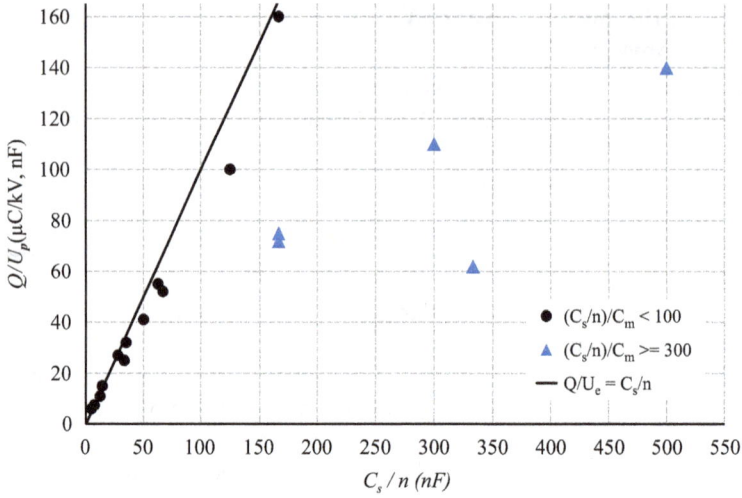

Figure 3.27   *Ratio of the transferred charge Q through the VI during a breakdown over the peak voltage U_p as a function of the charging capacitance C_s/n of various U_p tests sets compiled in [150], where C_s is the capacitance of each stage and n is the number of stages. Data are grouped by the ratio of the charging capacitance to the measurement capacitance.*

Table 3.6   *Formulas and the source of the formulas for calculating the peak of the recovery voltage u_c during capacitor switching tests*

Case	Ref.	u_c (kVpeak)	u_c (PU)
Three-phase	[73, Section 4.2.3]	$2.5 \cdot \left(\frac{\sqrt{2}}{\sqrt{3}} \cdot U_r\right)$	2.5
Single-phase $k_c = 1.0,\ 1.2,\ 1.4$	[70, Section 7.111.7]	$2.0 \cdot k_c \cdot \left(\frac{\sqrt{2}}{\sqrt{3}} \cdot U_r\right)$	$2.0 \cdot k_c$
Single-phase under earth fault $k_c = \sqrt{3} \approx 1.7$	[157, I_{ef2}][70, Annex J.2]	$2.0 \cdot k_c \cdot \left(\frac{\sqrt{2}}{\sqrt{3}} \cdot U_r\right)$	3.46

One HV dielectric test in HV that is used much less in MV is the chopped wave lightning impulse test [151, Section 5.3.5, Table 3.6]. The test consists of a lightning impulse test, where the voltage in the decaying portion of the voltage pulse is chopped to zero. This simulates the effect of a spark gap or coordination gap which flashes over to the ground after some tens of microseconds, removing the voltage across the interrupter. For certain types of older interruption techniques, the removal of the voltage during the decay allows the interrupter to hold off a higher voltage than during the standard lightning impulse test U_p. Despite the frequent claims that the standards are and should be technique independent, this special behavior was inserted into the IEEE standard.

Identifying the time when the voltage should be chopped is unfortunately ambiguous. Reference [151, Section 5.3.5] states, "The rated chopped wave impulse withstand voltage is the peak value of a standard lightning impulse voltage higher than the rated full wave impulse withstand voltage that a circuit breaker in new condition shall be capable of withstanding for a specified time, from the start of the wave at virtual time zero until flashover of a rod gap or coordinating gap occurs, when tested under specified conditions (see IEEE Std C37.09)," while [99, 4.5.7.c] states, "The time to the point of chop on the tail of the wave shall be no less than the times specified in IEEE Std C37.04." This leaves the readers themselves to break the circle.

The chopped wave test is only necessary according to [151] and is not part of [54]. For MV, the chopped wave is only required for outdoor circuit breakers. In HV, there is a chopped wave rating for both indoor and outdoor applications. Reference [151, Table 6] contains the interesting note that, "Special consideration should be addressed when performing chopped wave tests across open contacts of vacuum circuit breakers." Three main points should be considered. First, the U_p test and the chopped wave test are effectively equivalent for vacuum. The breakdown process in a vacuum is fast enough that the rising slope of the voltage and its ultimate peak determines the chance of breakdown. The decaying portion of the voltage pulse (and its removal after some tens of microseconds) has little or no effect on the breakdown behavior. Therefore, certainly, for development tests, the chop wave test can be replaced by the U_p test with the corresponding peak voltage. Second, in MV systems it is common in outdoor circuit breakers to test the chopped wave performance only with the VI contacts closed. This verifies chopped wave performance between the phases and from the phase to the ground. Achieving the chopped wave performance across the open contact gap for a vacuum would generally require moving to the next higher U_r rating. This significantly increases the expense, to cover a very rare condition of a voltage surge higher than the lightning impulse rating when the contacts are open. Third, because the breakdown process in vacuum for U_p and chopped wave tests are identical, the chopped wave tests should also follow the procedure outlined in [54, Section 7.2.6.1] and [99, Sec. 4.5.5], where preliminary impulses are allowed before formally starting the type test.

3.4.2.6 Common value versus isolating distance in standards

Tables of the dielectric requirements in the standards such as [54, Tables 1 and 2] give two values: the common value and across the isolating distance. The common value, columns 2 and 4 in [54, Table 1 and 2], is the correct value for circuit breakers, load break switches, capacitor switches, etc. The value across the isolating distance is only for disconnector applications, which have specific performance expectations that differ from circuit breakers and switches. Reference [94, Table 13.1] gives a clear definition of a disconnector:

"A mechanical switching device which provides, in the open position, an isolating distance in accordance with the specified requirements. A disconnector is intended to open or close a circuit under negligible current conditions or when there is no significant voltage change across the terminals of each of its poles. It is capable of carrying rated current under normal conditions and short circuit through currents for a specified time."

Unfortunately, customers particularly in certain markets incorrectly ask for the isolating distance dielectric values for circuit breaker applications. This only adds expense and/or difficulty in the VI and circuit breaker design, without automatically adding the expected behavior of a disconnector. Also, unless the higher dielectric value is also applied throughout the system to the switchgear cabinet, cabling, etc., any advantage from a higher value across the VI will be lost. Although researchers have explored the idea of using VI for disconnector applications [152–154], the application of dielectric values from across the isolating distance columns is incorrect except for specific disconnector applications.

3.4.3   Capacitive current switching

3.4.3.1   Basics and recovery voltage peak

Capacitive current (capacitor) switching is a strenuous duty for many interruption techniques, including vacuum [155]. Figure 3.28 shows an example of capacitive switching. Capacitor switching starts with a closing operation to connect the capacitance to the network. The worst case occurs when the closing operation is near the peak of the voltage across the circuit breaker. The HV across the contacts

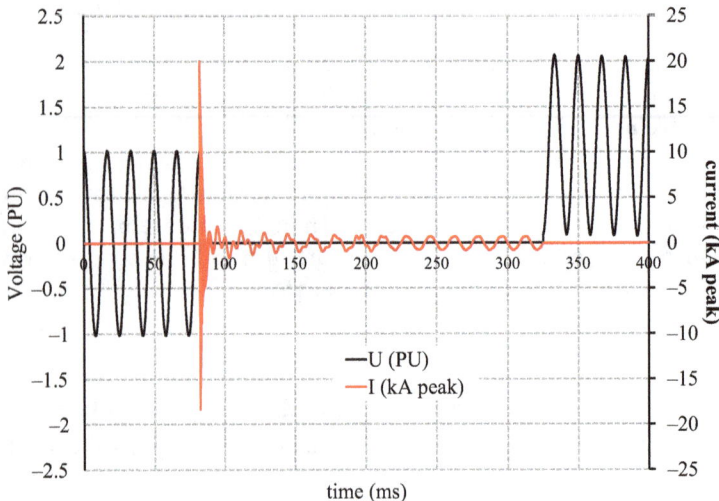

Figure 3.28   Example of a single-phase back-to-back capacitor switching operation plotting the voltage across the VI and current. The initially open contacts are closed at t = 83 ms. The in-rush current then begins, rising to 20 kA peak. The in-rush current and other harmonics then decay, so that the capacitor current of 540 A is interrupted at t = 325 ms. The voltage across the VI then oscillates with an offset voltage.

causes the contact gap to break down and the current to start flowing before the contact touch and is referred to as pre-strike arcing. This arcing is problematic because the short time and/or small contact gap reduce the effectiveness of the arc control method, leaving the arc constricted or with limited diffusion and little motion. The arc locally melts the contact surfaces, and these melted areas are then pushed together at contact closing, which on re-solidification leads to a weld. If only the power frequency current was involved, then the transferred charge during the pre-strike arcing would be limited due to the time it takes for the current to rise from zero at the start of the pre-strike. However, the discharging of the capacitor causes a high-frequency current with a high peak current. This current, referred to as the in-rush current, can lead to a significant transferred charge during the pre-strike arcing in comparison to power frequency currents [156]. The more transferred charge can lead to stronger contact welds.

The contact welds can then affect the opening process to later disconnect the capacitors. The first problem is when the weld is strong enough that the mechanism can no longer separate the contacts. This is an extreme failure, that is uncommon in well-designed circuit breakers. The more well-known problem is re-strikes. The circuit breaker interrupts the load current of the capacitor bank. The current can help to re-condition the contacts by melting the broken weld and smoothing the contacts. However, if the arcing time is short, then the re-conditioning of the contacts would be very limited. The recovery voltage across VI is simplest to understand in the case of a single-pole switch ($k_c = 1.0$). The capacitor is charged to 1 PU at CZ, and then the line-to-ground voltage of ± 1 PU appears across the VI. The voltage then oscillates between 0 PU and 2 PU across the VI, giving $u_c = 2$ PU. Other cases with different k_c for single-phase and three-phase cases follow a similar logic with a higher u_c, Table 3.6. Of note is the extremely high value for the single-phase under earth fault conditions. Here, the u_c value is almost as high as the peak of the U_d test, see Figure 3.14.

3.4.3.2   Type of capacitor switching and currents

The type tests in [70] and other standards for capacitive current switching look at four different cases. The first two are line charging and cable charging. These correspond to the charging of the stray capacitance in an outdoor line or indoor cable. Since all breakers must be connected to the network somehow, these are a very common requirement. The next two deal primarily with capacitors that are switched in and out of a network to help regulate the system voltage. An individual capacitor bank switched into the system is referred to as single-bank capacitor switching. Back-to-back capacitor switching occurs when one or more capacitor banks are already connected to the network and an additional bank is added or removed. Other applications where capacitance is switched into and out of a network can also be covered by these tests.

All four cases can have the power frequency voltage with an offset of $\geq 1\ PU$ after breaking the load current. All four cases also can have pre-strike arcing. The line and cable charging cases have insignificant in-rush currents; the single bank has a small to moderate in-rush of some kA peak, and the back-to-back case has the

Table 3.7   The breaking currents for various capacitor switching duties, according to different standards for MV and HV

Source	U_r (kV)	Line charging current (A)	Cable charging current (A)	Single/isolated bank current (A)	Back-to-back current (A)	In-rush current (kA peak)
[70, Table 1]	≤72	10	10–125	400	400	20
[151, Table 11, 15]	≤72	100	10–100	250–1600	630–2000	15–25/6
[110, Table 6]	≤38	2–5	10–40			
[157, Table 1, Sec. 5.106-108]	≤52	0.3–2.5	4–24	No preferred value	400	20
[70, Table 1]	>72	20–1300	125–500	400	400	20
[151, Table 21]	>72	160–900		1000–1200	700–800	16–25/6

most significant in-rush current, Table 3.7. Standard [151] specifies both a pre-ferred rating and an alternative value for the in-rush current. For the breaking current, line charging has the lowest values, with increasing breaking current for cable charging and the highest values for single- and back-to-back switching, also in Table 3.7. The values in Table 3.7 are preferred values, and the manufacturer is free to use other values if they are better suited to the application. The assumptions underlying the preferred values are subject to some criticism [158], and this will be discussed further regarding the in-rush current for back-to-back capacitor switching.

The type tests for capacitor switching have Test Duty 1 and Test Duty 2, referred to as BC1 and BC2 for single bank and back-to-back. BC1 is generally irrelevant for vacuum. Although the breaking current is only 10–40% of the rating, the tests are performed as opening operations, or closing with little/no in-rush current followed by the opening operation. Also, there are more operations at the minimum arcing time in BC2 which reduces the ability to re-condition the contacts with the breaking current. Standard [70] has two classes of capacitor switching, C1 and C2, while [151] has three classes, C0, C1, and C2. Class C2 has the hardest requirements and is the de facto requirement from customers [155], regardless of the details of the network or technical arguments. Therefore, this chapter will focus only on Class C2.

3.4.3.3   Switching process and re-strikes

Figure 3.29 illustrates the interactions between the capacitor switching duty and the chance for re-strikes in the opening operation, adopted from [159]. The first factor is the pre-strike arcing. One way to characterize the pre-strike arcing is to calculate the macroscopic electric field stress $|E_{pre}|$ when the pre-strike occurs

$$|E_{pre}| = \frac{U_{pre}}{d} \approx 5 - 15 \ \frac{kV}{mm} \tag{3.22}$$

where U_{pre} is the voltage across the contact gap d when the pre-strike arc starts and the range is from [160–162]. The average value and distribution of $|E_{pre}|$ can

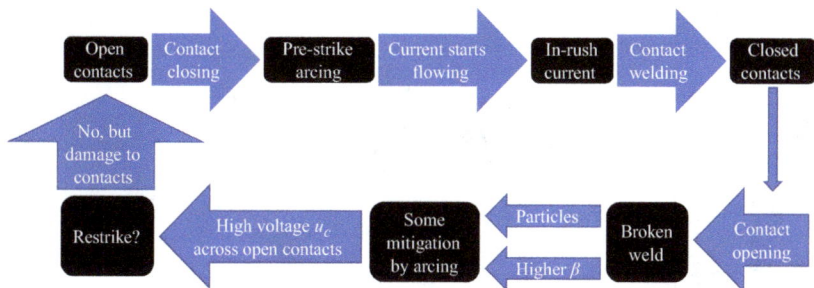

Figure 3.29   Flow chart for the back-to-back capacitor switching process including the sources of enhancement and mitigation

remain stable even over thousands of switching operations [163]. The primary methods to reduce the pre-strike arcing are to reduce the damage to the contacts from welding on previous operations and to shorten the pre-strike time by increasing the contact closing speed [49,164].

In addition to welding from pre-strike arcing, another potential problem is welding from contact bouncing on the closing operation. The high-frequency current again increases the transferred charge when the contacts are separated, and therefore leads to stronger welds. Reference [165] observed an increase in the percentage of re-strikes for higher closing speeds for $U_r = 72.5$ kV applications, with contact bouncing as the most logical source of the increase. The contrast with the results from [164] highlights the complex balance between reducing pre-strike arcing and minimizing contact bouncing.

The second factor is the peak value of the in-rush current. Not having an in-rush current reduces the re-strike rate [159] and is supported by the empirical experience that line and cable charging are the easiest duties, followed by single bank and back-to-back as the most difficult. The lower in-rush current during single bank testing leads to fewer re-strikes and less contact welding than the 20 kA peak value for back-to-back [162]. The in-rush current can be combined with the pre-strike arcing time to calculate the energy transferred into the welding process. The re-strike frequency increases significantly with this energy [166], and the electrical life is almost a step function of this energy with a factor of five reduction in the number of operations [163]. The contact material plays a key role in the welding behavior of the contacts. For the commonly used Cu–Cr contacts, increasing the Cr percentage and/or modifying the contact production method can significantly reduce the welding from the in-rush current [167]. Although 20 kA peak in-rush current is often the focus of research, there is some work at higher voltages. Reference [168] described tests with 400–1,000 operations with 30−59 kA peak in-rush current.

The third factor is the breaking current during the opening operation. The breaking current can re-condition the contacts by melting the portions damaged by welding. Figure 3.30 plots the Weibull cumulative distribution fits the re-strike

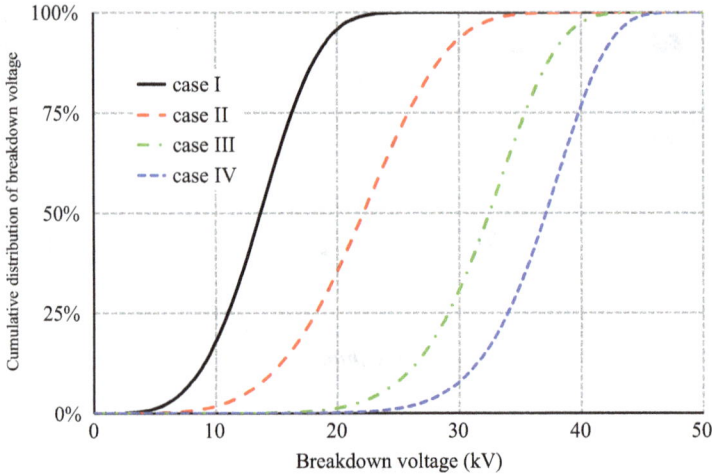

Figure 3.30   Cumulative Weibull distributions of the re-strike probability during capacitor switching with the in-rush current of 2 kA peak fitted to the data of [169] for four cases: I is opening without current; II and III have a breaking current of 180 A; IV has a breaking current of 340 A

breakdown voltages data for four cases described in [169]. Case I is with the making operation followed by an opening without current. This is, as expected, the worst case as there is no current conditioning of the welded spots. Cases II and III are opened with 180 A and Case IV with 340 A, with $t_{arc} = 5$–6 ms for cases II–IV. Current breaking operations increased the peak voltage that could be withstood, and higher current led to further improvements. A higher load current would generally help the operation in the field but would not help much in the type test where the focus is on short arcing times for the breaking current.

The final factor is the electrical field after breaking the load current. Reference [160] observed a clear reduction in the frequency of re-strikes when increasing the contact gap for a $U_r = 24$ kV circuit breaker from $d = 8$ mm to $d = 14$ mm. Increasing the electric field stress by increasing the k_c value also produced the expected increase in re-strikes [165]. This behavior suggested that the breakdowns were driven by the field emission of electrons from the cathode, leading to a breakdown of the contact gap [170]. Such theories drove successful efforts to measure the field emission current during capacitor switching tests and had initial success at connecting the field emission to re-strikes [171]. Models of the breakdown process also were extended to cases where the field emission drives local melting on the anode and breakdown [172]. Unfortunately, later tests observed no connection between the field emission values and the re-strike behavior [156]. New work observing particles between vacuum contacts (see Section 3.6.4) helped drive an increasing focus on particle-based models of breakdown and attempts to measure the existence and behavior of these particles [173].

3.4.3.4 In-rush current for back-to-back capacitor switching

Given the importance of the peak in-rush current to back-to-back capacitor switch testing, it makes sense to carefully review this value. A recent survey of end users found that ~90% of the users in MV had an in-rush current of ≤ 10 kA peak [155, Figure 3.31]. This result also agrees with an analysis of the effect of the inductance on peak in-rush current [8]. The peak breaking current $\sqrt{2}I_{bb}$ is [174, Section 9.2.1.2].

$$\sqrt{2}\,I_{bb} = 2\,\pi f_r\, C\, \frac{\sqrt{2}}{\sqrt{3}}\, U_r \qquad (3.23)$$

where f_r is the rated frequency, C is the capacitance of the individual bank, and $(\sqrt{2}/\sqrt{3})\, U_r$ is the peak of the line to earth voltage across the capacitor. Solving for C gives

$$C = \frac{\sqrt{3}\,I_{bb}}{2\,\pi f_r\, U_r} \qquad (3.24)$$

When connecting a new bank and n identical banks with identical inductance L between the switch are already connected, the peak in-rush current I_{bi} is [174, Table 44].

$$I_{bi} = \frac{\sqrt{2}}{\sqrt{3}}\, U_r\, \sqrt{\frac{C_{eq}}{L_{eq}}} \qquad (3.25)$$

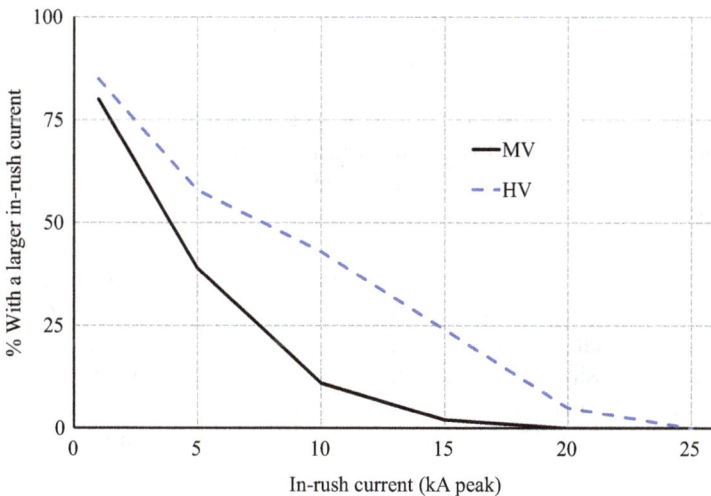

Figure 3.31   Percentage of back-to-back capacitor switching applications where the in-rush current is larger than the specified value for MV and HV systems [155]

where L_{eq} and C_{eq} are the inductance and capacitance from the already connected identical banks

$$L_{eq} = \frac{L}{\left(\frac{n}{n+1}\right)} \tag{3.26}$$

$$C_{eq} = \left(\frac{n}{n+1}\right)C \tag{3.27}$$

Inserting (3.26) and (3.27) into (3.25) solving for L gives

$$L = \frac{2}{3}\, U_r^2 \left(\frac{n}{n+1}\right)^2 \frac{1}{I_{bi}^2}\, C \tag{3.28}$$

and inserting (3.24) into (3.28) and specifying the units gives [8]

$$L/[H] = \left(\frac{n}{n+1}\right)^2 \frac{1}{\pi\,\sqrt{3}}\, \frac{\frac{I_{bb}}{[A]}\frac{U_r}{[V]}}{\frac{f_r}{[Hz]}\left(\frac{I_{bi}}{[A\,peak]}\right)^2} \tag{3.29}$$

Converting (3.29) into commonly used units gives

$$L/[\mu H] = 1{,}000 \left(\frac{n}{n+1}\right)^2 \frac{1}{\pi\,\sqrt{3}}\, \frac{\frac{I_{bb}}{[A]}\frac{U_r}{[kV]}}{\frac{f_r}{[Hz]}\left(\frac{I_{bi}}{[kA\,peak]}\right)^2} \tag{3.30}$$

and as a unitless formula (all RMS/effective values converted into peak values)

$$L = \left(\frac{n}{n+1}\right)^2 \frac{1}{2\,\pi\,\sqrt{3}}\, \frac{I_{bb}\,U_r}{f_r\,I_{bi}^2} \tag{3.31}$$

Taking a typical case of connecting a second bank for $U_r = 12$ kV, $I_{bb} = 400$ A, $f_r = 50$ Hz, and $I_{bi} = 10$ kA peak gives $L = 44$ µH. Since the inductance per length of the busbar [174, Table 45] and cables [155] is ~ 1 µH/m, this corresponds to 44 m between the switch and the bank. Outdoor capacitor banks easily reach this value, with most connections between 30 and 50 m [8]. Therefore, in-rush currents in MV above $I_{bi} > 10$ kA peak are difficult to achieve unless the breaker is very close to the bank. This behavior also appears in the type testing of back-to-back capacitor switching for MV. Here, the test object must be placed as close as safely possible to the capacitor bank, otherwise, the test lab cannot reach the I_{bi} requirement. For HV, higher in-rush currents are possible. Figure 3.31 shows that values out to $I_{bi} = 25$ kA peak occur, as can even higher values. In (3.29), the inductance L required to limit I_{bi} increases linearly with U_r, therefore the inductance of the connection between the bank and switch alone is not enough to significantly limit the in-rush current. The in-rush current in back-to-back capacitor switching will be a practical testing challenge for HV vacuum circuit breakers.

3.4.3.5   Perspectives on MV and HV capacitor switching

Capacitor switching is a challenging problem for VI, with a considerable body of technical research. However, in MV, it needs to be understood that capacitor switching, including back-to-back, is a solved problem. Every major VI and vacuum circuit breaker manufacturer can offer switches fully type tested according to [70] or [151] with years of field experience. While the capacitor switching performance needs to be actively considered during the development of a new VI or switch, the manufacturers have internal guidelines for what is required to pass the capacitor switching tests. This does not mean that every MV vacuum circuit breaker and switch can pass every capacitor switching duty according to the standards. Rather when a design cannot reach a specific duty, there is at a minimum an empirical understanding of why, knowledge of how the performance could be improved, and an alternative option.

References [49, Section 5.1.1] and [175] contain an extensive series of experiments on capacitor switching, with a range of different parameters and several measured quantities. Although the work is an excellent example of what tests to perform and how to analyze the results, the general application of the results is limited by the fact that the VI had no voltage or current conditioning before the tests. Eliminating this key production step significantly complicates the comparison of the experimental results to practical applications. The low voltage of $U_r = 7.2$ kV also limits comparison to most practical cases.

For HV, capacitor switching could be a challenging duty. Single VI at $U_r = 126-170$ kV have passed line/cable charging tests [48,79,80]. Placing two or more VI in a series would presumably enable higher U_r values for line/cable charging applications. Single bank and back-to-back capacitor switching applications would be harder. For HV, the higher U_r leads to a higher L to achieve a certain in-rush current I_{bi}, (3.29). Therefore, the inductance of cables between the bank and the switch will be less effective at limiting the in-rush current. This supports the tendency in Figure 3.31 for HV applications to have higher in-rush currents than MV. There is also a lack of technical publications describing single-bank and back-to-back capacitor switching tests on HV VI. As for MV, the expectation is that this problem can be solved; the interesting question will be what trade-offs will be necessary (going from a single VI to two VI in a series, for example [149], or controlled switching).

3.4.3.6   Controlled switching for capacitor switching

Controlled switching is a method of closing and/or opening the circuit breaker at a specific time range on the power frequency cycle. It requires measurement of the voltage and/or current, combined with detailed information on the circuit breaker's mechanical behavior. Controlled switching is already widely known and used in HV [176,177]. With capacitor switching, the first step would be during the closing operation. Here, the goal would be to choose a time window on the voltage cycle where the voltage across the contacts at contact touch would be minimum (to avoid pre-strike arcing) and/or to minimize the magnitude of the in-rush current. Reference [178] found a 4 ms window where the in-rush current could be reduced from $I_{bi} = 20\,kA$ peak $\rightarrow$ $10\,kA$ peak for a three-phase MV application. This application of controlled switching is well-known in HV. Controlled switching could also be applied to the opening operation. Here the goal would be to maximize the arcing time, and thereby maximize the re-conditioning of the contacts by the load current. Controlled switching is used much less in MV, nevertheless, the same logic also applies. The primary risks with controlled switching are ensuring that the control system can deal with irregularities in the inputs (noise on the signal, harmonics/non-sinusoidal behavior of the voltage, etc.) and can compensate for changes in the mechanical behavior with temperature, wear, etc.

3.4.4 Switching over-voltages
3.4.4.1 Irrelevance of chop current
The act of switching on and off an electrical power system can produce over-voltages beyond what could occur for a perfect ideal switch (zero resistance when on, instantaneous infinite resistance when off). VI have three main possible sources of over-voltages during switching operations. The first is chop current. Chop current has been erroneously connected with over-voltages in VI for decades [20]. Very early VI used pure Cu or Cu–Bi (0.5%) contacts, which had high chop values and led to over-voltage issues [27,179]. Cu–Cr-based contact materials were introduced in 1972 [180] which solved the chop current problem. Extensive comparisons focusing on HV applications showed no disadvantage of a vacuum relative to SF_6 interrupters regarding chop current with modern contact materials [20].

3.4.4.2 Multiple re-ignitions and virtual current chopping
The second source is multiple re-ignitions. These are a side effect of the excellent di/dt performance of vacuum [87]. When the arcing time is less than t_{min}, the VI can fail to interrupt the current at CZ. The power frequency current restarts, however, a high-frequency current from the stray capacitance in the system is superimposed on the power frequency current. The high-frequency current can be large enough to have another CZ with $di/dt \sim 100-300\,A/\mu s$. Vacuum can interrupt these currents. Because the increase in the contact gap is very small on these time scales, the VI is likely to fail to interrupt again, and the process then repeats until the contact gap is large enough to withstand the recovery voltage or the power frequency current rises enough so that early CZ no longer occur. Each re-ignition puts more energy into the load inductance and can increase the voltage magnitude and the steepness of the recovery voltage du/dt, both of which can place the load at risk. This behavior is not unique to vacuum; SF_6 breakers also see the same phenomena [20]. However, multiple re-ignitions are more common for vacuum [26].

The third source is virtual current chopping. Despite the name, virtual current chopping is unrelated to the current chopping and is instead a side effect of multiple re-ignitions [181,182]. Depending on the capacitive coupling between the phases and the coupling through the neutral point, re-ignition in one phase can induce a transient high frequency in one of the other phases. If the magnitude of the transient current is high enough, the sum of the power frequency and transient current can

Table 3.8   Key typical parameters for use in modeling the over-voltage behavior of VI

Voltage breakdown as a function of the gap			Eq. (3.12)
Maximum di/dt		450 A/µs	[87]
Material	Average chop current	Maximum chop current	Reference
Cu–Cr50	2.3 A	4.0 A	[187]
WC–Ag25	0.5 A	1.2 A	[187]
W–Cu30	2.8 A	5.3 A	[179,188]

create a CZ. The effect is similar to current chopping, except the magnitude of the virtual current chop is much higher, in the range of tens to a few hundred amperes. The resulting over-voltages can be significant. However, the critical factor is the coupling between the phases. Applications such as motor switching in MV can have coupling between the phases that makes virtual current chopping a valid concern. Test setups given in [183] were specifically designed to generate the worst cases from motor switching, including virtual current chopping. Even in this worst-case setup, the over-voltages from regular current chopping were insignificant [184]. For HV, the wider spacing and/or shielding between phases could reduce the likelihood of virtual current chopping, as well as the more efficient grounding of the neutral point. While modeling is recommended (see later in this section), no cases of virtual chop current in HV have been currently reported [20].

Multiple re-ignitions are first only a real concern when switching small (tens to a few hundreds of amperes) currents on inductive circuits. The presence and significance of multiple re-ignitions are a complex problem involving the interaction of the VI, circuit breaker, load, and the connection between the circuit breaker and the load, and possible interactions through stray capacitance. Therefore, modeling is the best way to evaluate the risk of multiple re-ignitions and the effect of mitigation techniques. References [185,186] provide an excellent background on the modeling of over-voltages from vacuum switching. Table 3.8 summarizes typical VI parameters for modeling.

3.4.4.3   Mitigation and controlled switching for over-voltages

In MV, extensive experience with the risk from multiple re-ignitions drove the development of standard recommendations to reduce the risk of damage from over-voltages. Table 3.9 summarizes one example from [189], which has various passive components that can be added when necessary to mitigate over-voltages. While different manufacturers may have slightly different recommendations [190], the general trend for MV VI applications is clear. The question would then follow whether the general recommendations from MV would also apply to HV. Reference [191] modeled the effect of passive RC-filter circuits, similar to the MV recommendations, on mitigating switching over-voltages from VI in some typical HV systems. The mitigation technique was successful, reducing the number of re-ignitions from >20 without the mitigation to an average number of 3 with

Table 3.9   Standard recommendations for over-voltage mitigation on MV systems when using VI [189]

Duty	Suggested countermeasures

Transformer switched under no-load conditions in normal service	No surge protector for normal applications
	Surge arresters are needed if:
	  – Connected to overhead lines
	  – Insulation level is List 1 of IEC 60071-1
	  – Insulation aged or unknown level
	  – High switching rate

Motor or generator	Surge arresters when starting current <600 A
	Add RC circuit if:
	  – Insulated does not follow IEC 60034-15
	  – Insulation aged or unknown level
	  – High switching rate
	  – At neutral of starting transformer

Converter transformer short-circuit limiting reactor Petersen coil	Surge arresters
Shunt reactor	RC circuits and surge arresters if current <600 A
Arc furnace transformer	RC circuits and surge arresters
Capacitive switching	None if Class C2
Back-to-back capacitor switching	Use Class C2 only. Recommend limiting in-rush current and consider thermal effects from harmonic currents
Filter circuit	Use Class C2 only. Consider the voltage rise from harmonics

mitigation. This gives the first evidence that the recommendations from MV could be transferred to HV.

As with capacitor switching, the switching of small inductive currents also has the option of using controlled switching. As with capacitor switching, the key is to identify an arcing time window where the problematic behavior is avoided. For multiple re-ignitions, the key is to avoid arcing times below t_{min}, the minimum arcing time that can be successfully interrupted. For example, [26] found a window from $t_{arc} = 4{-}10$ ms where $U_r = 145$ kV vacuum circuit breaker was free of re-ignitions. This wide arcing time window would allow a straightforward implementation of controlled switching to prevent over-voltages.

3.4.5   Applications at high external pressure

Some vacuum circuit breakers for HV used SF_6 as the insulating gas around the VI, which unfortunately undermined the advantage of using an SF_6-free interrupter. This was generally done either to simplify the testing of demonstration designs [79,192] or to gain the advantage of vacuum without the added challenges of using a different gas, typically for applications in Japan before the year 2000 [20]. Ultimately, since the drive to be SF_6-free is primarily a legal one, the value of reducing the amount of SF_6 in a breaker is questionable. Elimination is the primary

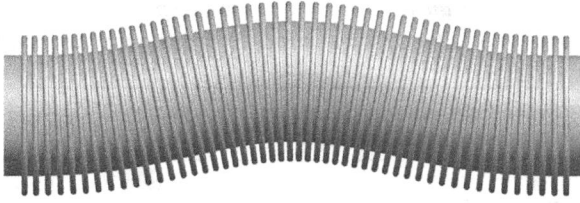

Figure 3.32   Example of bellows deformation due to high external gas pressure around the VI, leading to a reduction in the mechanical life

goal [193]. SF_6 alternative gases all offer lower dielectric insulation (see Chapter 4), therefore extra efforts are required to meet the dielectric requirements along the outside of the VI. The high U_p for HV ratings makes additional external insulation around the VI a de facto requirement. For example, for $U_r = 72.5$ kV has $U_p = 325$ kV which according to Figure 3.9 would require more than 650 mm of ceramic length, an impractical value. Therefore, all practical HV VI need external insulation, and the best insulating gas is no longer an option.

The use of pressurized dry air as the external insulation requires pressures up to 5 bar [23] and possibly even higher [141], and other replacement gases would require similar high external pressures around the VI. This pressure places a high risk on the bellows which is unstable to small perturbations and can "squirm" into a bent shape [194], Figure 3.32, which sharply reduces the mechanical life of the bellows and the VI. HV VI with long contact gaps would be particularly at risk. Potential solutions include separating the bellows from the high-pressure region [23] or providing guidance to the bellows [23,195]. While the bellows is the key concern, other parts of the VI can be at risk and could need additional support with increased pressure [196]. Finally, the higher external pressure could affect calculations of the VI lifetime based on leak rates, as the higher external pressure could push more gas into the VI through small, slow leaks [197].

3.5   How to use VI properly

3.5.1   Standard handling recommendations

3.5.1.1   Final tests in factory

When VI leaves the factory, each manufacturer performs certain tests on every individual VI and other tests on a batch basis. The two tests performed on each VI are a power-frequency withstand voltage test U_d [54] and a pressure measurement. The U_d measurement is often integrated as the last step of the voltage conditioning process. Although standards ask for a voltage application of 1 minute, this is a universal requirement to cover every insulation and interruption method. For vacuum, applying the U_d voltage for 15–30 s is perfectly sufficient to identify VI with a mid- to up-to-air pressure (see Section 3.5.2). Some VI are designed for use in insulating gases, oil, or solid insulation, which allows the use of shorter ceramic

Table 3.10   Typical procedure when performing two pressure measurements on VI in the factory, the first test just after brazing and a second later test [8]

First pressure measurement	Second pressure measurement after typically a few days	Comments
Fail	Fail	Large-scale leak. Reject
Fail	Pass	Poor getter activation or poor vacuum level in the furnace. Need to review.
Pass	Fail	Mid- to fast-pressure increase rate, see Table 3.12. Reject
Pass	Pass	Accepted

and total length. Reaching the rated U_d value may require extra insulation around the VI, both for the voltage conditioning and U_d test. Manufacturers typically use an insulating gas (possibly at a higher pressure) or an insulating oil to provide the additional insulation.

The pressure measurement using one of the magnetron methods (see Section 3.2.3), is normally performed at the end of the production process and covers the mid- to low-pressure range. The final pressure measurement may be combined with earlier magnetron gauge measurement after the VI come out of the brazing process. The time between these two measurements may include the time for the final processing of the VI (conditioning, additional tests, final assembly steps), or it may include storage for a fixed period, possibly in a noble gas. The difference between these two measurements can have four states with different outcomes, given in Table 3.10 [8]. Using two measurements allows the manufacturer to identify possible production issues before they become significant or help identify the potential size of the leak and aid in the identification of defects.

3.5.1.2   Steps for circuit breaker manufacturer

When the VI are received from the manufacturer, the first step is a visual inspection of the packaging for damage or contact with water during shipping. In general, it is best to leave the VI in the original packaging until ready to assemble the VI into the circuit breaker or switch. When removing the VI from the packaging, wear clean gloves and grasp the VI by the ceramic body to minimize the handling of the copper terminals [8]. The brazing process exceeds the annealing temperature of copper, so the copper terminals are softer than expected and can be bent or damaged by rough handling. The ceramics should be inspected for cracks, chips, or discolorations. Dust or other minor pollution can be removed from glazed ceramics using alcohol and a soft, dust-free cloth. Un-glazed ceramics should not be cleaned, and the manufacturer's instructions should be carefully followed. Minor discoloration on the flanges is not a concern. The current conducting area of the terminals can be cleaned with a metal brush for bare copper. Oxidation on silver-plated terminals can usually be easily removed with a soft cloth. If the manufacturer recommends the application of grease to the current conducting regions, make sure to only use the suggested lubricant to avoid the risk of corrosion on other parts of the VI or mechanism.

The bellows of the VI are made from a thin tube of stainless steel. Most designs have the bellows inside of the VI ceramic (Figure 3.3). However, some designs, particularly for LV applications, have an externally exposed bellows, Figure 3.33. For exposed bellows, extra care is necessary to not dent or otherwise damage the bellows as this will shorten the mechanical life of the VI. The thin walls of the bellows are vulnerable to chlorine corrosion. Even the levels of chlorine in public drinking water could pose a long-term risk to the bellows [198]. Therefore, it is important not to allow any liquids to become trapped in the bellows, and to only use alcohol or alcohol plus distilled water in the cleaning of the ceramics.

During the installation of the VI into the circuit breaker, pay close attention to the recommended torques for the connections to the terminals. The recommended torque is a function of the thread size, thread depth, and the material of the threads in the VI (generally the threads are either directly in the copper or have a stainless-steel insert/threads). What is attached to the terminals, as well as the washers, current paths, lubricants, and bolts all play a role in setting the necessary torque. While the recommended torque is a very good guideline, measuring the actual relationship between the applied torque and the resulting pre-load force on the parts is ideal. This curve should be measured with the actual planned setup and then can identify the torque region that applies an elastic force to the bolt and shows the torque value where plastic deformation begins (which should be avoided).

The torque applied to the moving terminal must be countered to leave no net torque on the moving terminal. Net torque on the moving terminal would be transferred to the bellows, which could damage it. Although the bearing of VI often

Figure 3.33   VI with external bellows, requiring careful handling to not damage the bellows and reduce the mechanical life. Courtesy of Siemens AG.

has an anti-twist feature to reduce the torque transmission to the bellows, these features allow far more torque to be transmitted to the bellows than would be expected. Therefore, it is very important to follow the manufacturer's recommendation regarding the support of the moving terminal during bolt attachment. Bearing designs do exist that can support considerable torque without transmitting the torque to the bellows, however, these designs have trade-offs in cost and complexity.

The torque to attach the external current path is best connected using internal threads in the VI terminals, or to external threads of a stainless-steel insert that sticks out of the copper terminal. Some VI designs do also have external threads on the copper terminals. The external threads in copper are generally not recommended. These threads are very vulnerable to handling damage, because of the fully annealed copper. In comparison, internal threads are much less exposed, and there are well-established options to increase their strength if necessary, such as stainless-steel inserts.

Once assembled in the circuit breaker, the next step is to carefully measure the contact gap and opening/closing speeds of the breaker mechanism [8]. Each VI has a datasheet from the manufacturer with a recommended range of values, and these values should be adhered to. Deviation from these values can lead to mechanical life problems, dielectric problems, and even failure to interrupt. Many VI are designed for more than one rating and can potentially operate at different combinations of gaps and speeds. However, the use of values not on the datasheet should be agreed upon with the manufacturer during the development phase and a new, possibly customer-specific, datasheet generated.

3.5.1.3   Steps in the field

VI are maintenance free and normally require no inspection during their service life. If the circuit breaker requires periodic inspection, then VI can also be inspected. The first step would be a visual inspection, looking for signs of excessive heating, contamination, or physical damage to the VI. The resistance over the pole can also be measured and compared to the commissioning results, and the U_d performance can also be verified. Testing 80% of U_d is generally sufficient to support the vacuum integrity of the VI [198]. Measurements on the mechanical parameters of the breaker can also include a measurement of the contact wear. The main point to keep in mind is that any inspection should minimize the handling of the VI and the breaker, to minimize the risk of damage from the inspection process itself.

Some circuit breaker designs either explicitly or implicitly allow for the refurbishing of the circuit breaker at the end of its service life, and this can include the replacement of the VI. This is an established process and has been successfully performed by both the original manufacturer and external firms. The key points to remember are that the replacement of the VI follows a similar process as in the original construction of the circuit breaker. However, this process lacks the assembly tools and experience used in the factory. Therefore, the use of reputable firms with a background in the specific circuit breaker design is vital.

Also, any replacement VI should be new or at least only be in storage. The long electrical and mechanical life of VI can create the temptation to re-use "lightly used" VI. However, the usage history from the field is not always reliable, and it can be difficult even for experts to assess the remaining life of a VI without efforts that rapidly exceed the cost of a new VI and even of a new circuit breaker.

3.5.2   Vacuum integrity and monitoring

3.5.2.1   Vacuum life of VI

The long electrical and mechanical life of VI can lead to circuit breakers being in service for many years and even some decades. This leads to growing interest in assessing the end-of-life of vacuum circuit breakers. Vacuum circuit breakers are complex, multi-component systems comprised of the VI, opening/closing mechanism, relays and control systems, and the insulation between the current path and ground. The mechanism itself is a multi-component system made of springs, motors and/or magnetic drives, insulation between the VI and the grounded components of the mechanism, etc. All these components must work in tandem for the circuit breaker to function. More than 61% of the failures in HV SF_6 circuit breakers stem from mechanical issues [199]. The designs of SF_6 and vacuum circuit breakers bear much in common [200] and a similar failure profile would be expected. Despite this, for vacuum circuit breakers the focus is often on the vacuum integrity of the VI.

While many people present vacuum integrity as if it was a new issue, in fact, this issue was addressed very early in the VI development. The initial vacuum sealing techniques were based on cathode ray tube assembly; an already established manufacturing process [201]. The vacuum level and dielectric performance were maintained after five years of field usage in HV systems in 1961 [16]. Later work by an MV VI manufacturer involved removing 12,500 VI out of the field after >10 years of service to refurbish the circuit breakers. The pressure level of these 12,500 VI was measured and observed that the remaining vacuum life could exceed 80 years, Table 3.11 [202].

Later experimental work delved into what gases were left behind in a sealed VI [203,204]. Residual gas analysis (RGA) agreed with pressure measurement by magnetron techniques, as well as identified the gases remaining inside the VI. The

Table 3.11   Data from [202] on the calculated remaining life of VI according to the pressure increase rate measured after more than 10 years of field service

Remaining vacuum life after >10 years in the field (1987)

Year made	Number	Life < 80 years	Life 80–160 years	Life > 160 years
1970–1971	9,000	0%	19.30%	80.70%
1972	500	0%	1.60%	98.40%
1973–1975	3,000	0%	1.50%	98.50%

gases comprised of H_2, CH_4, N_2/CO, CO_2, and Ar. These gases (other than Ar) are the typical outgassing components of stainless steel in a properly sealed and outgassed vacuum system. The typical RGA of a vacuum system with a leak would, by comparison, consist of H_2O and O_2. The Ar is the residual component of atmospheric air that cannot be pumped by the getter, from the gas remaining in the VI when it is sealed. The experiments in [204] also showed that the getter continues to pump after the initial sealing of the VI.

3.5.2.2   Mean time to failure (MTTF)

To help address the continuing questions about the vacuum integrity of VI, four VI manufacturers worked together on a joint assessment of the vacuum reliability of VI [205]. The key point from these manufacturers was the year-on-year increase in the number of VI produced, coupled with a lack of a corresponding increase in the number of VI returned to the manufacturer with failures due to leaks. Presently, several manufacturers produce more than 500,000 VI per year; this gives a large installed base of older VI, combined with large volumes of new VI. If vacuum leaks were a systematic problem, then it would appear in the number of VI returned to the manufacturer and identified as failures due to leaks. This perception can be quantified using the mean time to failure (MTTF) as a measurement of the product's lifetime [206]. The MTTF is calculated by

$$MTTF = \frac{\sum\limits_{15\ year\ period} (VI\ per\ year) \cdot (years\ in\ service)}{\sum\limits_{15\ year\ period} failures\ per\ year} \tag{3.32}$$

The MTTF of one manufacturer is plotted in Figure 3.34. The 15-year time length accounts for vacuum circuit breakers being removed from service for reasons unconnected with vacuum leaks. This includes the building or substation being

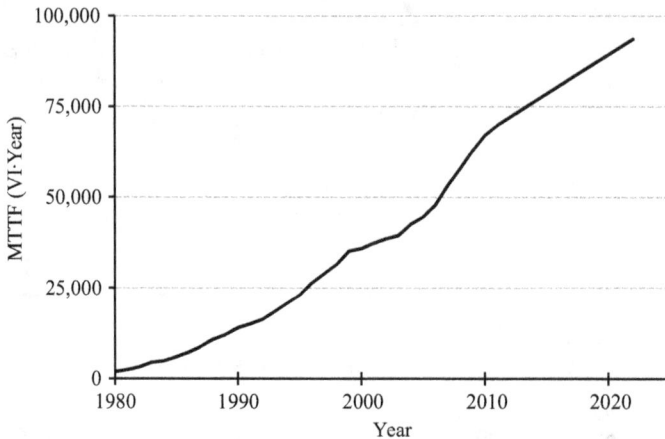

Figure 3.34   MTTF calculation using (3.32) over time from one VI manufacturer

torn down, replacement of old breakers with new ones, etc. Any VI returned during the designated period, even if older than 15 years, are included in the number of failures in the denominator. The key result of this figure is the increasing value of MTTF with time. This shows two primary points. First, if VI were approaching the end of the vacuum lifetime, then the MTTF should converge to a value. This behavior does not appear in the data. Second, if the end of the vacuum lifetime was reached, then a sharp increase in the number of returned VI would appear. This also did not appear in the data.

When attempting to determine the source of failure for a damaged piece of MV power systems, one key feature of VI and VCB designs is frequently overlooked. The circuit breaker system is arranged so that the stress is concentrated in the VI. Failures in one of the sub-systems making up the VCB (mechanical operating mechanism, relay controls, proper sizing/selection of the circuit breaker) can result in the damage or even destruction of the VI, without any damage to the source of the failure. In addition, failure to properly apply the IEC, IEEE, or other standards can result in damage to the VI [207]. The presence of damaged and/or VI without a vacuum does not automatically mean that vacuum loss was the root source of the failure. This is unfortunately a common mistake in poorly thought-out failure assessments.

3.5.2.3   Terminology for vacuum loss

A proper understanding of the potential source of pressure increases in VI is crucial to guide discussions of the vacuum lifetime. The risk of vacuum loss in VI is best evaluated by breaking the problem up into two key parts for each potential failure mode. The first is the type of vacuum loss, in particular the pressure increase rate (Q) and the ultimate pressure for the VI. Second is when the failure could occur. Related to this is whether the mode is a continuous process or is triggered by a specific event.

One note on terminology: the general term pressure increase rate Q will be used instead of the more widely used term leak rate. The problem with the term "leak rate" is that it implies that the pressure increase is due to a connection between the vacuum region and the external atmosphere. While this is a possible failure mode, it is not the only option, and quite possibly not the key failure mode. The term pressure increase rate Q serves as a reminder to also check the other features that will be summarized in Tables 3.12 and 3.13.

To determine the long-term vacuum integrity of VI, the only important potential failure modes are those with a slow Q, or that are triggered late in the VI lifetime. Modes with a faster Q or triggered early in the VI life would fall under the "infant mortality" potential failure modes, which are discussed and characterized in [198,205].

Table 3.12 summarizes the terminology that will be used to describe the vacuum loss rate and vacuum loss final pressure. The vacuum loss rate measures how fast the pressure increases in a VI. The values in Table 3.12 give the corresponding Q and time to the end of the VI life for a 0.5 L internal volume VI, a common mid-sized interrupter. Q and time to end of life for other volumes are

Table 3.12   Definitions for the pressure increase rate categories and final pressures

Pressure increase rate category	Time to end of life for 0.5 L VI internal volume	Pressure increase rate (Pa·L/s)
Fast	<1 year	$>1.6 \times 10^{-9}$
Mid	1–10 years	$1.6 \times 10^{-10} - 1.6 \times 10^{-9}$
Slow	>10 years	$<1.6 \times 10^{-10}$
Vacuum loss final pressure (hPa)		
Low	$<10^{-3}$	
Mid	10^{-3} to Paschen minimum	
Up-to-air	Paschen minimum to external pressure	

Table 3.13   Possible failure modes for vacuum loss of a VI [30], with the pressure increase rate and final pressures associated with the modes, along with when the situations could occur and whether the mode is triggered (sudden appearance of the pressure change) or continuous (gradual, constant change of pressure)

Failure mode	Vacuum loss		Occurrence	
	Rate	Final pressure	Where	Timing
Damage	Fast	Up to air	Field	Triggered
Corrosion	Mid to fast	Up to air	Field	Triggered
Operations	Fast	Low to mid	Field or testing	Triggered
Slow leaks	Slow	Low to up to air	Anywhere	Continuous
H/He perm.	Slow	Low to mid	Anywhere	Continuous
Outgassing	Slow to fast	Low to mid	Anywhere	Both
H outgassing	Slow	Low to mid	Anywhere	Continuous

Note: For terminology, see Table 3.12.

summarized in [27, Table 3.16]. A fast pressure increase rate Q_{fast} reaches the end of life in less than 1 year, and the slow rate Q_{slow} would theoretically take ten or more years. The mid value lies between Q_{slow} and Q_{fast}. The units of the Q measure the energy requires to move the gas across a plane in time, and therefore Pa·L/s = mW [36, Section 3.2]. In addition to different Q, different potential failure modes lead to different final pressures, and this is also summarized in Table 3.12. The low final pressure corresponds to a VI that retains its performance with a pressure $< p_{trans}$. The mid final pressure is between p_{trans} and $p_{Paschen}$, the pressure for the Paschen minimum breakdown voltage. From [1], the minimum breakdown voltage for atmospheric air is $p_{Paschen} \cdot d = 7$ hPa · mm, where d is the contact gap. For a common contact gap in HV of $d = 40$ mm gives

$p_{Paschen} = 0.2$hPa. Pressures above $p_{Paschen}$ only occur with a large external leak and will go up to air (or be filled with the surrounding insulating gas). Therefore, this case is referred to as up-to-air final pressure.

3.5.2.4 Potential vacuum failure modes

Table 3.13 lists the key potential failure modes for long-term loss of vacuum and the associated Q and final pressure. The key result is that every failure mode does not lead to all possible Q and final pressures, but rather the behavior of any pressure increase can be used to limit the possible sources. The first two modes are mechanical damage from inspections, handling, etc., and corrosion from the environment. Both could generally only occur in the field and could occur for any type of VI. Factors beyond the VI design and production method generally set these risks. The second mode is an increase in pressure due to the operations of the VI. Short-circuit switching, mechanical opening and closing, load current switching, capacitor switching, and passing short-circuit current with closed contacts could all theoretically lead to a pressure increase (however later data shows that they do not).

The next potential failure modes could occur for VI in any situation. The "anywhere" condition includes VI in storage, in breakers, and in the field. The first mode is slow venting leaks. This refers to leaks connecting the inside of the VI to the atmosphere at a rate Q_{slow}. A key feature of such leaks is the extremely small diameter of such a leak path. The formula for the vacuum conductance of a cylindrical leak path C is [36, Eq. (3.22)]

$$C = \frac{\pi \, v_{thermal} \, dia^3}{12 \, l} \tag{3.33}$$

where dia is the diameter of the hole, l is the length of the leak path (normally the thickness of the material with the leak), and $v_{thermal}$ is the thermal velocity of the gas molecules

$$v_{thermal} = \sqrt{\frac{8 \, k_B \, T}{\pi \, m}} \tag{3.34}$$

where $k_B = 1.38 \times 10^{-23}$ J/K is Boltzmann constant, T is the gas temperature, and m the mass of the gas molecules. The vacuum conductance is related to the Q and the pressure gradient across the leak Δp by the formula

$$C = \frac{Q}{\Delta p} \tag{3.35}$$

For VI, Δp is effectively the external pressure surrounding the VI of p_{ext}. Taking VI in atmospheric air $p_{ext} = 1$ atm at room temperature $T = 300$ K, and using the mass of nitrogen molecules $m = 23$g/mol, a pressure increase rate of Q_{slow} has a hole diameter of $dia = 40$ nm. Leaks with this rate and size cannot be detected by standard leak detection methods and are blocked from leaking by atmospheric gases [198,208]. For VI, these leaks can be detected by measuring a pressure change after exposure to a noble or other non-getterable gas.

Manufacturers developed effective tests by using different combinations of time, pressure, and temperature for exposure to a noble gas.

The next mode is hydrogen and helium permeation. This mode involves hydrogen and/or helium penetrating through material that is leak-tight to other, larger gas atoms and molecules. Insulating materials such as glass and ceramic are more vulnerable to this problem than metals. Glass, previously used for VI insulation housing, is much more vulnerable to permeation than alumina ceramics [36]. In general, VI are exposed directly to neither H_2 nor He. However, new applications for VI at cryogenic temperatures [209] could have exposure to He as a cooling medium. Also, increasing the use of hydrogen as a fuel could increase the chance of (likely accidental) H_2 exposure. The most likely source of H_2 exposure is the disassociation of water molecules on the surface. This process requires very large amounts of water vapor on the surfaces, and this quantity of water vapor is likely to lead to other problems within the circuit breaker and associated components well before any issues with hydrogen permeation appear.

The last two potential failure modes involve outgassing from parts inside the sealed VI. The first is the outgassing of atmospheric gases from the surface of the parts, from voids in the material, or the decomposition of oxides and other compounds into gases. Heating during the sealing process removes gases from the surfaces, and the higher temperatures used in Ag–Cu brazing help remove more of the gases from the VI. Operations by the VI can open voids and release gas, and arcing operations can decompose oxides into metal and gases. Alternately, the arcing could also expose clean metal surfaces, particularly Cr, which could have a gettering effect on the residual gases in the VI. Voids have a long history as a problem in vacuum systems in general and are controlled as a standard part of the quality control process [36]. Experiments in [210] suggest that outgassing, rather than slow leaks are the source of the pressure increase seen in certain VI after some decades of field service.

The outgassing process can be understood by looking at the time it takes to establish an equilibrium pressure. The gas adsorbed on a surface represents a balance between gas leaving the surface (outgassed) and gas returning to the surface (absorbed). The time a gas molecule spends on the surface τ is [211, Eq. (1.50)]

$$\tau = \tau_0 \, e^{E/(R\,T)} \tag{3.36}$$

where E is the energy of absorption for physical absorption or the free energy of formation for chemical absorption, $R = 8.314 \text{ J}/(\text{Kmol})$ is the gas constant, T is the temperature, and $\tau_0 = 10^{-13}$ s [211]. Some relevant values of E are summarized in Table 3.14.

τ is essentially a time for the surface outgassing to come into equilibrium with the gas inside the VI. Table 3.15 calculates τ for the values given in Table 3.14 at $T_{room} = 25°$, room temperature, and $T_{mid} = 450°C$, a temperature the VI would spend some hours at or above during the brazing process. Physio-absorbed gases are continually pumped away during the entire vacuum sealing process and are always in equilibrium between the surfaces and the VI volume. Therefore, they are

Table 3.14 Parameters for (3.36) for various gases physio-absorbed on the surface and key oxides

Material	Energy of absorption or free energy of formation (kJ/mol)	Reference
Physio-absorbed Ar, N_2, O_2, CO	12–17	[211]
CuO or Cu_2O	155–170	[212,213]
Cr_2O_3	1,128	[214]

Table 3.15 Time for the gases or oxides to come to an equilibrium pressure using (3.36) and the data from Table 3.14 at room temperature and a mid-point temperature typically seen during brazing of the VI

Equilibrium time T		25°C	450°C
Material	Physio-absorbed gases	95 ps	1.7 ps
	CuO	∞	0.2 s
	Cr_2O_3	∞	∞

an unlikely source of long-term outgassing effects. Copper oxides are very stable at room temperature but can rapidly dissociate and be pumped away during the vacuum sealing process. The pumping would be less effective than for physio-absorbed gases, as the time at high temperatures makes up only a portion of the total furnace time. Also, a major reason for the long furnace times is the need to equilibrate the temperature between VI directly visible to the furnace heaters, and VI (and VI joints) hidden from direct view, which would also reduce the effectiveness of copper oxide removal compared to physio-absorbed gas removal. Chromium oxide on stainless steel parts is extremely stable and unaffected by the sealing process and would not be expected to outgas over the lifetime of the VI. Although the value of τ_0 is subject to substantial discussion [211], the differences between the time on the surface for the different materials are large enough that the observations remain valid.

A key point to remember with outgassing is that the amount of gas is limited. Unlike a leak, where the external gas serves as an endless reservoir, in outgassing the amount of gas is limited by the quantity originally trapped inside the VI. To see the effect of HV VI designs on the risk of outgassing, the sources of outgassing can be divided into desorption from surfaces (including from near the surface) and outgassing from the bulk material. Outgassing from the bulk material is effectively eliminated through material choice and processing. For the surface outgassing, we can compare the vacuum volume of the VI to the internal surface area. Since we are interested in the general trend, we can simplify the problem. To compare the VI volume to the internal surface area, we can combine the average value of the VI diameter and length to a dimension , and treat the VI as a cube. Then, the

VI volume is

$$V = \Theta^3 \tag{3.37}$$

and the internal surface area a is

$$a = 6\chi\, \Theta^2 \tag{3.38}$$

where $\chi = 1 - 4$, a factor of the additional surfaces in the vacuum from the parts aside from the housing. If the VI was empty and the only surfaces are the VI housing, then $\chi = 1$. VI shields cover a significant portion of the interior of the VI and have two sides in the vacuum chamber. This adds in worse case 2 to χ, and raises it to $\chi = 3$. The other internal parts such as the terminals, contacts, etc., add some more surface and could raise the factor to $\chi = 4$. The key feature is the ratio of the volume over the area

$$\frac{V}{a} = \frac{\Theta^3}{6\ \chi\ \Theta^2} \sim \frac{\Theta}{\chi} \tag{3.39}$$

As the dimensions of the VI, Θ, increase with the voltage level, the internal volume increases faster than the surface area. Therefore, space for gas to occupy increases faster than the source of these gases, namely the internal surfaces. Also, χ, which is a rough measure of the part density inside the VI, would also go down for HV VI designs, given the wider part spacing necessary to achieve the dielectric requirements. This suggests that larger VI for HV applications present a lower risk from outgassing than the smaller MV VI. Based on the success of MV VI, this is a very encouraging result for HV VI designs.

The last mode is hydrogen outgassing. The source of hydrogen outgassing is separate from hydrogen permeation. Stainless steel has hydrogen dissolved in the metal, and this hydrogen can outgas into the vacuum chamber over time [215,216]. The level of hydrogen in stainless steel is much higher than in other materials used in VI such as alumina ceramics, copper [36], or Cu–Cr [217]. The high temperatures during brazing outgas the hydrogen from the steel, and in particular, the higher temperatures during Ag–Cu brazing outgas even more hydrogen [218].

The testing and/or operation of the VI could alter the internal pressure. In the field, VI operations can occur at any point during the VI lifetime. As a trigger mechanism for leaks, these operations could conceivably lead to a loss of vacuum integrity beyond the infant mortality period. The extensive development and certification testing performed on new VI and vacuum circuit breaker designs provide us with the opportunity to measure the effect of arcing, mechanical, and current-carrying operations on pressure. Earlier work examined a small number of VI after load and short-circuit switching and saw no significant change in the vacuum level or composition of residual gases inside the VI [203,219]. The internal pressure after testing was measured in 562 individual VI in [30]. The majority of the VI experienced short-circuit testing, while the remainder saw no-load switching, load current switching, capacitor switching, short-time current, and making operations testing. This group includes VI that saw particularly onerous test duties, the most

Table 3.16 Examples of extreme testing and field operations, after which no long-term increase in the VI pressures, were observed [30]

Operation	Examples
Short-circuit current	Use as test lab breakers
	50–100x breaking operations
	100–140% rated current
	2x Class E2 endurance
	Pressure measurement after >1 year
Mechanical life	1–7 million operations
Load current	10–550K operations
	Arc furnace switching
Capacitor switching	Back-to-back switching
	Pressure measurement after >1 year
Short-time current	4s STC test
Making current	14x closing ops.

interesting of which are listed in Table 3.16. No increase in pressure due to any operations appeared. A short-term reduction in the internal pressure after current interruption operations was seen in [220].

Table 3.13 and the above discussion illustrate the importance of knowing the potential pressure increase sources when attempting to understand a pressure measurement later in a VI lifetime. Simply taking the initial pressure measurement at the start of the VI life, combining it with a later pressure measurement after some time in the field, and then assigning a Q could be problematic. First, this estimated Q must be compared to the value where the leak would be blocked or significantly reduced by atmospheric air, particularly water. Second, some sources such as outgassing can end, once all the gas is released. This could happen quickly, or still be ongoing and would require further pressure measurements at later times to see if the process is continuing. The risks from permeation and outgassing/hydrogen outgassing have already been reduced by design and material improvements. The risk of damage to the VI can be increased through handling to measure the pressure in the field and/or the creation of dielectric weak points by the measurement equipment for continuous monitoring. For example, multiple VI were damaged during the process of removing them from circuit breakers to measure the pressure after several years in field service [221,222]. Thus, efforts to increase reliability through vacuum monitoring could introduce more risks than they eliminate. Finally, corrosion is a threat not just to the VI, but to multiple components of the circuit breaker, and requires general attention.

3.5.2.5 Pressure monitoring in the field

The general interest in circuit breaker monitoring prompted interest in also monitoring the vacuum level of VI in the field. Reference [223, Section 3.3.13] contains a detailed overview of options to monitor the vacuum level. Table 3.17 summarizes

Table 3.17   Options for measuring the pressure of VI after they have left the factory [223, Section 3.3.13], including the range of pressures measured (using the terminology from Table 3.12) and the output of the measurement

Method	Vacuum level	Output
Dielectric testing	Mid to up to air	Pass/fail
Magnetron	Low to mid	Pressure level
Mechanical pressure	Up to air	Pass/fail
Vacuum gauges	Low to mid	Pressure level
Partial discharge	Mid	Mixture
Field emission	Low	Pressure level
Current interruption	Mid	Pass/fail
Sound waves	Up to air	Pass/fail
Light emission	Up to air	Pass/fail

the options, the pressure range covered (using the same terminology as earlier), and the type of output. The first point from this table is that no single technique covers the full range of the vacuum level. Second, most techniques give only a pass/fail measurement, only three options give an actual pressure level. Third, for the three techniques producing a pressure level, only the magnetron method is technically mature. Section 3.2.3 introduced the magnetron method of pressure measurement.

For field/monitoring applications of the magnetron, some points must be considered. First, the magnetic field generation and the voltage application require some handling of the VI, either in the circuit breaker, for example in [224], or its removal from the breaker [221,222]. This handling introduces its own risk, which must be carefully balanced against the chance of finding a VI with elevated pressure. An FMEA (failure mode and effects analysis) would be the perfect tool for balancing these risks [225]. Second, a successful measurement requires a careful calibration curve for the specific VI design [32,43] and magnetron parameters [226,227]. Third, magnetron measurements are not simple and require expertise to properly interpret [30,228]. Fourth, the measurement process itself disturbs the system, and this must be accounted for in the measurement procedure [229,230]. Finally, as described earlier in this section, pressure measurement is only the start of the process for identifying at-risk VI and the timescale of this risk.

Interest in measuring the pressure of VI after reaching 20–40 years has recently appeared. The subtext of this work is often the interest from the end-users to extend the lifetime of the circuit breaker beyond the manufacturers' recommendations. Again, it must be re-emphasized that the vacuum integrity is only one issue of one component of the circuit breaker, and if all the components do not simultaneously function then the circuit breaker will not work as intended. In [221], the pressure of 54 VI was measured after 5–25 years of field service by the original manufacturer. The pressure was $\leq 2 \times 10^{-6}$ hPa for all the VI and had a distribution of pressures similar to or better than the current production of the same design. Reference [231] measured the pressure of 140 VI after 25–40 years of field service and found 61 with $>10^{-4}$ hPa. Later work [232] measured 207 VI after 25–40 years of field service and found 88 with $>10^{-4}$ hPa (it was not clear if some or all the previous measured VI in [231] were included in the later paper). Finally, work in [233] measured the pressure on 21 VI after >35 years of service, and 21 VI after ~20 years of service. All the VI had pressures $<10^{-4}$ hPa. Some VI had higher than expected pressures, but there was very little information on the measurement calibration and no effort to evaluate possible sources of the suspected pressure increase.

The massive effort of measuring 12,000 VI after >10 years of field service reported in [202] combined with the recent work in [221] show that VI vacuum integrity is very reliable out to ~25 years of field service. Beyond 25 years, more work could be justified to identify the source of higher pressures observed in [231] and [232]. Changes in the VI design from glass to ceramic insulation, and the switch from tubulation sealing (with a very limited temperature at the time of sealing) to sealing with Ag-Cu-based brazes could eliminate some of the failures for older designs [30]. Also, as already mentioned, Q only make sense when associated with a gas source, so that the ultimate possible pressure range can be identified. Finally, for circuit breakers and switchgear older than 25 years, every part, not just the VI, must work for the circuit breaker to function. The retrofitting of new circuit breakers into old enclosures is a well-established market with considerable experience in complex installations [234] and could present a more cost-effective option than attempting to measure the pressure of old VI.

3.5.3 X-ray emission
3.5.3.1 IEC and ANSI X-ray tests
Metal contacts in a vacuum with a voltage across them in the MV or HV range are a potential source of X-ray emission [27]. Field emission currents on the $nA - \mu A$ level involve electrons generated at the cathode that are accelerated across the vacuum gap and can generate X-ray emission when impacting the anode. It is very important to understand that VI does not emit X-rays when sitting in storage or when the contacts are closed, even when the circuit breaker is energized [235]. No X-rays are emitted when the contacts are opened without voltage nor during mechanical operations without HV on the circuit breaker. It is only when voltages on the order of U_r or U_d are applied across the open contact gap that X-ray emission needs to be considered. Technically X-rays could be emitted when performing U_d on a circuit breaker with closed contacts between the phases and ground. However, only a portion of the total voltage drop occurs within the VI, which sharply reduces the X-ray emission. Therefore, the only significant potential concern is HV application over the open contact gap.

The potential health risk of X-ray emissions is set by the exposure dose rate measured in $\mu Sv/h$. Both the IEC standard [54, Section 7.11] and the ANSI standard [236] specify very similar X-ray tests for VI. The test is applied to a VI with the contacts opened to the recommended contact gap. The voltage is applied across the contact gap, and the X-ray detector is set 1 m from the VI outer surface. If the

VI is normally used in insulating gas or solid insulation, the test can be performed in insulating oil, as long as the oil and housing do not significantly attenuate the X-ray measurement. If necessary, distances other than 1 m can be used by converting the measured dose rate $R(l)$ and the distance l to the equivalent measurement at 1 m of $R(1\ m)$ using the relationship

$$R(1\ m) = R(l) \cdot \left(\frac{l}{1\ m}\right)^2 \tag{3.40}$$

Two voltage levels are tested. The first is at U_r with a limit of $R(1\ m) \leq 5\ \mu\mathrm{Sv/h}$. The second test is at U_d for [54] and $0.75 \cdot U_d$ for [236] with a limit of $R(1\ m) \leq 150\ \mu\mathrm{Sv/h}$. For both standards, when $U_d > 160\ \mathrm{kV}$, if $R(1\ m) > 150\ \mu\mathrm{Sv/h}$ then the VI does not fail the test, but the manufacturer must state the measured value to their customers. Typically, the VI manufacturer would specify an additional safety distance beyond 1 m to bring the X-ray emissions down to $\leq 150\ \mu\mathrm{Sv/h}$. Since only a rather foolhardy field technician would be standing only 1 m away from a VI energized to $U_d > 160\ \mathrm{kV}$, the specification of an additional safety distance is not a significant problem in practice.

3.5.3.2   Additional X-ray requirements

The EU issued a directive to the national governments regarding the regulation of X-ray emissions [237]. Unfortunately, some national governments, such as Germany, applied limits on X-ray emission from VI based on previous regulations for cathode ray tube (CRT) monitors [238]. This regulation applies when the peak system voltage is $U_r > 30\ \mathrm{kV}$ peak [239]. For VI where $U_r \leq 52\ \mathrm{kV}$, the X-ray emissions at U_r cannot exceed $1\ \mu\mathrm{Sv/h}$ at $d - 0.1\ m$. For $U_r > 52\ \mathrm{kV}$, the manufacturer can specify a safety distance, and the measurement is then performed at the safety distance plus 0.1 m. Given the l^2 dependence of the emissions, this is clearly a much harder requirement. The requirements of [238] make perfect sense for CRT computer monitors, where people easily spend hours a short distance away from the X-ray source. If someone is truly spending hours only 0.1 m away from an open, energized VI then the risk of electrocution or causing an arc fault vastly outweighs the risk from the X-ray emissions. Nevertheless, legal requirements trump everything, including reason and the existence of well-established standards [54,236].

MV VI generally passes the X-ray tests, including the German regulation [238], without any problems [239]. These tests are necessary, but in MV are only done at the end of the development process to complete the test documentation. For HV VI, the X-ray emissions do need to be considered in the development process. The dose rate dH/dt [240] scales as

$$\frac{\mathrm{dH}}{\mathrm{dt}} = \frac{a_1}{l^2} \cdot \beta^2 \cdot \frac{1}{d^2} \cdot U^4 \cdot e^{-\frac{a_2\, d}{\beta\, U}} \tag{3.41}$$

where l is the distance at which the X-ray emissions are measured, d is the contact gap, U is the peak applied voltage, β is the field enhancement factor [241] associated with the material and its treatment, and a_1 and a_2 are constants associated with the material. The two key features of (3.41) are the sharp increase in dose rate with applied voltage, and the decrease with increasing contact gap. The $U_r = 145$kV VI design described in [240] meet the IEC and ANSI requirements, as well as the German requirements. In contrast, the $U_r = 126$ kV design described in [242] would met the IEC and ANSI requirements but would not have directly passed the German requirements. The $dH/dt = 3\ \mu Sv/h$ measured at $l = 1$ m would translate to $dH/dt = 300\ \mu Sv/h$ at $d = 0.1$ m. A safety distance of $l = 1.7$ m might be sufficient to bring the emissions below the German requirement limit. For $U_r > 145$ kV, more attention in the design phase would be needed to pass the X-ray tests. For these voltages, using two VI in series would cut the voltage on each VI in roughly half, and (3.41) would result in a dramatic reduction in the X-ray emission.

The IEC and ANSI tests deal with VI in field applications, and not with the potential for X-ray emission in manufacturing or type testing. The main situations where X-ray emission from VI are a concern are in the manufacturing of the VI and test labs. In manufacturing, the conditioning process can generate significant X-ray emissions, and every manufacturer uses a combination of shielding, distance, and exposure monitoring to reduce employee exposure. Similarly, test labs can see higher X-ray emissions when testing above the rating for development tests, or in cases of problems with the design. Again, every test lab uses a combination of shielding, distance, and exposure monitoring to manage the exposure. The two areas that require attention are when a sub-supplier or contractor applies U_r and/or U_d to the VI as part of a production process, and the circuit breaker/switchgear manufacturer when performing U_d tests on completed breakers. In both these cases, the sub-supplier or circuit breaker/switchgear manufacturer must be informed that VI could emit X-rays, and that they are responsible to adhere to any local laws and regulations regarding X-ray emissions.

3.5.4 Non-sustained disruptive discharge

Non-sustained disruptive discharge(s) (NSDD) are phenomena currently only measured in vacuum during switching operations in test labs [243]. The NSDD occurs after the CZ, during the application of the recovery voltage. When the collapse of the voltage across the VI to zero leads to the flow of either the load current either the power frequency current for short-circuit tests or the in-rush current for capacitor tests, then the breakdown is considered a re-strike. If neither the power frequency nor the in-rush current flows after the voltage collapse, then the event is an NSDD. A very high-frequency current can follow the NSDD, but this is very hard to measure in practice.

The concern over NSDD came from the effect of the NSDD on the other two phases of the system. The magnitude of the voltage reduction to zero on the phase with the NSDD is ΔU_n, and is a function of when the NSDD occurs on the voltage cycle. The NSDD produces a constant shift in the voltage on the other two phases of ΔU_{NSDD} [73, Eq. (4.3)]

$$\text{for } k_{pp} = 1.5 : \ \Delta U_{NSDD} = \Delta U_n$$
$$\text{for } k_{pp} = 1.3 : \ \Delta U_{NSDD} = 0.43\ \Delta U_n \tag{3.42}$$

The concern is that this offset voltage ΔU_{NSDD}, when superimposed on the recovery voltage, could lead to a re-strike. In fact, the VI generally withstands the recovery voltage plus ΔU_{NSDD} without trouble. The rare instances where NSDD leads to a re-strike are when multiple NSDD occur during the same operation, and the ΔU_n from each NSDD has the same polarity. In this case, the recovery voltage can be shifted enough to lead to a re-strike.

The concern in testing is that if you performed some more tests, then you would see an NSDD leading to a re-strike. The issue then becomes whether enough tests are performed to see whether this case can occur. NSDD can happen at all current levels [243], so for short-circuit tests, we can look at all the tests where the full recovery voltage is applied. Reference [48] gives a typical example of a $U_r = 145$ kV VI, where 59 operations were performed. The long electrical life of a vacuum encourages users to ask for various electrical life tests, which add even more operations; up to 274 for Class E2 list 1 in [70]. Capacitor switching tests have a similar story, where 44–120 operations are performed in test duty 2 [70, Section 7.111.9.4]. For HV VI, the systems are normally $k_{pp} = 1.3$, which reduces the effect of the NSDD on the other two phases. Therefore, there is enough evidence to show that NSDD is not a general problem, and if the NSDD were frequent enough to warrant concern, then they would lead to re-strikes during the standard type testing.

In the author's experience, multiple NSDD on a single operation are most frequently due to mechanical issues with the circuit breaker design. A distant second source of NSDD is manufacturing issues within the VI, particularly particles and cleanliness. Finally, if there was a specific concern about the risk from multiple NSDD during the same operation on a capacitor bank, there are straightforward methods to reduce the over-voltage across the bank [244].

3.6 Evolving research

3.6.1 Modeling of AMF (and RMF/TMF) designs

3.6.1.1 Basis of modeling

Several different research groups focused on the numerical modeling of the vacuum arc under an AMF. The key advantage of numerical modeling is the ability to reduce the amount of high-power testing required to evaluate the short-circuit performance of a VI design. The key disadvantage is the complexity of the multi-component models and the exact connection between the calculated parameters and the ultimate interruption performance. Reference [245] provides an excellent summary of research into modeling vacuum arc behavior in an AMF, and [246] contains a clear outline of the main outputs of the modeling.

The basis of the bulk of the models is a multi-fluid MHD (magneto-hydrodynamics) model. Some basic features of the plasma model are first that the AMF confines the electrons to the magnetic field lines. The ions are then in turn confined by the electrostatic interaction with the electrons. Both the electrons and ions have thermal motion, and for the ions, a directed velocity from the cathode to the anode is superimposed on the thermal motion. Plasma is primarily generated at the cathode; some models do consider plasma generation at the anode as well. A voltage sheath exists between the cathode and plasma and another sheath with a lower voltage drop exists between the anode and the plasma. The AMF is externally supplied and ranges from a constant value throughout the model area to a radial profile based on magnetic field modeling of the actual coil structure. Light emissions can be a significant loss term for energy, especially as the arc constricts from the diffuse mode. These basic behaviors drive the choices for the model complexity and for the boundary conditions.

The models have some key features in common. First, they all ignore the drawn arc and transition mode between the contact separation and formation of the arc over the entire contact [128]. They focus on the fully evolved arc at a specific contact gap, current level, and AMF, typically using the highest current and the biggest contact gap as a worst case. All or most models use commercial modeling programs run on high-end PC computers, as opposed to supercomputers. The main comparison to experiments is through arc visualization, with Figure 3.35 showing an example. Here the transitions from the diffuse to a columnar mode in the arc visualization experiments at two AMF strengths in [127] are compared to the transition point as a function of a critical arc voltage U_{cr}^{eff} calculated from an AMF arc model [125]. Hopefully, some of the arc parameter measurements outlined in Section 3.6.2 will enable a deeper comparison in the future. Finally, the equations and the common assumptions of the models are very similar to and often directly based on the work in [247] and [248]. These two references give a complete list of the equations and a broad discussion of the meaning of the various terms.

Two key boundary conditions for the modeling are at the cathode and the anode. The cathode is the source of the plasma and the directed motion of the ions. The complex behavior of the cathode spots is not directly included in the model,

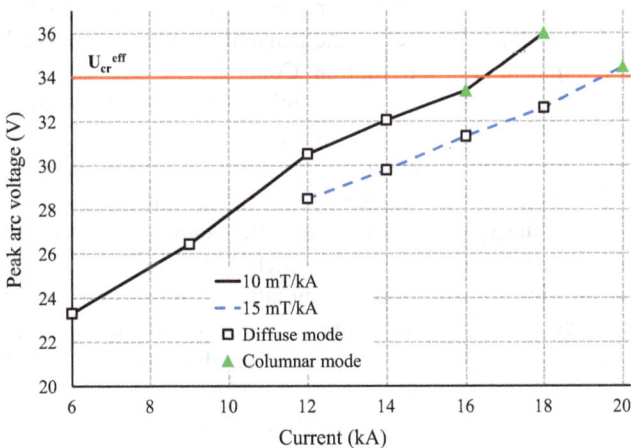

Figure 3.35   One example of an AMF arc model (based on changes in the arc voltage) successfully predicting changes in the arc behavior [125]

rather the plasma just above the spots serves as the boundary. Simpler models have a uniform plasma generation along the boundary, while more complex models look at assigning a spatial distribution of the plasma based on experiments or another modeling. Finally, the long-term goal is to couple the results of advanced models of the cathode spot behavior such as in [249] to the boundary conditions for the AMF model. The simplest boundary conditions for the anode are to treat it as passive. This condition is certainly violated when an anode spot forms, and some theories suggest that the anode is not passive even when there is no anode spot [131]. More complex models added anode evaporation to the model, and the long-term goal would be to link advanced models of the behavior of the molten anode including the motion of the molten pool [250] to the AMF model.

The models outlined in [247] and [248] contain several scalar values. Some such as the average/mean charge of the ions are assigned clear values in the papers. Others such as the conductivity of the plasma parallel and orthogonal to the magnetic field, or the plasma viscosity are left vague at best. I used the term "scalar" instead of "constant" because the values may not be constant – some can and are expected to depend on the values calculated in the model such as the ion and/or electron temperature. While assumptions and simplifications about the value of these scalars are fully acceptable, more clarity is needed about what values are used and what are the assumptions to help future researchers and the article readers understand how to apply the results.

3.6.1.2   Future work

Although considerable progress has been made, some significant challenges remain. One key issue is that the value or values most relevant to the current interruption are unclear. As discussed in Section 3.4.1, the transition from a presumably "good" state to a presumably "bad" state is reasonably clear. Calculated parameters such as the anode temperature, constriction in ion density, or changes in the anode sheath are likely candidates for predicting the transition between success and failure in interruption. However, the calculated values have not been clearly connected to the transition in interruption. Coupled with this issue is the modeling focuses on the highest current level while ultimately the interruption is dictated by what happens at CZ. The behavior at high current influences the situation at CZ, but the connection remains empirical.

Despite the limitations of the models, modeling demonstrated useful qualitative comparisons with experiments. One specific example in HV is [126], where a change in the convergence of the model predicted the transition to a constricted arc mode in visualization experiments. Quantitative comparisons also show initial success. Reference [246] contains some comparisons of the radial profile of the electron density and the effect of the AMF strength on the ion and electron temperatures between the modeling and experiments. There was very reasonable agreement in the magnitude and general trends. Hopefully, the work described in Section 3.6.2 will give more data to compare to modeling.

The empirical success of AMF models is encouraging development in two main areas. The first is in concrete applications such as the effect of the magnetic fields from neighboring phases on the vacuum arc on AMF contacts [251]. The second is to add additional effects to the AMF model. For example, [252] adds the ionization and re-combination behavior along with the effect of imposed non-axial fields on the arc behavior. This points to a useful and exciting future for AMF modeling. One of the key features of future work will be the detailed comparison to experiments, and a greater understanding of which terms in the model are significant, and which can safely be deleted to keep the model complexity within reason.

Modeling of rotating arc contact designs (RMF) is much more limited in comparison to the AMF case. One basic challenge is the distinct boundary between the high-density plasma and the vacuum background. Although there is some transition in the plasma density, it is much sharper than in the AMF case. This compels the model to either force the arc to be constricted or attempt the difficult task to have the constricted arc appear as part of the solution. With RMF contacts, arc rotation is the main purpose of the contact design, and therefore this motion is a vital output of the model. Reproducing this time-based evolution of the arc adds an additional challenge over the AMF modeling. Two-dimensional modeling in [253,254] reproduced the behavior where a minimum contact gap seems to be necessary to start arc rotation. These works also had interesting results modeling how the arc jumps the gaps in the contacts; a behavior that is empirically fully established, but was difficult to understand from theory. The model was later extended to three dimensions [255].

For MV, RMF/TMF contacts are very important and have even penetrated countries and companies that were previously only interested in AMF contacts [256,257]. For HV, as mentioned in [20], few applications exist where RMF contacts are used. Greater usage of vacuum in HV, particularly at the lower voltages like $U_r = 72.5$ kV, could open more opportunities for RMF contacts and increase interest in modeling RMF designs with larger contact gaps. Nevertheless, the focus of RMF modeling is likely to remain on MV applications.

3.6.2   Detailed in situ measurements of arc parameters

Arc visualization techniques using high-speed cameras to record the behavior of the arc in visible light are well-established and extremely useful techniques [27]. The typical experiment setup consists of a vacuum chamber with a viewport. VI contacts are placed inside and opened under short-circuit current. Movies of the resulting arc behavior allow insightful comparisons between different contact materials, contact designs, and externally applied magnetic fields. While arc visualization was, is, and will continue to be a key technique to study VI, it does have the limitation that the results are largely empirical.

Light emissions from plasma are proportional to the temperature and density, and light from the contacts appears when they are hot and is also proportional to the temperature. This enables the measurement of key fundamental quantities in situ. However, there are some significant difficulties with extracting temperature and density measurements from light emissions. First, there are multiple sources of light emissions including the hot/molten contact surface, various plasma species, and neutral vapor in the contact gap. Second, the mathematical relationships are complex and sometimes involve constants with poorly or not measured values. Third, many techniques require a separate calibration to return absolute measurements, and these calibrations are themselves complex. There are numerous examples of efforts to get temperature and/or density data from optical emissions in vacuum arcs in the literature [27].

Recently there was a renaissance in this field after many years of limited activity. The first group of techniques focuses on measuring the temperature of the contact surface, measuring the anode behavior [258,259] or cathode behavior [260]. The techniques include near-infrared, two-color pyroscopy, high-speed thermography [261], and the combination of a visible high-speed camera with filters to remove line emission. The second group consists of video spectroscopy [262,263] and absorption spectroscopy [263,264], and focuses on the temperature and density of the neutral and ionized metal. Table 3.18 summarizes these techniques. An exciting feature of this recent work is the combination of two or more of these techniques in the same experiment. One other interesting technique uses lasers to measure the electron density and neutral vapor density in intense vacuum arcs [265]. So far, the technique has only been applied to small contacts at relatively low currents, such as $D = 6$ mm and ≤ 4 kA in [266], for example. If the technique can be expanded to bigger contacts and higher currents, it could become a very useful tool.

The application of arc visualization and optical in situ techniques has expanded greatly over the past few years. Table 3.19 summarizes some recent papers on in

Table 3.18   Compares the main in situ techniques for measuring vacuum arc parameters in terms of the current level, output, location, and time resolution of the measurement

Technique	When	What	Where	How long
Near-infrared spectrometry	After CZ $>$ 700°C	Surface temperature	Point	1–2 ms
Two-color pyroscopy	High current and after CZ to $>$ 1,200°C	Surface temperature	Point	$<$100 μs
High-speed camera plus filtering of line emission	High current	Surface temperature (relative, absolute when calibrated with other techniques)	Area	100 μs
Video spectroscopy	High current	Neutral and ionized metal	Along a line	$>$50 μs
Absorption spectroscopy	Around CZ	Neutral metal	Along a line	one point
High-speed thermography	High current and after CZ to $>$ 700°C	Surface temperature	Area	$<$200 μs

Table 3.19   *Recent references on in situ measurements of the arc behavior, whether the measurement is of the plasma or the contact surface, and the output of the measurement along with which contact design analyzed*

Reference	AMF	RMF	High-current	CZ	Location	Measurement
[269]	X		X		Plasma	Radiation power
[260]	X			X	Surface	Temperature
[261]	X			X	Surface	Temperature
[267]	X			X	Surface	Temperature
[270]	X			X	Surface	Temperature
[271]		X	X		Plasma	Cu I (neutral)
[69]		X	X		Plasma	Arc rotation
[259]		X		X	Surface	Temperature
[268]		X		X	Surface	Temperature
[272]		X		X	Surface	Temperature

Note: High-current refers to measurements around the peak of the current, and CZ to measurements near to and/or after the current zero.

situ techniques dealing with Cu or Cu–Cr contacts with currents of ≥ 10 kA. While most are optical, there is also an example of a method using external magnetic probes to measure the arc rotation inside commercial VI [69]. The recent work includes both AMF and RMF contacts and even includes some work specifically focusing on the change in behavior at the interruption limit [267] or at the end of the short-circuit electrical life [69]. One paper focused on the effect of contact welding on the later behavior of the arc [268], further demonstrating the capability of these techniques to answer practical VI questions. These techniques are fully applicable to HV VI, and it will be exciting to see these same techniques applied to the larger contact gaps relevant to HV.

3.6.3   Breakdown experiments and modeling

3.6.3.1   General overview

Despite practical progress in increasing breakdown voltages, many challenges to theoretically and practically understanding vacuum breakdown remain. An article from 1965 describes many of the issues that remain today [273]: empirical importance of conditioning, lack of significant improvement from surface treatment beyond an initial improvement, uncertainty over the actual source of field emission, etc. For many decades, the arguments between field emission and particle-based models of contact breakdown in a vacuum resembled a theological debate more than a scientific evaluation. The reason for this state is the difficulty in observing the behavior before and during the breakdown. Most analyses developed a theoretical model based on field emission or particle-based breakdown, and then fit the model to a specific set of data. Applications of the models to different setups are very rare. The possibility that different models could apply at different contact gaps and/or conditioning states is also rarely considered. Although this state could be discouraging, progress has happened. The key is to appreciate the difficulty of the problem, and temper expectations to focus on incremental improvements.

3.6.3.2   Field emission of electrons as a source of breakdown

Three applications critically depend on vacuum breakdown: field emission arrays (FEA) in micro-electronics, VI, and linear accelerators in particle physics and nuclear fusion. Although vacuum breakdown is a problem in all three fields, only a few publications looked at comparing models and experimental data between different regimes. FEA uses micrometer-sized sharp tips as a cathode together with a nearby anode to maximize the electric field stress on the tips, Figure 3.36. These tips can then be grouped array of 10s, 100s, and even 1000s of these tips to generate a useable field emission current when a voltage is applied. VI and other macroscopic contacts in a vacuum also generate a field current with voltage across the gap. Figure 3.37 plots a wide range of FEA and VI-related data for the field emission current as a function of voltage [241]. The current-voltage $I - V$ data can be converted into current density-electric field stress $j - |E|$ data using the relations

$$j = I/A_e$$
$$|E| = \beta V/d \tag{3.43}$$

where β is the field enhancement factor, A_e is the emission area, and d is the contact gap. β and A_e are calculated by fitting the $I - V$ data to the Fowler–Nordheim equation

$$I - \frac{1.5 \times 10^{-6}}{\varphi} \exp\left(\frac{10.4}{\sqrt{\varphi}}\right) A_e \left(\frac{\beta V}{d}\right)^2 \exp\left(-\frac{6.44 \times 10^9 d}{\beta V} \varphi^{3/2}\right) \tag{3.44}$$

with the current $I[\text{A}]$, voltage $V[\text{V}]$, work function of the metal φ [eV], contact gap $d[\text{m}]$ to calculate the emission area $A_e[\text{m}^2]$ and field enhancement factor

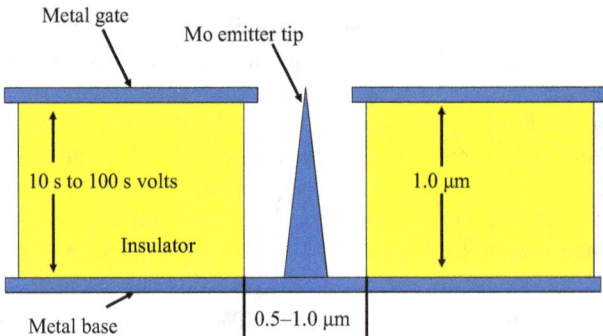

Figure 3.36   Typical example of a single-tip field emitter array (FEA)

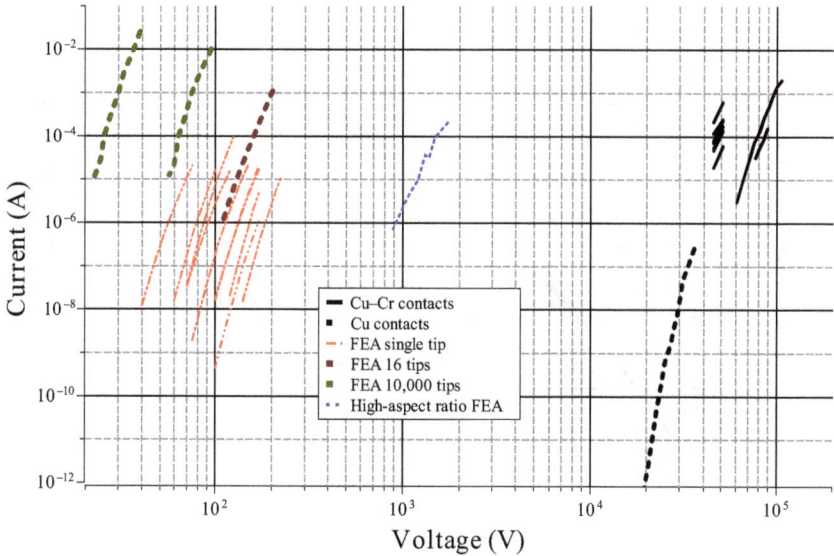

Figure 3.37   Comparison of field emission currents as a function of voltage for various FEA, a microscopic high-β feature, and macroscopic contacts of Cu and Cu–Cr from multiple sources summarized in [241]

β [unitless]. The Fowler–Nordheim equation in the $j - |E|$ form is

$$j = \frac{1.5 \times 10^{-6}}{\varphi} \exp\left(\frac{10.4}{\sqrt{\varphi}}\right) \; |E|^2 \exp\left(-\frac{6.44 \times 10^9}{|E|} \; \varphi^{3/2}\right) \qquad (3.45)$$

Figure 3.38 plots the data from Figure 3.37 in terms of $j - |E|$ together with (3.45) using the commonly used value of $\varphi = 4.5$ eV. Here the diverse range of data, from FEA with 10s, 100s, and even 1,000s of tips, to extremely high β microtips, to macroscopic Cu contacts and actual VI all fall along the same curve. The results show that a narrow range of electric field stress at the electron emission point of $|E| = 2 \times 10^9 - 1.2 \times 10^{10} \frac{V}{m}$ generates a massive range of current densities of $j = 10 - 10^{14} \frac{A}{m^2}$. Even if the interpretation of the electric field and current density is subject to criticism, the fact remains that (3.44) with two free parameters (β and A_e) can describe an extremely wide range of electron emission sources.

3.6.3.3 Particle-based breakdown models

This success at integrating these diverse data sets is tempered by the problem that (3.44) or (3.45) alone does not translate into a breakdown model. The second problem is that the field enhancement factor β calculated from fitting (3.44) gives extremely high values. These values for β of 100s to 1,000s would at first glance require very sharp, tall, microscopic points. Reference [274] shows an example

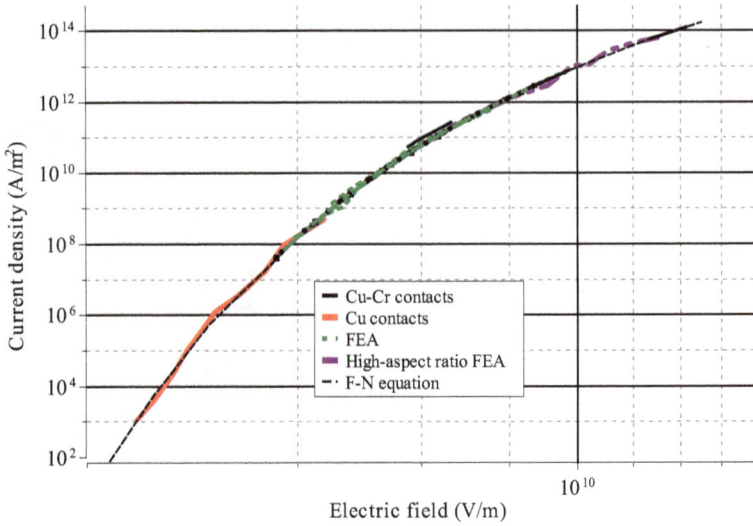

Figure 3.38   Data from Figure 3.37 converted into current density and electric field stress plotted together with (3.45) [241]

where such $\beta \sim 2,600$ features with a height of 42 μm and a tip radius of < 10 *nm* were deliberately made on a surface and the corresponding $I - V$ curve measured. Such structures are simply not seen on contact surfaces.

The main competing theory is that particles from the surface (or even surrounding parts) drive the breakdown process [275]. For the particle-based theory, the key value is the parameter W which is calculated along a possible path of a particle between the cathode and the anode

$$W = |E_C|^{\alpha} V^{\beta} |E_A|^{\gamma} \tag{3.46}$$

where $|E_C|$ is the electric field stress at the cathode end of the path, $|E_A|$ is the electric field stress at the anode end of the path, V is the voltage drop along the path. The exponential powers can be fit as part of the comparison to experiments, from the theory they are $\alpha = 1$, $\beta = 1$, $\gamma = 2/3$ [275]. When W rises above a threshold value, the probability of breakdown becomes non-zero. This model is used in the development of multiple-stage accelerator design for 1 *MV* [276], and at the other end of the spectrum was also applied to breakdown data from commercial VI [277,278].

3.6.3.4   Relevant work from linear accelerators

The difficulties with the high β values and the details of the breakdown process prompted several groups to look in detail at surface behavior that could drive the field emission leading to breakdown for linear accelerators. The first set involved detailed experimental measurements of the localized β, A_e, and φ values on real surfaces [279] and measurements of the exact location and sequence of the breakdowns for a later detailed evaluation of the surface and material underneath [280,281]. The second focused on breakdown modeling driven by the motion of dislocations in the bulk material altering the surface and driving field emission [282] with the experimental comparison in [283]. A third modeled the evolution of sharp nanotips on a surface which more than doubled in height over a <10 ns period and would give a basis for high β values [284]. Finally, a model assuming the presence of a dielectric layer on the metal contacts reproduced the behavior of (3.44) seen in actual experiments [285]. This model could be interesting for materials such as stainless steel and aluminum that are known to have very strong oxide layers on the surface. These works demonstrate the interesting progress being made in field emission models of vacuum breakdown.

On the practical side, linear accelerators and VI share very similar requirements regarding part cleaning and voltage conditioning. In addition to the common need for vacuum-compatible materials [36,211], both linear accelerators and VI require careful part cleaning for the voltage performance. Many different cleaning techniques exist, each with their own advantages and disadvantage and a long history of research [286,287]. VI have the advantage of heating up to $\geq 800°$ during the brazing process, which is very effective in outgassing the components and removing H_2O from the surfaces. Accelerators are limited in the temperatures they can apply and therefore must rely on additional techniques. Ultimately, both VI and accelerators depend on voltage conditioning to reach their desired breakdown performance. Despite sometimes heroic efforts in accelerators to use diamond surface polishing and other techniques to optimize the surfaces, all practical devices require voltage conditioning.

The conditioning behavior is very similar between VI [288] and accelerators [289], with a rapid increase in the voltage withstands with the early voltage applications, followed by a slower rise to the ultimate voltage limit. Recent work began looking at monitoring the X-ray emissions during the conditioning process of stainless-steel contacts in a large vacuum chamber to identify when the process has saturated [290]. The aim of accelerator design for a breakdown voltage above 1 MV [276] compares well to the goal of using VI for $U_r = 245 - 300 \, kV$, with $U_p = 850 - 1,050 \, kV$ peak [54]. As discussed in Section 3.4.2, the breakdown distributions of VI and accelerators do overlap, suggesting that the high breakdown values seen in accelerators may be achievable in VI.

3.6.4 *Role of particles and molten droplets*

3.6.4.1 **Generation and breakdown process for particles**

Capacitor switching in Section 3.4.3 presents a particular challenge to contact breakdown models. The peak of the offset voltage across the VI u_c is below the peak voltages of the dielectric tests U_d and U_p, so some degradation of the contact withstand voltage must happen for a re-strike to occur. The slow rise of the recovery voltage in comparison to current interruption tests rules out the direct effects of arcing over the bulk of the contact, which would otherwise be seen in T10 tests where the first-pole-to-clear has $u_c = 2.55 \, PU$, Table 3.3, especially in synthetic tests where the voltage remains at a constant value. During single-bank and back-to-back capacitor switching, breaking the welds from the in-rush current is presumably the source of re-strikes, either through field emission or particles from the broken weld. For line and cable charging, the source of re-strikes is harder to identify.

The possibility of particles as a source of breakdown, in general, and for capacitor switching specifically, prompted interest in studying the behavior of small particles between contacts at HV. The first idea was to controllably add particles to the contact gap and then make detailed observations of the behavior. This behavior can then be fed into models and even potentially be connected to breakdowns. Interest appeared early on to designing such a system [291] with some early measurements of particles under HV in [292], however, more recent work looked at the particle behavior in-depth [293,294]. Both Cu and Al_2O_3 particles were added to the contacts in [293], and an AC 50 kV peak voltage was applied between Cu contacts. Figure 3.40 compares the size of particles observed and/or added to contacts for breakdown studies. Two main particles appeared: the first bouncing between the contacts and the second charged particles that would attach to the cathode and move to the other contact when the voltage changed polarity.

Later work looked at adding particles in three size groups between 38 and 212 μm to a 5 mm gap of 30 mm Cu contacts with a voltage of 50 kV DC across the gap [294]. This gave an electric field stress of

$$|E_{pre}| = 10 \; \frac{kV}{mm} \tag{3.47}$$

which is in the middle of the value from (3.22). This arrangement allowed many breakdowns to be observed, with the results in Table 3.20. Cases 1–3 were the most common, making up roughly $\geq 75\%$ of the observed breakdowns. The particles were either flying towards or already attached to the cathode before the breakdown. Light emission, believed to indicate the start of the breakdown process, occurred either between the cathode and the particle or from the tip of the particle. Although both [293] and [294] involved particles artificially added to the contact gap, they give critical data on the real behavior of particles and can help guide model

Table 3.20   Motion before and after breakdowns of particles added to Cu contacts, and the sources of light emission during the breakdown process [294]

Case	Particle before breakdown	Location of light emission	Particle after breakdown
1	Move towards cathode	Between cathode and particle	Move away from the cathode
2	Attached to cathode	Between cathode and particle	Attached to cathode
3	Attached to cathode	Tip of particle	Attached to cathode
4	Move towards cathode	Between cathode and particle	Attached to cathode
5	Move towards cathode	Tip of particle	Move away from the cathode
6	Attached to cathode	Between cathode and particle	Move away from the cathode

development. The key observation is that many breakdown paths are possible, but not all with equal probability.

The second idea was to observe the effect of mechanical operations on particle generation and breakdown. A voltage was applied to the Cu–Cr contacts after a mechanical close and open operation, and the number of particles measured, Figure 3.39 [293]. The voltage is compared to the value that the gap withstood before the mechanical operations. The first operation had the highest number of particles and broke down at 50–70% of the previously withstood voltage. The number of particles decreased with the subsequent operations and the gap did not break down. When the closing velocity was doubled, the process restarted.

The final idea was to look at the behavior after the current interruption. Reference [293] gave one example where the motion of a particle in between the contacts correlated with the breakdown. However, only this one case was reported. Later work looked at small Ag–WC contacts with a small diameter of $D = 10$ mm [295]. Ag–WC material combined with the small contact size drove the formation of particles during the arcing. Many particles were generated, but two types led to breakdowns. The first was particles with a diameter of $\sim 20-40$ μm and a speed of ~ 15 m/s, and the second was particles with a diameter of ~ 100 μm and a speed of ~ 1 m/s.

All these experiments were specifically designed to maximize the production of particles and thereby enable the researchers to gather data. Therefore, the direct application to Cu–Cr contacts in capacitor switching tests is complex, where a design that fails the test may have only 2 re-strikes over several tens of operations. Nevertheless, the research provides vital guidance to theories of particle-based vacuum breakdown for capacitor switching. Interesting work also looked at

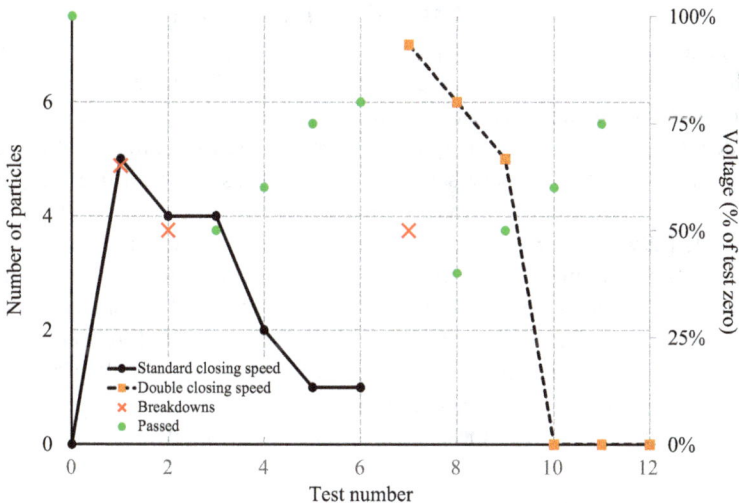

Figure 3.39   Number of particles observed and the breakdown behavior of an open Cu–Cr contact gap after closing operations at different speeds [293]

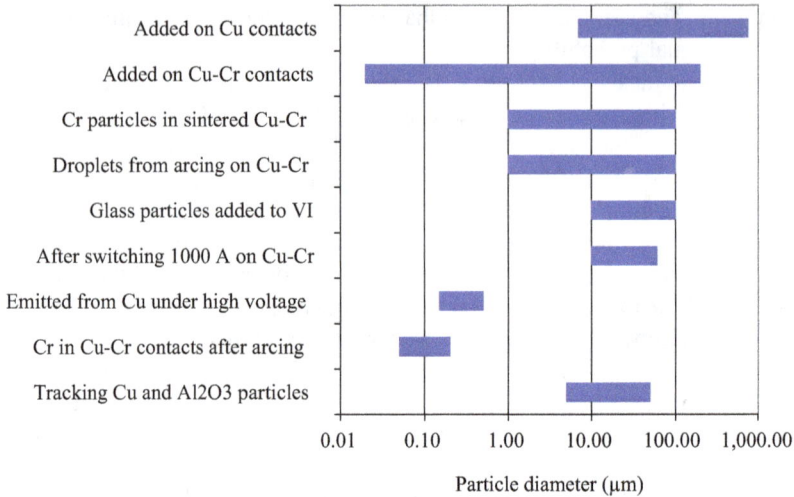

Added on Cu contacts
Added on Cu-Cr contacts
Cr particles in sintered Cu-Cr
Droplets from arcing on Cu-Cr
Glass particles added to VI
After switching 1000 A on Cu-Cr
Emitted from Cu under high voltage
Cr in Cu-Cr contacts after arcing
Tracking Cu and Al2O3 particles

0.01   0.10   1.00   10.00   100.00   1,000.00

Particle diameter (μm)

Figure 3.40   Comparison of the diameter of particles added to or observed in various experiments [241,293]

measuring the influence of the in-rush current on particle generation for Cu–Cr10 and Cu–Cr50 contacts [296]. The number of particles increased linearly with the in-rush current, for the in-rush currents above 4 kA peak. Hopefully, this research into the breakdown process during capacitor switching will continue, as this is a key challenge for HV VI.

3.6.4.2   Generation of droplets and contact erosion

The cathode spots are a source of particle generation from molten droplets ejected from the spots. These droplets were extensively studied for vacuum arcs with a few cathode spots for the vacuum coating application [297]. The droplets created defects in the deposited coatings and many ideas were explored to eliminate them. In VI, molten droplets are the primary source of contact erosion in diffuse arcs, see for example [298], and the major source of contact material deposition on the arc-facing components [299]. Numerous publications over the years measured various droplet parameters. Some recent work made very detailed measurements of the size, angular distribution, and velocity of droplets from vacuum arcs on Cu contacts [300,301]. Modeling of the droplet ejection process is also proceeding, with recent work modeling the behavior of the cathode spot including the ejection of molten droplets at a shallow angle relative to the contact surface [249]. It will be very interesting to see if detailed comparisons between these experiments and the modeling can be made.

3.6.5   Contact material

3.6.5.1   Cu–Cr and contact material production

For MV, the dominant contact material is Cu–Cr [4]. Other materials are used, specifically Cu–W for load break and capacitor switches [302] and Ag–WC for contactor applications [4], and there is some revived interest in Cu–WC [303], and some exploration of Cu–W/WC [184]. Nevertheless, to the author's knowledge all MV VI manufacturers use Cu–Cr for circuit breaker applications. The reason for this situation is that the contacts must perform several duties well. Table 3.21 summarizes the different properties and compares the key contact materials [304]. In addition to the dielectric and switching performance, the contact material must work in the manufacturing process. It must be vacuum compatible, reasonable to machine especially for complex spiral contact designs, and compatible with the brazing assembly process. Hence the limited number of options.

Because of the trauma from the high chop current for Cu–Bi contacts, some work did look at using Ag–WC for MV circuit breakers to take advantage of the low chop current [305], despite extensive prior experience with Cu–Cr at this manufacturer [306,307]. Some efforts were even spent looking at the possibility to combine the best of Cu–Cr and Ag–WC contacts into a new material, Ag–Cr [299]. However, the MV contact material of choice remained Cu–Cr for circuit breakers. HV vacuum circuit breakers have focused on Cu–Cr as the contact material.

Currently, all practical contact materials and materials under investigation are composites, where two or more metal components do not alloy. Therefore, the manufacturing of Cu–Cr and all other contact materials, regardless of the exact method, start with separate powders of the individual components. The details of the powder manufacture and powder size play a role in the performance of the Cu–Cr contacts [308–310].

Three main techniques are used to manufacture contacts [311]. The first method is infiltration [312], where the higher melting point component is compacted under high mechanical pressure into a skeleton. The other component is then melted and allowed to penetrate the skeleton and re-solidify. The second method is sintering. Here, the two (or more) powders are mixed, mechanically pressed, and then heated in a vacuum or a neutral gas oven. In solid phase sintering, the temperature is below, but close to, the melting point of the lower melting point powder [313]. Although the powder does not melt, the heat gives enough energy for the

Table 3.21   Comparison of the performance of three modern (Cu–Cr, WC–Ag, and W–Cu) contact materials and one obsolete material (Cu–Bi) [304]

Material	Current interruption	Electrical life	Dielectric strength	Chop current	Weld resistance
Cu–Cr	••••	••••	••••	•••	••
WC–Ag	•	•••	•••	••••	••••
W–Cu	•	••	••••	•••	•••
Cu–Bi	••	•	•	•	•••

• poor
•• acceptable
••• good
•••• excellent

powder grains to change shape and bind. In liquid phase sintering, the temperature exceeds the melting point of one of the powders [314]. This is like infiltration; however, it generally gives a more uniform distribution of the second component and can often use smaller powder grains. Sintering techniques for Cu-Cr advanced to the point that even complex RMF contacts can be directly shaped without or with limited machining of the final part [315].

The final technique is casting [316]. Here, the temperature rises above the melting point of all the powders. Although the component powders do not alloy, this method allows for a fine dispersion of small Cr particles in the copper [317]. The temperature can be provided through heating or by an arc which locally melts the powders. Infiltration is often used for tungsten-based contacts while sintering and casting are the main methods for Cu–Cr.

3.6.5.2   Effect on performance

Within the Cu–Cr monopoly for vacuum circuit breakers, the two key issues are the Cr percentage and the manufacturing method. The Cr percentage typically ranges from 20% to 50%. Higher and lower values are possible but tend to be challenging to manufacture and have negative effects on performance. The interruption performance increases with lower Cr % [93,318]. For capacitor switching, the manufacturing methods influence the performance [319,320] while increasing the Cr percentage and/or modifying the contact production method can significantly reduce the welding from the in-rush current [167]. Some limited data support the theory that cast contacts can be better than sintered contacts at capacitor switching [321].

Given the importance of the contact material on the performance, many authors examined the changes in short-circuit current interruption performance from Cr content, manufacturing methods, O_2 content, powder manufacturing techniques, etc. Although some tests used MV VI at high short-circuit current with the corresponding arc control method [93,318], the expense and difficulty with securing enough testing time for a meaningful comparison prompted others to look for simpler test methods. Synthetic test methods were used, together with small diameter contacts from 12 to 35 mm that then required modest short-circuit currents of 3.5–11 kA [311,322]. Further work looked at using a high-frequency current to compare contact materials [323]. These test methods with lower short-circuit current or high-frequency current have the advantage of reducing the size of the capacitor bank supplying the current. The disadvantage is that aside from very low short-circuit currents of 4–10 kA, all VI for circuit breaker applications use arc control. Translating results when the arc is in the natural diffuse mode or a stationary constricted arc to the arc modes seen with AMF or RMF contact designs at high currents is not straightforward. Also, while current interruption is a very important performance parameter, it is not the only duty required of a VI and the other duties must be checked when making significant changes to the contact material.

3.6.5.3   Additives to Cu–Cr contacts

Given the success of Cu–Cr, many efforts looked at improving the performance with additives of other materials to the contacts. Reference [304] summarizes several different additives and their proposed advantages. A quick review of the patent literature on VI Cu–Cr contacts will identify numerous attempts to improve Cu–Cr with additives. One example serves to illustrate the challenges [324]. The patent summarizes some previous work adding a metal with a higher vapor pressure to reduce the chopping current. Since chop current is not the real source of over-voltages, the value of this "improvement" is questionable, perhaps they also hoped to reduce the occurrence of multiple re-ignitions. Nevertheless, while the high vapor pressure material did indeed reduce the chopping current, it also reduced the current interruption performance and even made brazing of the contacts difficult. This patent attempted to get around these problems by applying the high vapor pressure material only to the arc-facing portion of the contact, and not through the entire material. Ultimately this idea was not used in practice. More recently, one of the materials used in the patent, Te, was re-evaluated in very small amounts to try and improve the capacitive switching performance [325,326]. Another interesting description of the challenges of optimizing all the required performance with Cu–Cr–Nb contacts is described in [327].

HV VI could change the dynamic for additives in Cu–Cr. The general issue remains that adding enough of a third material to significantly improve one behavior generally degrades another behavior. The different challenges facing HV VI could make trade-offs feasible, such as improving the dielectric and capacitor switching duties, perhaps with a trade-off in interruption duty and/or manufacturing complexity. Nevertheless, while the author does not want to discourage further research into contact materials in general and additives specifically, it is vital to look at the broader picture when testing new materials. Improved performance in one test duty can only be regarded as a preliminary result until other duties are evaluated, especially ones that could be degraded by an improvement in the first duty.

3.7 Summary

"Vacuum. In high voltage systems. I am not sure about this," paraphrases one comment overheard by the author at a technical meeting. Hopefully, this chapter helps show the exciting possibilities to use VI in HV systems. The application of VI in HV is not new, with about a half-century of field experience. Although VI have so far only been a minor player in HV, there are clear examples of a variety of successful applications. Furthermore, the HV VI is fundamentally based on the MV VI, giving a massive research and application background to draw upon.

The most pressing challenge will be the dielectric performance, particularly in U_p and capacitor switching. Other challenges will also need to be faced, such as continuous current ratings and current interruption at very large contact gaps. Nevertheless, the recent developments in understanding the vacuum arc and vacuum breakdown will help push the performance boundaries of VI, together with new extensions of the MV experience to HV.

This chapter provides many references for readers interested in tracking down the details of the work mentioned here. For readers who want to dive deeper into

Table 3.22　Key references for further reading

Reference	Description
[27]	Encyclopedic reference on MV VI, including extensive and detailed summaries of the published literature.
[20]	Cigre technical report on HV VI, with extensive information on field applications and developments by different manufacturers. Updated information especially on commercial products in [117].
[49]	Summarizes the broad range of HV VI research particularly in China, with a rare chapter on circuit breaker–VI interaction.
[73]	A deep review of type testing of circuit breakers.
[15,45,59,86,328]	Review papers and chapters covering many aspects of MV VI in a compact form.
[329]	Older book with information on the circuit breaker side of vacuum circuit breakers.
ISDEIV conference	Main conference for vacuum arc, vacuum breakdown, and VI research. Proceedings contain many key papers, and many references for this chapter come from these proceedings.

HV VI on a more general level, Table 3.22 provides a summary of some key references.

References

[1] T. W. Dakin, G. Luxa, G. Oppermann, J. Vigreux, G. Wind, and H. Winkelnkemper, "Breakdown of gases in uniform fields: Paschen curves for nitrogen, air and sulfur hexafluoride," *Electra*, no. 32, p. 61, 1974.

[2] K. Becken, D. de Graaf, C. Elsner, *et al.*, *Avoiding Fluorinated Greenhouse Gases: Prospects for Phasing Out, German Federal Environment Agency (Umweltbundesamt)*, Dessau-Roßlau, Germany, 2010.

[3] P. O'Connell, F. Heil, J. Henriot, *et al.* Taillebois, "SF$_6$ in the Electric Industry, Status 2000," *Electra*, no. 200, *p.* 16, 2002.

[4] P. G. Slade, "High current contacts: a review and tutorial," in: *Proceedings of the ICEC 21st International Conference on Electrical Contacts*, 2002.

[5] F. Y. Chu, "SF$_6$ decomposition in gas-insulated equipment," *IEEE Trans. Elect. Insul.*, vol. EI-21, no. 5, *p.* 693, 1986.

[6] B. Belmadani, J. Casanovas, A. M. Casanovas, R. Grob, and J. Mathieu, "SF$_6$ decomposition under power arcs physical aspects," *IEEE Trans. Dielect. Elect. Insul.*, vol. 26, no. 6, *p.* 1163, 1991.

[7] C. T. Dervos and P. Vassiliou, "Sulfur hexafluoride (SF$_6$): global environmental effects and toxic byproduct formation," *J. Air Waste Manag.*, vol. 50, p. 137, 2000.

[8] E. D. Taylor, D. Gentsch, and M. Leusenkamp, "Field experience and handling of vacuum switching devices," *Cigre Tutorial, Paris Session*, 2022.

[9]   G. A. Farrall and J. D. Cobine, "Recovery strength measurements in arcs from atmospheric pressure to vacuum," *IEEE Trans. Power App. Syst.*, vol. PAS-86, no. 8, p. 927, 1967.

[10] R. Hackam, "Effects of voltage polarity, electric current, external resistance, number of sparkings, supply frequency, & addition of hydrogen and air on electrical breakdown in vacuum," *J. Appl. Phys.*, vol. 46, no. 9, p, 3789, 1975.

[11] R. W. Sorensen and H. E. Mendenhall, "Vacuum switching experiments at California Institute of Technology," *Trans. AIEE*, p. 1102, 1926.

[12] P. A. Redhead, "The ultimate vacuum," *Vacuum*, vol. 53, p. 137, 1999.

[13] J. D. Cobine, "Research and development leading to the high-power vacuum interrupter – an historical review," *IEEE Trans. Power App. Syst.*, vol. 82, no. 64, p. 201, 1963.

[14] T. H. Lee, A. Greenwood, D. W. Crouch, and C. H. Titus, "Development of power vacuum interrupters," *Trans. AIEE Part III: Power App. Syst.*, vol. 81, no. 3, p. 629, 1963.

[15] R. Renz, *Vacuum Interrupters. In Vacuum Electronics*: *Components and Devices, Chapter 8*, Springer, 2008.

[16] H. C. Ross, "Vacuum power switches: 5 Years of field application and testing," *Trans. AIEE Part III: Power App. Syst.*, vol. 71, p. 758, 1961.

[17] H. Ross, "Switching in vacuum – Interruption of high voltage power, at moderate currents to 4000 amperes and possibly 10,000 amperes or more," in: *Proceedings of International Research Symposium on Electrical Contact Phenomena*, 1961, p. 345.

[18] P. G. Slade, P. O. Wayland, R. E. Voshall, *et al.*, "The development of a vacuum interrupter retrofit for the upgrading and life extension of 121 kV–145 kV oil circuit breakers," *IEEE Trans. Power Del.*, vol. 6, no. 3, p. 1124, 1991.

[19] L. Falkingham and M. Waldron, "Vacuum for HV applications – Perhaps not so new? – Thirty years service experience of 132kV vacuum circuit breaker," in: *Proceedings of 22nd ISDEIV International Symposium on Discharges and Electrical Insulation in Vacuum*, 2006.

[20] R. Smeets, Cigre Working Group A3.27, "The Impact of the Application of Vacuum Switchgear at Transmission Voltages," *Cigre Technical Report #589*, 2014.

[21] Y.Matsui, K.Nagatake, M.Takeshita, *et al.*, "Development and technology of high voltage VCBs; brief history and state of art," in: *Proceedings of 22nd ISDEIV International Symposium on Discharges and Electrical Insulation in Vacuum*, 2006.

[22] R. E. Voshall, C. W. Kimblin, P. G. Slade, and J. G. Gorman, "Experiments on vacuum interrupters in high voltage 72 kV circuits," *IEEE Trans. Power App. Syst.*, vol. PAS-99, no. 2, p. 658, 1980.

[23] H. Komatsu, K. Saito, and T. Furuhata, "Progress of vacuum interrupt (VI) and recent technical trends," *Meiden Rev. Ser.*, vol. 163, no. 1, p. 8, 2015.

[24] R. Renz, "High voltage vacuum interrupters – technical and physical feasibility versus economical efficiency," in: *Proceedings of 22nd ISDEIV*

International Symposium on Discharges and Electrical Insulation in Vacuum, 2006.

[25] www.eea.europa.eu. "Regulation (EU) No 517/2014 of the European Parliament and of the Council of 16 April 2014 on fluorinated greenhouse gases," accessed 16 Dec. 2022. Available: https://www.eea.europa.eu/ds_resolveuid/b471e1af4e06431c8048970f6c992099.

[26] F. Richter, D. Makareinis, J. Teichmann, F. Reincke, R. Mross, and S. Giere, "Comparison of switching behavior of 145 kV vacuum and SF6 circuit-breakers in the case of switching off shunt reactor currents," in: *Proceedings of Cigre*, paper 432, 2013.

[27] P. G. Slade, *The Vacuum Interrupter: Theory, Design, and Application*. New York, NY, USA: CRC Press, 2021.

[28] Y. T. Sasaki, "A survey of vacuum material cleaning procedures: A sub-committee report of the American Vacuum Society Recommended Practices Committee," *J. Vac. Sci. Technol. A*, vol. 9, no. 3, p. 2025, 1991.

[29] I. Nuta, C. Chatillon, F. Chombart, and A. Moreau, "Thermodynamic evaluation of oxidation during brazing process of medium-voltage electrical circuit breakers," *Metall. Res. Technol.*, vol. 113, p. 510, 2016.

[30] E. D. Taylor, A. Lawall, and D. Gentsch, "Long-term vacuum integrity of vacuum interrupters," in: *Proceedings of 26th ISDEIV International Symposium on Discharges and Electrical Insulation in Vacuum*, 2014, p. 433.

[31] S. Kobayashi, "Recent experiments on vacuum breakdown of oxygen-free copper electrodes," *IEEE Trans. Dielectr. Electr. Insul.*, vol. 4, no. 6, p. 841, 1997.

[32] W. Kuhl and K. Wiehl, "Device for measuring the internal pressure of vacuum interrupters," *Siemens Rev.*, vol. 45, no. 2, p. 87, 1978.

[33] K. Hencken, T. Kaufman, D. Gentsch, and T. Delachaux, "Modeling and simulation of magnetron discharges inside a vacuum interrupter as a method to analyze the vacuum status," in: *Proceedings of 27th ISDEIV International Symposium on Discharges and Electrical Insulation in Vacuum*, 2016, p. 529.

[34] J. C. Helmer and R. L. Jepsen, "Electrical characteristics of a penning discharge," in: *Proceedings of IRE*, vol. 49, no. 12, p. 1920, 1961.

[35] G. Barnes, J. Gaines, and J. Kees, "Relative sensitivity and pumping rate of the redhead magnetron gauge," *Vacuum*, vol. 12, no. 3, p. 141, 1962.

[36] J. F. O'Hanlon, *A User's Guide to Vacuum Technology*, Hoboken: John Wiley, 2003.

[37] K. Kageyama, "Magnetically confined low-pressure gas discharge generated in a vacuum switch," *J. Vac. Sci. Technol. A*, vol. 1, no. 3, p. 1522–1528, 1983.

[38] J. A. Fedchak, P. J. Abbott, J. H. Hendricks, P. C. Arnold, and N. T. Peacock, "Review article: recommended practice for calibrating vacuum gauges of the ionization type," *J. Vac. Sci. Technol. A*, vol. 36, no. 3, 030802-1, 2018.

[39] www.pfeiffer-vacuum.com. "5.1.2 Indirect, gas-dependent pressure measurement," accessed 29 Dec. 2022. Available: https://www.pfeiffer-vacuum.com/en/know-how/vacuum-measuring-equipment/fundamentals-of-total-pressure-measurement/indirect-gas-dependent-pressure-measurement/

[40] K. Kageyama, "Properties of a series crossed-field discharge in a vacuum switch," *J. Vac. Sci. Technol. A*, vol. 1, no. 3, p. 1529–1532, 1983.

[41] A. Vesel, M. Mozetic, M. Zumer, V. Nemanic, and B. Zajec, "Pressure/current characteristics of a magnetron cold cathode gauge," *Vacuum*, vol. 78, p. 13, 2005.

[42] R. N. Peacock and N. T. Peacock, "Plots of gauge constant vs pressure for analyzing cold-cathode gauge calibration data," *Vacuum*, vol. 45, no. 10–11, p. 1055, 1994.

[43] X. Godechot, C. Nicolle, M. Hairour, S. Olive, and P. Picot, "Investigation and optimization of magnetron discharge in a vacuum switch," in: *Proceedings of 25th ISDEIV International Symposium on Discharges and Electrical Insulation in Vacuum*, 2012, p. 465.

[44] R. Renz, "On criteria of optimized application of AMF and RMF contact systems in vacuum interrupters," in: *Proceedings of 19th ISDEIV International Symposium on Discharges and Electrical Insulation in Vacuum*, 2000, p. 176.

[45] P. Picot, "Vacuum switching," *Schneider Electric, Cahier Technique* no. 198, 2000.

[46] H. Urbanek, K. R. Venna, and N. Anger, "Vacuum circuit breakers – promising switching technology for pumped storage power plants up to 450MVA," in: *Proceedings of the 4th ICEPE International Conference on Electric Power Equipment – Switching Technology*, 2017, p. 107.

[47] S. Yanabu, S. Souma, T. Tamagawa, S. Yamashita, and T. Tsutsumi, "Vacuum arc under an axial magnetic field and its interrupting ability," *Proc. IEE*, vol. 126, no. 4, p. 313, 1979.

[48] S. Giere, T. Heinz, A. Lawall, C. Stiehler, E. D. Taylor, and S. Wethekam, "Control of diffuse vacuum arc using axial magnetic fields in commercial high voltage switchgear," *Plasma Phys. Technol.*, vol. 6, no. 1, p. 19, 2019.

[49] Z. Liu, J. Wang, Y. Geng, and Z. Wang, *Switching Arc Phenomena in Transmission Voltage Level Vacuum Circuit Breakers*. Singapore: Springer Nature, 2021.

[50] R. K. Smith, R. W. Long, and D. L. Burmingham, "Vacuum interrupters for generator circuit breakers: They're not just for distribution breakers anymore," in: *Proceedings of CIRED 17th International Conference on Electrical Distribution*, Session 1, Paper No. 5, 2003.*

[51] S. Yanabu, E. Kaneko, H. Okumura, and T. Aiyoshi, "Novel electrode structure of vacuum interrupter and its practical application," *IEEE Trans. Power App. Syst.*, vol. PAS-100, no. 4, p. 1966, 1981.

[52] E. Dullni, E. Schade, and W. Shang, "Vacuum arcs driven by cross-magnetic fields (RMF)," *IEEE Trans. Plasma Sci.*, vol. 31, no. 5, p. 902, 2003.

[53] P. G. Slade and R. E. Voshall, "Effects of arc shield proximity to the electric contacts on the current interruption capability of vacuum inter- rupters," *Elektrotec. Inf.*, vol. 107 Jg., H. 3, p. 138, 1990.

[54] High-Voltage Switchgear and Controlgear—Part 1: Common Specifications for Alternating Current Switchgear and Controlgear, IEC Standard 62271-1, ed. 2.1, 2021.

[55] W. Li, L. G. Campbell, G. Balasubramanian, and L. D. Loud, "Measuring external dielectric strength of the vacuum interrupter envelope," in: *Proceedings of 28th ISDEIV International Symposium on Discharges and Electrical Insulation in Vacuum*, 2018, p. 531.

[56] C. T. Lav, D. B. Staley, and T. W. Olsen, "Practical design considerations for application of GIS MV switchgear," *IEEE Trans. Ind. Appl.*, vol. 40, no. 5, p. 1427, 2004.

[57] M. B. J. Leusenkamp, J. H. L. A. Hilderink, and K. Lenstra, "Field calcu- lations on epoxy resin insulated vacuum interrupters," in: *Proceedings of 17th ISDEIV International Symposium on Discharges and Electrical Insulation in Vacuum*, 1996, p. 1065.

[58] B. H. Krafft and K. Mascher, "Vacuum circuit-breaker and a method for controlling the same," U.S. Patent Application 2004/0061504, 1 April 2004.

[59] K. Fröhlich, H. C. Kärner, D. König, M. Lindmayer, K. Möller, and W. Rieder, "Fundamental research on vacuum interrupters at technical uni- versities in Germany and Austria," *IEEE Trans. Electr. Insul.*, vol. 28, no. 4, p. 592, 1993.

[60] H. Schellekens, M.-F. Devismes, and C. Nicolle, "Vacuum interrupters: Design optimization requires reliable experimental data," in: *Proceedings of 1st ICEPE International Conference on Electric Power Equipment – Switching Technology*, 2011, p. 56.

[61] E. D. Taylor, M. Eiselt, R. Tschiesche, U. Suresh, and T. Brauner, "Electrical life of vacuum interrupters for load current switching," *IEEE Trans. Plasma Sci.*, vol. 50, no. 9, p. 2642, 2022.

[62] M. Schlaug, L. Dalmazio, U. Ernst, and X. Godechot, "Electrical life of vacuum interrupters," in: *Proceedings of 22nd ISDEIV International Symposium on Discharges and Electrical Insulation in Vacuum*, 2006.

[63] H. A. Fohrhaltz, "Load tap changing with vacuum interrupters," *IEEE Trans. Power App. Syst.*, vol. PAS-86, no. 4, p. 422, 1967.

[64] D. Dohnal, *On-Load Tap-Changers for Power Transformers: A Technical Digest*. Regensburg, Germany: Maschinenfabrik Reinhausen, 2009.

[65] Ted Olsen, "Arc-furnace switching applications," Siemens TechTopics, No. 65, Accessed 22 Nov. 2022. Available: https://new.siemens.com/us/en/ products/energy/techtopics/techtopics-65.html.

[66] A. Bianco, B. Brewer, M. Stefanka, and M. Riva, "High performance smart MV apparatus for arc furnace applications," in: *Proceedings of 25th International Conference on Electricity Distribution CIRED*, 2019, paper 0883.

[67] W. R. Wilson, "High-current arc erosion of electrical contact materials," *Trans. AIEE III, Power App. Syst.*, vol. 74, no. 3, p. 657, 1955.

[68] T. Rettenmaier, "Untersuchung der bewegungscharakteristik vonSschaltlichtbögen in handelsüblichen Vakuumschaltröhren der Mittelspannungsebene," Ph.D. dissertation, Fachbereich Elektrotechnik und Informationstechnik, Darmstadt, Germany, 2015.

[69] H. Janssen, V. Hinrichsen, E. D. Taylor, and T. Rettenmaier, "Comparison of arc motion in different lifetime sealed vacuum interrupters with RMF contacts," *IEEE Trans. Plasma Sci.*, vol. 47, no. 8, p. 3525, 2019.

[70] High-Voltage Switchgear and Controlgear—Part 100: Alternating-Current Circuit-Breakers, IEC Standard 62271-100, ed. 3.0, 2021.

[71] High-Voltage Alternating-Current Circuit-Breakers, GB Standard 1984, 2014.

[72] Working Group 13.08, "Life management of circuit-breakers," CIGRE, Paris, France, Tech. Rep. 165, 2000.

[73] R. Smeets, L. Sluis, M. Kapetanovic, D. F. Peelo, and A. Janssen, *Switching in Electrical Transmission and Distribution Systems*, Chichester: Wiley, 2015.

[74] A. E. Geisler and N. Wenzel, "Metal vapor deposition in vacuum interrupters: Utilization of DSMC methods and considerations on vapor sources," in: *Proceedings of 28th ISDEIV International Symposium on Discharges and Electrical Insulation in Vacuum*, 2018, p. 479.

[75] P. G. Slade and R. K. Smith, "Electrical switching life of vacuum circuit breaker interrupters," in: *Proceedings of 52nd IEEE Holm Conference on Electrical Contacts*, Sep. 2006, p. 32.

[76] E. D. Taylor, S. A. Baus, and A. Lawall, "Increase in contact resistance of vacuum interrupters after short-circuit testing," in: *Proceedings of 27th ICEC International Conference on Electrical Contacts*, 2014, p. 203.

[77] E. Dullni, D. Gentsch, W. Shang, and T. Delachaux, "Resistance increase of vacuum interrupters due to high-current interruptions," *IEEE Trans. Dielectr. Electr. Insul.*, vol. 23, no. 1, p. 1, 2016.

[78] U. Ernst, K. Cheng, X. Godechot, and M. Schlaug, "Dielectric performance of vacuum interrupters after switching," in: *Proceedings of 22nd International Symposium on Discharges and Electrical Insulation in Vacuum*, 2006, p. 17.

[79] X. Yao, J. Wang, Y. Geng, J. Yan, Z. Liu, and J. Yao, "Development and type test of a single-break 126-kV/40-kA–2500-A vacuum circuit breaker," *IEEE Trans. Power Del.*, vol. 31, no. 1, p. 182, 2016.

[80] J. Teichmann, S. Kosse, M. Koletzko, *et al.*, "145/170 kV vacuum circuit breakers and clean-air instrument transformers – product performance and first installations in AIS substations," in: *Proceedings of Cigre* Session, 2018, paper A3-311.

[81] I. Benfatto, A. Machio, and S. Manganaro, "DC breaking tests up to 55 kA in a single vacuum interrupter," *IEEE Trans. Power Del.*, vol. 3, no. 4, p. 1732, 1988.

[82] I. Benfatto, A. DeLorenzi, A. Maschio, W. Weigand, H. P. Timmert, and H. Weyer, "Life tests on vacuum switches breaking 50 kA unidirectional current," *IEEE Trans. Power Del.*, vol. 6, no. 2, p. 824, 1991.

[83] T. Bonicelli, A. DeLorenzi, D. Hrabal, *et al.*, "The European development of a full scale switching unit for the ITER switching and discharging networks," *Fusion Eng. Des.*, vol. 75–79, p. 193, 2005.

[84] R. Piovan, L. Zanotto, and T. Bonicelli, "Vacuum breaker for high DC current: experimental performances and operational limits," *IEEE Trans. Plasma Sci.*, vol. 37, no. 1, p. 229, 2009.

[85] A. Zamengo, "Operational and performance overview of the 50 kA–35 kV RFX-Mod DC-current interruption system," *IEEE Trans. Plasma Sci.*, vol. 47, no. 11, p. 5198, 2019.

[86] E. Schade, "Physics of high-current interruption of vacuum circuit breakers," *IEEE Trans. Plasma Sci.*, vol. 33, no. 5, p. 1564, 2005.

[87] E. D. Taylor, K. R. Venna, D. Gentsch, A. Lawall, and S. Wethekam, "Behavior of vacuum interrupters during switching operations with a high rate of rise of recovery voltage (RRRV)," in: *Proceedings of 29th ISDEIV International Symposium on Discharges and Electrical Insulation in Vacuum*, 2020, p. 367.

[88] High-voltage switchgear and controlgear – Part 37-013: Alternating-current generator circuit-breakers, IEC/IEEE Standard 62271-37-013, Edition 2.0, 2021.

[89] R. P. P. Smeets and W. A. van der Linden, "Current-zero measurements of vacuum circuit breakers interrupting short-line faults," *IEEE Trans. Plasma Sci.*, vol. 31, no. 5, p. 852, 2003.

[90] A. Lee and L. S. Frost, "Interruption capability of gases and gas mixtures in a puffer-type interrupter," *IEEE Trans. Plasma Sci.*, vol. PS-8, no. 4, p. 362, 1980.

[91] W.G. Heinmiller, R. W. Katterhenry, S. R. Lambert, and T. W. Stringer, "Transient recovery voltage failures of two 15 kV indoor oilless circuit breakers," *IEEE Trans. Power App. Syst.*, vol. PAS-102, no. 8, p. 2578, 1983.

[92] M. B. Schulman and R. K. Smith, "Better interpretation of vacuum interrupter tests," *IEEE Industry Appl. Mag.*, vol. 5, no. 2, p. 18, 1999; Errata: vol.5, no. 4, p. 8, 1999.

[93] T. Delachaux, F. Rager, R.A. Simon, T. Schmoelzer, M. Boehm, and D. Gentsch, "Development of a simple procedure for screening current interruption capability of vacuum interrupter contact materials," in: *Proceedings of 26th ISDEIV International Symposium on Discharges and Electrical Insulation in Vacuum*, 2014, p. 413.

[94] C. R. Bayliss and B. J. Hardy, *Transmission and Distribution Electrical Engineering*, Oxford: Elsevier, 2012.

[95] Working Group A3.11, "Guide for the application of IEC 62271-100 and IEC 62271-1 Part 2 – Making and Breaking Tests," *Cigre Technical Brochure* 305, 2006.

[96]   E. D. Taylor, J. Oemisch, M. Eiselt, and M. Hinz, "Performance of vacuum interrupters in electrical power systems with an effectively earthed neutral," in: *Proceedings of 27th ISDEIV International Symposium on Discharges and Electrical Insulation in Vacuum*, 2016, p. 513.

[97]   E. D. Taylor, A. Lawall, and D. Gentsch, "Single-phase short-circuit testing of vacuum interrupters for power systems with an effectively earthed neutral," in: *Proceedings of 28th ISDEIV International Symposium on Discharges and Electrical Insulation in Vacuum*, 2018, p. 579.

[98]   E. D. Taylor, A. Lawall, J. Genzmer, and T. Heydenreich, "Effect of arcing time on the current breaking capability of vacuum interrupters," in: *Proceedings of 2nd ICEPE International Conference on Electric Power Equipment – Switching Technology*, 2013, p. 1.

[99]   Test Procedures for AC High-Voltage Circuit Breakers with Rated Maximum Voltage Above 1000 V, IEEE Standard C37.09, 2018.

[100] High-voltage switchgear and controlgear – Part 101: Synthetic testing, IEC Standard 62271-101, ed. 3.0, 2021.

[101] D. Dufournet and G. Montillet, "Three-phase short circuit testing of high-voltage circuit breakers using synthetic circuits," *IEEE Trans. Power Del.*, vol. 15, no. 1, p. 142, 2000.

[102] W. Shi, B. Zhang, J. Wang, *et al.*, "Minimum arcing interrupting capability and opening velocity of vacuum interrupters: an impact of magnetic field," in: *Proceedings of 28th ISDEIV International Symposium on Discharges and Electrical Insulation in Vacuum*, 2018, p. 271.

[103] W. Shi, B. Zhang, C. Guan, *et al.*, "Influence of opening velocity on arcing time windows of fast vacuum circuit breaker in duties of terminal fault test T100s," in: *Proceedings of 5th ICEPE International Conference on Electric Power Equipment – Switching Technology*, 2019, p. 140.

[104] C. Guan, X. Yao, J. Zheng, *et al.*, "A relationship between vacuum arc characteristics and short-circuit current interrupting capability at minimum arcing times," *IEEE Trans. Plasma Sci.*, vol. 50, no. 9, p. 2670, 2022.

[105] J. Panek, "Synthetic methods for testing of vacuum breakers," *IEEE Trans. Power App. Syst.*, vol. PAS-97, no. 4, p. 1328, 1978.

[106] K. R. Venna, A. Lawall, D. Gentsch, and M. Leusenkamp, "Suitability of vacuum circuit breakers for generator switching applications according to dual logo standard IEC/IEEE 62271-37-013," in: *Proceedings of 5th ICEPE International Conference on Electric Power Equipment – Switching Technology*, 2019, p. 130.

[107] cesi.it. "CESI acquires KEMA Labs," accessed 23 Nov. 2022. Available: https://www.cesi.it/news/2019/cesi-acquires-kema-labs/

[108] cesi.it. "IPH Berlin," accessed 23 Nov. 2022. Available: https://www.cesi.it/about-us/overview/iph-gmbh/.

[109] B.-C. Kim, S.-T. Kim, K.-Y. Ahn, and J.-H. Lee, "Study on electrical lifespan of VI by means of calculation of arc energy during arcing time in synthetic tests," in: *Proceedings of 26th ISDEIV International Symposium on Discharges and Electrical Insulation in Vacuum*, 2014, p. 153.

[110] High-voltage switchgear and controlgear – Part 111: Automatic circuit reclosers for alternating current systems up to and including 38 kV, IEC Standard 62271-111/IEEE Standard C37.60, ed. 3.0, 2019.

[111] www.gwelectric.com. "The World's First High-Voltage Pole Top Recloser," accessed 24 Nov. 2022. Available: https://www.gwelectric.com/products/distribution-reclosers-and-overhead-switches/introducing-viper-hv-recloser/.

[112] P. G. Slade and R. K. Smith, "The use of vacuum interrupters to control short circuit currents and the probability of interruption at the first current zero after contact part," in: *Proceedings of 9th International Conference on Switching Arc Phenomena*, Lodz, Poland, 2001, p. 54.

[113] E. D. Taylor, A. Lawall, J. Genzmer, and T. Heydenreich, "Current interruption performance of axial and radial magnetic field vacuum interrupters," in: *Proceedings of 28th ISDEIV International Symposium on Discharges and Electrical Insulation in Vacuum*, 2018, p. 571.

[114] A. Henon, T. Altimani, P. Picot, and H. Schellekens, "3D finite element simulation and synthetic tests of vacuum interrupters with axial magnetic field contacts," in: *Proceedings of 20th ISDEIV International Symposium on Discharges and Electrical Insulation in Vacuum*, 2002, p. 463.

[115] M. B. J. Leusenkamp, "Vacuum interrupter model based on breaking tests," *IEEE Trans. Plasma Sci.*, vol. 27, no. 4, p. 969, 1999.

[116] B. Bridger Jr., T. A. Burse, and M. W. Wactor, "Design considerations for 38 kV metal-clad switchgear using vacuum interrupting technology," in: *Proceedings of IEEE/PES Transmission and Distribution Conference*, 1994, p. 15.

[117] R. P. P. Smeets, Working Group A3.41, "Current interruption in SF6-free switchgear," *Cigre Tech. Rep.*, 871, 2022.

[118] X. Godechot, S. Chakraborty, A. Girodet, and P. Vinson, "Design and tests of vacuum interrupters for high voltage circuit breakers," in: *Proceedings of 26th ISDEIV International Symposium on Discharges and Electrical Insulation in Vacuum*, 2014, p. 417.

[119] K. Ma, X. Yao, S. Ai, *et al.*, "Development and test of a 252 kV multi-breaks Bus-tie fast vacuum circuit breaker," in: *Proceedings of 5th ICEPE International Conference on Electric Power Equipment – Switching Technology*, 2019, p. 590.

[120] X. Yao, C. Guan, C. Wang, *et al.*, "Design and verification of an ultra-high voltage multiple-break fast vacuum circuit breaker," *IEEE Trans. Plasma Sci.*, vol. 37, no. 5, p. 3436, 2022.

[121] M. B. Schulman, P. G. Slade, and J. V. R. Heberlein, "Effect of an axial magnetic field upon the development of the vacuum arc between opening electric contacts," *IEEE Trans. Comp., Hybrids, Manufact. Technol.*, vol. 16, no. 2, p. 180, 1993.

[122] H. C. W. Gundlach, "Interaction between a vacuum arc and an axial magnetic field," in: *Proceedings of 8th ISDEIV International Symposium on Discharges and Electrical Insulation in Vacuum*, 1978, p. A2-1.

[123] A. M. Chaly, A. A. Logatchev, K. K. Zabello, and S. M. Shkol'nik, "High-current vacuum arc in a strong axial magnetic field," *IEEE Trans. Plasma Sci.*, vol. 35, no. 4, p. 939, 2007.

[124] M. Keidar and M.B. Schulman, "On the effect of an axial magnetic field on the high-current vacuum arc," *IEEE Trans. Plasma Sci.*, vol. 28, no. 1, p. 347, 2000.

[125] M. Keidar and E. D. Taylor, "A generalized criterion of transition to the diffuse column vacuum arc," *IEEE Trans. Plasma Sci.*, vol. 37, no. 5, p. 693, 2009.

[126] N. Wenzel, A. Lawall, U. Schümann, and S. Wethekam, "Combined experimental and theoretical study of constriction threshold of large-gap AMF vacuum arcs," in: *Proceedings of 26th ISDEIV International Symposium on Discharges and Electrical Insulation in Vacuum*, 2014, p. 193.

[127] E. D. Taylor, "Visual measurements of plasma arc modes in a high-current vacuum arc with an axial magnetic field," in: *Proceedings of 49th Holm Conference Electrical Contacts*, 2003, p. 70.

[128] E. D. Taylor, P. G. Slade, and M. B. Schulman, "Transition to the diffuse mode for high-current drawn arcs in vacuum with an axial magnetic field," *IEEE Trans. Plasma Sci.*, vol. 31, no. 5, p. 909, 2003.

[129] B. Fenski and M. Lindmayer, "Vacuum interrupters with axial field contacts: 3-d finite element simulation and switching experiments," *IEEE Trans. Dielectr. Electr. Insul.*, vol. 4, no. 4, p. 407, 1997.

[130] Z. Zhang, Z. Liu, H. Ma, *et al.*, "Anode input heat flux density in high-current vacuum arcs as anode spot formation," *IEEE Trans. Plasma Sci.*, vol. 50, no. 10, p. 3732, 2022.

[131] S. M. Shkolnik, "Processes on contacts and thermal state of contact surfaces in AMF-contact systems of medium-voltage vacuum interrupters when switching the limiting current: a review," *IEEE Trans. Plasma Sci.*, vol. 50, no. 9, p. 2634, 2022.

[132] E. D. Taylor and J. Genzmer, "Performance of axial magnetic field vacuum interrupters at various transient recovery voltages and contact gaps," in: Proceedings of 27th ISDEIV International Symposium on Discharges and Electrical Insulation in Vacuum, 2016, p. 165.

[133] W. Shang, H. Schellekens, and J. Hilderink, "Experimental investigations into the arc properties of vacuum interrupters with horseshoe electrode, four pole electrodes, and their application," *IEEE Trans. Plasma Sci.*, vol. 21, no. 5, p. 474, 1993.

[134] H. Li, Z. Wang, Y. Geng, Z. Liu, and J. Wang, "Arcing contact gap of a 126-kV horseshoe-type bipolar axial magnetic field vacuum interrupters," *IEEE Trans. Plasma Sci.*, vol. 46, no. 10, p. 3713, 2018.

[135] H. Schellekens, "50 Years of TMF contacts design considerations," in: *Proceedings of 23rd ISDEIV International Symposium on Discharges and Electrical Insulation in Vacuum*, 2008.

[136] C. Korasli, "Statistical inference for breakdown voltage in SF_6 GIS from first breakdown data," *IEEE Trans. Dielectr. Electr. Insul.*, vol. 5, no. 4, p. 596, 1998.

[137]   E. D. Taylor, J. Genzmer, and A. Schulz, "Vacuum breakdown voltage distributions between different contact arrangements and materials," in: *Proceedings of 5th ICEPE International Conference on Electric Power Equipment – Switching Technology*, 2019, p. 157.

[138]   A. Grudiev, S. Calatroni, and W. Wuensch, "New local field quantity describing the high gradient limit of accelerating structures," *Phys. Rev. ST Accelerators Beams*, vol. 12, p. 102001, 2009.

[139]   A. Descoeudres, T. Ramsvik, S. Calatroni, M. Taborelli, and W. Wuensch, "DC breakdown conditioning and breakdown rate of metals and metallic alloys under ultrahigh vacuum," *Phys. Rev. ST Accelerators Beams*, vol. 12, p. 032001, 2009.

[140]   H. Okubo, H. Kojima, K. Kato, N. Hayakawa, and M. Hanai, "Advanced electrical insulation techniques for higher voltage vacuum interrupters," Cigre, Study Committee B3, Brisbane, Australia, Sept. 2013, paper 218.

[141]   H. Schellekens and G. Gaudart, "Compact high-voltage vacuum circuit breaker, a feasibility study," *IEEE Trans. Dielectr. Electr. Insul.*, vol. 14, no. 3, p. 613, 2007.

[142]   T. Betz and D. König, "Influence of grading capacitors on the breaking capability of two vacuum circuit-breakers in series," *IEEE Trans. Dielectr. Electr. Insul.*, vol. 6, no. 4, p. 405, 1999.

[143]   T. Fugel and D. König, "Switching performance of two 24kV vacuum interrupters in series," *IEEE Trans. Dielectr. Electr. Insul.*, vol. 9, no. 2, p. 164, 2002.

[144]   T. Fugel and D. König, "Influence of grading capacitors on the breaking performance of a 24-kV vacuum breaker series design," *IEEE Trans. Dielectr. Electr. Insul.*, vol. 10, no. 4, p. 569, 2003.

[145]   Y. Shiba, N. Ide, H. Ichikawa, Y. Matsui, M. Sakaki, and S. Yanabu, "Withstand voltage characteristics of two series vacuum interrupters," *IEEE Trans. Plasma Sci.*, vol. 35, no. 4, p. 879, 2007.

[146]   L. Min-fu, D. Xiong-ying, Z. Ji-yan, F. Xing-ming, and S. Hui, "Dielectric strength and statistical property of single and triple-break vacuum interrupters in series," *IEEE Trans. Dielectr. Electr. Insul.*, vol. 14, no. 3, p. 600, 2007.

[147]   Z. Xiang, X. Zhang, R. Huang, Z. Wu, M. Liao, and X. Zhang, "Research on voltage distribution of double-break vacuum circuit breakers with different interrupters in series," in: *Proceedings of 28th ISDEIV International Symposium on Discharges and Electrical Insulation in Vacuum*, 2018, p. 635.

[148]   M. Liao, X. Duan, X. Cheng, Z. Huang, and J. Zou, "Property of 126kV vacuum circuit breaker based on three 40.5kV fiber-controlled vacuum interrupter modules in series," in: *Proceedings of 25th ISDEIV International Symposium on Discharges and Electrical Insulation in Vacuum*, 2012, p. 22.

[149]   L. T. Falkingham, K. W. Cheng, and W. J. Molan, "The design of a 245kV vacuum circuit breaker," in: *Proceedings of 27th ISDEIV International*

Symposium on Discharges and Electrical Insulation in Vacuum, 2016, p. 497.

[150] M. Leusenkamp, "Impulse voltage generator design and the potential impact on bacuum interrupter de-conditioning," in: *Proceedings of 25th ISDEIV International Symposium on Discharges and Electrical Insulation in Vacuum*, 2012, p. 453.

[151] IEEE Standard for Ratings and Requirements for AC High-Voltage Circuit Breakers with Rated Maximum Voltage Above 1000 V, IEEE Standard C37.04, 2018.

[152] E. D. Taylor and P. G. Slade, "High voltage breakdown performance and circuit isolation capability of vacuum interrupters," in: *Proceedings of the 22nd International Symposium on Discharges and Electrical Insulation in Vacuum*, 2006, p. 208.

[153] H. Schellekens, A. Henon, and P. Picot, "Vacuum switch-disconnectors: 1. Dielectric behaviour," in: *Proceedings of the 23rd International Symposium on Discharges and Electrical Insulation in Vacuum*, 2008.

[154] H. Schellekens, I. Gall, and D. Goulielmakis, "Vacuum switch-disconnectors: 2. Compliance with insulation coordination," in: *Proceedings of the 23rd International Symposium on Discharges and Electrical Insulation in Vacuum*, 2008.

[155] E. Dullni, B. Baum, D. Desmond, and C. Heinrich, "The performance of in-service shunt capacitor switching devices as investigated by Cigre WG A3.38," in: *Proceedings of the 25th International Conference on Electricity Distribution – CIRED*, 2019, paper 1312.

[156] R.P.P. Smeets, R. Wiggers, S. Chakraborty, H. Bannink, S. Kuivenhoven, and G. Sandolache, "The impact of switching capacitor banks with very high inrush current on switchgear," in: *Proceedings of the Cigre Session*, 2012, A3-201.

[157] High-voltage switchgear and controlgear – Part 103: Alternating current switches for rated voltages above 1 kV up to and including 52 kV, IEC Standard 62271-103, Edition 2.0, 2021.

[158] A. M. Chaly and I. N. Poluyanova, "Relevancy of IEC requirements related to switching cable and line charging currents for medium voltage vacuum circuit breakers (VCB)," in: *Proceedings of the 22nd International Symposium on Discharges and Electrical Insulation in Vacuum*, 2006.

[159] T. Kamikawaji, T. Shioiri, T. Funahashi, Y. Satoh, E. Keneko, and I. Ohshima, "An investigation into major factors in shunt capacitor switching performances by vacuum circuit breakers with copper-chromium contacts," *IEEE Trans. Power Del.*, vol. 8, no. 4, p. 1789, 1993.

[160] E. Dullni, W. Shang, D. Gentsch, I. Kleberg, and K. Niayesh, "Switching of capacitive currents and the correlation of restrike and pre-ignition behavior," *IEEE Trans. Dielect. Elect. Insul.*, vol. 13, no. 1, p. 65, 2006.

[161] F. Körner, M. Lindmayer, M. Kurrat, and D. Gentsch, "Contact behavior in vacuum under capacitive switching duty," *IEEE Trans. Dielect. Elect. Insul.*, vol. 14, no. 3, p. 643, 2007.

[162]   G. Sandolache, U. Ernst, X. Godechot, S. Kantas, M. Hairour, and L. Dalmazio, "Switching of capacitive current with vacuum interrupters," in: *Proceedings of the 24th International Symposium on Discharges and Electrical Insulation in Vacuum*, 2010, p. 129.

[163]   S. Griot and A. Moreau, "Vacuum circuit breaker's electrical life for shunt capacitor switching," in: *Proceedings of the 24th International Symposium on Discharges and Electrical Insulation in Vacuum*, 2010, p. 194.

[164]   T. Delachaux, F. Rager, and D. Gentsch, "Study of vacuum circuit breaker performance and weld formation for different drive closing speeds for switching capacitive current," in: *Proceedings of the 24th International Symposium on Discharges and Electrical Insulation in Vacuum*, 2010, p. 241.

[165]   S. Giere, R. Renz, F. Richter, and N. Trapp, "Capacitive current switching capability of 72.5 kV high-voltage vacuum interrupters," in: *Proceedings of the 25th International Symposium on Discharges and Electrical Insulation in Vacuum*, 2012, p. 217.

[166]   A. Juhasz and W. Rieder, "Capacitive switching with vacuum interrupters," in: *Proceedings of the 10th International Conference on Gas Discharges and their Applications*, 1992, p. 62.

[167]   X. Godechot and P. Novak, "New trends in capacitive switching current," in: Proceedings of the *28th International Symposium on Discharges and Electrical Insulation in Vacuum*, 2018, p. 599.

[168]   R. Alexander and E. Dullni, "Melting pattern of vacuum interrupter contacts subjected to high inrush currents at high frequencies," *IEEE Trans. Power Del.*, vol. 37, no. 2, p. 710, 2022.

[169]   Z. Zalucki, "Restrike and reignition voltages of a short contact gap during capacitance switching using vacuum circuit breakers," in: *Proceeding of the 19th International Symposium on Discharges and Electrical Insulation in Vacuum*, 2000, p. 56.

[170]   W. T. Diamond, "New perspectives in high voltage insulation. 1. The transition to field emission," *J. Vac. Sci. Technol. A*, vol. 16, no. 2, p. 707, 1998.

[171]   M. K. Zadeh, V. Hinrichsen, R. Smeets, and A. Lawall, "Field emission currents in vacuum breakers after capacitive switching," *IEEE Trans. Dielect. Elect. Insul.*, vol. 18, No. 3, p. 910, 2011.

[172]   P. G. Slade and E. D. Taylor, "Calculations on the potential role of emission currents on restrikes after capacitor switching using vacuum interrupters," in: *Proceeding of the 27th International Symposium on Discharges and Electrical Insulation in Vacuum*, 2018, p. 177.

[173]   B. Surges, V. Hinrichsen, and E. D. Taylor, "Simultaneous measurement of high and low frequency pre-breakdown currents on HV vacuum interrupters," in: *Proceeding of the 27th International Symposium on Discharges and Electrical Insulation in Vacuum*, 2016, p. 72.

[174]   High-voltage switchgear and controlgear – Part 306: Guide to IEC 62271-100, IEC 62271-1 and other IEC standards related to alternating current circuit-breakers, IEC Standard 62271-306, edition 1.1, 2018.

[175] H. Yang, Y. Geng, Z. Liu, Y. Zhang, and J. Wang, "Capacitive switching of vacuum interrupters and inrush currents," *IEEE Trans. Dielect. Elect. Insul.*, vol. 21, no. 1, p. 159, 2014.

[176] Cigre Working Group A3.07, "Controlled switching of HVAC circuit breakers: planning, specification and testing of controlled switching systems," *Cigre Technical Report* No. 264, 2004.

[177] D. Goldsworthy, T. Roseburg, D. Tziouvaras, and J. Pope, "Controlled switching of HVAC circuit breakers: application examples and benefits," in: *Proceedings of the 61st Annual Conference for Protective Relay Engineers*, 2008, p. 520.

[178] H. Schellekens and M. Bono, "Synchronous capacitor bank switching with vacuum circuit breakers," in: *Proceeding of the 28th International Symposium on Discharges and Electrical Insulation in Vacuum*, 2018, p. 567.

[179] E. D. Taylor, P. G. Slade, and W. Li, "High chop currents observed in vacuum arcs between tungsten contacts," in: *Proceedings of the 23rd International Conference on Electrical Contacts*, 2006, p. 1.

[180] P. G. Slade and M. F. Hoyaux, "The effect of electrode material on the initial expansion of an arc in vacuum," *IEEE Trans. Parts, Hybrids, Packag.*, vol. PHP-8, no. 1, p. 35, 1972.

[181] J. Panek and K. G. Fehrle, "Overvoltage phenomena associated with virtual current chopping in three phase circuits," *IEEE Trans. Power App. Syst.*, vol. PAS-94, no. 4, p. 1317, 1975.

[182] S. F. Farag and R. G. Bartheld, "Guidelines for the application of vacuum contactors," *IEEE Trans. Ind. Appl.*, vol. IA-22, no. 1, p. 102, 1986.

[183] High-voltage switchgear and controlgear – Part 110: Inductive load switching, IEC Standard 62271-110, Ed. 4.0, 2017.

[184] E. D. Taylor, K. Niemeyer, and C. Pietsch, "Generation of overvoltages by chop current on Ag-WC and Cu-W/WC contacts in vacuum," in: *Proceedings of the 64th IEEE Holm Conference on Electrical Contacts*, 2018, p. 242.

[185] J. F. Perkins, "Evaluation of switching surge overvoltages on medium voltage power systems," *IEEE Trans. Power App. Syst.*, vol. PAS-101, no. 6, p. 1727, 1982.

[186] J. F. Perkins and D. Bhasavanich, "Vacuum switchgear application study with reference to switching surge protection," *IEEE Trans. Ind. Appl.*, vol. 1A-19, no. 5, 1983.

[187] P. Halbach, "Einfluss des Prüfkreises auf das Abreißstromverhalten von Vakuumschaltern unter Berücksichtigung spezieller Netzkonfigurationen in der Mittelspannung," Ph.D. dissertation, Elektrotechnik und Informationstechnik, Technischen Universität Darmstadt, Darmstadt, Germany, 2013.

[188] P. Frey, N. Klink, R. Michal, and K. E. Saeger, "Metallurgical aspects of contact materials for vacuum switching devices," *IEEE Trans. Plasma Sci.*, vol. 17, no. 5, p. 734, 1989.

[189] A. Müller and D. Sämann, "Switching phenomena in medium voltage systems – good engineering practice on the application of vacuum circuit-breakers and contactors," in: *Proceedings of the Petroleum and Chemical Industry Conference Europe* (PCIC EUROPE), 2011, p. 1.

[190] E. Dullni, J. Meppelink, and L. Liljestrand, "Vacuum circuit breaker, switching interactions with transformers and mitigation means," in: *Proceedings of the Cigre Session*, 2014, Paper A3 302.

[191] K. Trunk, A. Lawall, E. D. Taylor, *et al.*, "Small inductive current switching with high-voltage vacuum circuit breakers," in: *Proceedings of the 29th ISDEIV International Symposium on Discharge and Electrical Insulation in Vacuum*, 2020, p. 331.

[192] J. Ryu, Y.-G. Kim, J. Choi, and S. Park, "The experimental research of 170 kV VCB using single-break vacuum interrupter," in: *Proceedings of the 25th ISDEIV International Symposium on Discharge and Electrical Insulation in Vacuum*, 2012, p. 493.

[193] Y. Ohki, "Japan AE power systems develops high-voltage, vacuum circuit breaker," *IEEE Elect. Insul. Mag.*, vol. 23, no. 1, p. 48, 2007.

[194] Y. Matsui, K. Nagatake, M. Takeshita, *et al.*, "Development and technology of high voltage VCBs; brief history and state of art," in: *Proceedings of the 22nd ISDEIV International Symposium on Discharge and Electrical Insulation in Vacuum*, 2006.

[195] T. Tsutsumi, Y. Kanai, N. Okabe, E. Kaneko, T. Kamikawaji, and M. Homma, "Dynamic characteristics of high-speed operated, long-stroke bellows for vacuum interrupters," *IEEE Trans. Power Del.*, vol. 8, no. 1, p. 163, 1993.

[196] L. Baron, F. Graskowski, A. Lawall, and C. Stiehler, "Overpressure-resistant vacuum interrupter tube," U.S. Patent 11,289,292, 29 March 2022.

[197] M. B. J. Leusenkamp, "Quality assurance of vacuum interrupters intended for transmission voltage level applications," in: *Proceedings of the 29th ISDEIV International Symposium on Discharge and Electrical Insulation in Vacuum*, 2020, p. 339.

[198] P. G. Slade, W. Li, S. Mayo, R. K. Smith, and E. D. Taylor, "Vacuum interrupter, high reliability component of distribution switches, circuit breakers and contactors," *J. Zhejiang Univ. Sci. A*, vol. 8, no. 3, p. 335, 2007.

[199] Cigre WG 13.06, "High voltage circuit breaker reliability data for use in system reliability studies," in: *Proceedings of the Cigre Symposium on Electrical Power System and Reliability*, 1991, paper 2-01.

[200] R. D. Garzon, *High Voltage Circuit Breakers: Design and Applications*, New York, NY: Marcel Dekker, 1997.

[201] L. J. Bayford, "Some aspects of modern cathode-ray-tube manufacture," *Proc. IEE – Part IIIa: Television*, vol. 99, p. 514, 1952.

[202] M. Okawa, T. Tsutsumi, and T. Aiyoshi, "Reliability and field experience of vacuum interrupters," *IEEE Trans. Power Del.*, vol. PWRD-2, no. 3, p. 799, 1987.

[203] W. Li, R. L. Thomas, and P. G. Slade, "Residual gas analysis of vacuum interrupters," in: *Proceedings of the 3rd International Conference on Electrical Contacts, Arcs, Apparatus and Their Applications*, 1997, p. 491.

[204] D. Gentsch and T. Fugel, "Measurement by residual gas analysis inside vacuum interrupters," *IEEE Trans. Plasma Sci.*, vol. 37, no. 8, p. 1484, 2009.

[205] R. Renz, D. Gentsch, H. Fink, P. Slade, and M. Schlaug, "Vacuum interrupters – sealed for life," in: *Proceedings of the CIRED 19th International Conference & Exhibition on Electricity Distribution*, 2007, Paper 0156.

[206] E. L. Grant and R. S. Leavenworth, *Statistical Quality Control*, New, York, NY: McGraw-Hill, 1988.

[207] R. J. Deaton and R. L. Sellers Jr., "Analysis of the failure of a vacuum circuit breaker applied on a consumer-utility interface," *IEEE Trans. Ind. Appl.*, vol. IA-22, no. 4, p. 610, 1986.

[208] T. H. Lee, D. R. Kurtz, and D. J. Veras, "Problems in detecting leaks with long time constants in long-time vacuum devices with sealed glass envelopes," *Vacuum*, vol. 13, no. 5, p. 167, 1963.

[209] K. Golde, V. Hinrichsen, D. Gentsch, A. Lawall, and E. D. Taylor, "Short-circuit current interruption in liquid nitrogen environment," in: *Proceedings of the 27th International Symposium on Discharges and Electrical Insulation in Vacuum*, 2016, p. 533.

[210] R. Reeves, and L. T. Falkingham, "A measurement of intrinsic outgassing rates in vacuum interrupters," in: *Proceedings of the 4th ICEPE International Conference on Electric Power Equipment – Switching Technology*, 2017, p. 863.

[211] G. F. Weston, *Ultrahigh Vacuum Practice*, London: Butterworths, 1985.

[212] en.wikipedia.org. "Copper(I) oxide." Accessed 15 Nov. 2022. Available: https://en.wikipedia.org/wiki/Copper(I)_oxide.

[213] en.wikipedia.org. "Copper(II) oxide." Accessed 15 Nov. 2022. Available: https://en.wikipedia.org/wiki/Copper(II)_oxide.

[214] en.wikipedia.org. "Chromium(III) oxide." Accessed 15 Nov. 2022. Available: https://en.wikipedia.org/wiki/Chromium(III)_oxide

[215] J. R. J. Bennett, S. Hughes, R. J. Elsey, and T. P. Parry, "Outgassing from stainless steel and the effects of the gauges," *Vacuum*, vol. 73, p. 149, 2004.

[216] J. K. Fremerey, "Residual gas: traditional understanding and new experimental results," *Vacuum*, vol. 53, p. 197, 1999.

[217] Y. Koyatsu, H. Mikl, and F. Watanabe, "Measurements of outgassing rate from copper and copper alloy chambers," *Vacuum*, vol. 47, p. 709, 1996.

[218] L. Westerberg, B. Hjoervarsson, E. Wallen, and A. Mathewson, "Hydrogen content and outgassing of air-baked and vacuum-fired stainless steel," *Vacuum*, vol. 48, p. 771, 1997.

[219] M. E. Arthur and M. J. Zunick, "Useful life of vacuum interrupters," *IEEE Trans. Power App. Syst.*, vol. PAS-97, no. 1, p. 1, 1978.

[220] M. Weuffel, D. Gentsch, and P. G. Nikolic, "Influence of current interruption operations on internal pressure in vacuum interrupters," *IEEE Trans. Plasma Sci.*, vol. 45, no. 8, p. 2144, 2017.

[221] M. Leusenkamp, M. Binnendijk, B. T. Hedde, and T. Van Rijn, "A field check on the condition of vacuum interrupters after long periods of service," in: *Proceedings of the 23rd International Conference on Electricity Distribution* – CIRED, 2015, paper 1461.

[222] C. Holdsworth, G. Wood, B. Pentecost, and S. Hennell, *Performance Review of Vacuum Interrupters in UK Distribution Networks, Energy Network Assoc., Engineering Report* 135, Issue 1, 2015.

[223] Joint Working Group A3.32/CIRED, "Non-intrusive methods for condition assessment of distribution and transmission switchgear," Cigre Technical Report No. 737, 2018.

[224] D. Eichhoff, D. Gentsch, M. Weuffel, and A. Schnettler, "Magnetron-based on-site measurement of the internal pressure in vacuum interrupters," in: *Proceedings of the 26th ISDEIV International Symposium on Discharges and Electrical Insulation in Vacuum*, 2014, p. 461

[225] Potential Failure Mode and Effects Analysis in Design (Design FMEA) and Potential Failure Mode and Effects Analysis in Manufacturing and Assembly Processes (Process FMEA) and Effects Analysis for Machinery (Machinery FMEA), SAE Standard J-1739, 2021.

[226] L. T. Falkingham, R. Reeves, C. H. Gill, and S. Mistry, "Studies in inverse magnetron discharges of vacuum interrupters: Part 1 – variations in electric field," in: *Proceedings of the 23rd ISDEIV International Symposium on Discharges and Electrical Insulation in Vacuum*, 2008.

[227] L.T. Falkingham, R. Reeves, C. H. Gill, and S. Mistry, "Studies in inverse magnetron discharges of vacuum interrupters: Part 2 – variations in magnetic field," in: *Proceedings of the 24th ISDEIV International Symposium on Discharges and Electrical Insulation in Vacuum*, 2010, p. 214

[228] L. T. Falkingham, R. Reeves, and S. Mistry, "Studies in Inverse Magnetron Discharges of Vacuum Interrupters – Part 3 – Anomalies," in: *Proceedings of the 25th ISDEIV International Symposium on Discharges and Electrical Insulation in Vacuum*, 2012, p. 1.

[229] R. Reeves and L. T. Falkingham, "The effect of magnetron discharge pressure measurement on the actual pressure in vacuum interrupters," in: *Proceedings of the 3rd International Conference on Electric Power Equipment – Switching Technology*, 2015, p. 31.

[230] D. Eichhoff, D. Gentsch, M. Weuffel, and A. Schnettler, "Influence of the residual gas composition and the getter-ion effect on the measurement of the internal pressure in vacuum interrupters," in: *Proceedings of the 26th ISDEIV International Symposium on Discharges and Electrical Insulation in Vacuum*, 2014, p. 457.

[231] L. T. Falkingham, R. Reeves, S. Mistry, and C. H. Gill, "A study of vacuum levels in a sample of long service vacuum interrupters," in: *Proceedings of the 25th ISDEIV International Symposium on Discharges and Electrical Insulation in Vacuum*, 2012, p. 181.

[232] R. Reeves and L. T. Falkingham, "An appraisal of the insulation capability of vacuum interrupters after long periods of service," in: *Proceedings of the*

2nd International Conference on Electric Power Equipment – Switching Technology (ICEPE-ST 2013), 2013.

[233]   W. Hopper, "One Mill's experience using MAC testing to evaluate vacuum interrupter integrity in 15 kV vacuum circuit breakers," *IEEE Trans. Ind. Appl.*, vol. 53, no. 1, p. 774, 2017.

[234]   J. Bowen and T. A. Burse, "Medium-voltage replacement breaker projects," *IEEE Trans. Ind. Appl.*, vol. 38, no. 2, 2002.

[235]   R. Kirkland Smith and R. William Long, "Vacuum interrupters and x-rays," *Electrical World*, p. 36, 2001.

[236]   Alternating-Current High-Voltage Power Vacuum Interrupters—Safety Requirements for X-Radiation Limits, ANSI Standard C37.85, 2020.

[237]   Council Directive 96/29/EURATOM, 13 May 1996.

[238]   Announcement of Bundesamt für Strahlenschutz (BfS), 'Permission of type of construction and operation of vacuum interrupters according to §§5, 8 and attachment 2 No. 5 of the German X-ray regulation (RöV)', 9th of August 2005, Berlin.

[239]   R. Renz and D. Gentsch, "Permissible X – ray radiation emitted by vacuum – interrupters / – devices at rated operating conditions," in: *Proceedings of the 24th ISDEIV International Symposium on Discharges and Electrical Insulation in Vacuum*, 2010, p. 133.

[240]   S. Giere, T. Heinz, A. Lawall, *et al.*, "X-Radiation emission of high-voltage vacuum interrupters: dose rate control under testing and operating conditions," in: *Proceedings of the 28th ISDEIV International Symposium on Discharges and Electrical Insulation in Vacuum*, 2018, p. 523.

[241]   E. D. Taylor, "Application of research in field emitter arrays to the breakdown of contacts in vacuum," in: *Proceedings of the 25th ISDEIV International Symposium on Discharges and Electrical Insulation in Vacuum*, 2012, p. 41.

[242]   J. Yan, Z. Liu, S. Zhang, Y. Geng, Y. Zhang, and G. He, "X-Ray radiation of a 126 kV vacuum interrupter," in: *Proceedings of the 25th ISDEIV International Symposium on Discharges and Electrical Insulation in Vacuum*, 2012, p. 461.

[243]   R. P. P. Smeets, A. G. A. Lathouwers, and L. T. Falkingham, "Assessment of non-sustained disruptive discharges (NSDD) in switchgear. Test experience and standardization status," in: *Proceedings of the Cigre Session*, 2004, Paper A3-303.

[244]   H. Schellekens, "Capacitor bank switching with vacuum circuit breakers," in: *Proceedings of the 23rd ISDEIV International Symposium on Discharges and Electrical Insulation in Vacuum*, 2008.

[245]   L. Wang, X. Zhang, J. Deng, Z. Yang, and S. Jia, "Research progress of model of vacuum arc and anode activity under axial magnetic fields," *IEEE Trans. Plasma Sci.*, vol. 47, no. 8, p. 3496, 2019.

[246]   L. Wang, S. Jia, Z. Shi, and M. Rong, "High-current vacuum arc under axial magnetic field: Numerical simulation and comparisons with experiments," *J. Appl. Phys.*, vol. 100, p. 113304, 2006.

[247]   E. Schade and D. Shmelev, "Numerical simulation of high-current vacuum arcs with an external axial magnetic field," *IEEE Trans. Plasma Sci.*, vol. 31, no. 5, *p. 890, 2003; Errata*: vol. 32, no. 2, p. 829, 2004.

[248]   L. Wang, S. Jia, L. Zhang, *et al.*, "Current constriction of high-current vacuum arc in vacuum interrupters," *J. Appl. Phys.*, vol. 103, p. 063301, 2008.

[249]   H. T. C. Kaufmann, M. D. Cunha, M. S. Benilov, W. Hartmann, and N. Wenzel, "Detailed numerical simulation of cathode spots in vacuum arcs: interplay of different mechanisms and ejection of droplets," *J. Appl. Phys.*, vol. 122, p. 163303, 2017.

[250]   L. Wang, S. Jia, Y. Liu, B. Chen, D. Yang, and Z. Shi, "Modeling and simulation of anode melting pool flow under the action of high-current vacuum arc," *J. Appl. Phys.*, vol. 107, 113306, 2010.

[251]   X. Huang, H. Zhu, M. Yang, *et al.*, "Numerical investigation of magnetic field in AMF vacuum arc switch with transverse magnetic field from three-phase structure," *IEEE Trans. Plasma Sci.*, vol. 48, no. 7, p. 2558, 2020.

[252]   L. Wang, M. Luo, J. Deng, *et al.*, "Numerical simulation of multispecies vacuum arc subjected to actual spatial magnetic fields," *IEEE Trans. Plasma Sci.*, vol. 49, no. 11, p. 3652, 2021.

[253]   T. Delachaux, O. Fritz, D. Gentsch, E. Schade, and D. L. Shmelev, "Simulation of a high current vacuum arc in a transverse magnetic field," *IEEE Trans. Plasma Sci.*, vol. 37, no. 8, p. 1386, 2009.

[254]   D. L. Shmelev and T. Delachaux, "Physical modeling and numerical simulation of constricted high-current vacuum arcs under the influence of a transverse magnetic field," *IEEE Trans. Plasma Sci.*, vol. 37, no. 8, p. 1379, 2009.

[255]   O. Fritz, D. Shmelev, K. Hencken, T. Delachaux, and D. Gentsch, "Results of 3D numerical simulations of high-current constricted vacuum arcs in a strong magnetic field," in: *Proceedings of the 29th ISDEIV International Symposium on Discharges and Electrical Insulation in Vacuum*, 2010, p. 359.

[256]   W. Shi, L. Wang, R. Lin, Y. Wang, J. Ma, and S. Jia, "Experimental investigation of drawing vacuum arc under different TMF contacts in vacuum interrupter," *IEEE Trans. Plasma Sci.*, vol. 47, no. 4, p. 1827, 2019.

[257]   A. Daibo, Y. Niwa, N. Asari, W. Sakaguchi, Y. Sasaki, and J. Kondo, "Analysis of arc behavior and electrode surface temperature in spiral electrode of vacuum interrupter," in: *Proceedings of the 6th International Conference on Electric Power Equipment - Switching Technology*, 2022, p. 63.

[258]   R. Methling, S. Franke, S. Gortschakow, *et al.*, "Anode surface temperature determination in high-current vacuum arcs by different methods," *IEEE Trans. Plasma Sci.*, vol. 45, no. 8, p. 2099, 2017.

[259]   T. Pieniak, M. Kurrat, and D. Gentsch, "Surface temperature analysis of transversal magnetic field contacts using a thermography camera," *IEEE Trans. Plasma. Sci.*, vol. 45, no. 8, p. 2157, 2017.

[260] A. A. Logachev, I. N. Poluyanova, K. K. Zabello, Y. A. Barinov, and S. M. Shkol'nik, "Cathode surface state and cathode temperature distribution after current zero of different AMF-contacts," *IEEE Trans. Plasma Sci.*, vol. 47, no. 8, p. 3516, 2019.

[261] H. Li, R. Methling, S. Franke, *et al.*, "Spectroscopic analysis of anode surface thermal emission with single and dual vacuum arc columns," *IEEE Trans. Plasma Sci.*, vol. 47, no. 11, p. 5204, 2019.

[262] A. Khakpour, S. Gortschakow, D. Uhrlandt, *et al.*, "Video spectroscopy of vacuum arcs during transition between different high-current anode modes," *IEEE Trans. Plasma Sci.*, vol. 44, no. 10, p. 2462, 2016.

[263] A. Khakpour, S. Franke, S. Gortschakow, *et al.*, "Investigation of anode plume in vacuum arcs using different optical diagnostic methods," *IEEE Trans. Plasma Sci.*, vol. 47, no. 8, p. 3488, 2019.

[264] A. Khakpour, S. Popov, S. Franke, *et al.*, "Determination of Cr density after current zero in a high-current vacuum arc considering anode plume," *IEEE Trans. Plasma Sci.*, vol. 45, no. 8, p. 2108, 2017.

[265] Y. Inada, T. Kamiya, S. Matsuoka, A. Kumada, H. Ikeda, and K. Hidaka, "Intense-mode vacuum arc characterization by using 2-D electron and vapor density image," *IEEE Trans. Plasma Sci.*, vol. 45, no. 1, p. 129, 2017.

[266] R. Kikuchi, Y. Hirano, H. Ejiri, *et al.*, "Neutral atom density at current zero in interruption of intense-mode vacuum arc," in: *Proceedings of the 29th ISDEIV International Symposium on Discharges and Electrical Insulation in Vacuum*, 2020, p. 135.

[267] K. K. Zabello, I. N. Poluyanova, A. A. Logachev, D. I. Begal, and S. M. Shkol'nik, "Anode surface state and anode temperature distribution after current zero for different AMF-contact systems," *IEEE Trans. Plasma Sci.*, vol. 47, no. 8, p. 3563, 2019.

[268] S. Gortschakow, D. Gonzalez, R. Methling, *et al.*, "Influence of ignition position on the properties of vacuum arc generated by switching RMF contacts," in: *Proceedings of the 29th ISDEIV International Symposium on Discharges and Electrical Insulation in Vacuum*, 2020, p. 127.

[269] Y. A. Barinov, K. K. Zabello, A. A. Logachev, *et al.*, "Radiation power of a short high-current vacuum arc at high current densities," *IEEE Trans. Plasma Sci.*, vol. 50, no. 9, p. 2729, 2022.

[270] K. K. Zabello, I. N. Poluyanova, A. A. Logachev, Y. A. Barinov, and S. M. Shkolnik, "Thermal effect on AMF electrodes at different opening speeds," *IEEE Trans. Plasma Sci.*, vol. 49, no. 4, p. 1454, 2021.

[271] Z. Wang, Y. Li, Y. Pan, Y. Du, Z. Zhou, and L. Sun, "Time-dependent 3D reconstruction of arc shape in a TMF contact system," in: *Proceedings of the 29th ISDEIV International Symposium on Discharges and Electrical Insulation in Vacuum*, 2020, p. 214.

[272] A. Daibo, Y. Niwa, N. Asari, W. Sakaguchi, Y. Sasaki, and T. Yoshida, "Measurement of transient temperature distribution at anode surface after current zero in vacuum interrupter," in: *Proceedings of the 29th ISDEIV*

International Symposium on Discharges and Electrical Insulation in Vacuum, 2020, p. 143.

[273] R. P. Little and S. T. Smith, "Electrical breakdown in vacuum," *IEEE Trans. Electron Devices*, vol. 12, no. 2, p. 77, 1965.

[274] I. W. Rangelow and St. Biehl, "Fabrication and electrical characterization of high aspect ratio silicon field emitter arrays," *J. Vac. Sci. Technol. B*, vol. 19, no. 3, p. 916, 2001.

[275] N. Pilan, P. Veltri, and A. DeLorenzi, "Voltage holding prediction in multi electrode–multi voltage systems insulated in vacuum," *IEEE Trans. Dielect. Elect. Insul.*, vol. 18, no. 2, p. 553, 2011.

[276] G. Chitarin, A. Kojima, M. Boldrin, *et al.*, "Strategy for vacuum insulation tests of MITICA 1 MV electrostatic accelerator," *IEEE Trans. Plasma Sci.*, vol. 50, no. 9, p. 2755, 2022.

[277] N. Marconato, A. De Lorenzi, N. Pilan, *et al.*, "Prediction of lightning impulse voltage induced breakdown in vacuum interrupters," *IEEE Trans. Dielect. Elect. Insul.*, vol. 24, no. 6, p. 3367, 2017.

[278] N. Marconato, T. Patton, P. Bettini, *et al.*, "Application of the voltage holding prediction model to floating and fixed shield vacuum interrupters," in: *Proceedings of the 29th ISDEIV International Symposium on Discharges and Electrical Insulation in Vacuum*, 2020, p. 302.

[279] C. H. Moore, A. Jindal, E. Bussmann, *et al.*, "Progress in micron-scale field emission models based on nanoscale surface characterization for use in PIC-DSMC vacuum arc simulations," in: *Proceedings of the 29th ISDEIV International Symposium on Discharges and Electrical Insulation in Vacuum*, 2020, p. 272.

[280] X-D. Qiu, J.-C. Su, Y. Zhang, *et al.*, "Investigation of explosive electron emission sites on surface of polished cathodes in vacuum," *IEEE Trans. Dielect. Elect. Insul.*, vol. 25, no. 6, p. 2018.

[281] I. Profatilova, X. Stragier, S. Calatroni, A. Kandratsyeu, E. Rodriguez Castro, and W. Wuensch, "Breakdown localization in a pulsed DC electrode system," *Nuclear Inst. Methods Phys. Res. A*, vol. 953, p. 163079, 2020.

[282] E. Z. Engelberg, A. B. Yashar, Y. Ashkenazy, M. Assaf, and I. Popov, "Theory of electric field breakdown nucleation due to mobile dislocations," *Phys. Rev. Acclerators Beams*, vol. 22, p. 083501, 2019.

[283] E. Z. Engelberg, J. Paszkiewicz, R. Peacock, S. Lachmann, Y. Ashkenazy, and W. Wuensch, "Dark current spikes as an indicator of mobile dislocation dynamics under intense dc electric fields," *Phys. Rev. Acclerators Beams*, vol. 23, p. 123501, 2020.

[284] V. Jansson, E. Baibuz, A. Kyritsakis, *et al.*, "Growth mechanism for nanotips in high electric fields," *Nanotechnology*, vol. 31, no. 35, p. 13, 2020.

[285] E. Spada, A. De Lorenzi, L. Lotto, N. Pilan, S. Spagnolo, and M. Zuin, "New development of BIRD model," *IEEE Trans. Plasma Sci.*, vol. 50, no. 9, p. 2763, 2022.

[286] M. E. Cuneo, "The effect of electrode contamination, cleaning and conditioning on high-energy pulsed-power device performance," *IEEE Trans. Dielectr. Electric. Insul.*, vol. 6, no. 4, p. 469, 1999.

[287] H. Schellekens, F. Chombart, R. Mottin, and F. Vianna, "Dielectric strength of vacuum interrupters influence of manufacturing process," in: *Proceedings of the 24th ISDEIV International Symposium on Discharges and Electrical Insulation in Vacuum*, 2010.

[288] J. Ballat, D. Konig, and U. Reininghaus, "Spark conditioning procedures for vacuum interrupters in circuit breakers," *IEEE Trans. Elect. Insul.*, vol. 28, no. 4, p. 621, 1993.

[289] A. Degiovanni, W. Wuensch, and J. G. Navarro, "Comparison of the conditioning of high gradient accelerating structures," *Phys. Rev. Accelerators Beams*, vol. 19, p. 032001, 2016.

[290] S. Spagnolo, N. Pilan, A. DeLorenzi, *et al.*, "Characterization of X-ray events in a vacuum high voltage long-gap experiment," *IEEE Trans. Plasma Sci.*, vol. 50, no. 11, p. 4788, 2022.

[291] M. M. Menon and K. D. Srivastava, "Electrostatic detection of large microparticles (>10 μm) in a high-voltage vacuum insulated apparatus," *IEEE Trans. Elect. Insul.*, vol. EI-9, no. 4, p. 142, 1974.

[292] D. A. Eastham and P. A. Chatterton, "The detection of microparticle-induced breakdowns using a twin-beam laser scattering system," *IEEE Trans. Elect. Insul.*, vol. EI-18, no. 3, p. 209, 1983.

[293] H. Ejiri, K. Abe, Y. Kikuchi, *et al.*, "Motion and production of microparticles in vacuum interrupter," *IEEE Trans. Dielect. Elect. Insul.*, vol. 24, no. 6, p. 3374, 2017.

[294] H. Ejiri, A. Kumada, K. Hidaka, T. Donen, and K. Kokura, "Motion and breakdown related to microparticles in vacuum gap," *IEEE Trans. Plasma Sci.*, vol. 47, no. 8, p. 3384, 2019.

[295] H. Ejiri, A. Kumada, K. Hidaka, *et al.*, "Late breakdowns caused by microparticles after vacuum arc interruption," *IEEE Trans. Plasma Sci.*, vol. 47, no. 8, p. 3392, 2019.

[296] W. Yan, Z. Wang, Z. Cao, H. Li, L. Sun, and J. Wang, "Particle generation, their behavior and effect on post-arc breakdowns after capacitive-current vacuum arc interruption," *IEEE Trans. Power Del.*, vol. 37, no. 5, p. 4303, 2022.

[297] A. Anders, *Cathodic Arcs: From Fractal Spots to Energetic Condensation*, New York: Springer, 2008.

[298] D. T. Tuma, C. L. Chen, and D. K. Davies, "Erosion products from the cathode spot region of a copper vacuum arc," *J. Appl. Phys.*, vol. 49, no. 7, p. 3821, 1978.

[299] M.B. Schulman, P.G. Slade, L.D. Loud, and W. Li, "Influence of contact geometry and current on effective erosion of Cu-Cr, Ag-WC and Ag-Cr vacuum contact materials," *IEEE Trans. Comp. Packag. Technol.*, vol. 22, no. 3, p. 405, Sept. 1999. Errata: *IEEE Trans. Comp. Packag. Technol.*, vol. 23, no. 1, p. 194, 2000.

[300] P. Siemroth, M. Laux, H. Pursch, *et al.*, "Diameters and velocities of dro-plets emitted from the Cu cathode of a vacuum arc," *IEEE Trans. Plasma Sci.*, vol. 47, no. 8, p. 3470, 2019.

[301] P. Siemroth, M. Laux, H. Pursch, *et al.*, "Investigation of vacuum arc droplets from copper, titanium, and tungsten by means of light scattering," *IEEE Trans. Plasma Sci.*, vol. 50, no. 9, p. 2736, 2022.

[302] K. Yokokura, M. Matsuda, K. Atsumi, *et al.*, "Capacitor switching cap-ability of vacuum interrupter with CuW contact material," *IEEE Trans. Pow, Del.*, vol. 10, no. 2, p. 804, 1995.

[303] M. Muir, C. R. Belton, S. W. J. Scully, K. Matuszak, P. G. Klaja, and J. E. Cunningham, "Compact Vacuum Interruption with Microsecond Accuracy," in: *Proceedings of the 28th ISDEIV International Symposium on Discharges and Electrical Insulation in Vacuum*, 2018, p. 185.

[304] P. G. Slade, "Advances in material development for high power, vacuum interrupter contacts," *IEEE Trans. Comp., Packag., Manufact. Technol. A*, vol. 17, no. 1, p. 96, 1994.

[305] E. Kaneko, K. Yokokura, M. Homma, *et al.*, "Possibility of high current interruption of vacuum interrupter with low surge contact material: improved Ag-WC," *IEEE Trans. Power Del.*, vol. 10, no. 2, p. 797, 1995.

[306] S. Yanabu, M. Okawa, E. KIaneko, and T. Tamagawa, "Use of axial magnetic fields to improve high current vacuum interrupters," *IEEE Trans. Plasma Sci.*, vol. PS-15, no. 5, p. 524, 1987.

[307] M. Okawa, S. Yanabu, E. Kaneko, and K. Otobe, "The investigation of copper-chromium contacts in vacuum interrupters subjected to an axial magnetic field," *IEEE Trans. Plasma Sci.*, vol. PS-15, no. 5, p. 533, 1987.

[308] W. Li, R. L. Thomas, R. E. Haskins, and A. M. Rubel, "Outgassing mea-surements of Cu powder, Cr powder, and their mixture," in: *Advances in Powder Metallurgy and Particulate Materials*, 2001, Metal Powder Industries Federation, p. 253.

[309] W. Li and A. M. Rubel, "Outgassing characteristics of Cr powders," in: *Proceedings of the World congress on Powder Metallurgy & Particulate Materials*, Part 13: Sintering, 2002.

[310] W. F. Rieder, M. Schussek, W. Glatzle, and E. Kny, "The influence of composition and Cr particle size of Cu/Cr contacts on chopping current, contact resistance, and breakdown voltage in vacuum interrupters," *IEEE Trans. Compon. Packag. Technol.*, vol. 12, no. 2, p. 273, 1989.

[311] M.-F. Devismes, H. Schellekens, P. Picot, *et al.*, "The influence of CuCr25 characteristics on the interruption capability of vacuum inter-rupters," in: *Proceedings of the 21st ISDEIV International Symposium on Discharges and Electrical Insulation in Vacuum*, 2004, p. 359.

[312] D. M. Atkinson, "Improvements in or relating to the manufacture of contact elements for vacuum interrupters," Great Britain Patent 1,459,475, 23 May 1974.

[313] R. E. Gainer, "High-density high-conductivity electrical contact material for vacuum interrupters and method of manufacture," U.S. Patent 4,190,753, 26 Feb. 1980.

[314] H. Hassler, H. Kippenberg, and H. Schreiner, "Composite metal as a contact material for vacuum switches," U.S. Patent 4,032,301, 28 June 1977.

[315] C. Kowanda, M. Hochstrasser, A. Schwaiger, and F. E. H. Müller, "Net shape manufacturing of CuCr for vacuum interrupters," *IEEE Trans. Dielect. Elect. Insul.*, vol. 18, no. 6, p. 2131, 2011.

[316] R. Müller, "Arc-melted CuCr alloys as contact materials for vacuum interrupters," *Siemens Forsch.-Entwickl.-Ber.*, vol. 17, no. 3, p. 105, 1988.

[317] P. Li, X. Yao, S. S. Zhang, X. Wang, L. Gang, and Z. Liu, "Comparison of Electrical Performance of CuCr Contact Materials Manufactured by Two Methods," in: *Proceedings of the 28th ISDEIV International Symposium on Discharges and Electrical Insulation in Vacuum*, 2018, p. 607.

[318] W. Li, R. L. Thomas, and R. K. Smith, "Effects of Cr content on the interruption ability of CuCr contact materials," *IEEE Trans. Plasma Sci.*, vol. 29, no. 5, p. 744, 2001.

[319] H. Wang, Y. Geng, Z. Liu, J. Lin, X. Li, and Y. Li, "Prestrike characteristics of arc-melted CuCr40 and infiltration CuCr50 contact materials in 40.5 kV vacuum interrupters under capacitive making operations," *IEEE Trans. Dielect. Elect. Insul.*, vol. 24, no. 6, p. 3357, 2017.

[320] G. A. Kumichev and I. N. Poluyanova, "Investigation of welding characteristics and NSDD probabilities of different contact materials under capacitive load conditions," in: *Proceedings of the 27th ISDEIV International Symposium on Discharges and Electrical Insulation in Vacuum*, 2016, p. 189.

[321] S. Zhang, W. Wang, P. Li, *et al.*, "Influences of different manufacture ways on the anti-welding ability of electrical contacts," in: *Proceedings of the 29th ISDEIV International Symposium on Discharges and Electrical Insulation in Vacuum*, 2020, p. 380.

[322] E-D. Wilkening, M. Lindmayer, and U. Reininghaus, "A test method for 50 Hz capability of vacuum interrupter materials," *IEEE Trans. Dielect. Elect. Insul.*, vol. 4, no. 6, p. 854, Dec. 1997.

[323] T. Delachaux, F. Rager, R.A. Simon, and D. Gentsch, "Testing procedure for the current interruption capability of vacuum interrupter contact materials," in: *Proceedings of the 25th ISDEIV International Symposium on Discharges and Electrical Insulation in Vacuum*, 2012, p. 209.

[324] H. Kippenberg, R. Müller, H. Schnödt, I. Paulus, and R. Hess, "Contact pieces for vacuum switchgear, and method for the manufacture thereof," U. S. Patent 4,749,830, 7 June 1988.

[325] B. Miao, J. Xie, J. He, G. Liu, W. Wang, and X. Wang, "Effects of trace Te on the anti-welding property of Cu-30CrTe alloy contact material," in: *Proceedings of the 25th ISDEIV International Symposium on Discharges and Electrical Insulation in Vacuum*, 2012, p. 189.

[326] H. Yang, Y. Geng, Z. Liu, *et al.*, "Back-to-back capacitor bank switching performance of vacuum interrupters: comparison of three contact materials," in: *Proceedings of the 2nd ICEPE International Conference on Electric Power Equipment – Switching Technology*, 2013.

[327] W. Li, M. Leusenkamp, and D. Ellis, "Testing a Cu-8Cr-4Nb contact material in vacuum interrupters," in: *Proceedings of the 26th ISDEIV International Symposium on Discharges and Electrical Insulation in Vacuum*, 2014, p. 409.

[328] P. Malkin, "The vacuum arc and vacuum interruption," *J. Phys. D: Appl. Phys.*, vol. 22, p. 1005, 1989.

[329] A. Greenwood, *Vacuum Switchgear, Stevenage*, UK: Institution of Engineering and Technology, 1994.

Chapter 4

Environmentally friendly high-voltage AC switching technology: gas circuit breakers with SF$_6$ alternative gases

Martin Seeger[1]

4.1 Introduction to high-voltage gas circuit breakers and interruption processes

4.1.1 Introduction

Circuit breakers (CB) are important elements in the power system that are needed to ensure safe power flow under various network conditions [1–5]. Requirements are defined in the relevant standards e.g., [6,7] and in Chapter 2 of this book. IEC [6] defines a CB as 'mechanical switching device capable of making, carrying and breaking currents under normal circuit conditions and also making, carrying for a specified time and breaking currents under specified abnormal circuit conditions such as those of short-circuit'. Currents of interest range from a few 10 A to several 10–100 kA. CBs are required to operate reliably over a wide temperature range from –50°C to 60°C. The safe operation needs to be ensured over a typical lifetime of at least 30 years.

In distribution networks, below 1,000 V air CB are commonly used. At the medium-voltage level, up to 52 kV, mainly vacuum circuit breakers (VCB) are used at present; they have mostly replaced SF$_6$ CB in this voltage range. For higher voltage, from 52 kV to more than 1,000 kV, modern CBs to date mainly use SF$_6$ as the interruption medium. There are still CBs in older installations in the field that use air or oil as the quenching and insulation medium. All these technologies use an electric arc to interrupt the current at a current zero (CZ)-crossing. In AC CBs such zero-crossings occur naturally, whereas in DC CBs, a zero-crossing must be generated artificially. The difference in interruption performance of the various CB types is related to the medium in which the plasma is formed, i.e., the material properties. Differences in material properties need to be taken into account when designing CBs. The control of arc conductance around CZ is a crucial aspect of a design. To ensure high insulation strength, the insulation properties of the gas and insulating materials used must be optimised.

[1]Hitachi Energy Switzerland Ltd, Central European Research Center, Baden-Dättwil, Switzerland

This section will focus on the introduction of CB on the example of present HV AC CBs that use SF_6. It serves as a starting point for later discussion of SF_6 alternative gases. Section 4.1.2.1 gives a short overview of the types of circuit breakers. The interrupter unit, where the arc interruption occurs, is explained in Section 4.1.2.2 and the operation of circuit breakers is shown in Section 4.1.2.3.

The fundamental processes and important parameters of insulation and current interruption are introduced in Section 4.1.3. In Section 4.2, these parameters will be compared for the various SF_6 alternative gases and available information on the interruption and insulation performance will be shown in Section 4.3. In Section 4.4, service issues, such as material compatibility, contact and nozzle erosion, gas consumption and leakage of the new gases will be addressed.

4.1.2   CB design and operation

4.1.2.1   Types of circuit breakers

Examples of different types of circuit breakers are shown in Figure 4.1. Different enclosures are used depending on the application. The enclosure can be made of insulating material in the so-called live-tank CBs, where the enclosure is on potential and is primarily composed of porcelain or composite insulators, the latter consisting of fibre-reinforced epoxy covered by silicon sheds. Other enclosure types consist of a grounded metal housing with bushings. Such enclosures are employed, for example, in the so-called 'dead tank CBs' (DTB) or in 'gas insulated switchgear' (GIS). For the highest currents, but for voltages in the medium voltage rather than in the high-voltage range, generator CBs are used; they are located

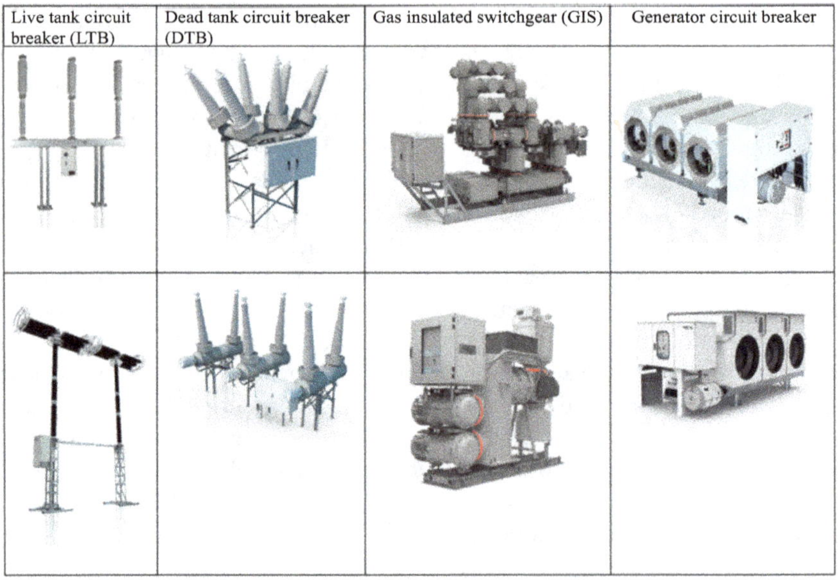

Live tank circuit breaker (LTB)	Dead tank circuit breaker (DTB)	Gas insulated switchgear (GIS)	Generator circuit breaker

Figure 4.1   Types of circuit breakers (courtesy of Hitachi Energy Ltd.)

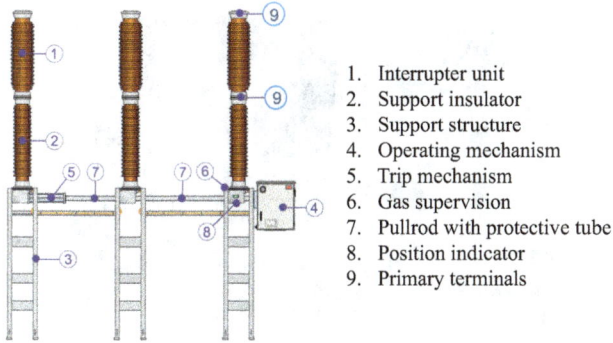

1. Interrupter unit
2. Support insulator
3. Support structure
4. Operating mechanism
5. Trip mechanism
6. Gas supervision
7. Pullrod with protective tube
8. Position indicator
9. Primary terminals

Figure 4.2   LTB CB sketch, adapted from [7]

between the generator and the step-up transformer in power plants and typically have nominal voltage ratings of 10–30 kV.

To give an overview of the main components of a circuit breaker, a sketch of an LTB CB is shown in Figure 4.2. Arc interruption occurs in the interrupter unit, which is connected to the operating mechanism via a pull-rod. For three-pole-operated CBs, one operating mechanism is used for all three phases, whereas in a single-pole-operated CB, one operating mechanism is used per phase. The latter is used more frequently for transmission voltages at and above 245 kV due to the operating mechanism requirements. The entire CB including support insulators is mounted on a support structure. The CB is connected to the network at the primary terminals. Gas supervision allows to control the fill pressure in the CB and the position indicator shows if contacts are separated (open) or connected (closed).

4.1.2.2   Types of interrupter units

The interrupter chamber is located within the interrupter unit, see Figure 4.3. Inside this interrupter chamber a pressure build-up is created that is used to extinguish the arc between the arcing contacts at a natural current zero (CZ) of the applied AC current. This principle is used for SF_6 as well as for SF_6-alternative gases.

A sketch with the details of such an interrupter chamber is shown in Figure 4.4. It consists of the tulip and plug arcing contacts, which are surrounded by the nominal current contacts. Nozzles made of PTFE are used for insulation between the contacts, which are denoted the auxiliary and insulating nozzles. There is a so-called moving contact, which is connected to the operating mechanism on the one side and the fixed contact, which is connected to the opposite side. Note that the expression 'fixed contact' stems from technologies where the plug-arcing contact was static. Today, movement of the plug contact is often utilised and achieved by means of a gear connected to the insulating nozzle and the plug. This so-called double-move system enables higher relative contact velocities, resulting in higher interruption performance and shorter arcing time for some switching cases. Self-blast (SB) CB, see Figure 4.4 (top), is characterized by the presence of a so-called heating chamber, which remains unchanged in volume during the operation of a

145 kV, 40 kA

Figure 4.3   *Interrupter chamber located in an LTB CB (courtesy of Hitachi Energy Ltd.)*

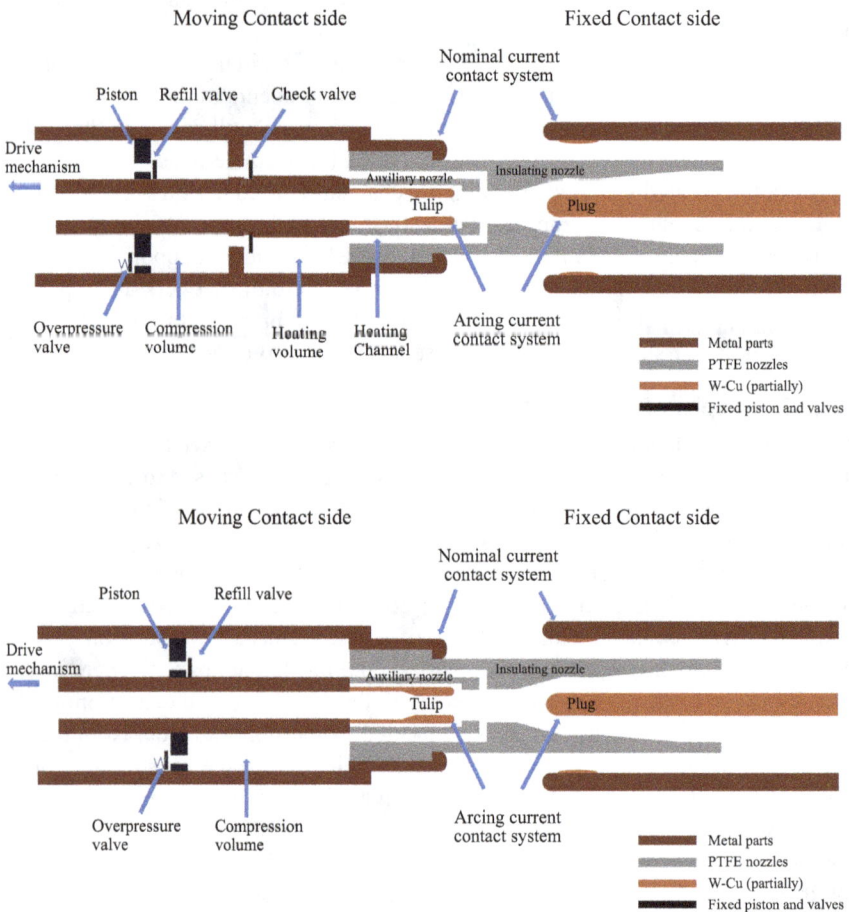

Figure 4.4   *(Top) Sketch of SB CB interrupter chamber cross-section in open position; (bottom) sketch of puffer CB interrupter chamber cross-section in open position*

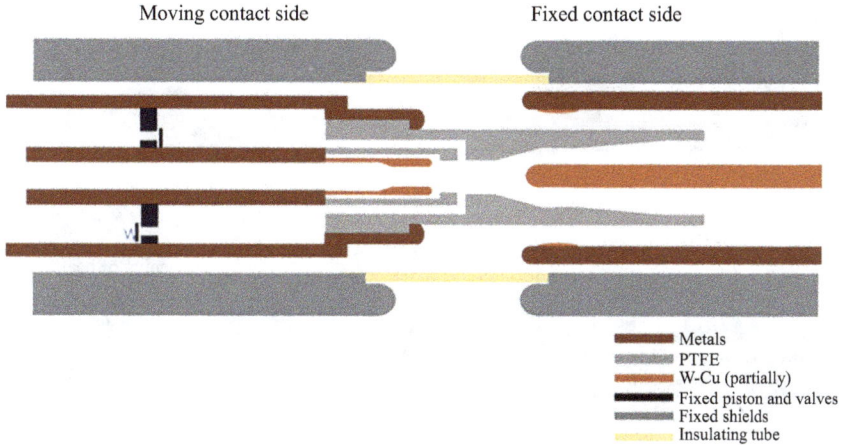

Figure 4.5   Sketch of a typical puffer interrupter cross-section as employed in metal-enclosed CB, with additional shields and insulating tube

CB. In this volume, pressure build-up is created during the interruption of high short-circuit currents above 10–20 kA, typically. This pressure build-up stems from the ablation of nozzle material (PTFE) and clogging of the nozzles by the arc, which is denoted as 'back-heating' in the following. The heating chamber is connected via a check valve to the compression chamber, which, as the name indicates, is used for gas compression during operation. This mechanical gas compression is required for the interruption of load and low short-circuit currents, below 10–20 kA, typically. In the compression chamber, the piston is static and there are valves for refilling gas during closing operation (refill valve) and an overpressure valve to avoid excessive pressure build-up acting on the operating mechanism during an opening operation. Excessive pressure could lead to slowing down of the movement of the moving contact and should be minimised. In a so-called 'puffer' CB, no heating chamber is used, and the compression chamber is directly connected to the arc zone (Figure 4.4, bottom). This type of CBs does rely less on the pressure build-up from nozzle ablation and nozzle clogging than SB CBs.

In a metal-enclosed CB, the interrupter unit is placed inside a grounded metal tank. Additional shields towards the tank are usually used and an insulator tube to connect both sides for higher stability is typically also implemented; see Figure 4.5 for a view of the interrupter chamber.

4.1.2.3   Operation of CBs

After a trip signal is delivered to the operating mechanism, the moving contact starts to move, and the contacts separate. For simplicity, this will only be explained on the example of a single-move, self-blast CB at large short-circuit current. The sequence of movements is shown in Figure 4.6, together with a possible short-circuit current at the given instant. It should be noted that the phase angle of the current is random and here just shown for illustration.

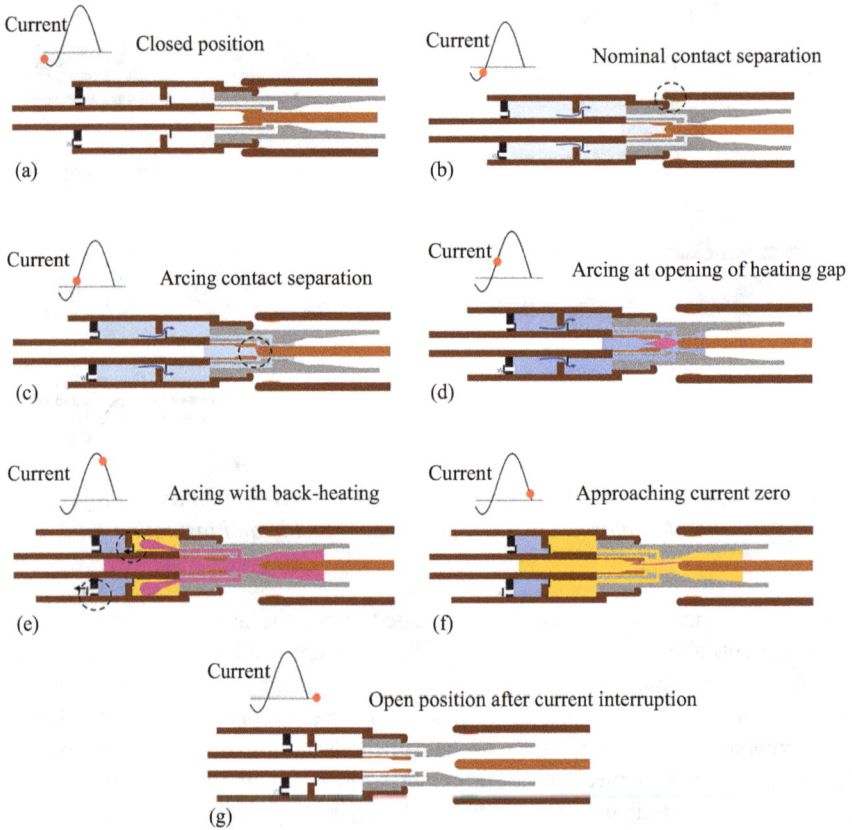

Figure 4.6   Operating sequence of a SB circuit breaker with back-heating

First, the nominal contacts separate. Due to a small arc voltage created at this contact separation, the current through the CB commutes to the arcing contacts, which have not separated yet. After the separation of the arcing contacts, an arc develops in the arc zone. At this time, some pressure is created in the compression chamber, due to the ongoing reduction in volume of the compression chamber. This creates a flow through the check-valve to the heating chamber and from there to the arc zone. This flow can push out residual material generated as a result of contact separation, such as metal droplets and vapor, provided the current is still low. Before the plug contact opens the heating channel gap, the arc is only weakly blown and no significant flow of hot gas and plasma from the arc zone to the heating chamber, i.e., back-heating, can occur, even if the alternating current is near a peak at this instant. As the contacts separate further and if the current is sufficiently high, back-heating can occur, i.e., ablated PTFE vapor from the nozzles flows to the heating chamber and increases by mass and energy transfer the pressure in this chamber. During this phase, the pressure in the nozzles is higher than the pressure in the heating chamber and can reach several

MPa. Towards CZ the pressure in the arc zone drops, due to the reduction of current and increased outflow cross-section due to the shrinking of the arc. Eventually, when the pressure in the arc zone is lower than in the heating chamber, the so-called 'flow reversal' occurs and relatively cold gas from the heating chamber starts to flow to the arc zone and removes the residual gases and materials from the previous high-current phase. At CZ, if arc cooling is sufficient (this will be explained later), the current is interrupted and the transient voltage from the network, the transient recovery voltage (TRV) develops across the CB contacts. The present case was explained based on the example of a high short-circuit current. In the case of nominal currents, no back-heating occurs, and the arcing time is usually much shorter than for a short-circuit current case. At closing, the opposite movement occurs, and fresh gas that has not been exposed to the arc flows into the compression chamber via the refill valve.

The operation of an interrupter chamber is the same for SF_6 and SF_6-alternative gases. However, as will be explained in the later sections, due to the change in the material properties, adaptions are needed for optimisation of performance.

4.1.3   Fundamental processes for insulation and interruption

In the present section, the relevant processes and parameters for insulation and current interruption will be presented. Since the current interruption needs gaseous insulation in gas CB, Section 4.1.3.1 will give the concepts of gaseous insulation, followed by the interruption process in Section 4.1.3.2.

4.1.3.1   Gaseous insulation

The relevant processes for gaseous insulation will only briefly be summarized in this section. Details can be found in many textbooks [8–11] and these will not be further referenced or explained here in detail. A dielectric breakdown generally occurs through a series of several steps, including an electron avalanche which can develop into a 'streamer' if sufficient charge is released by collisions of electrons with atoms, molecules or ions in the head of the avalanche. This is the so-called avalanche to streamer transition. A streamer is characterized by low, close to ambient temperature in the streamer channel. Such streamers can propagate at sufficiently high background fields through the gap. A streamer can also do transition into a so-called 'leader' by thermal heating of a streamer channel, as will be explained below. After a streamer or leader crosses the gap, a high current flows through the discharge channel, leading to significant channel heating followed by the collapse of the applied voltage. This process is denoted as spark transition.

Electron avalanches occur at electric fields exceeding the critical field E_{cr}, which is defined by the zero crossing of the so-called effective ionisation coefficient $\alpha_{eff} = \alpha-\eta$, the net difference between the ionisation coefficient α and the electron attachment coefficient η. Typical field dependencies of the effective ionisation coefficient are shown in Figure 4.7 for air and SF_6. Note that usually density-reduced values are used, i.e., α_{eff}/n and E/n, where n is the particle number density. Due to the much higher electron attachment in SF_6, the effective ionisation coefficient is negative over a large range of electric fields and crosses zero only at

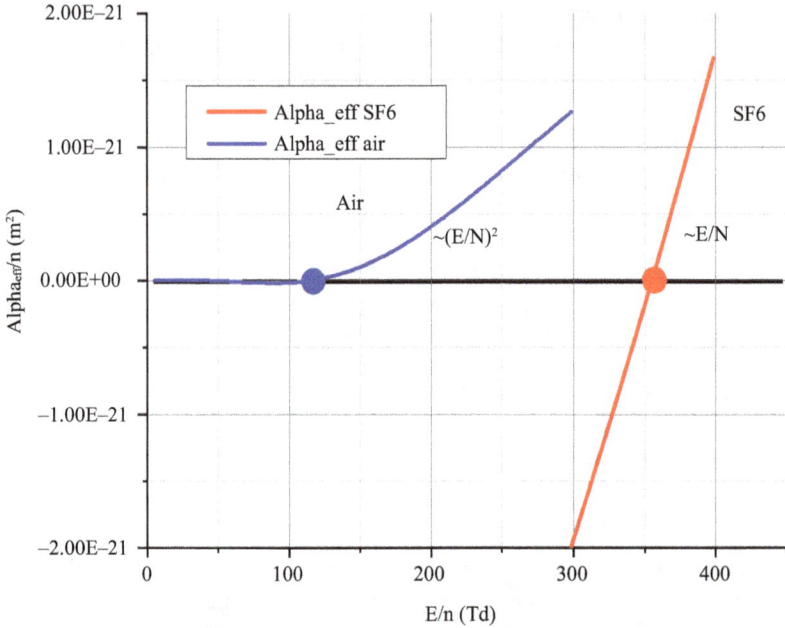

Figure 4.7   Effective ionisation coefficients for air and SF₆ vs electric field. Particle density reduced values are shown for effective ionisation coefficient and electric field, with 1Td = 10⁻²¹ V·m², which is the typical unit for density reduced electric fields.

fields much higher than that for air. This explains the much higher insulation performance of SF₆ compared to air.

Since insulation strength scales proportional to particle number density, the density-reduced values are used in the figure. A convenient unit for density-reduced field is 'Townsend' with $1Td = 10^{-21}$ V·m². From the figure, it can be seen that the critical field E_{cr} is about 112 Td and 355 Td for air and SF₆, respectively. This is denoted by the solid circles in the figure. With one Td corresponding to about 2.5×10^4 V/m at a pressure of 0.1 MPa and 20°C, this translates to the well-known pressure-reduced critical fields of 28 V/(m·Pa) and 89 V/(m·Pa) for air and SF₆, respectively.

As mentioned before, an avalanche can develop into a streamer if a sufficient charge is released due to ionisation, leading to sufficient field distortion for further ionisation and possibly streamer propagation. This transition is described by the streamer inception criterion, which follows from the effective ionisation coefficient as

$$K = \int_0^{lcr} \alpha_{eff} \cdot dz \tag{4.1}$$

where the length l_{cr} is the distance at which the electric field drops to the critical field along an electric field line and the ionisation integral is $K = 10.5$ for SF₆. A streamer can undergo a transition to a leader via thermal heating of a streamer channel; see, for example, [12–14]. This lowers the electric field in the channel and increases the electric field at the tip of the leader accordingly, which eases discharge propagation through the gap. Typically, leaders occur in most gases at high gas densities at pressures above 0.2 MPa at ambient temperature [15] and are, therefore, the dominant discharge propagation mechanism for CB applications.

For streamer inception, small surface imperfections (surface roughness) can locally enhance the electric field [16–20]. This needs to be taken into consideration in streamer inception calculations, particularly at the high pressures of about 0.6 MPa used in HV CBs. As can be seen from Figure 4.7, the rise of the effective ionisation coefficient in air as a function of reduced electric field is lower than that of SF_6. This has an impact on the 'roughness sensitivity', e.g., [9]. For the streamer integral, larger field enhancements or lengths l_{cr} over which the electric field is above the critical field is needed in air than in SF_6 to fulfil the criterion. This explains that air insulation at atmospheric pressure is less sensitive to small surface protrusions. At increased pressures ($p > 0.5$ MPa) as used e.g., in gas-insulated switchgear, this difference is less pronounced and a significant roughness sensitivity can be seen also in air, as will be explained in Section 4.2. Due to the much lower critical field of air compared to SF_6, breakdown fields in air remain always significantly below those of SF_6, even at pressures well above 1 MPa.

It should be noted that the critical field, as a minimum breakdown condition, depends on several parameters [21], which in a simplified way can be expressed as:

$$E_{cr} = E_{cr,0} \cdot \frac{T_0}{T} \cdot p \cdot f(T,p) \tag{4.2}$$

where $E_{cr,0}$ = 89 V/(m·Pa) for SF_6 [22] denotes the critical field at standard conditions, e.g., at a temperature of e.g., T_0 = 300 K and a pressure of 0.1 MPa. The pressure p and temperature T in this expression describe the particle number density n, which is $n = p/(kT)$ for an ideal gas, where k = 1.380649 × 10^{-23} J/K is the Boltzmann constant. The function f includes the pressure and temperature dependence of the critical field, which due to gas decomposition affects the kinetic properties of the gas. This function is not as easy to determine for gases at elevated temperatures in comparison to ambient temperatures, since composition and kinetic calculations become increasingly complex with rather large uncertainties, see e.g., [23,24] and references therein. For the calculation of the effective ionisation and attachment coefficients, the knowledge of the electron energy distribution function and electron impact cross-sections (see e.g. [9]) are required for each species in addition to the composition of the gas. This information is usually incomplete and hence assumptions and simplifications need to be made, such as employing the local thermodynamic equilibrium and local chemical equilibrium (LTE/LCE) approach or making assumptions regarding the impact cross sections. Precise measurements at elevated temperatures, as required for CB applications, are on the other hand also very challenging [13,25,26]. An example of the calculated pressure and temperature dependence of the function f for SF_6 is shown in Figure 4.8. Above the decomposition temperature of about 1,500 K, SF_6 decomposes into smaller

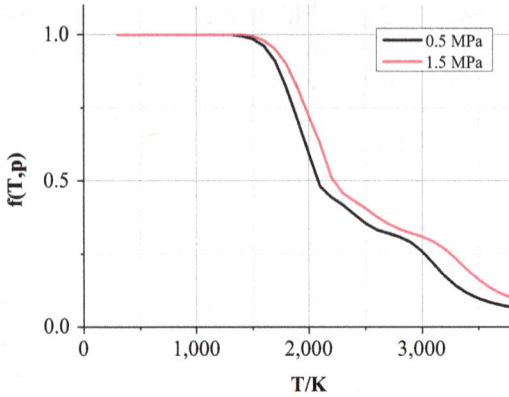

Figure 4.8   *Pressure and temperature dependence of the material function* f*(T,p),*
data from [12]

molecules, such as SF_4 and SF_2. At 2,100 K, the function f is reduced to about 50% of the value at ambient temperature for a pressure of 0.5 MPa. Below the decomposition temperature, the particle density reduced critical field is constant, i.e., the critical field would simply scale as $1/T$ for constant pressure. A higher pressure delays the reactions and therefore shifts the function to higher temperatures. This relation is very important for the comparison of the dielectric recovery performance of various gases, as will be discussed in Section 4.2.

4.1.3.2   Current interruption

A sketch of typical current and voltage waveshapes for a terminal fault are shown in Figure 4.9. A zoom-in of the CZ phase is also shown. For the interruption of

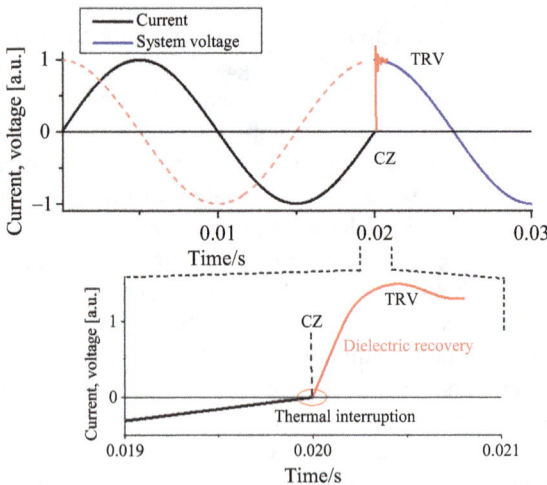

Figure 4.9   *Sketch of current and voltage waveshapes, adapted from [38]*

short-circuit currents, thermal interruption at CZ and dielectric recovery in the interrupter are crucial for the successful clearing of a fault event. Both phases are indicated in Figure 4.9. Since short circuits in the power grid are typically inductive in nature, the voltage and the current have a phase difference. This leads to the most severe conditions when the voltage rapidly rises to the maximum value of the TRV within a short-time interval of some 10–100 µs after CZ. There are other switching cases, such as switching of inductive and capacitive loads. These low-current switching cases are described in the standards of IEC and ANSI [6,7] and Chapter 2. The main challenge for these cases is not the current interruption but the withstand of the TRV, i.e., for these cases, the dielectric withstand is decisive. Therefore, the focus in this section is on the interruption of short-circuit currents. The different switching cases can be found in the relevant standards and many textbooks [1–7] and will not be discussed in detail, since the physical principles for short-circuit current inter-ruption are similar for the different short-circuit current switching cases.

A typical pressure rise in the heating chamber that results from the mechan-isms explained above is shown in Figure 4.10. After the pressure is maximum, the pressure decreases due to the outflow of gas from the heating or compression chamber to the arc zone via the heating channel, see Figure 4.11. This is the 'flow reversal' phase, which sets the stage for the current interruption process. During this phase, the arc mode switches from ablation controlled [27], where the arc burns in the nozzle vapor (typically PTFE), to axially blown [28,29], where the arc is convectively stabilised by the imposed flow from the heating chamber. In the axially blown arc mode, the axial pressure and velocity profiles in the arc zone between the contacts are of great importance [30]. These are shown in Figure 4.12. As can be shown theoretically for a parabolic pressure profile, the arc is of

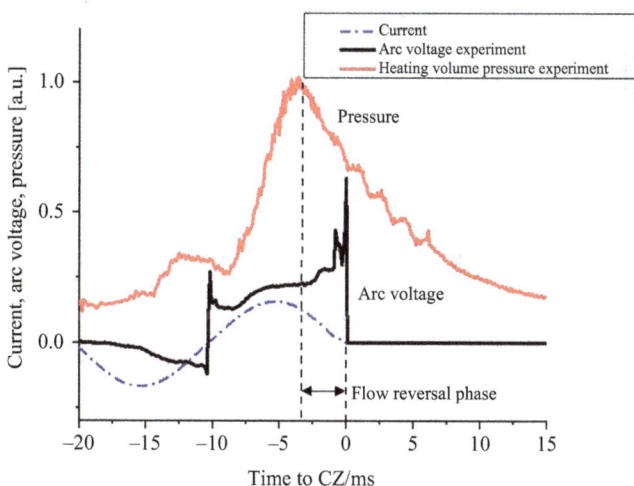

Figure 4.10   Current, arc voltage and pressure during arcing in an SF$_6$ SB CB, adapted from [38]

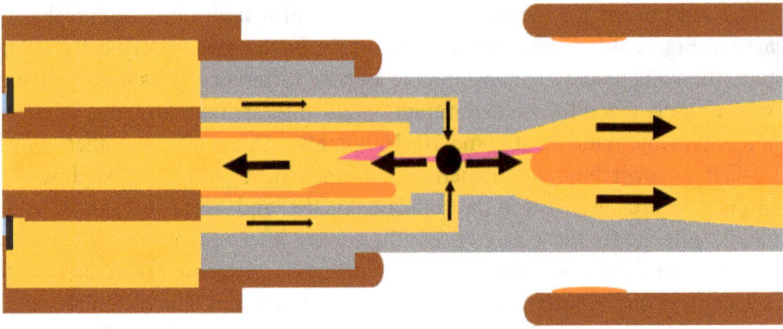

Figure 4.11   *Sketch of the arc zone. The arrows indicate the flow from the heating chamber before CZ. The stagnation point is indicated by the full circle.*

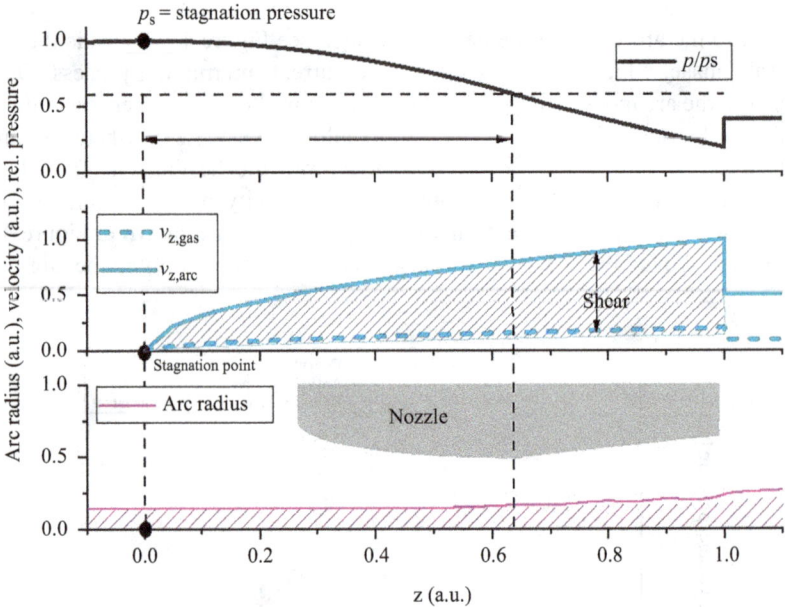

Figure 4.12   *Sketch of axial distribution of pressure, gas and plasma velocity and arc radius, adapted from [30]*

cylindrical shape [31]. This is the ideal case, which is displayed here only as an example. The contacts are omitted for simplicity and only one side of the flow is shown from the stagnation point, which refers to the point where the velocity v on the axis is zero. Due to the pressure profile, the arc plasma and surrounding gas are accelerated, which keeps the arc diameter small. For similar Mach numbers v/c

(c = velocity of sound), different absolute flow velocities will result in both inside and outside the arc. This creates vorticity, which leads to turbulent heat exchange. This is the dominant cooling mechanism at CZ, since radiation and convective cooling become negligible at CZ [32,33]. Due to turbulent cooling by large eddies entering the arc from the outside, its shape deviates from cylindrical in the diffuser region after the nozzle throat. Further downstream a shock might occur (position $z = 1$ in Figure 4.12). Shock formation is typically due to changes in flow cross-sections, which leads to a pressure rise and decrease of velocity in the diffuser. As a result, the arc widens in the diffuser region.

4.1.3.2.1   Thermal interruption

For the interruption at CZ, the arc needs to be cooled extensively to achieve a sufficiently low electric conductivity. This process is denoted as a thermal interruption. As described in [34,35], the most efficient arc cooling occurs in the region of the nozzle where the velocity is high, and the arc cross-section is small. The smallest cross-section usually occurs around the nozzle throat, which is, as a result, the most important region for current interruption at CZ. In this region, the highest arc resistance is built up, see Figure 4.13. The computational fluid dynamic (CFD) simulations from [34] for the case of a short-line fault with 90% of the rated short-circuit current (SLF90) show the differential resistance distribution along the axis of a typical CB geometry at different times before CZ. About 100 μs before CZ, the stagnation point still has the highest resistance, but at about 200 ns before CZ, resistance peaks in the nozzles have built up. This was also observed in simulations described in [36]. A high resistance at CZ reduces the current after CZ, the so-called post-arc current (PAC). The interruption

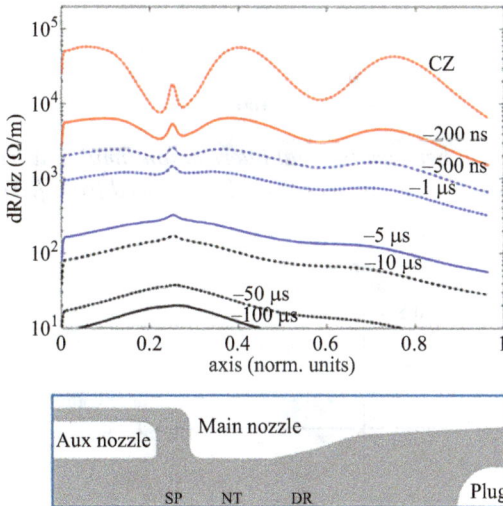

Figure 4.13   *Top: Differential arc resistance distribution in a CB geometry adapted from [34]. Bottom: Geometry used in the CFD simulation, SP, stagnation point; NT, nozzle throat; DR, diffuser region.*

performance is usually described by the slope of the short-circuit current di/dt at a zero-crossing, which can be interrupted for a given line surge impedance Z. In [37], it was shown that interruption performance can be correlated with the arc resistance 200 ns before CZ (Figure 4.14). The higher the resistance, the higher the di/dt that can be interrupted for a given line surge impedance. Note that this relation holds regardless of the details of the CB design. The required resistance value is only a function of the line surge impedance Z and possibly the gas type, as will be discussed later.

If the cooling power in the decision phase of interruption after CZ exceeds the power input from the PAC, the current is successfully quenched. In the opposite case of insufficient cooling, the arc channel can slowly heat up, leading to thermal runaway. This explains why this phase is called 'thermal' interruption phase. The current at and after CZ for such situations is shown schematically in Figure 4.15. In case of interruption, the PAC decreases to zero over the span of a few 100 ns for SF_6. In case of a failure, the PAC increases after CZ. Thus, the thermal interruption

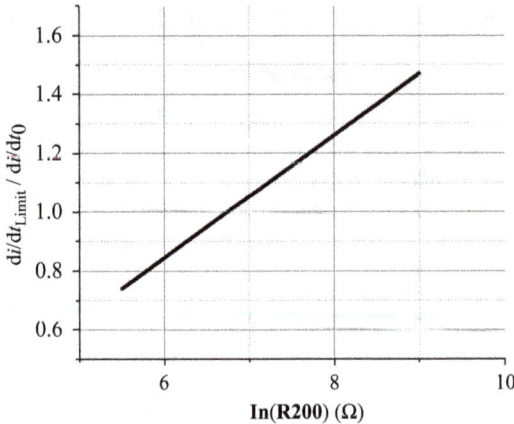

Figure 4.14   Relation between thermal interruption limit di/dt (relative units) and arc resistance 200 ns before CZ deduced from [37] for SF_6 CB

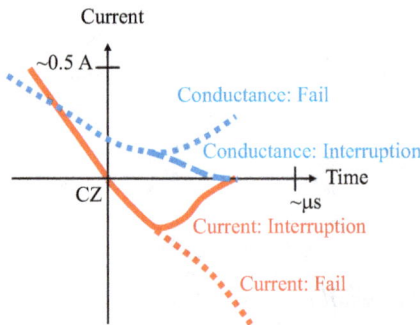

Figure 4.15   Sketch of PAC and conductance decay at CZ

is typically decided within a few 100 ns to a few microseconds after CZ, depending on quenching gas and test circuit parameters. The level of PAC at which the decision occurs depends mainly on the gas, switching case and possibly design and is typically around a few 10 mA to a few 100 mA for SF_6. The material properties that describe the thermal interruption performance will be presented in Section 4.2.3 for SF_6 alternative gases and mixtures and compared to those of SF_6. It must be noted that thermal and chemical non-equilibrium effects significantly influence the properties of the gas around CZ, as discussed in [33,38–42].

As explained above, the axial pressure profile is an important, design-dependent parameter. Additionally, the absolute value of the stagnation pressure is one of the most decisive parameters for thermal interruption. As shown in [43], the current interruption performance di/dt_{limit} depends on the stagnation pressure p_{stag} roughly according to the relation: $di/dt_{limit} \sim \sqrt{p_{stag}}$. This is an important nearly gas-independent relation that will be discussed later in the context of comparing SF_6 to alternative gases. Qualitatively, this relation can be understood by realizing that the arc diameter decreases with increasing pressure, a thin arc is easier to cool, and the pressure depends on the material properties.

Simplified equations for the description of the arc cooling of a cylindrical arc section of length L at CZ are given in [44]. The turbulent heat conduction can be written by $\kappa_t = 2 \cdot \rho \cdot c_p \cdot \epsilon$, where ρ is the density and $\epsilon \approx f_t \cdot R \cdot c_a$ is a heat exchange coefficient that scales with a characteristic length and velocity. For the characteristic length and velocity, the arc radius R and velocity of sound in the arc c_a, respectively, are chosen. The latter is valid in the relevant arc section in the nozzle throat where interruption is decided. f_t is an empirical factor that can be deduced from experiments. The energy balance at CZ can be written in a simplified form, neglecting radiative heat transfer as

$$\frac{d}{dt}\left(\pi \cdot R^2 \cdot L \cdot \rho \cdot h\right) = -2 \cdot \pi \cdot R \cdot L \cdot \kappa_t \cdot (T_a - T_0) \tag{4.3}$$

with the arc core temperature T_a and the surrounding gas temperature $T_0 \ll T_a$. Thus, the energy content of the arc is only reduced by the radial turbulent heat exchange across the arc boundary in this simplified representation and can be written in pressure-normalized form using the relations for κ_t

$$\frac{d}{dt}\left(\frac{\rho \cdot h}{p}\right) \approx -4 \cdot \left[\frac{\rho \cdot c_p}{p}\right] \cdot f_t \cdot c_a \cdot T_a \tag{4.4}$$

Accordingly, the change in the energy content of the arc is proportional to the material parameter $\frac{\rho \cdot c_p}{p}$, the velocity of sound and the arc core temperature. The parameter $\frac{\rho \cdot c_p}{p}$ is mainly temperature and gas dependent. For SF_6, it is shown in Figure 4.16 as an example. The high values and large peak around the dissociation temperature of 2,000 K combined with low values above 4,000 K leads to the most efficient heat transfer at temperatures where the electric conductivity is negligible. In other words, turbulent heat exchange will rapidly cool the arc core to temperatures where electric conductivity is low. This efficient heat exchange explains the

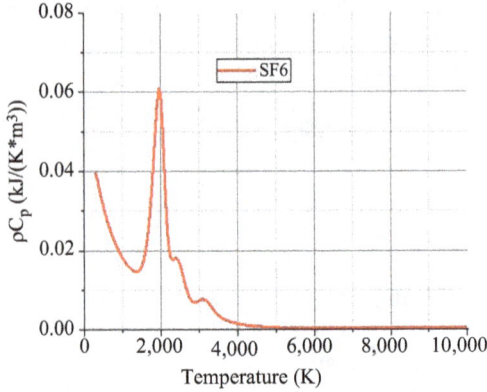

Figure 4.16   *Material parameter $\rho \cdot c_p/p$ for SF_6 at a pressure of 0.8 MPa*

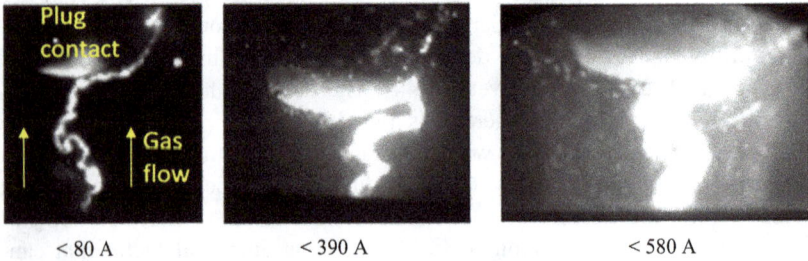

| < 80 A | < 390 A | < 580 A |

Figure 4.17   *Images of the arc shortly before CZ (< 50 μs) after current peaks of 5 kA, 25 and 37 kA, i.e., at currents less than 80 A, 390 A and 580 A, respectively, in SF_6. Details of the experiments can be found in [21].*

good thermal interruption properties of SF_6. This will be discussed and compared to SF_6 alternative gases in Section 4.2.

It should be noted that the arc structure close to CZ is characterized by 3D, mainly helical instabilities. This can be observed in Figure 4.17, particularly in the diffuser region, as already indicated in Figure 4.12. This affects the total arc resistance and partially explains the scatter of thermal interruption performance in CB tests.

4.1.3.2.2   Dielectric recovery

After the current is interrupted, the TRV rises across the arcing contacts, as shown in Figure 4.9. During this rise, the CB recovers its dielectric strength. Breakdown might occur if the dielectric strength between the contacts or to the ground is not sufficient to withstand the applied TRV stress. Since arc temperatures are below 5,000 K during this phase, the electric conductivity is generally low enough so that thermal runaway is less likely to occur than during the thermal interruption phase. However, the insulation strength can break down via an avalanche-to-streamer or leader transition as discussed in Section 4.1.3.1 on dielectric breakdown above. For this reason, this phase is referred to as the dielectric recovery phase. The breakdown during this time is characterized by a fast collapse of the applied voltage within typically a few 10–100 ns. As shown in [21], two phases can be distinguished, the early and the late dielectric recovery. During the early dielectric recovery, the breakdown occurs in what remains of the arc channel. This phase has a typical duration of 100 µs or less and is simply denoted as early dielectric recovery in the following text.

The most important quantity to predict breakdown to a first approximation is the critical field, see (4.2) above. If the critical field is exceeded, avalanches and possibly streamers or even leaders can start in what remains of the arc channel, as shown, for instance, in [45]. In that investigation, the early dielectric recovery in an air gas flow could be described well by the streamer inception criterion using the pressure, temperature and electric field on the axis from CFD simulations of the arc zone, see Figure 4.18. The recovery characteristics for different filling pressures are shown normalized to the breakdown level at ambient temperatures (=cold

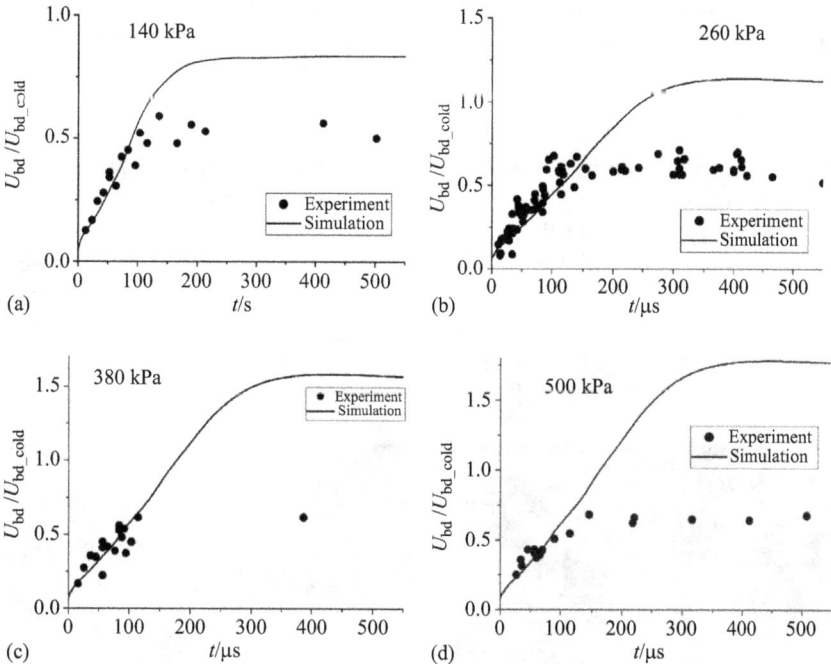

Figure 4.18 *Dielectric recovery in a test CB filled with air at (a) 140 kPa, (b) 260 kPa, (c) 380 kPa, and (d) 500 kPa reproduced after [45]. The simulations were done by CFD calculations and agree reasonably well in the early recovery phase. The relative recovery characteristics are similar for different filling pressures, indicating that the dielectric recovery scales with pressure.*

breakdown). The cold breakdown voltages increase with pressure as shown in [45]. The similarity of these characteristics indicates that the dielectric recovery scales with filling pressure. For this example, the dielectric recovery in air was chosen since it is much slower than in SF_6 and can, therefore, be studied more easily. It should be noted that the assumption of capacitive electric field distribution on the axis shortly after CZ might not be correct, since at the beginning of the recovery process, a high ion concentration will lead to significant field distortions in the channel. However, at elevated temperatures, the streamer length will be quite large, as shown in [21], which might reduce the dependencies of detailed field distribution in the channel. Since the pressure distribution in a CB does not change significantly within the first 100 µs after CZ, the voltage rise of the early dielectric recovery is mainly determined by the temperature decay of the arc channel. In addition, related non-LTE and non-LCE effects will play a role, since they affect the chemical composition and electron temperature compared to heavy particle temperatures [42]. This influences the critical field as well as the electric field and temperature distribution in the channel.

After the arc channel has been cooled, sufficient breakdown at dielectrically weak spots can occur outside the arc channel within the arc zone, see Figure 4.19. The properties of such weak spots characterize the breakdown level during the late recovery. In other words, breakdown depends on the distribution of temperature, pressure, electric field and gas composition. Similar breakdown models for ambient temperature gas can also be applied to hot gas as shown for SF_6 in [21], see Figure 4.20. In that work, it was shown that while the streamer inception calculations were quite pessimistic, the predictions done using a leader model were in reasonable agreement with the measured recovery in a test CB. Note that in the investigation, the early dielectric recovery in SF_6 could not be measured since it was much faster than e.g. in the case of air shown in Figure 4.18. This method was also successfully used for commercial CB tests [46] and achieved good agreement between predictions and the results of power tests.

Figure 4.19   Breakdown at a CB plug around CZ in SF_6 after a peak current of 23 kA and 240 kV were applied. The direction of the flow is indicated by the arrows. The breakdown after CZ does not occur in the previous arc channel; adapted from [13].

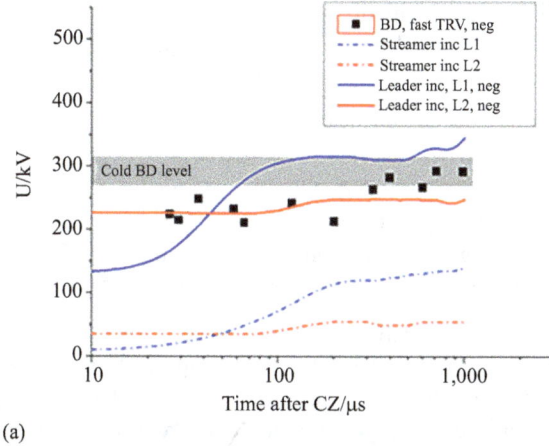

(a)

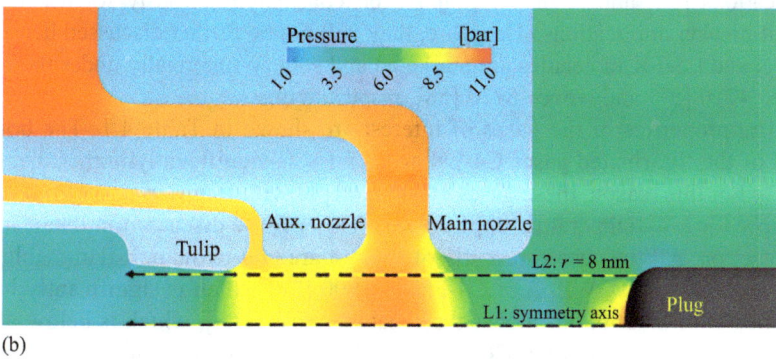

(b)

Figure 4.20   Dielectric recovery in SF_6 in a test CB adapted from [21] after the application of a short-circuit currents in the range of 20–30 kA. (a) Experimental results compared to simulations with streamer and leader models, shown by the curves for a line on the symmetry axis (L1) and another line (L2) with a radial offset of 8 mm from L1, i.e., off-axis. (b) CFD pressure calculation with indicated lines L1 and L2. The evaluations of streamers and leaders always started at the plug.

4.2   Material properties of SF_6 alternative gases compared to SF_6

4.2.1   Overview of gases of interest

Numerous investigations in the last decade have identified the most promising gases to be used for replacing SF_6 [47–52]. Electric insulation and interruption properties, global warming potential for a 100-year horizon [52,53], referred to as GWP_{100}, toxicity, chemical stability under operating conditions, flammability, material compatibility, heat dissipation capability and ozone depletion potential (ODP) were considered. This led to the selection of the following molecules as most suited for HV circuit breaker applications:

- CO_2
- O_2
- Heptafluoroisobutyronitrile, $((CF_3)_2\text{-}CF\text{-}CN$; molecular formula C_4F_7N; designation in HV community: C4-FN
- 1,1,1,3,4,4,4-Heptafluoro-3-(trifluoromethyl)-2-butanone, $CF_3C(O)CF(CF_3)_2$; molecular formula $C_5F_{10}O$, designation in HV community: C5-FK

Figure 4.21 shows the chemical structures of these molecules. Other molecules were also investigated, as e.g., HFO-1234ze for the application in MV switchgear or CF_3I, which was mostly studied in academia [54–56]. These two gases have several drawbacks regarding decomposition and toxicity in high-current switching applications [52] and have so far not been considered seriously for use in HV circuit breaker applications. Therefore, they will not be further discussed here. The C4-FN and C5-FK molecules are marketed and sold commercially under the name Novec 4710 [57] and Novec 5110 [58], respectively.

The properties of the gases of interest are shown in Table 4.1. The boiling point of the fluorinated gases C4-FN and C5-FK is significantly higher than that of SF_6. For this reason, these gases must be mixed with a carrier gas, such as CO_2, to allow for low-temperature applications. This will be discussed in detail in the next section. All the gases presently of interest are non-toxic, non-flammable and have no ODP. The GWP_{100} of C4-FN is 2,100 [57] which is significantly lower than SF_6 with 25,200, but higher than that of C5-FK [58], which is below unity. Note that in the coming IPCC report [53], these numbers might be slightly adapted. The boiling point of C4-FN of $-4.7°C$ is much lower than that of C5-FK of $26.9°C$, which allows for lower temperature applications of the former. The dielectric withstand of pure C4-FN and C5-FK is about twice that of SF_6 at a reference pressure of 0.1 MPa. CO_2 has a dielectric withstand comparable to air. As will be explained in the following sections, the fluorinated compounds are most suitably used in a mixture with CO_2/O_2 [51,52], but there is also one

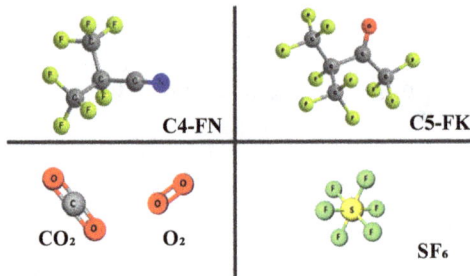

Figure 4.21   Molecular structures of SF_6 and SF_6 alternatives

Table 4.1 Properties of pure gases compared to SF₆, adapted from [51]

	CAS number[c]	Molecular weight (g/mol)	Density (kg/m³) at 0.1 MPa, 20°C	Boiling point/°C	GWP	ODP	Flammability	Toxicity LC^{50} (4 h) ppmv	Occupational exposure limit TWA[a] ppmv	Electric strength/p.u. at 0.1 MPa	References
SF₆	2551-62-4	146	6.17	−64[b]	25,200	0	No	8e5	1e3	1	14,23,52,53,61
CO₂	124-38-9	44	1.98[d]	−78.5[b]	1	0	No	≥1.6e5	5e3	≈0.3	60,62,63
C5-FK	756-12-7	266	10.7	26.9	<1	0	No	≥1.4e4	225	≈2	59,57,61
C4-FN	42532-60-5	195	7.9	−4.7	2,100	0	No	≥1e4	65	≈2	48,58,61,63
O₂	7782-44-7	32	1.43[d]	−182.96	0	0	No	–	4e5	≈0.3	6,16
N₂	7727-37-9	14	1.25[d]	−195.8	0	0	No	–	N.A.	≈0.3	16,24

[a]The occupational exposure limit is given by a time-weighted-average (TWA), 8-h, which is derived based on airborne exposure during a complete occupational working lifetime (30 years, 52 weeks per year, 5 days a week).
[b]Sublimation point.
[c]A unique numerical identifier assigned to every chemical substance described in the open scientific literature.
[d]https://www.engineeringtoolbox.com/gas-density-d_158.html

manufacturer that does not add any O_2 to the mixture [67]. The O_2 concentrations typically used are in the range of molar fractions of 10–13% [51,52], as will be explained later. Note that in the following, always molar fractions are used when giving concentrations, if not otherwise stated. For very low-temperature applications of −50°C only CO_2/O_2 mixtures are used [68,69]. The properties of the mixtures of interest are given in Table 4.2. For a filling pressure of 700 kPa and 3.5 mol% C4-FN in CO_2/O_2 a dielectric strength comparable to SF_6 with a GWP_{100} of about 293 is achieved [57,70]. This mixture can be used down to −30°C and is favoured for low-temperature CB applications. Mixtures based on C5-FK have a negligible GWP_{100} but can be used only for indoor applications with minimum temperatures of above −5°C. Due to the reduced maximum molar concentration of C5-FK permitted for such applications, the dielectric strength of the mixture is lower than for a −30°C $CO_2/C4\text{-}FN/O_2$ mixture. For this reason, most manufacturers have moved towards using $CO_2/C4\text{-}FN/O_2$ or $CO_2/C4\text{-}FN$ as the preferred mixtures for HV CB applications, see e.g., [52,70–72] and references therein.

4.2.2   Condensation of C4-FN and C5-FK gases and mixtures

Circuit breakers are exposed to a relatively wide temperature range, typically ranging from −50°C to 60°C. This requires the gases or gas mixtures used in circuit breakers to have sufficiently low liquefaction temperatures; otherwise, the gas density, and with it the insulation strength, would be reduced when the temperature drops below a certain level and condensation begins to occur. The most important parameter in this regard is the dew point, or, correspondingly, the vapor pressure of a gas or gas mixture. These parameters determine the maximum density of a gas or gas mixture that can be used in a given piece of equipment. Condensation is a phase change that is described by the equation of state (EOS) of a gas, or a gas mixture, which gives the relations between pressure, temperature and volume. Typically, these values are determined by measurement or from calculations using real gas properties, like Van der Waals or Peng-Robinson EOS, see e.g. [52,73–76] for further details. Measurements can, for example, be performed by slowly cooling a given amount of high-purity gas in a vessel at a slow rate in a climate chamber, while simultaneously recording the pressure and temperature precisely, e.g., [77]. The dew point can then be determined from a change in the slope of pressure decay with decreasing temperature.

An example of the pressure-temperature relation for SF_6 is shown in Figure 4.22 at two gas densities, defined by the pressure and temperature at gas filling. When the temperature is decreased, the gas pressure decreases accordingly until the dew point is reached. For instance, Figure 4.22 marks a density of 40.3 kg/m^3 corresponding to 0.62 MPa at 20°C. Condensation occurs at the dew point of −30°C. It should be noted that condensation does not necessarily occur exactly when the saturation curve is reached, but the gas can be in the supersaturated state depending on the availability of condensation sites.

Table 4.2 Properties of gas mixtures compared to SF_6, adapted from [51]

	C_{ad}[a]	p_{min}/MPa[b]	T_{min}/°C[c]	GWP	Electric strength[d]	Toxicity LC_{50} (4 h) ppmv	References
SF_6	–	0.43...0.6	−41...−31	25 200	0.81...1	–	20
CO_2/O_2	–	0.6...1	<−50[f]	1	0.4...0.7	>3e5	17,26,20,50
CO_2/C5-FK/O_2	≈4–6.5	0.7	−5...+5[g]	1	≈0.75...0.86	>2e5	2,15,28,64
CO_2/C4-FN and CO_2/C4-FN/O_2	≈3.5...6	0.7...0.9	−30...−25[g]	293...500	≈0.87...0.96	>80,000	14,20,31,47, 52,65,66

[a]Concentration of admixture is in mole % referred to the gas mixture.
[b]Typical lock out pressure range.
[c]Minimum operating temperature for p_{min}.
[d]Electric strength compared to SF_6 at 0.55 Mpa.
[e]Compared to SF_6 at 0.13 Mpa, measurements were for a mixture at −15°C.
[f]Calculations with Refprop: https://www.nist.gov/srd/refprop.
[g]Lower temperature with pressure reduction.

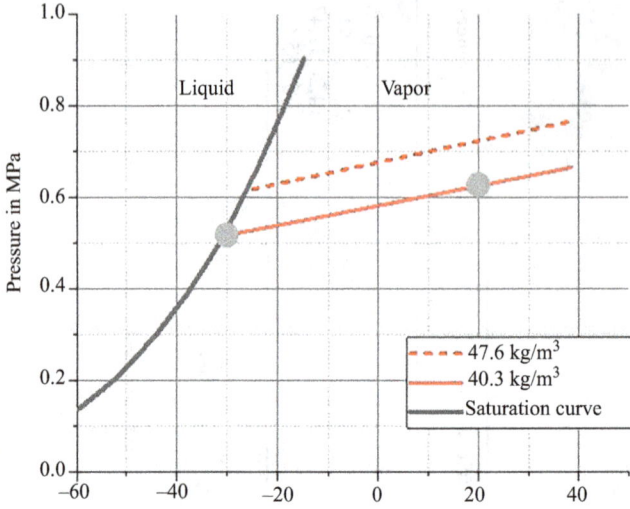

Figure 4.22   Pressure vs. temperature for SF$_6$ with saturation curve

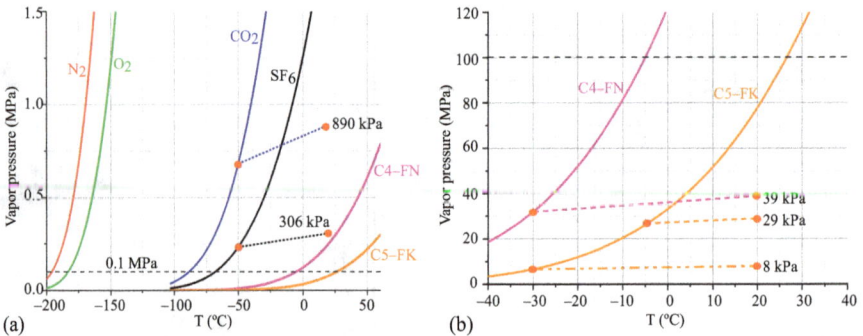

Figure 4.23   (a) Vapor pressures for pure substances using data from [52] and (b) zoom for C4-FN and C5-FK

Vapor pressure curves for SF$_6$ and alternative gases of interest are compared in Figure 4.23. Here the properties of pure gases are shown.

It can be seen that N$_2$ and O$_2$ and CO$_2$ have higher vapor pressures than SF$_6$ at the same temperature, allowing significantly lower temperature applications (at the same filling pressure) than with SF$_6$. The dew points of SF$_6$ and CO$_2$ at 0.1 MPa are −64°C and −78.5°C, respectively, as given in Table 4.1. As mentioned above, for a filling pressure of 0.62 MPa at 20°C with SF$_6$, the dew point is −30°C. For lower operating temperature, the fill pressure of SF$_6$ at 20°C needs to be decreased. To this end, a gas with low dew point is sometimes added to SF$_6$, e.g., N$_2$ or CF$_4$ (not shown here). Typically, the minimum SF$_6$ fill pressure at 20°C needs to be reduced to about 306 kPa to allow for $-50°C$ applications, as indicated in Figure 4.23(a) by the black dashed line. A linear ideal gas pressure scaling was used to draw this line.

Similarly, it can be deduced for pure CO_2 that a $-50°C$ requirement allows for fill pressures of about 890 kPa, shown by the blue dashed line in Figure 4.23(a). With the addition of O_2 to avoid soot formation [69], a minimum functional pressure of about 1 MPa absolute at 20°C CO_2/O_2 can be deduced for $-50°C$ applications, see Table 4.2 and [51].

Other promising SF_6 alternative gases use a mixture of CO_2/O_2 with C4-FN or C5-FK compounds. From Figure 4.23(b), it can be seen that the vapor pressure of the fluorinated components is much lower than for SF_6, i.e., these gases can only be used at a reduced partial pressure. For 0.1 MPa, the dew point for C4-FN and C5-FK is $-4.7°C$ and $26.9°C$, respectively, as given in Table 4.1. Figure 4.23(b) shows that for $-30°C$ applications a partial pressure of about 39 kPa and 8 kPa at 20°C can be used with C4-FN and C5-FK, respectively. For $-5°C$ applications and C5-FK, a partial pressure of about 29 kPa can be used at 20°C. This explains that in HV applications, such gases cannot be employed in their pure form like SF_6 but need to be mixed with a carrier gas with suitable material properties. For CB applications, the best-suited carrier gas is CO_2 with some O_2 admixture, as mentioned before.

As the first approximation, the partial pressures of the components (Dalton's law) can be used to estimate the admixture of fluorinated compounds to CO_2 or CO_2/O_2 for a given temperature application [78,79]. Calculations of dew temperatures with C4-FN concentration for various fill pressures with CO_2 as carrier gas are shown in Figure 4.24 [79]. From this figure (or Figure 2 of [78]), the approximate percentage of C4-FN can be deduced for a given pressure and temperature requirement. According to this approximation, for 0.9 MPa, absolute at

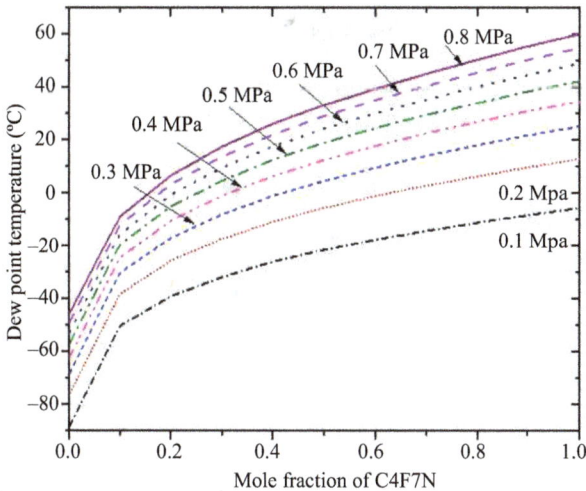

Figure 4.24   Dew temperatures of C4-FN in various gas mixtures and absolute pressures (pressures not temperature corrected) with CO_2 [78]

−30°C up to 3.5% C4-FN can be admixed. Note that it is common to give filling pressures temperature corrected to +20°C. This correction has not been applied here.

It should be noted that due to interaction (Van der Waals forces) between the different molecules in a mixture, the condensation temperature will usually deviate from that of the pure gas; see, for example [52,80]. This deviation depends on the concentrations of the admixture, the type of carrier gas and the pressure. These real gas effects can lead to deviations from the simple scaling as shown in Figure 4.24. For C4-FN/CO_2 mixtures, this translates into dew temperature increase of a few K. For C5-FK/CO_2 mixtures, this effect is similar but less pronounced. Using N_2, the effect is positive for C4-FN and C5-FK mixtures, i.e., higher concentrations than expected from the simple scaling can be used in a mixture, which decreases the dew temperature. Additionally, it should be noted that the condensation itself is a complex process that can deviate from the simple curves shown here, since condensation can be delayed due to supersaturation. Interactions between the compounds in liquid and vapor phase need additional consideration, as mentioned in [52,75]. The correct vapor pressure curves for mixtures can be calculated, e.g., using calculations with REFPROP [75]. The fluorinated substances C4-FN and C5-FK are not yet included as a standard there. Thus, precision condensation measurements have been performed [80]. The resulting curves can be parametrised, e.g., with Antoine or Wagner-type equations. Examples of vapor pressure curves for C4-FN/CO_2/O_2 mixtures are shown in Figure 4.25.

From such figures, the maximum usable lockout pressure for a given concentration of fluorinated additives in O_2 and CO_2 can be roughly deduced for a given temperature application. For −30°C, 3.5% C4-FN and 13% O_2 admixture,

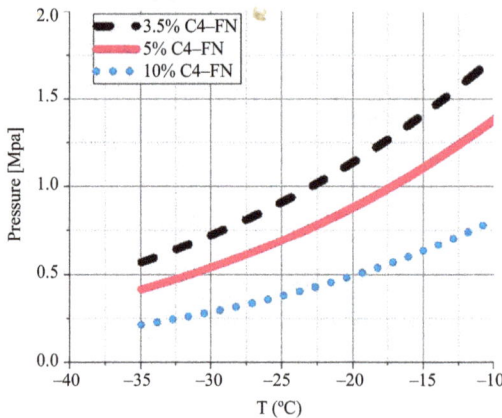

Figure 4.25   Pressure vs. dew temperature for C4-FN/CO_2/O_2 mixtures with 13% O_2 content based on the empirical correlation from [80] using the Antoine equation (pressures not temperature corrected, i. e. corresponding filling pressures at 20°C will be higher)

the lockout pressure at 20°C can be estimated to be slightly below 0.9 MPa, applying the ideal gas law.

4.2.3   Material properties

In the present section, the fundamental material properties of SF_6 and alternative gases, including gas mixtures based on CO_2, will be compared. Such properties are important for the performance of HV CB such as current interruption, insulation performance or heat transfer. These material properties include the composition of pure gases and gas mixtures as a function of temperature, thermodynamic and transport properties, and additional properties particularly important for circuit breaker applications, such as $\rho \cdot c_p$, the adiabatic coefficient, and the speed of sound, which can all be deduced from the thermodynamic properties. Additionally, radiative properties will be briefly compared and discussed.

It should be noted that such material properties are mostly based on calculations, since the experimental determination of parameters at elevated temperatures and pressures, and the short-time scales of processes in HV CB is very challenging. The uncertainties of such calculations are still quite large due to assumptions made.

4.2.3.1   Composition calculations

Plasma composition calculations are the basis for the calculation of thermophysical and transport parameters of arc plasmas. The common carrier gas for HV CB applications is CO_2. Calculations for mixtures of C4-FN and C5-FK with CO_2 were shown by various authors [81–89] and will not be repeated here. Differences could be observed in the results presented by these publications, which are due to the different assumptions made in the models and in the species included. Usually, the models assume local chemical and thermodynamic equilibrium in the plasma, denoted as LCE and LTE, respectively. Under the assumption of LTE, all species are locally in thermodynamic equilibrium and have the same temperature. In LCE, the chemical composition is the one obtained after an infinitely long time, i.e., neglecting reaction rates. The equilibrium calculations use the mass-action law via the Saha and Guldberg–Waage laws if only gaseous species are considered, or minimization of the Gibbs free energy which is preferred for arc plasmas and for considering non-gaseous phases, as solid and liquids [86]. It should be noted that the composition calculations available today need further investigations that take into account condensed phases and non-equilibrium effects, i.e., chemical kinetic simulations, that take into account reaction rates and pathways.

4.2.3.2   Thermodynamic parameters and transport coefficients

From the compositions, the thermophysical properties, transport coefficients and radiation properties of the arc plasma can be deduced. These parameters are the basis for arc models and thermal plasma behaviour. From the composition, the thermodynamic parameters are obtained by using the thermodynamic properties of the species and taking into account the interaction between charged species via the Debye–Hueckel correction. Transport properties are obtained from the Boltzmann equation using the Chapman–Enskog approach; see e.g., [32,81–84] for further details and references on the methods.

4.2.3.2.1   C4-FN/CO₂ mixtures

Since it is difficult to find calculations in the literature for the new gases and mixtures mentioned above, the same approach as described before was used to perform own calculations using software from [77] for SF_6, CO_2, CO_2/O_2, C4-FN and C4-FN/CO_2. For pure C4-FN, the chemical species from [81] were used. For mixtures of C4-FN with CO_2 and O_2, the relevant species were added as was done in [81]. Similarly, calculations were done for all the different gases and mixtures. For SF_6, the typical species, e.g., as given in [90], were used. For simplicity, in the following, a 10% C4-FN and 10% O_2 admixture to CO_2 is used, which is close to typical values in practical applications.

Figures 4.26 and 4.27 show the thermodynamic parameters mass density, enthalpy, and specific heat and the transport coefficients thermal conductivity, viscosity and electrical conductivity for C4-FN pure gas and mixtures compared to those of CO_2.

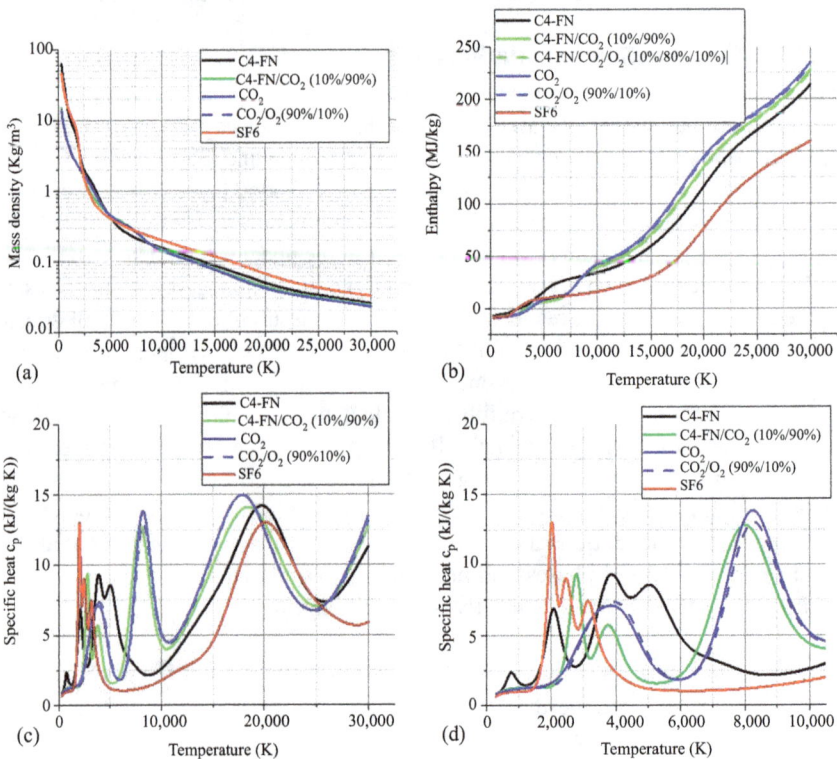

Figure 4.26   Thermodynamic parameters of C4-FN and mixtures with CO_2 and O_2 at 0.8 MPa. (a) Mass density, (b) enthalpy, (c) specific heat (300–30,000 K), and (d) specific heat (300–10,000 K).

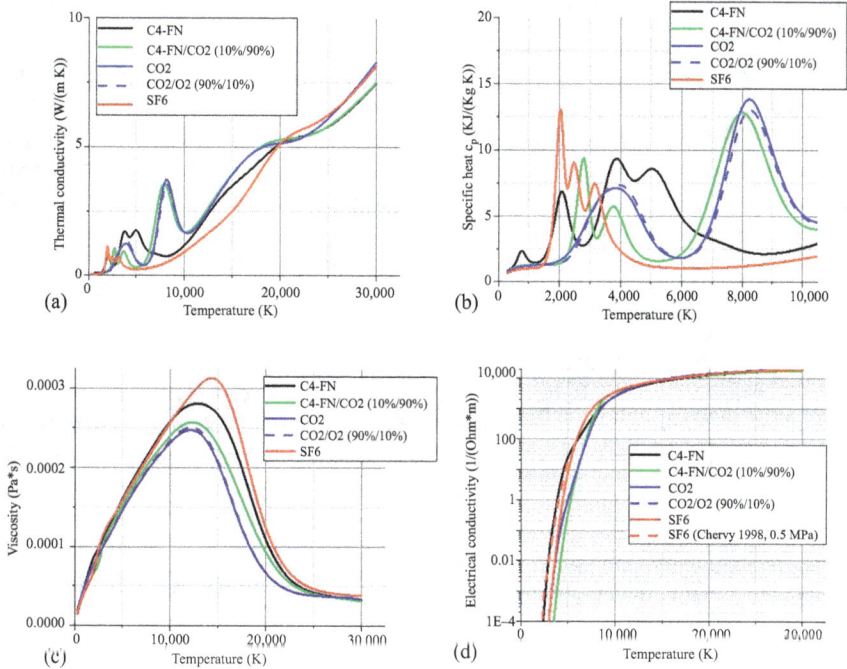

Figure 4.27   Transport coefficients of C4-FN and mixtures with CO_2 and O_2 compared to SF_6 at 0.8 MPa. (a) Thermal conductivity (300–30,000 K). (b) Specific heat (300–30,000 K). (c) Viscosity. (d) Electrical conductivity. For the electrical conductivity of SF_6 below 5,000 K, the data from [154] was used.

The mass density of pure C4-FN (Figure 4.26(a)) is similar to pure SF_6 and higher than CO_2 due to the larger molecular weight below 2,500 K. A 10% C4-FN admixture to CO_2 does not significantly change the mass density compared to pure CO_2. In the figure, the curves are overlapping. Also, a CO_2/O_2 mixture with 10% O_2 has a similar density as pure CO_2. The same is also the case for the enthalpy of the CO_2-based mixtures in Figure 4.26(b). Compared to the CO_2-based mixtures, SF_6 has a significantly lower enthalpy above 7,000 K. The specific heat is shown in Figure 4.26(c) together with a zoom to temperatures below 10,000 K (Figure 4.26(d)). This quantity is an indicator for decomposition reactions and is important for the interruption performance, as will be discussed later in Section 4.2.3.4. The peak at about 800 K in pure C4-FN corresponds to the decomposition of the C4-FN molecule. This peak is absent for the mixture of C4-FN with CO_2, due to the low admixture concentration. The specific heat peak at about 2,000 K for pure C4-FN is mainly caused by the dissociation of CF_3CN, C_2F_4 and CF_4 and is very close to the specific heat peaks of SF_6. Further peaks between 3,000 K and 5,000 K are due to the decomposition of CF_2 and CF.

Above 7,000 K, the C4-FN molecule is completely decomposed into atoms and above 10,000 K the peaks are due to the ionization of atoms.

For 10% C4-FN in CO_2, below 2,000 K, the differences from pure CO_2 are very small. The narrow peak in the specific heat for the 10% C4-FN admixture at around 2,800 K (Figure 4.26(d)) is due to the decomposition of large polyatomic molecules like CF_4 and COF_2. The peak at about 3,800 K is due to the decomposition of CO_2. Interestingly, this peak is more narrow and smaller for the CO_2/C4-FN mixtures than for pure CO_2. This was also observed by [81].

Figure 4.26(c and d) also shows for comparison the specific heat of SF_6, which has peaks in the range of 2,000–3,000 K and which is well known to be related to the good interruption performance of SF_6 [32]. These peaks are at lower temperature and are larger than for the CO_2-based mixtures.

A similar behaviour as for the specific heat can be seen for the thermal conductivity; see Figure 4.27(a and b). However, the thermal conductivity peak for C4-FN and mixtures with CO_2 is similar to that of SF_6. For temperatures below 5,000 K, there is no significant difference in the viscosity (Figure 4.27(c)). At plasma temperatures of 10,000–15,000 K, the viscosity is lower for CO_2-based mixtures than for SF_6. This is the temperature range of an arc approaching CZ. Above 20,000 K, which is the typical temperature range of a high current arc, the differences between SF_6 and CO_2/C4FN mixtures become small. The electrical conductivity is similar for all gas mixtures above 5,000 K, see Figure 4.27(d). In the range of the arc at CZ, i.e., 5,000–10,000 K, typically, the electrical conductivity of CO_2-based mixtures is lower than for SF_6. This might be beneficial for interruption performance.

4.2.3.2.2   C5-FK/CO_2 mixtures

Mixtures of C5-FK with CO_2 at 0.8 MPa were investigated by Li *et al* in [84]. The resulting thermophysical and transport parameters are shown in Figures 4.28 and 4.29, respectively.

Mass density (Figure 4.28(a)) is not significantly affected by the mixing ratio in these calculations, which is due to the recombination calculation approach used in this case. Increasing the fraction of C5-FK increases the enthalpy (Figure 4.28(b)) of C5-FK/CO_2 mixtures below 8,400 K and decreases it above, which is due to mass density changes. With decreasing C5-FK mixing ratio, the specific heat peaks (Figure 4.28(c)) at the lower temperatures of 1,200 K and 3,000 K disappear and the properties become more similar to those of pure CO_2. This is due to the lower C5-FK and CF_4 content at these temperatures. The characteristic peaks of CO_2 around 8,200 and 17,700 K are due to dissociation of CO, and ionization of C and O. With increasing C5-FK content, these peaks shift slightly. It must be noted that the data for the specific heat was collected from different references and at different pressures. For this reason, the comparability might not be well given. For the CO_2-based mixtures with C5-FK additive a lower specific heat peak around 2,000 K can be observed compared to C4-FN. This might indicate a lower cooling performance for such mixtures. However, it is unclear how much this is produced by differences in the calculation methods and assumptions.

The transport properties for C5-FK/CO_2 mixtures are shown in Figure 4.29. The thermal conductivity peaks below 3,000 K disappear with decreasing C5-FK

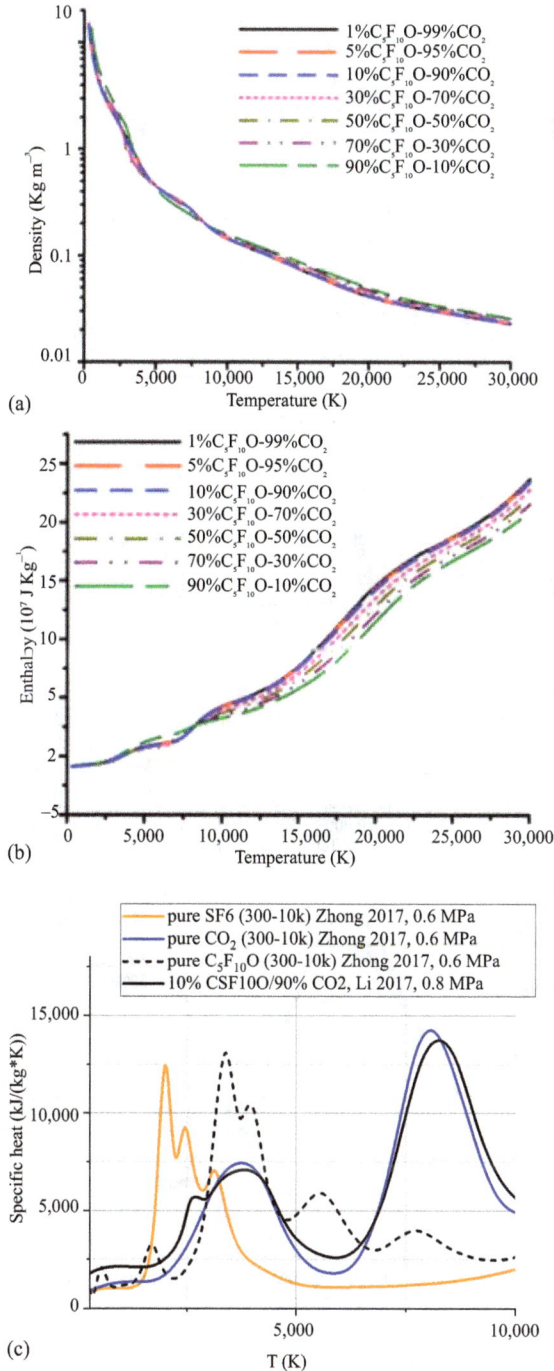

Figure 4.28   *Thermodynamic parameter for C5-FK/CO$_2$ mixtures. (a) Mass density from [88] at 0.8 MPa. (b) Enthalpy from [88] at 0.8 MPa. (c) Specific heat from Zhong [82] and Li [88] at 0.8 and 0.6 MPa.*

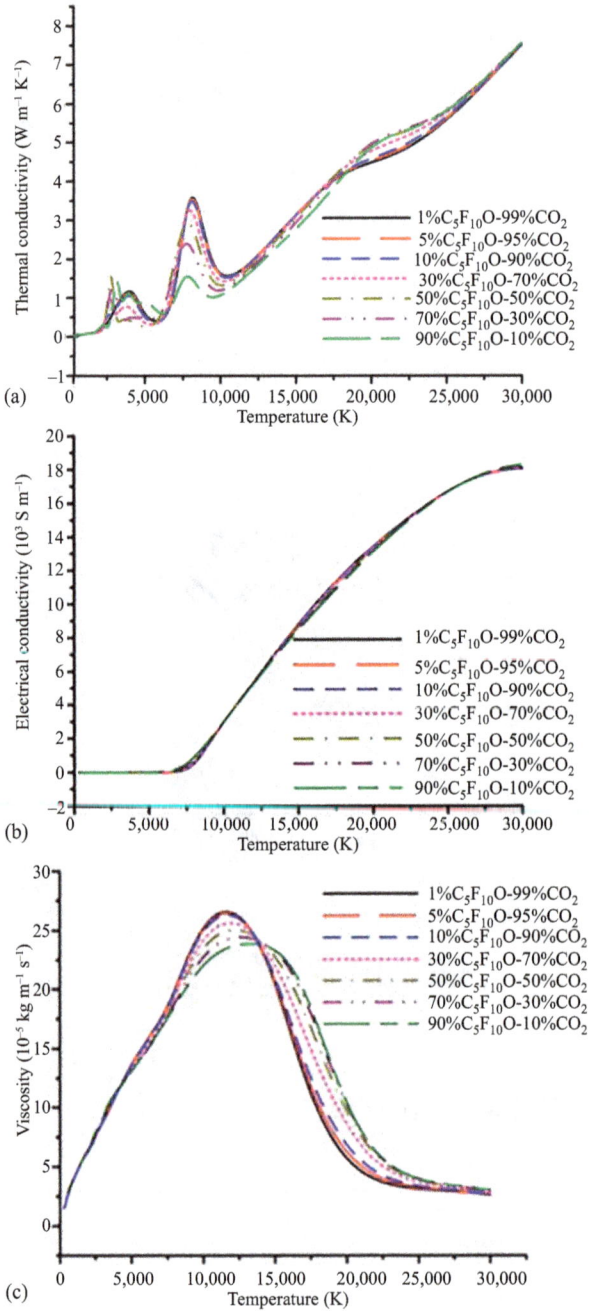

Figure 4.29   Transport coefficients of C5-FK/CO₂ mixtures at 0.8Mpa from [88].
(a) Thermal conductivity. (b) Electrical conductivity. (c) Viscosity.

content, as observed for the specific heat. Viscosity changes for different admixture ratios become significant only above 7,500 K. Similar to the C4-FN/CO_2 mixtures discussed previously, the viscosity peaks at 12,000–15,000 K are lower than for SF_6.

For the electrical conductivity, no significant differences are observed and for the viscosity differences become significant only above 7,500 K.

4.2.3.3 Additional important material parameters

4.2.3.3.1 $\rho \cdot c_p$ and $\rho \cdot c \cdot h/p$

For the interruption performance, the product of specific heat c_p and mass density ρ is important, since it describes the radial turbulent heat transport [32], which is important for thermal interruption at CZ. Figure 4.30(a) compares this property for C4-FN admixtures to CO_2, pure CO_2 and SF_6.

For a high thermal interruption performance, it is important to reach high values at temperatures below the temperature where electrical conductivity rises, i.e., about 4,000 K, and low values above. This leads to a constricted arc core and

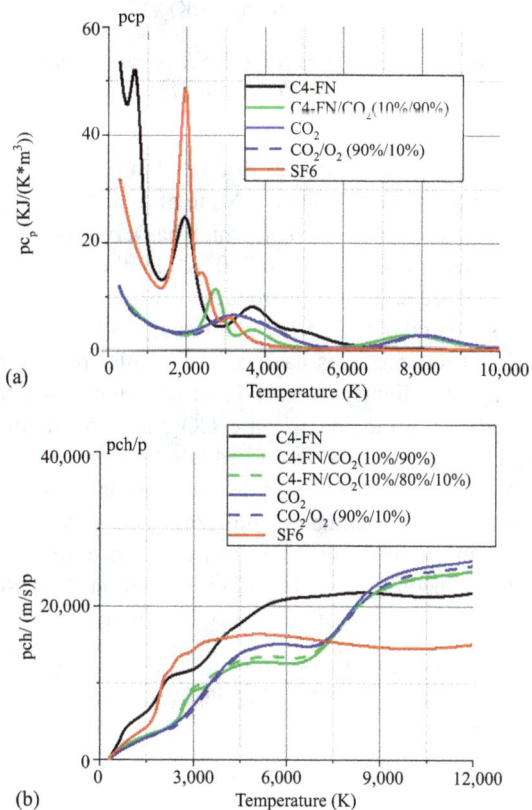

Figure 4.30 Material parameters $\rho \cdot c_p$ (a) and $\rho \cdot c \cdot h/p$ (b) for C4-FN mixtures in comparison to SF_6 and CO_2 at 0.8 MPa

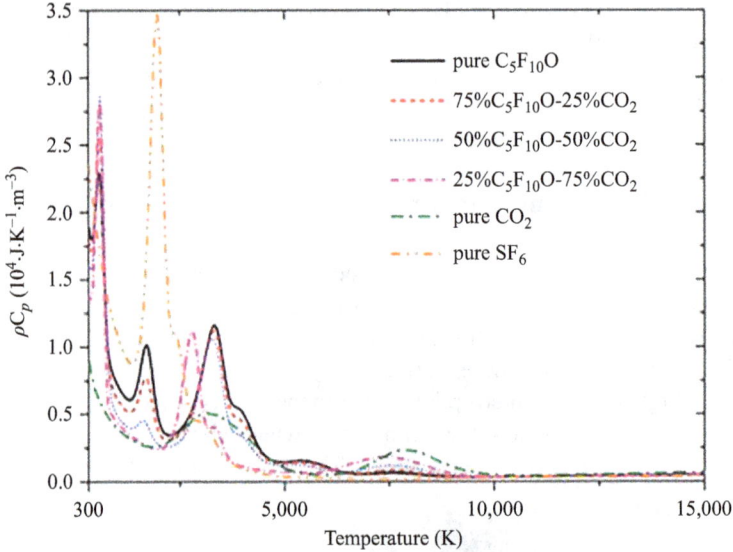

Figure 4.31   Material parameter ρ·c_p for C5-FK mixtures at 0.6 MPa in comparison to SF6 and CO2 from [155]

large heat dissipation in the arc fringes, as discussed in [32,87], which are optimum conditions for cooling and decay of the arc. None of the CO_2-based mixtures match the performance of SF_6. Even pure C4-FN, which has a $\rho \cdot c_p$ peak at around 2,000 K has significantly higher values above 4,000 K. In the mixture with CO_2, the low-temperature peaks are significantly reduced and only a small peak at around 2,800 K is left. This is similar for C5-FK admixtures to CO_2, see Figure 4.31.

Pure CO_2 and CO_2/O_2 mixtures have only a smaller peak at around 3,000 K. Hence, one might expect a slightly increased thermal interruption capability for C4-FN and C5-FK mixtures compared to pure CO_2, but still a lower performance compared to SF_6. This topic will be discussed later.

Another important quantity is the combined material property $\rho \cdot c \cdot h/p$, which describes the axial convective enthalpy transport. The axial convective enthalpy transport can be seen in the equation for the power balance of a cylindrical arc with a rectangular radial temperature profile and cross-section A_{arc}. Note that this is a strongly simplified representation of an axially blown arc:

$$(1 - \vartheta) \cdot U \cdot I = \frac{(1 - \vartheta) \cdot L \cdot I^2}{\sigma \cdot A_{arc}} = \rho_{ex} \cdot h_{ex} \cdot v_{ex} \cdot A_{arc}$$

$$\approx \left[\frac{\rho \cdot c \cdot h}{p} \right]_{stag} \cdot p_{stag} \cdot f \cdot A_{arc} \qquad (4.5)$$

where J is the fraction of the input power (Ohmic losses) that leaves the arc via transparent radiation, U is the arc voltage, I is the current, L is the arc length, s is the electrical conductivity, ρ is the mass density, h is the enthalpy and v the plasma flow velocity. The indices 'ex' and 'stag' denote values at the exit of the arc cross-section and at the upstream stagnation point, respectively. The dimensionless factor f is the flow function as introduced in [52] and gives the ratio of the actual value of the mass flow density in the exit cross-section to the sonic mass flow density $\rho(p_{stag}, T_{stag})$ at the stagnation values of pressure and temperature. For sonic flow, f is about 0.59. The arc cross-section for sonic flow can then be expressed as:

$$A_{arc} = I \cdot \sqrt{\frac{(1 - \vartheta) \cdot L}{\sigma \cdot \left[\frac{\rho \cdot c \cdot h}{p}\right]_{stag} \cdot p_{stag} \cdot f}} \qquad (4.6)$$

It depends proportionally on the current and is inversely proportional to the square-root of stagnation pressure p_{stag}. The quantity $(\rho \cdot c \cdot h)/p$ is a fundamental, pressure-reduced material property that describes the arc cross-section of the convectively stabilized arc. It mainly depends on the temperature. The larger $\frac{\rho \cdot c \cdot h}{p}$ and the stagnation pressure p_{stag}, the smaller the arc cross-section. Figure 4.30 shows $\frac{\rho \cdot c \cdot h}{p}$ for the different pure gases and mixtures. Again, there is a significant difference between SF_6 and the CO_2-based mixtures. Below 7,000 K, the values for CO_2 mixtures are below SF_6, whereas above this temperature, they are significantly higher. This is due to the effect of higher enthalpy and speed of sound for the CO_2 mixtures. Since close to CZ the arc core temperature drops to below 7,000 K [33], typically, an increased arc core cross-section due to convection might be expected for the CO_2 mixtures. This can be seen also in the simulations shown in Figure 4.35 which will be discussed below. Also in the arc fringes, applying the same reasoning a significantly larger cross-section of the CO_2 mixtures would result, again indicating a larger total arc cross-section compared to SF_6. Note that here we neglected the possible differences in arc core temperature and radiation losses between SF_6 and CO_2 mixtures.

4.2.3.3.2 Adiabatic coefficient and speed of sound

For use in circuit breakers, the adiabatic coefficients and speed of sound c of a gas or gas mixture are also very important. The adiabatic coefficient, defined by the ratio $\gamma = c_p/c_v$, is important for the pressure rise, e.g., in exhaust and heating volumes, since for a volume V, an energy input Q (adiabatic conditions) will lead to a pressure rise Δp:

$$\Delta p = \frac{Q \cdot (\gamma - 1)}{V} \qquad (4.7)$$

The speed of sound for an ideal gas follows from $c = \sqrt{\gamma \cdot p/\rho}$, which shows the dependencies; due to the decrease in mass density ρ the speed of sound increases with temperature. The speed of sound is of importance for flow processes inside the CB. This will be further discussed later.

These quantities are deduced from the thermodynamic parameters and are shown in Figure 4.32 for C4-FN mixtures with CO_2 and CO_2/O_2 and for SF_6 for comparison.

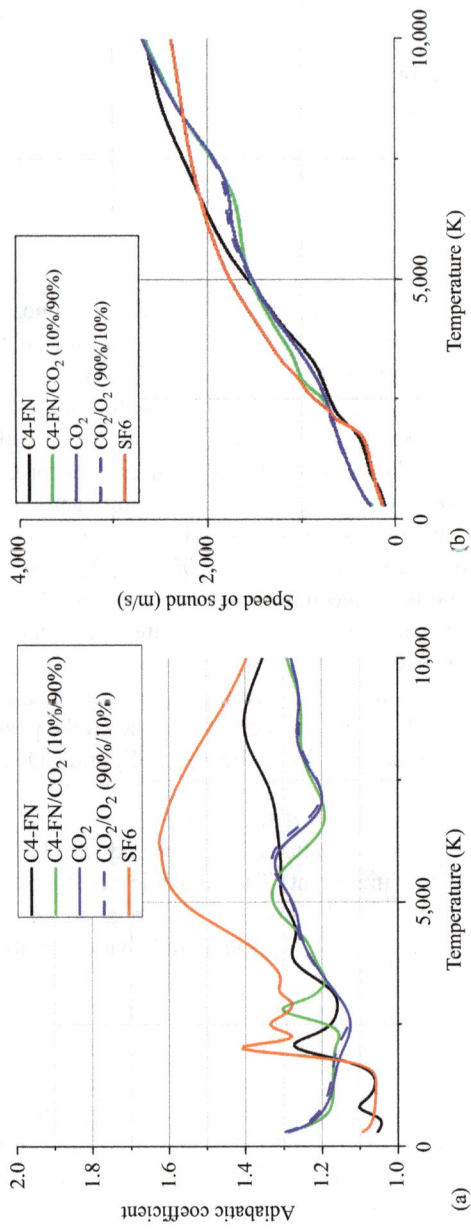

Figure 4.32   Adiabatic coefficient (a) and speed of sound (b) for C4-FN mixtures at 0.8 MPa

Below 1,500 K, SF_6 and pure C4-FN have significantly lower adiabatic coefficients g compared to CO_2-based mixtures. This changes above 1,800 K, where SF_6 has a higher adiabatic coefficient than the other gases. Thus, based on (4.7) at lower temperatures, SF_6 will lead to lower pressure build-up compared to CO_2-based mixtures. This fact is important for the pressure rise in exhaust or thermal volumes, which are typically below such temperatures. Above 1,800 K, the adiabatic coefficient depends more sensitively on the temperature with SF_6 having systematically higher values than CO_2-based mixtures. Only around 2,800 K do C4-FN/CO_2 mixtures have similar values of the adiabatic coefficient as SF_6.

The speed of sound is lower for SF_6 than for CO_2-based mixtures below 2,200 K and higher above. At about 1,000 K, it is about twice as high for CO_2-based mixtures compared to SF_6. Flow processes in CB, which are usually at sonic or even supersonic speeds, are hence faster in CO_2-based mixtures.

4.2.3.3.3 Radiative properties

Radiative properties are obtained from the radiation spectra for the different gases and gas mixtures, taking into account transition probabilities, line broadening mechanisms and continuum radiation [91,92]. Several investigations in recent years have calculated net emission coefficients (NEC) for mixtures of C4-FN and CO_2 taking into account only emission and neglecting absorption, e.g. [82,87]. For this purpose, the arc is assumed to be an isothermal sphere or cylinder of radius R and the corresponding NEC is calculated as a function of arc radius and temperature [93,94]. A comparison is shown in Figure 4.33.

In that work, both atomic and molecular radiations were considered, the latter being important at lower temperatures. SF_6 has lower NEC below about 8,000 K compared to CO_2-based mixtures in these calculations. It should be noted that the absolute levels at such temperatures are quite small, and these differences might be less important for CB applications, especially since in [87], the bump at low temperatures disappears for the mixtures. At higher temperatures in the range of 10,000–17,000 K, the CO_2-based mixtures have higher NECs than SF_6, whereas the opposite occurs at even higher temperatures.

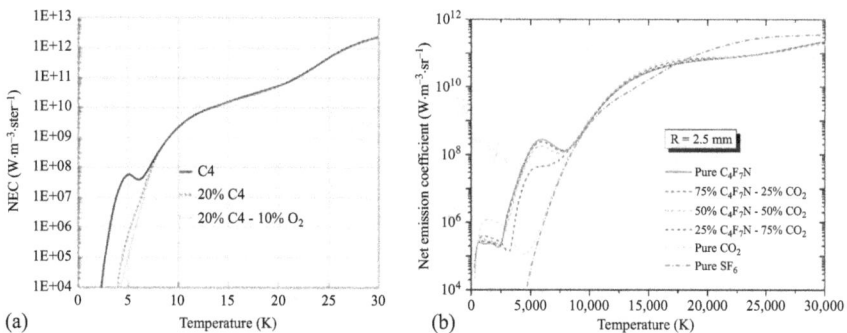

Figure 4.33 NEC for C4-FN, C4-FN/CO_2 mixtures, CO_2 and SF_6: (a) at 0.1 MPa from [87] and (b) at 0.6 MPa from [82]

However, as it is well known, net emission can only describe the radiation losses in the arc core. For CB interruption performance, the arc temperature profile is decisive, and this profile depends on all the power input and loss mechanisms. Radiation absorption in the arc boundary and at the nozzle walls is very important. This needs to be calculated by more advanced radiation models with suitable band-averaging methods, as described, e.g., in [87,95]. Accordingly, the radiative power flux can be compared for the different gases and mixtures of interest by comparing the divergence of the radiative power flux over the arc radius, which is computed for a predefined temperature profile. An example is shown in Figure 4.34.

Positive values denote net emission and negative ones net absorption of radiation, which occurs in the arc fringes. Significant differences can be observed for SF_6 and CO_2-based mixtures in the centre of the arc, indicating less emission in the centre of the SF_6 arc for the given temperature profile. Also, in the arc fringes, significant differences in the location of the re-absorption zones can be seen. However, as mentioned this is calculated for a given temperature profile. More precise calculations need to be done self-consistently. This can only be done in spatially resolved models, such as CFD models. Then, the temperature profile would be different for the different gases and mixtures. This was shown in [87] for SF_6, air and C4-FN/CO_2/O_2 mixtures in a simplified geometry for a free burning 10 A arc, see Figure 4.35.

For the same current, the SF_6 arc is narrower than for the CO_2-based mixtures. O_2 admixture does not significantly affect the arc diameter. The steeper arc temperature profile leads to increased thermal conduction losses, which are relevant for the laminar flow conditions investigated in the reference. Additionally, the arc fringes show a more pronounced arc mantle for the CO_2-based mixtures, which is in qualitative agreement with the findings discussed in the previous subsection on (4.5). Due to the larger arc radius, turbulent eddies cannot reach the conductive arc

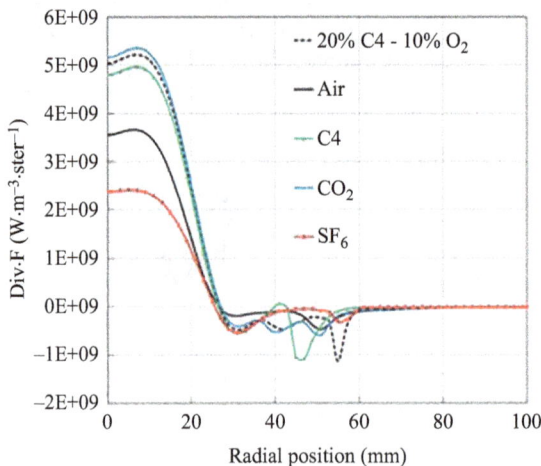

Figure 4.34   Divergence of radiative energy flux at 0.1 MPa from [87]

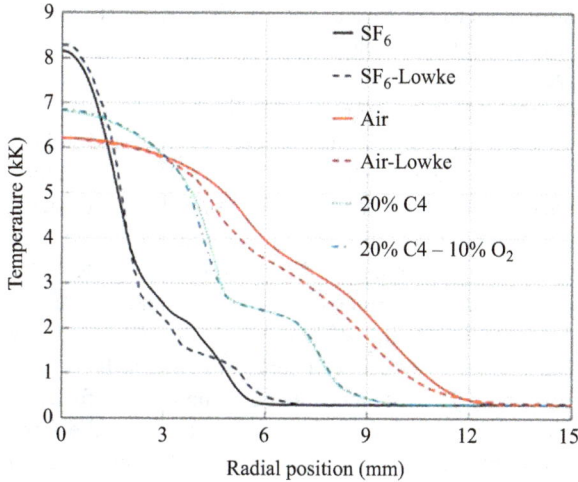

Figure 4.35 Calculated temperature profile for a 10 A arc at the midpoint between electrodes at 0.1 MPa from [87]

core and arc conductance decay is reduced. Taken together, these differences are expected to reduce interruption performance compared to SF_6. Air has an even larger arc radius compared to SF_6 and CO_2-based mixtures, which is also reflected in the interruption performance, as will be discussed later in Section 4.3. In [45], this is explained by the higher thermal diffusivity of air and CO_2-based mixtures at temperatures below 25,000 K (compared to SF_6), which leads to more spreading of heat in the non-SF_6 gases. Note that radiation around CZ for the case shown is negligible due to the low temperatures. In CB applications, the forced convection, which was not considered in [87], leads to turbulent heat exchange and is very important and might change the temperature profile significantly. However, the qualitative conclusions given here will possibly hold even in that case.

4.2.4 Gaseous insulation

Gaseous insulation of SF_6 alternative gases has been addressed in many publications over the last decade. Overviews are given in CIGRE TB 730 for dry air, N_2, CO_2 and N_2/SF_6 mixtures [96]. For the new SF_6 alternative gas mixtures containing fluorinated additives, an extensive comparative study was done by CIGRE WG D1.67 and summarised in the technical brochure 849 [97]. Results were also discussed in the technical brochure 871 from CIGRE WG A3.41 [52], and a recent review [98] gives further useful references for the prediction of dielectric properties of alternative gases.

4.2.4.1 Effective ionisation coefficients and critical fields

The fundamental properties to be investigated first are the effective ionisation coefficients as explained in Section 4.1.3.1. For the new CO_2-based gas mixtures with C4-FN and C5-FK admixtures, these properties were investigated by performing swarm experiments: see [97,99–102], respectively. In such experiments, a swarm of charged particles drifts through a gas gap with a defined gas mixture, pressure and applied electric field. From the measured displacement currents, the attachment and ionisation rate coefficients can be deduced. The authors of [73] used the Pulsed Townsend Method, whereas the authors of [100,101] used the Steady State Townsend method. Details of the methods and experimental procedures are described in the references. Effective ionisation rate coefficients obtained from these experiments are shown in Figure 4.36 as a function of the applied particle density reduced electric field for C4-FN and C5-FK. Note that in the following text, 'particle density reduced' values will be simply denoted 'reduced' values. The figure shows that with increasing admixture of fluorinated additives to CO_2 the reduced effective ionisation rate coefficients shift towards higher fields. As discussed in Section 4.1.3.1, avalanches occur when the effective ionisation rate coefficient is positive. The electric field that corresponds to the zero-crossing of the effective ionization coefficient is denoted the critical electric field, referred to as a critical field in the following text. Around the critical field, the reduced effective ionization coefficients vary approximately linearly with the reduced field, as discussed in Section 4.1.3.1 for SF_6; see Figure 4.37 on the example of C4-FN/CO_2 mixtures. The effective ionisation coefficient and slope of the effective ionization coefficient vs. electric field curve near the critical field for CO_2 mixtures containing 20 % (molar) C4-FN is similar to the one for SF_6. This is significantly different from CO_2 and air, where the slope is not constant and increases with the electric field, as shown in Section 4.1.3.1. As discussed, a similar slope of the effective ionisation coefficient indicates a similar roughness sensitivity. Hence for the new fluorinated gas mixtures, a similar roughness sensitivity as for SF_6 is expected.

The dependence of the reduced critical field on the admixture of C4-FN and C5-FK to CO_2 is shown in Figure 4.38.

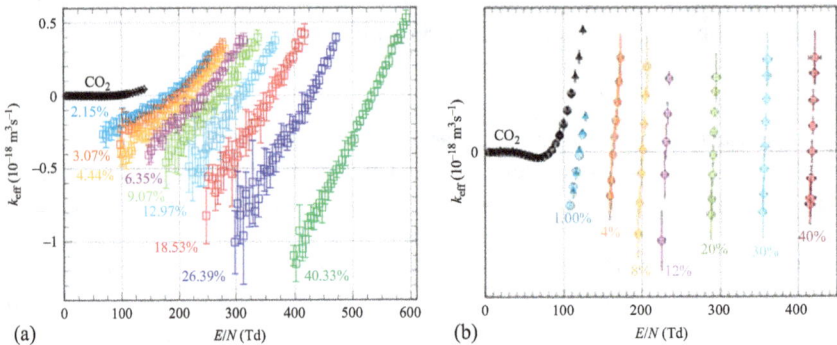

Figure 4.36   *Reduced effective ionisation rate coefficients for (a) C4-FN [156]*
and (b) C5-FK [102]. Note that the rate coefficient is related to the
reduced effective ionisation coefficient by, $k_{eff}=(\alpha_{eff}/n)\cdot v_{dr}$ using the
drift velocity v_{dr}.

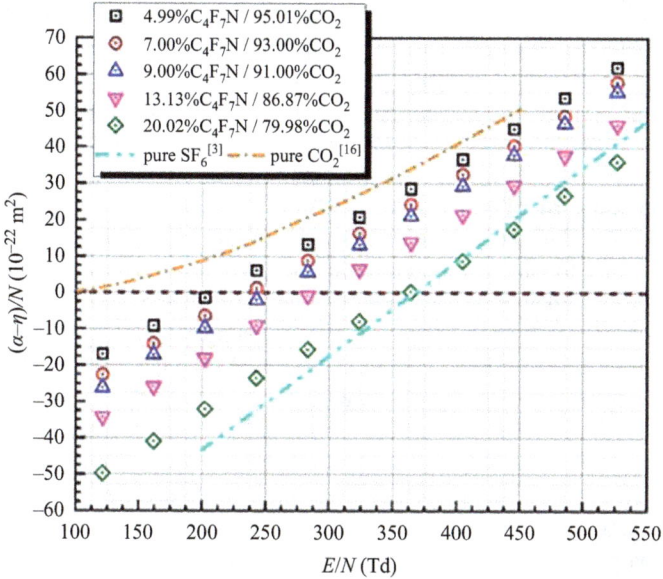

Figure 4.37   Reduced effective ionisation coefficients for C4-FN CO$_2$ mixtures, CO$_2$ and SF$_6$ from [101]

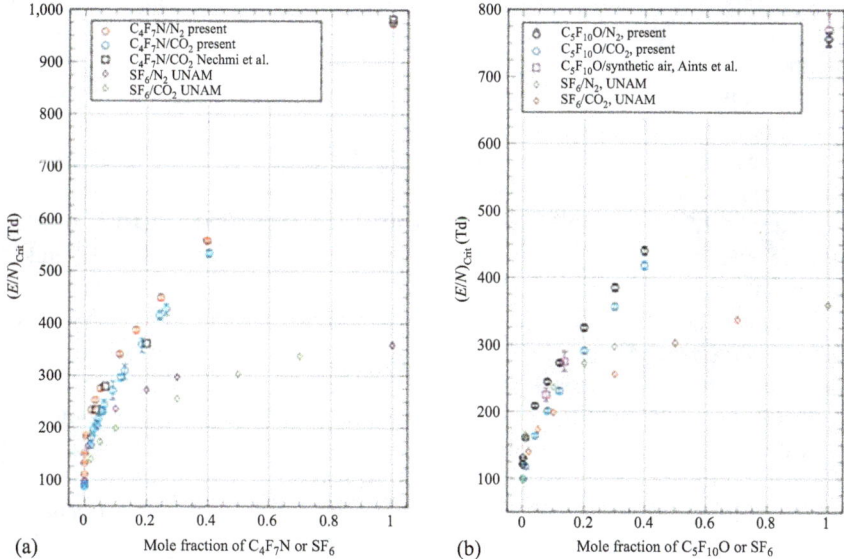

Figure 4.38   Reduced critical fields vs. C4-FN (a) and C5-FK (b) content in CO$_2$ from [102,156]. For comparison, SF$_6$/N$_2$ and SF$_6$/CO$_2$ mixtures are also shown. The critical field of SF$_6$ is indicated by the horizontal dashed line.

From the figure, a significant so-called 'synergy' [97] can be observed, i.e., a small amount of fluorinated admixture leads to a large increase in the critical electric field. Without synergy, a linear increase of the critical field with admixture mole fraction would be expected. This effect is well known for SF_6/N_2 mixtures [96,103], an example of which is also shown in Figure 4.38 for comparison. Due to this synergy, a small amount of fluorinated additive can be used to significantly increase the dielectric performance of such gas mixtures.

It can also be seen that N_2 mixtures have a slightly higher critical field compared to CO_2 mixtures for the same admixture concentration. This is due to a stronger electron energy moderation in N_2. Differences can be seen for C4-FN and C5-FK mixture at the same admixture mole fraction, showing significantly lower values for the C5-FK mixtures compared to C4-FN mixtures. At about 15–20% and 25–30% for C4-FN and C5-FN, respectively, the critical field of SF_6 is reached.

The experimental data shown were performed at reduced pressure (compared to the typical pressure in electrical equipment) and are strictly valid only for that pressure range. Particle-density-reduced critical fields can be scaled to higher pressures of several 100 kPa, as for example used in gas-insulated switchgear, assuming that they scale only with particle density and are otherwise independent of pressure. However, the particle-density-reduced critical field may depend on the pressure as a result of complex ion kinetic processes. Studies on the ion kinetics in pure C4-FN by [104] up to 69 kPa showed that electron detachment from negative ions cannot be neglected and that there are additional kinetic processes that can affect the pressure dependence. Through experiments and calculations, it was shown that such processes lower the reduced critical field of pure C4-FN to about 785 ± 5 Td which is 175–205 Td lower than the results from the experiments at low pressures in Figure 4.38. This explains why the breakdown voltage of pure C4-FN was found in previous investigations to be around twice that of SF_6; see, for example, Table 4.1 in Section 4.2.1.

As was shown in [104], the change in field dependence of the effective ionisation rates also affects predictions made using the streamer criterion as explained in Section 4.1.3.1. By taking into account detachment when determining the effective rate coefficients, excellent agreement with breakdown experiments in uniform fields was obtained for C4-FN.

Recently the effect of ion conversion reactions and detachment on the critical field was investigated also in CO_2/O_2 mixtures [105]. The change of the critical field due to these mechanisms is only relevant for increased O_2 concentrations, above 40% typically, and at low pressures of less than 0.1 MPa. For switchgear applications that use O_2 concentrations of less than 40% and total pressures of several 100 kPa, this influence can probably be neglected.

An overview of measured reduced critical fields from various references is shown for C4-FN mixtures from [101] in Figure 4.39 stating that 13.13% and 9% C4-FN admixture corresponds to 80% and over 70% of the reduced critical field of SF_6, respectively. Pure CO_2 has only about 25% of the reduced critical field of SF_6. Thus, relatively small amounts of C4-FN or C5-FK admixture can significantly increase the insulation performance of the new SF_6 alternative gas mixtures containing fluorinated additives.

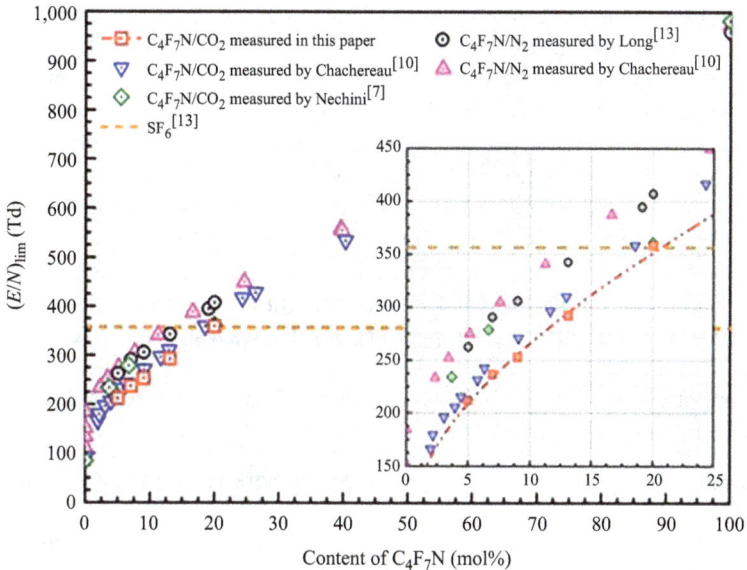

Figure 4.39   Reduced critical fields vs. C4-FN content in CO_2 from the comparison [101]

4.2.4.2   Breakdown fields

In real applications of HV CB, the experimental confirmation of gaseous insulation is crucial, i.e., not only the critical fields but also breakdown fields at the operating pressure are decisive for judging the insulation performance. To understand breakdown fields, mechanisms including streamer inception and propagation, transition to a leader and finally spark transition are relevant, as explained in Section 4.1.3.1. The breakdown field depends not only on the gas type but also on the details of the gap (size and surface roughness of electrodes, gap distance, non-uniformity, etc.), the pressure and the applied voltage waveform (AC, DC, LI, SI, VFT); i.e. a large parameter space needs to be covered by experiments.

As mentioned above, CIGRE WG D1.67 [97] did a literature survey and an extensive experimental campaign for the determination of breakdown fields in representative simplified geometries using uniform (Rogowski electrode), weakly non-uniform (sphere), and strongly non-uniform (needle) arrangements at various test laboratories for comparison (round robin test). Realistic pressures as used in equipment were tested. Due to the large parameter space investigated, only a few results will be shown here. For further details, refer to the reference.

4.2.4.2.1   C4-FN breakdown fields

The results of the literature survey from [97] can be summarised for C4-FN:

- In pure form, C4-FN has more than twice the breakdown strength of SF_6.
- In mixtures with CO_2 or N_2, an equivalent breakdown strength to SF_6 is reached for about 20% admixture in uniform and weakly non-uniform fields. For weakly

non-uniform fields, negative polarity at the contact with higher electric field is most critical. This behaviour is similar to that of SF_6, and the polarity effect can be explained by the availability of a first electron for discharge inception.

- Positive polarity becomes critical for strongly non-uniform fields and at high pressure. This behaviour is also similar to SF_6 and can be explained by the importance of streamer-leader transition, which is favoured at positive polarity.
- Typical C4-FN admixtures to CO_2 or CO_2/O_2 in the literature references in [97] are given in the range 3.5–10%. For such mixtures, slightly lower performance compared to SF_6 can be compensated by an increase in pressure. For a 3.7% C4-FN admixture in uniform and weakly non-uniform fields, SF_6 equivalence is reported at about 60% higher pressure for LI waveshape, for example, see Table 7-4 in [97].
- The impulse ratio, i.e. the breakdown field ratio of LI/AC, is cited to be comparable to that of SF_6.

These findings were confirmed in the benchmark test campaign as shown in [97]. The measurement of breakdown voltages in a uniform field accurately reflected the synergy expected from the results of the critical field experiments, see Figure 4.40. For an admixture of 20%, C4-FN similar breakdown field strength as for SF_6 is obtained. Note that an effect of O_2 admixture, which might be expected to have a slight positive influence [106,107], was not seen in these experiments (not shown here).

For judging practical applications, in the benchmark tests, CO_2 mixtures with about 5% C4-FN admixture were tested at 600 kPa with AC and lightning impulse voltage applied in a weakly non-uniform and in a strongly non-uniform field arrangement, see [97] for the details. The results are compared with breakdown results in SF_6 under identical conditions, see Figure 4.41. It can be seen that for weakly non-uniform fields the performance of the C4-FN mixture is about 25% lower than for SF_6 at a given pressure. This can be compensated by a higher pressure, as mentioned before. The negative LI breakdown voltage is lower than that for positive LI in weakly non-uniform gaps, and, for a strongly non-uniform gap, this relation changes and positive LI is more critical; both effects are also seen for SF_6. The breakdown voltages for strongly non-uniform arrangements are similar to those of SF_6.

From the median of the breakdown voltages in Figure 4.41, the impulse ratios $U_{bd}(LI)/U_{bd}(AC)$ for weakly non-uniform fields were determined to be 1.4 and 1.05 for positive and negative LI, respectively. This is very similar to SF_6, for which this experiment resulted in impulse ratios of 1.4 and 1.15 for positive and negative LI, respectively. It can be concluded from the experimental benchmark campaign that C4-FN mixtures show similar trends to those seen with SF_6. The statistical distribution of breakdown voltages is comparable, and the impulse ratios are nearly identical to SF_6. As an important consequence, existing test and design rules developed for SF_6 can be applied also to C4-FN mixtures.

Given the similarities of SF_6 and C4-FN mixtures in terms of electric breakdown, a concept of SF_6 equivalency can be defined [70] that eases the application

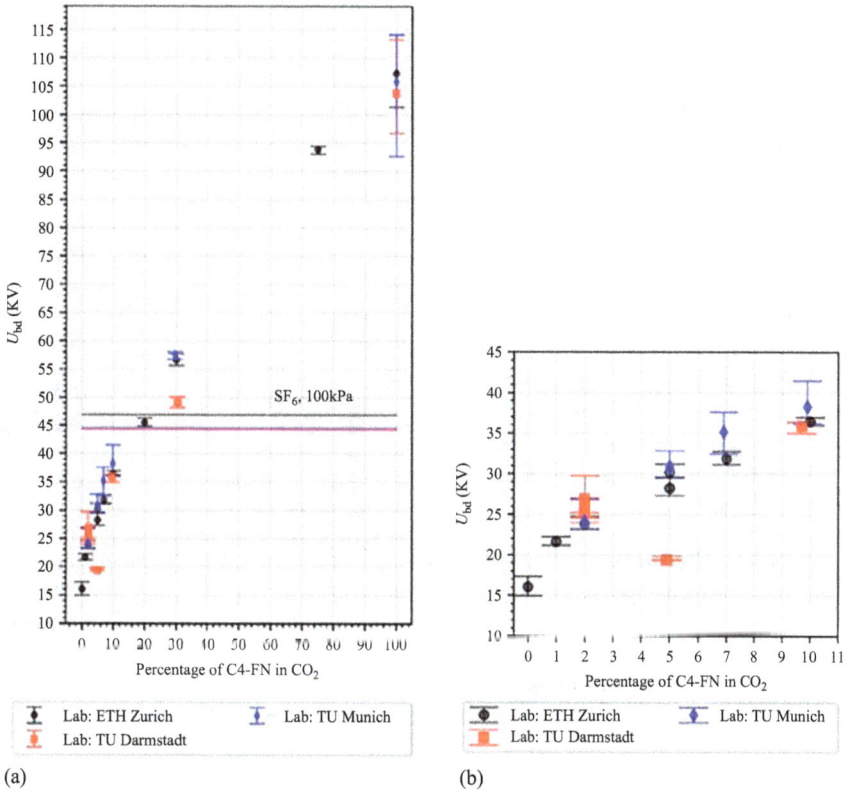

(a)   (b)

Figure 4.40   *Synergy scan of a mixture of C4-FN in CO$_2$ at 100 kPa compared with*
SF$_6$ at 100 kPa (horizontal lines) for AC breakdowns in a uniform
arrangement in a 5 mm gap: (a) 0–100% and (b) focused view 0–10%
from [97], reprinted with permission from CIGRE, @2021

of SF$_6$ design rules to the new alternative gases. This is due to the observation that
with increasing pressure, the curve of effective ionisation coefficient vs. electric
field for C4-FN/CO$_2$ mixtures increases in slope around the critical field. At fields
close to the critical field of SF$_6$, the slope of the curve for the mixtures and the
curve for SF$_6$ become similar. As explained in Section 4.1.3.1, this leads to a
similar roughness sensitivity, since the streamer inception criteria would give
similar streamer inception fields. Consequently, comparison can be done based on
the critical field only. In other words, breakdown fields should be similar for
pressures that result in the same critical field. SF$_6$ equivalency for a given C4-FN/
CO$_2$ mixture is then defined by the pressure at which SF$_6$ would yield the same
critical field. Note that for pure CO$_2$ or air, a slight correction of some percent is
necessary, since the slope of the effective ionisation coefficient is then lower than
that of SF$_6$, which is due to the lower attachment (see Figure 4.7 and Section
4.1.3.1). Based on this concept, as shown in [70], various SF$_6$ alternative gases can

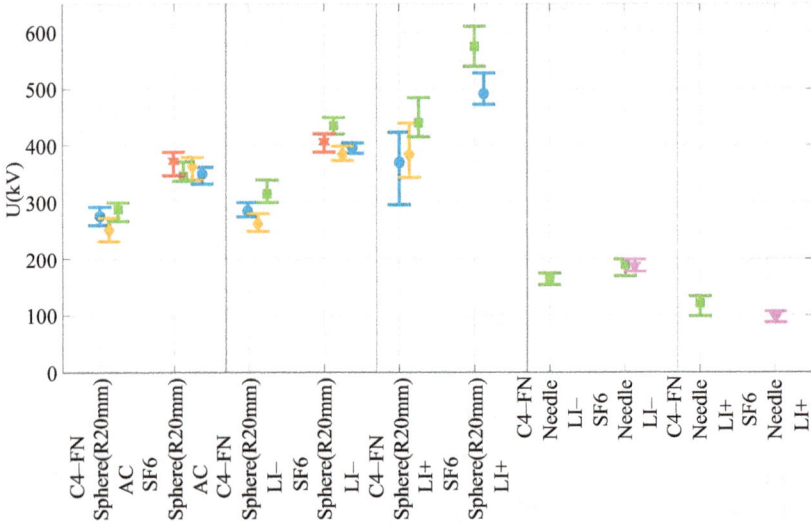

Figure 4.41   *Breakdown voltages in sphere-plane (weakly non-uniform) and needle-plane (strongly non-uniform) configuration, for a mixture of C4-FN/O$_2$/CO$_2$ (5%/5%/90%) mixture at 0.6 MPa absolute pressure compared to SF$_6$ adapted from [97]. Different colours show the results of different research groups for the same arrangement, reprinted with permission from CIGRE, @2021.*

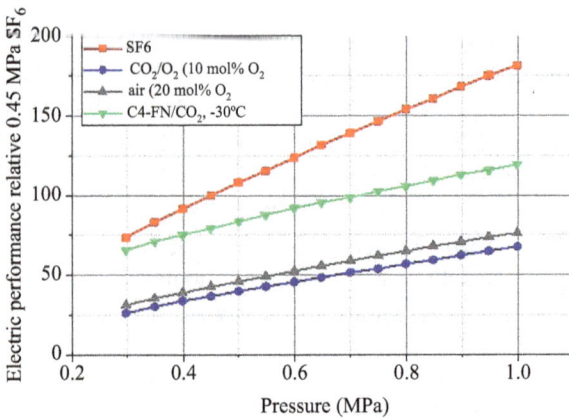

Figure 4.42   *Relative insulation performance of SF$_6$ alternative gases and mixtures, compared to SF$_6$ at 0.45 MPa adapted from [70]*

be benchmarked, see Figure 4.42. For the same pressure, the C4-FN/CO$_2$/O$_2$ gas mixture for −30°C application has about 20% lower electric strength than SF$_6$, as stated before from [97]. Increasing the pressure to 700 kPa leads to the same performance as for SF$_6$ at 450 kPa. CO$_2$/O$_2$ mixtures at 450 kPa without fluorinated additives have slightly less than 40% of the electric strength of SF_6 at the same pressure. For similar performance as SF_6 at 450 [109] sensitivity at such high pressures, breakdown fields will tend to saturate, see Figure 4.43. The saturation is indicated by streamer inception calculations using various assumptions on

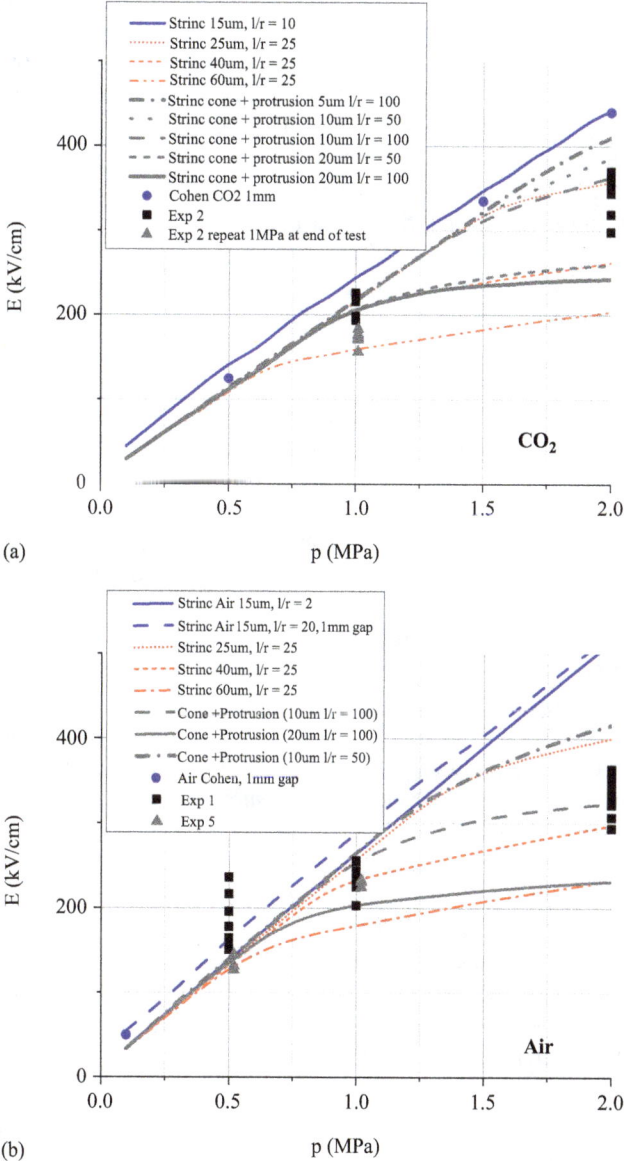

(a)

(b)

Figure 4.43   Breakdown fields in (a) CO_2 and (b) synthetic air vs. pressure adapted from [110]. Measurements are shown by symbols and streamer inception calculations for various assumptions by the curves.

protrusion shapes and confirming the measurements within the large uncertainties. Hence, in natural origin gases like CO_2 or synthetic air, further design adaptions are needed to achieve similar insulation performance as that of SF_6, i.e., a pressure increase alone is not sufficient.

4.2.4.2.2   C5-FK breakdown fields

Similar to C4-FN mixtures in the previous section, the C5-FK mixtures were investigated for synergy in [97], see Figure 4.44. The results are consistent with previous investigations on synergy [101,109], i.e., small concentrations of C5-FK can significantly increase the breakdown voltage. C5-FK mixtures with air showed a slightly higher withstand compared to CO_2 as background gas (not shown here). With 30% C5-FK admixture to CO_2 a performance similar to that of SF_6 is reached, i.e., slightly lower withstand than measured for C4-FN mixtures, as seen in Figure 4.40.

The breakdown voltage in C5-FK mixtures was investigated, e.g. in [49,110–112]. A comparison of various mixtures with SF_6 adapted from [49] is shown in Figure 4.45.

For LI and AC waveshape and uniform field, a similar electric strength as SF_6 with 450 kPa is reached at 700 kPa for a C5-FK/CO_2/O_2 mixture with 5.6% C5-FK and 11.2% O_2 content. It should be considered, however, that this mixture is only suitable down to 0°C. CO_2 at 700 kPa has about 40% and 30% lower electric strength than SF_6 at 450 kPa for LI and AC, respectively. In strongly non-uniform fields positive LI is much lower than negative LI for all gases, similarly for SF_6 and C5-FK mixtures. AC breakdown fields are only slightly lower than positive LI for all gases.

In the round-robin benchmark test of [69], the practical insulation performance was tested at 600 kPa with C5-FK/CO_2/O_2 mixtures for weakly and strongly non-uniform field configurations, see Figure 4.46 with similar results as reported from [49]. AC and LI-breakdown voltage are similar, with the LI level being slightly higher than AC. Compared to SF_6, the C5-FK/CO_2 mixtures show roughly 60% of the breakdown voltage of SF_6 at the same pressure. For positive LI, the results from different test labs vary widely and are on average larger than for negative LI, as for SF_6. The LI/AC impulse ratio was found to be identical to SF_6. Also, with strongly non-uniform fields, the results indicate a similar behaviour with similar breakdown voltages as those measured for SF_6.

The important conclusion was given in [97] that, similar to C4-FN/CO_2 mixtures, existing test and design rules can be used also C5-FK mixtures when scaled with the ratio of electric strength. Assumptions on impulse ratios as used in the definition of IEC standard type and routine test voltages are valid for C5-FK as well, such that these values can be used also for the new SF_6 alternative gas mixtures.

4.2.4.2.3 O_2 admixture influence

There are various, partially quite differing statements on the influence of O_2 admixture. Reference [108] reports the influence of O_2 concentration in C4-FN mixtures and for AC breakdown voltages relevant for MV application, see

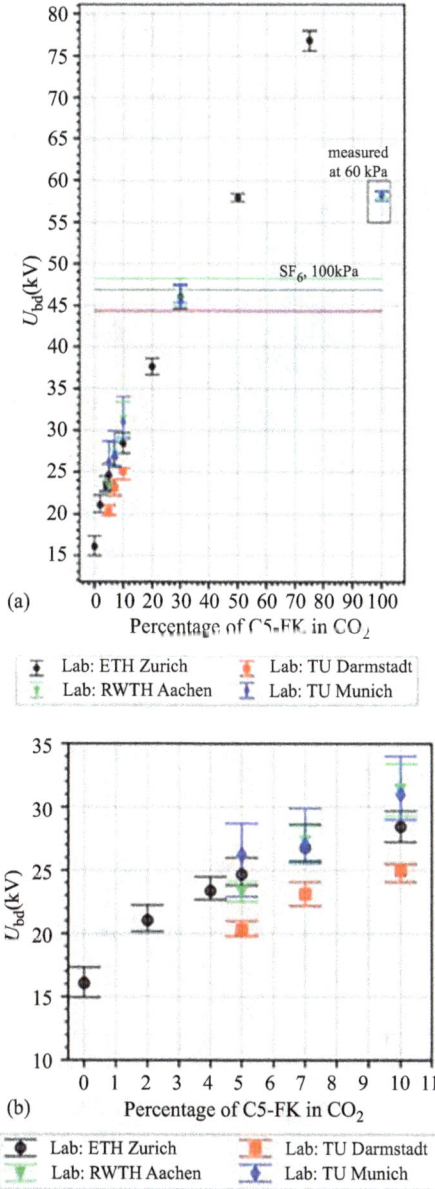

Figure 4.44 *Synergy scan of a mixture of C5-FK in CO_2 at 100 kPa compared with SF_6 at 100 kPa (horizontal lines) for AC breakdowns in a uniform field arrangement of 5 mm gap (a) 0–100% and (b) focused view 0–10% from [89], reprinted with permission from CIGRE, @2021.*

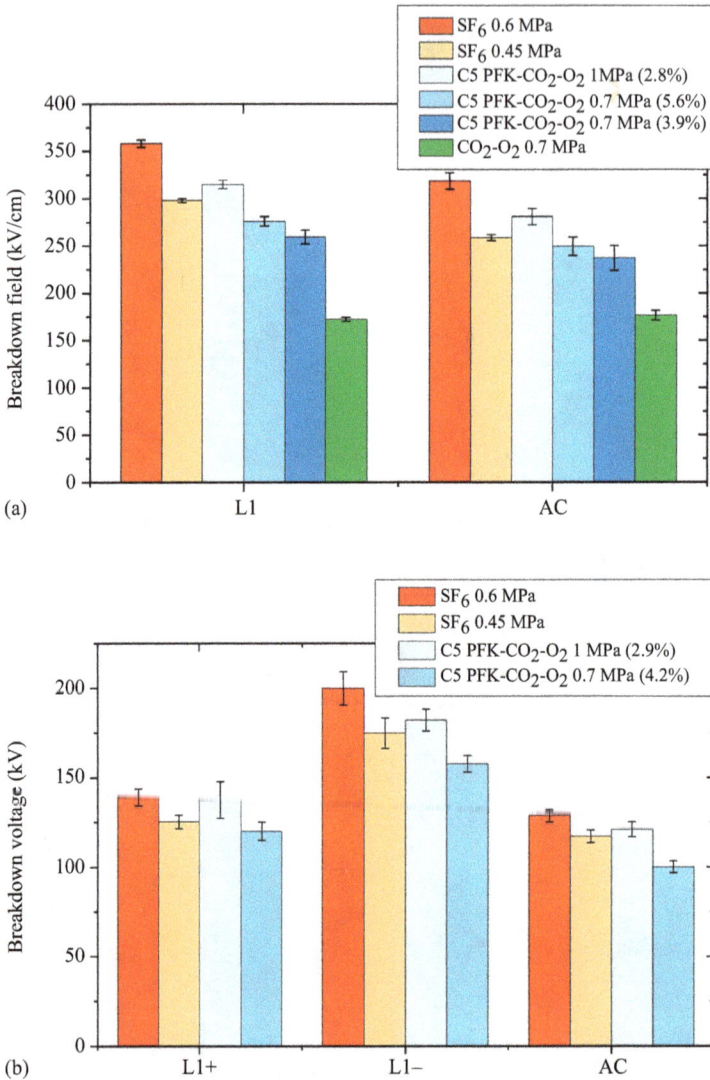

Figure 4.45   Breakdown voltage of mixtures of C5-FK with CO_2 and O_2, under different experimental gas mixtures. The figure is taken from [97]. The molar ratio of C5-FK used in each measurement is indicated in parentheses. In all cases, the partial pressure of O_2 was fixed to twice the partial pressure of C5-FK. For measurements using a lightning impulse, the polarity corresponds to the polarity at the needle. (a) Uniform field: Rogowski electrodes with 7.5 mm gap. (b) Non–uniform field: 1.2 mm needle fixed at the centre of one of the two Rogowski electrodes. The inter–electrode distance is fixed to 10 mm, reprinted with permission from CIGRE, @2021.

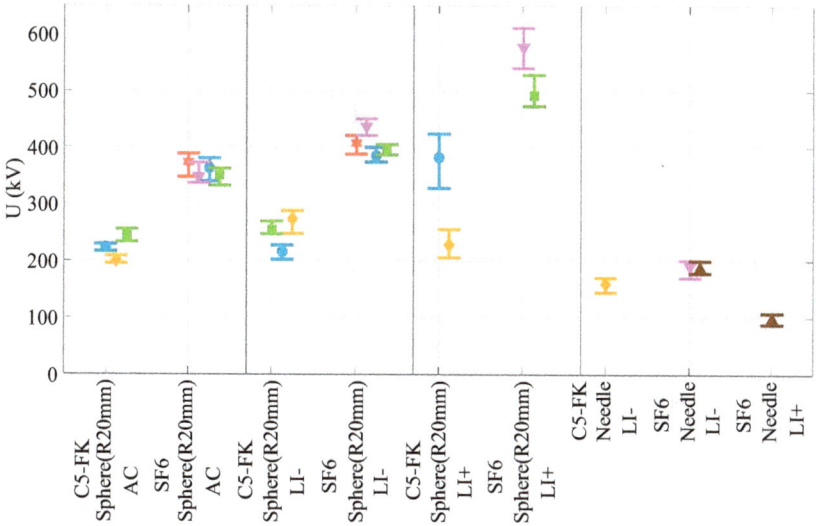

Figure 4.46 Breakdown voltages in sphere-plane (weakly non-uniform) and needle-plane (strongly non-uniform) configuration, for a mixture of C5-FK/O_2/CO_2 (5.5%/11%/83.5%) mixture at 0.6 MPa absolute pressure compared to SF_6, adapted from [97]. Different colours show the results of different research groups for the same arrangement, reprinted with permission from CIGRE, @2021.

Figure 4.47. In these experiments, an increase of at most 7.7% is reported; it occurs for an O_2 concentration of 6%. The gas pressure was 0.14 MPa and 15% C4-FN molar admixture was used, which is quite different from HV applications with significantly higher pressures and lower C4-FN concentrations.

In the benchmark test of [97] using AC waveshape, it was noted that the differences at 100 kPa and 5% O_2 admixture were too small for a clear conclusion regarding the influence of O_2, given that any differences were within the uncertainty of the measurement. Since the uncertainties were in the range of the expected increase from [112], this result is not unexpected.

In the effective ionisation coefficients of the pulsed Townsend experiments of [113], no significant effects were seen with up to 10% O_2 admixture in 5% C4-FN or 5% C5-FK mixtures with CO_2. For the C5-FK/N_2/O_2 mixture, however, the critical field strength dropped by 6.5% when using up to 19.5% O_2.

In CO_2/O_2 mixtures, an increase in electric strength is reported, e.g., in [106], see Figure 4.48. With 10% and 30% O_2 admixture, an increase of dielectric strength of about 6% and 13%, respectively, compared to pure CO_2 is obtained. Reference [107] reported in calculations a 22% increase in the critical field when adding 20% O_2 to CO_2 at room temperature. This is higher than the curve shown in Figure 4.48, where only about 10% increase would be expected.

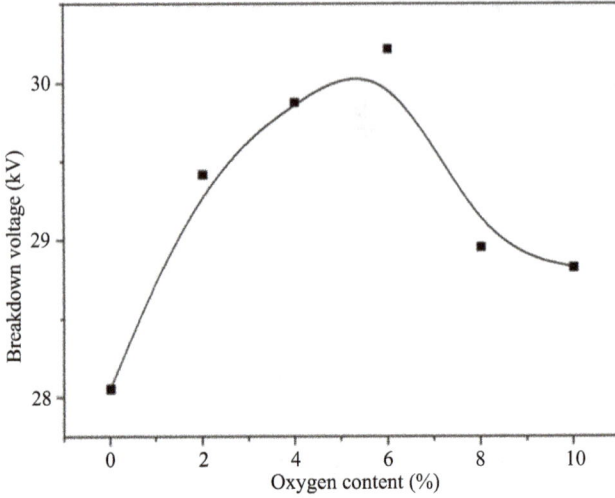

Figure 4.47   AC breakdown voltages for different O_2 concentrations in C4-F7N/ CO_2 mixtures from [108]. The electrode arrangement was a sphere-sphere gap (radius 25 mm and 3 mm gap).

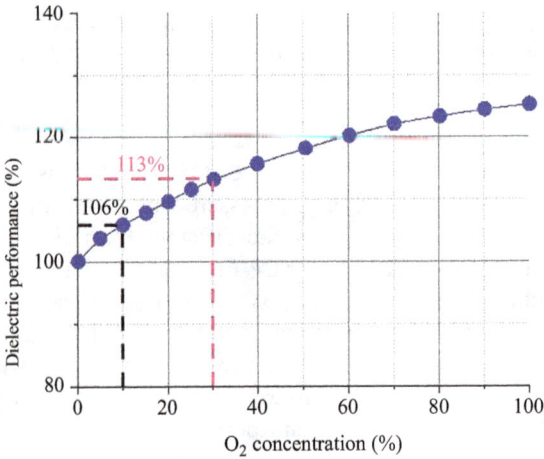

Figure 4.48   Dielectric performance in CO_2/O_2 mixtures vs. O_2 concentration adapted from [106]

4.3   Switching properties

In the present section, the switching properties of SF_6 alternative gases will be compared to those of SF_6. There is still not exhaustive information on direct comparisons. Such work is currently ongoing, e.g. [114]. An overview of the present publicly available information was given in [52]. The section will be organized as follows: first, load current switching properties, especially those of capacitive and small inductive currents will be addressed in Section 4.3.1. Thermal interruption information will be compared in Section 4.3.2. This will include a comparison of indicators like conductance decay, post-arc currents (PAC) and extinction peaks in addition to the thermal interruption performance at CZ. Section 4.3.3 addresses the dielectric recovery performance.

4.3.1   Load current interruption

For load currents, such as capacitive or inductive currents, in HV CB, the interruption challenge in SF_6 is mainly of dielectric nature, i.e., the insulation strengths at a given contact distance, pressure and electric field distribution. The reduction of dielectric strength due to the load current is usually negligible. Due to this, the dielectric strength variation with stroke is usually called a cold characteristic [52]. For SF_6 alternative gases and mixtures, there is only little information so far on this topic; e.g. [115,116] report successful capacitive switching test duties for modified commercial SF_6 CB's using CO_2/O_2/C5-FK and CO_2/O_2/C4-FN mixtures, respectively. This agrees with the expectation that for similar dielectric strength of a gas mixture compared to SF_6 a similar capacitive switching performance should result. Such successful use of alternative gases and mixtures was reported also by several manufacturers using CO_2-based mixtures [69,71,117,118]. To the author's knowledge, there is no direct comparison of the performance of SF_6 and alternative gas mixtures in HV CB under capacitive or inductive switching conditions available to date. This should be addressed in the near future.

Reference [114] states that test procedures prescribed by the international standards are expected to be applicable also for SF_6 alternative gases, since breakdown scatter is similar to SF_6.

4.3.2   Thermal interruption performance
4.3.2.1   Interruption limits

As explained in Section 4.1.3.2, thermal interruption is defined by the current slope di/dt that can be interrupted for given circuit conditions (e.g. surge impedance) and depends on fundamental properties of the employed gas and also on design parameters. An important design-related parameter is the blow pressure at CZ. The blow gas temperature might play also a role [72,119]. Other important parameters are flow conditions in the arc zone, as explained in Section 4.1.3.2.

To judge interruption performance, benchmark tests need to be done [114]. For such tests, it is important to test the performance under the same conditions, which include:

- Test circuit parameters, including line surge impedance, charging voltage of injection circuit, current decay rate di/dt, rate of recovery voltage rise du/dt, time delay of the TRV, and parallel and stray capacitances.
- Blow pressure, i.e., the pressure in the heating volume at CZ.

- Blow gas temperature.
- Contact stroke and arcing time.
- No metal vapor, since this could affect different gases differently.
- Negligible or similar amount of PTFE vapor in the heating volume.
- Similar exhaust pressure.
- Liquefaction of additive gases like C4-FN or C5-FK in the blow gas before reaching the arc zone must be avoided for a well-defined concentration of the gas mixture.

In the literature, there are only few publications available that might approximately fulfil these requirements. In the following, such comparative tests will be discussed.

Thermal interruption in CO_2 and air was compared to SF_6 by Stoller *et al.* [68] in a so-called two-pressure device, where pressurized gas at ambient temperature is released to the arc zone sufficiently early before CZ to establish quasi-stationary flow conditions. Only low currents of about 2 kA peak were tested, thereby keeping metal vapor or nozzle vapor negligible. The test circuit was a synthetic Weil–Dobke circuit with the circuit elements adjusted for an equivalent surge impedance of 450 Ω. The pressure range covered was between 1.2 and 1.9 MPa. The relative interruption performance compared to SF_6 is shown in Figure 4.49 by the black symbols. It can be deduced from the tested conditions that air has about 30% of the interruption performance of SF_6 and CO_2 has about 67% of the interruption

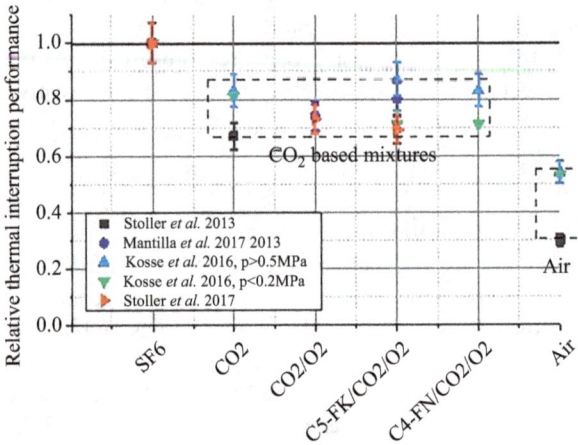

Figure 4.49   Relative thermal interruption performance of synthetic air, CO_2, CO_2/O_2, C5-FK/CO_2/O_2, C4-FN/CO_2/O_2 compared to and SF_6 from various authors [49,68,120–122]. The performance in each experiment was determined based on the interrupted di/dt. The uncertainties were estimated from the respective publications where measured data was shown. For [121], the same uncertainty was assumed as for the other references.

performance of SF_6. Typical uncertainties of these ratios are estimated to about ±7% based on the scatter of the results shown in the publication.

A comparative study including SF_6, CO_2/O_2 and C5-FK/CO_2/O_2 was carried out by Mantilla et al. [120] using a 245 kV self-blast CB. The C5-FK additive concentration was given in the range up to 5%. From the pressure-reduced di/dt values for C5-FK/CO_2/O_2 and CO_2/O_2, a performance of ∼80% and ∼74% of that of SF_6, respectively, was determined, see the red symbols in Figure 4.49.

Another comparative test was done by Kosse et al. [121] using a two-pressure test device, comparing SF_6, N_2, air, CO_2 and CO_2/C4-FN and CO_2-C5-FK mixtures. The mixtures with fluorinated additives were chosen for −30°C application [122]. Since molar concentrations were not given in the publications, it can be assumed that typical partial pressures were in the range as discussed in Section 4.2.2. The limiting di/dt for interruption was determined for upstream pressures of >0.5 MPa and <0.2 MPa. The device was placed in a tank filled with ambient pressure with the gas under test. The results are shown in Figure 4.49 by the blue and green symbols. The fluorinated gas mixtures showed an interruption capability in the range of 83–87% of SF_6, which is similar to the 83% observed with CO_2. Air showed only about 54% and 53% of the performance of SF_6, respectively. N_2 was found to be similar to air in its interruption capability (not shown in Figure 4.49).

As discussed in [1], benchmark tests with two-pressure devices may lead to erroneous results if the high boiling point component liquifies during compression. This could be the case for the C4-FN and C5-FK additives in the study of [121]. However, the similar result for the C5-FK mixture shown in [121] (where a two-pressure device was not used and liquefaction thus not an issue) suggests that this was not an issue, possibly due to a limited upstream pressure in the high-pressure volume. This is confirmed by the similar interruption capabilities using an upstream pressure of below 0.2 MPa, where liquefaction of chosen mixtures should not occur due to the low partial pressures of the additives, see Figure 4.49.

Stoller et al. [49] used a modified 145 kV SF_6 SB CB and tested SF_6, C5-FK/CO_2/O_2 (8%/84%/8%) and CO_2/O_2 (92%/8%) mixtures and compared the interruption performance directly to SF_6. Similar values as those of the other investigations were found, see Figure 4.49.

Comparing all the results from the different studies of Figure 4.49, it can be concluded that all CO_2-based mixtures show a similar thermal interruption performance in the range 67–87% compared to SF_6, within the scatter and uncertainties of the different experiments. The performance of air is significantly lower and varies between 30% and 54%. These ranges are indicated by the dashed boxes in Figure 4.49. It needs to be considered that the interruption performance depends on the design and test circuit details. This might lead to differences between the different experiments.

The change in interruption performance brought about by adding O_2, CH_4, N_2 and He admixtures to CO_2 was investigated by Uchii et al. [123]. This investigation was done in a puffer circuit breaker using a synthetic test circuit with 450 Ohm surge impedance. The filling pressure of the CB was 0.7 MPa. The pressure in the heating chamber at CZ was not given, but it was stated that test conditions were

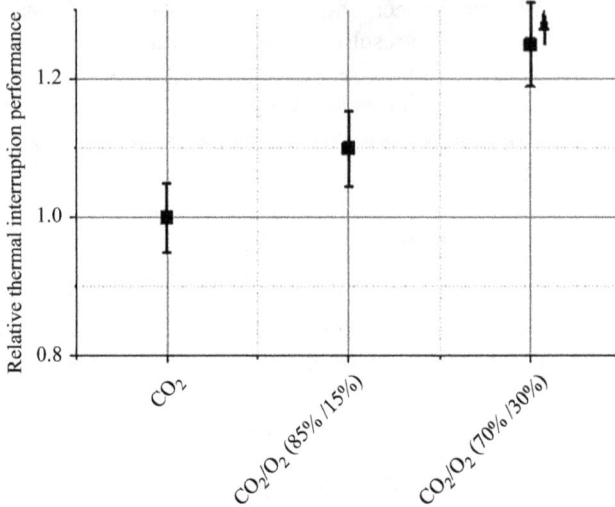

Figure 4.50   Relative thermal interruption performance of different O_2 admixture to CO_2 from [123]

almost equivalent. An interesting additive for practical applications is O_2, as explained in Section 4.2.3, which is beneficial for reducing switching by-products and possibly also increasing dielectric strength. The results for O_2 admixtures are shown in Figure 4.50. With 15% O_2 addition, which is a typical value for practical applications, about 8% improvement was obtained. With 30% O_2 addition the limit could not be found but at least 25% improvement was reported. Based on these results, a 30% O_2 addition to CO_2 was preferred, which was reported also later as a promising mixture [124]. CH_4 also showed an improved interruption performance (not shown in Figure 4.50) in [123] but could be less suited due to flammability and moisture generation issues.

The various experiments so far available can be summarized:

(a) The qualitative ranking of thermal interruption is:

$SF_6 > CO_2 + F$-Additive $\geq CO_2 >$ air

(b) The fluorinated additives do not significantly improve the thermal interruption performance of CO_2. The performance increase compared to CO_2 for the investigated mixtures is within the uncertainties. Only moderate effect is expected based on a comparison of material parameters, e.g., the $\rho \cdot c_p$ values discussed in Section 4.2.3. Hence, the experimental results confirm such a moderate effect for small additive concentrations in the range 3–5%.

(c) CO_2 has about 17–33% lower interruption performance compared to SF_6. This corresponds roughly to one current rating. It can be deduced: CO_2-based CB have intrinsically roughly one current rating lower thermal interruption performance than SF_6. To improve this performance, specific design adaptions in the interrupter unit are required for CO_2-based CB to reach the performance of SF_6 CB.

(d) Different quantitative results of the different investigations can possibly be explained by differences in the details of the test circuit, test procedures and designs. For example, could a decrease in charging voltage – a means to achieve a lower di/dt in the injection circuit when searching for a limit – lead to a less stiff circuit due to arc-network interaction? This could ease the relative interruption performance for gases with significantly lower performance, such as air or N_2, and might explain the higher performance of air in [121] compared to [68]. Unfortunately, such details are usually not given in the publications.

Since the experimental conditions of the above-given tests are not fully comparable, additional experiments with well-defined and controlled parameters are needed. Such experiments are currently ongoing at ETH-Zürich in Switzerland, where all gases of interest will be tested in the same test device and circuit. The test device design and procedures are described in [114]. These tests should give more precise numbers for comparison of thermal interruption performance of the various CO_2-based mixtures to that of SF_6.

4.3.2.2   Arc conductance decay and post arc currents

As discussed in Section 4.1.3.2, arc conductance or resistance shortly before CZ are important parameters to judge interruption performance. The arc conductance decay for CO_2/O_2 is shown in Figure 4.51. For a successful interruption, the conductance 200 ns before CZ (g_{200}) should be less than half the value required for successful interruption in SF_6 and is estimated to be less than 0.5 mS. Figure 4.52

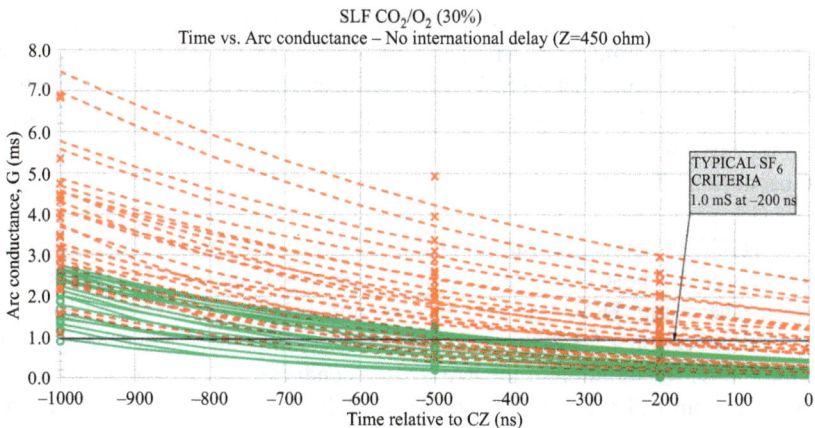

Figure 4.51   Arc conductance decay in CO_2/O_2 (70/30%) under SLF conditions from [124]. The red symbols and curves denote interruption failure, whereas the green symbols and curves denote successful interruption.

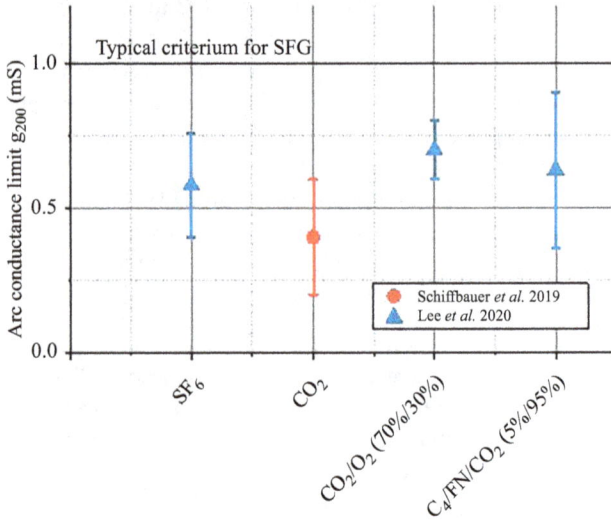

Figure 4.52   Limit values for the arc conductance 200 ns before CZ from Schiffbauer et al. [124] and Lee et al. [125]

shows this value in comparison to other tests with CO_2/O_2, SF_6 and C4-FN/CO_2 mixtures from [125].

Within the uncertainties, all these values are similar, but significantly lower than the limit typically found for SF_6 [126,127], indicated by the horizontal line in the figure. To understand the differences, it needs to be considered that not only different gas mixtures were tested but also under different conditions of fill pressures and du/dt after CZ [125], which influences the result and, hence, the critical g_{200} for a successful interruption. Additional parameters which might be different are the arc-network interaction and timing of the current injection before CZ. It can be concluded that the arc conductance g_{200} needs to be about a factor two lower in SF_6 alternative gases compared to SF_6 for achieving successful interruption. This can be attributed to the lower cooling power of the arc in CO_2-based gases before CZ.

After CZ the post arc current (PAC) is important for judging interruption performance, the post arc current peak and duration are parameters of special interest. The rising TRV will lead to a current in the residual arc channel. As explained in Section 4.1.3.2, the stronger the arc cooling, the smaller the amplitude and the shorter the duration of the PAC. If arc conductance during this phase is too high, the current will rise during the rising TRV, leading to reheating of the residual arc channel followed by thermal interruption failure. Figure 4.53 shows measured current signals from unsuccessful tests with various gases under the same conditions from Stoller *et al.* [68]. Within the measurement accuracy of 0.5 A, no post-arc current could be observed in the decision phase from CZ to restarting of current flow, indicating a low PAC level.

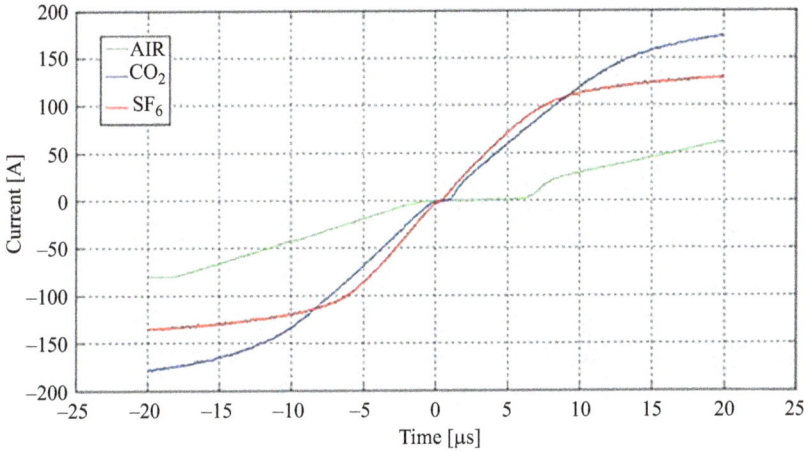

Figure 4.53   Current measurement around CZ for unsuccessful interruptions in SF_6, CO_2 and air [68]. Note that the saturation of the current signal is produced by the current measuring system.

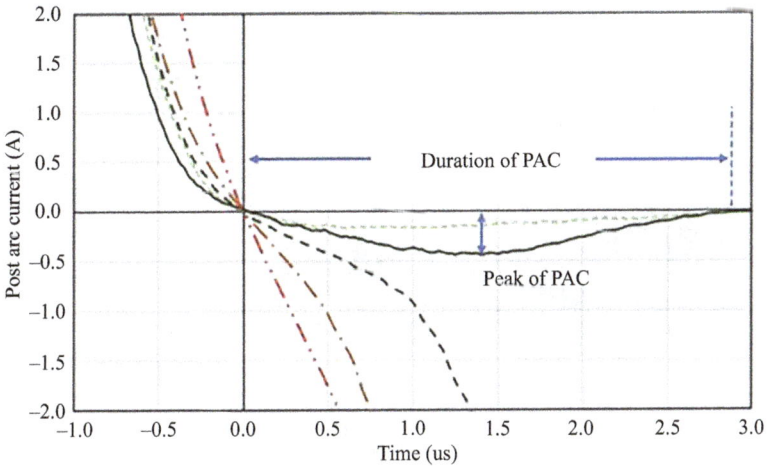

Figure 4.54   PAC measurement in an L90 test with $CO_2/O_2/C5$-FK SB CB for 145 kV, 40 kA, 50Hz, adapted from [115]. The various curves are for different applied pressures leading to increasing and decreasing PAC at interruption failure and success, respectively.

This phase is much larger for air and slightly larger for CO_2, compared to SF_6. PAC measurements in SLF90 tests are also shown in [115] for a $CO_2/O_2/C5$-FK mixture (82%/11.4%/5.7%), see Figure 4.54. Different arcing time and pressure conditions lead to different PAC evolutions. The PAC peak and duration at interruption are less than about 0.5 A and 3 μs, respectively, in the shown example.

(a)

(b)

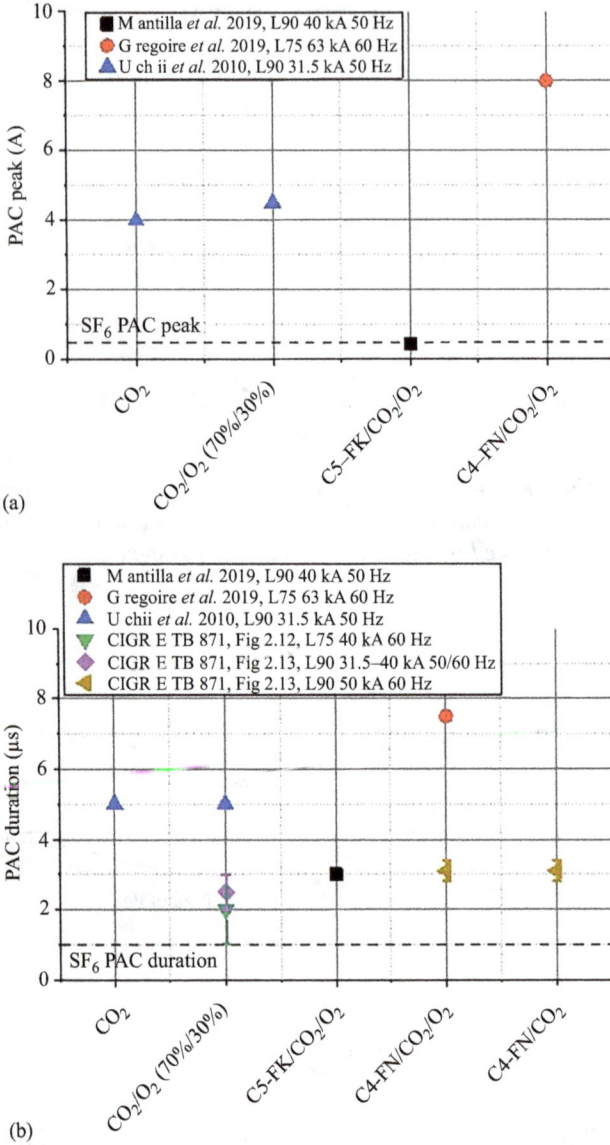

Figure 4.55   PAC peaks and durations from various authors [114,115,123,128] at different SLF conditions

Figure 4.55 compares available data from various short-line fault tests for SLF75 (L75) and SLF90 (L90) conditions. A long post-arc current duration phase of several microseconds and PAC peak of several amperes was seen in the tests of Uchii *et al.* [123] with CO_2 and Gregoire *et al.* [128] with $CO_2/O_2/C4$-FN. PAC peaks and durations of up to 8 A and 8 µs were observed, respectively. This is much larger than observed in Figures 4.53 and 4.54. With SF_6, PAC peaks are below 0.5 A and durations are below 1 µs, typically [129]. The post arc current peak and duration depend on the conductance of the arc at CZ, i.e., blow conditions, on arc zone design, on the gas, and on circuit parameters like di/dt, surge impedance and arc-circuit interaction, as discussed in Section 4.1.3.2. This might explain the differences in the experiments shown. The main conclusion from the comparison is then that in CO_2-based gases the 'critical' PAC peak and duration is higher than for SF_6. This can be expected from the material property discussion in Section 4.2.3, since the arc is wider in CO_2-based mixtures and cooling is less efficient.

The longer PAC duration might have an influence on the criticality of different SLF conditions. The IEC standard defines SLF60, SLF75 and SLF90, with the latter usually the most critical since the short-circuit current is the largest and accordingly the di/dt before CZ. However, the TRV line peaks of the different test duties occur at different times that increase with increasing distance of the fault, i.e., with decreasing short-circuit current. Due to the longer PAC duration in CO_2-based SF_6 alternative gas mixtures, it might be beneficial if the first TRV line peak occurs during the PAC duration, since it also leads to a change in du/dt applied to the CB. On the other hand, it might be more critical for the interruption if the TRV rises without such a change of du/dt. This could be the case for L75 where the line peak occurs for a 170 kV, 50 kA rating at about 6.4 µs in the SLF75, compared to 2.5 µs in the SLF90 case [67]. In [128,130], this was reported to be more critical than SLF90. However, there is still no general consensus on this topic yet [67,114].

The pressure build-up peak in the thermal or compression chamber in CO_2 and CO_2-based mixtures is usually higher than in SF_6 [68,114,130,131], see e.g., Figure 4.56, where the pressure build-up is compared between SF_6 and a CO_2/C4-FN

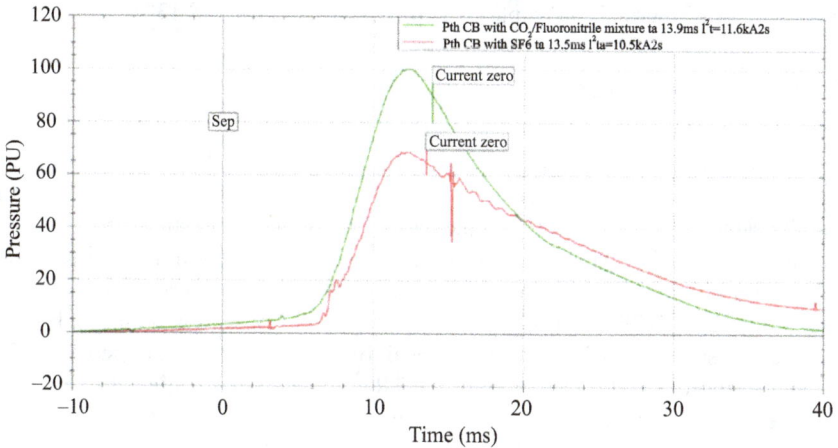

Figure 4.56 Pressure rise in the thermal chamber for SF$_6$ and C4-FN/CO$_2$ mixture from [131], reprinted with permission from CIGRE, @2018

mixture. This is due to the adiabatic coefficient, which is higher for CO_2-based mixtures compared to SF_6, as discussed in Section 4.2.3.3. However, it should be noted that the energy input to the thermal chamber is not necessarily the same for the different gases, even if the short-circuit current to be interrupted is the same. In an SB CB, for example, the arc burns in PTFE vapor and the pressure in the arc zone is defined by the nozzle geometry. In the case of back-heating at the high short-circuit currents relevant for SLF tests, the amount of energy flowing to the thermal chamber is then defined by the pressure difference between arc zone and thermal chamber. A faster pressure rise in the thermal chamber, e.g., due to the higher adiabatic coefficient of CO_2, will then reduce the energy flow to the thermal chamber [68]. This might lead to lower pressure than expected simply from (4.7) based on the adiabatic coefficient. It should be noted that the process is of transient nature and depends on the design, contact movement and applied current. A higher pressure would in principle be beneficial for thermal interruption, since interruption performance rises with pressure as shown in Section 4.1.3.2. However, due to the higher speed of sound of CO_2-based gas mixtures the pressure decay after the peak is faster, as can be seen also in Figure 4.56. The same behaviour is observed for CO_2/C5-FK mixtures [49]. The addition of C5-FK to the CO_2 does not result in any significant change in the pressure build-up.

It is important to consider these differences in pressure build-up in the design of the CB, e.g., by proper control of flow cross-sections and contact travel characteristic. Additionally, the pressure rise in terminal faults, such as T100a, needs to be considered to ensure that the mechanical strength of the interrupter chamber arc zone is not exceeded. The mechanical requirements arising from the pressure rise in terminal faults can indirectly set limits for the maximum possible pressure build-up in SLF tests. Hence, simply replacing SF_6 in an existing CB by a CO_2-based alternative gas mixtures could lead to larger de-ratings than if an optimized design is used. A performance close to that of SF_6 can be reached with SF_6-alternative gases, especially when considering a limited additional parallel capacitor for improvement of thermal interruption, as discussed, e.g. in [128,130].

4.3.3   Dielectric recovery

After a successful interruption of the current around CZ, the TRV develops across the CB contacts. As explained in Section 4.1.3.2, the early dielectric recovery phase within a few 10 µs is characterized by the removal of the arc plasma, whereas the late dielectric recovery phase is characterized by the removal of hot gas surrounding the arc [21]. Measurement of the dielectric recovery in a modified commercial SB circuit breaker under nearly identical conditions was shown by [49] comparing SF_6, C5-PFK/CO_2/O_2 (mixtures and pure CO_2, see Figure 4.57). For this purpose, a fast-rising TRV with time to peak of about 30 µs was applied after CZ. By the variation of the trigger time for the TRV application, the dielectric recovery characteristic could be determined, as explained in Section 4.1.3.1 and [21]. At low and medium current, the C5-FK admixture to CO_2/O_2 leads to significantly faster dielectric recovery compared to the CO_2/O_2 mixture. SF_6 recovers much faster than both gas mixtures, however. At such currents where significant back-heating

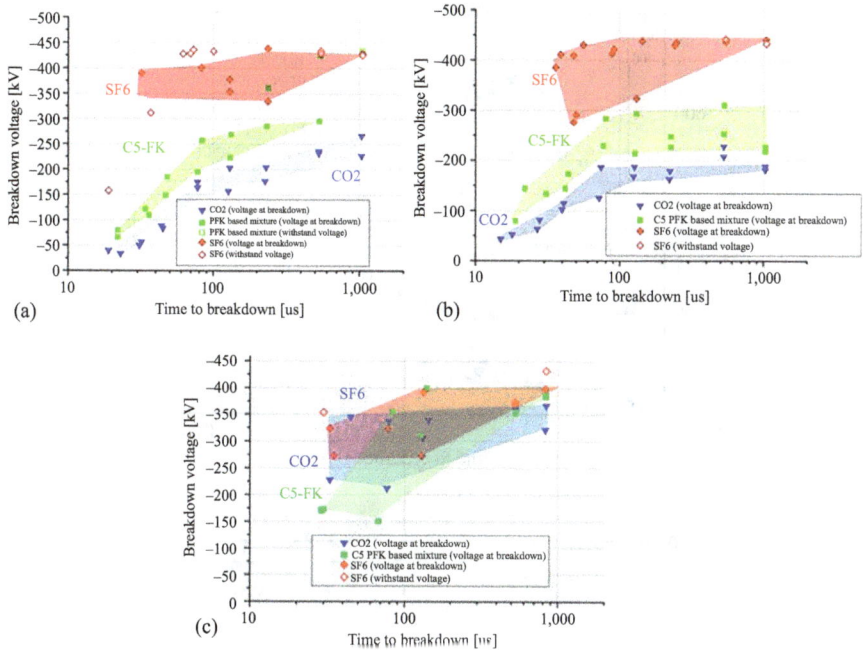

Figure 4.57 *Dielectric recovery at (a) low (4 kA rms), (b) medium (12 kA rms) and (c) high currents (30 kA rms) in a SB CB using C5-FK/CO₂/O₂ (10%/80%/10%) and CO₂/O₂ (90%/10%) mixtures in comparison to SF₆ adapted from [49]. Filling pressure was 0.4 MPa (abs).*

(see Section 4.1.2) still does not occur, the recovery characteristics of the different gases are clearly separated, which is indicated by the coloured areas in Figure 4.57. In the late recovery, the difference between SF_6 and CO_2 is more than a factor of two for these currents (Figure 4.57(a and b)). A similar difference between SF_6 and CO_2 was observed in [68]. The C5-FK mixture is in between SF_6 and CO_2 at such currents. At higher currents (Figure 4.57(c)), where back-heating occurs, the recovery characteristics overlap within the large scatter of the experiments. This might be due to pressure build-up and decomposition of the gases due to the back-heating. The steep rise of the early recovery could not be seen in these tests, i.e., it is faster than the minimum possible trigger time delay.

Another measurement of the dielectric recovery in SF_6 alternative gases was shown in [132], using a test device based on a single pole of a HV CB. Tested currents corresponded to T10 and T30 test duties for a 31.5 kA short-circuit current rating. Dielectric recovery characteristics of CO_2/O_2 (90%/10%), C5-FK/CO_2/O_2 (1.5%/88.5%/10%) and C4-FN/CO_2/O_2 (2.8%/87.2%/10%) at a total filling pressure of 0.7 MPa (abs) were compared. For the C5-FK and C4-FN mixtures, this corresponds to a dew point of about −25°C and −40°C, respectively. The resulting dielectric recovery characteristics are shown in Figure 4.58. For T10 currents

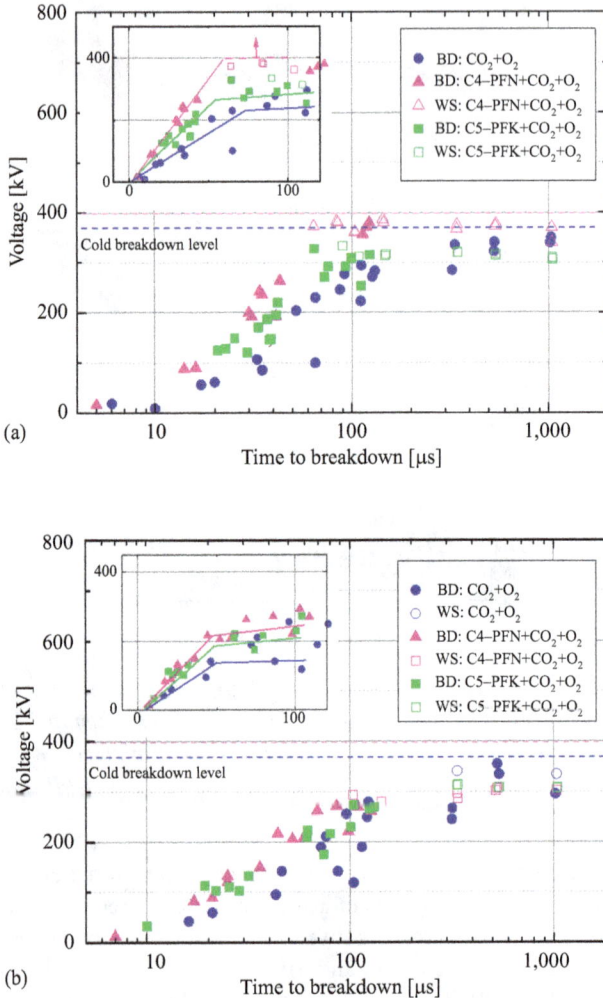

Figure 4.58   *Dielectric recovery characteristics at (a) low currents, T10 and (b) medium currents T30 in a SB CB adapted from [132] for CO_2/O_2 (90%/10%), C5-FK/CO_2/O_2 (1.5%/88.5%/10%) and C4-FN/CO_2/O_2 (2.8%/87.2%/10%) mixtures. The total filling pressure was 0.7 MPa (abs). The inserts show the early dielectric recovery phase on a linear time scale. The solid lines are a guide to the eye for the minimum level of the early dielectric recovery. The horizontal dashed lines indicate the cold breakdown voltage (see text) for C5-FK and CO_2/O_2 mixtures. For C4-FN mixtures the cold breakdown voltage was higher than 325 kV but could not be determined due to limitations on the test voltage.*

(Figure 4.58(a)), the recovery is significantly faster with fluorinated admixtures in comparison to CO_2/O_2, despite the low admixture concentrations used. C4-FN mixtures show a faster recovery compared to C5-FK mixtures, which is probably due to the higher admixture concentration. The inset in Figure 4.58(a) shows that the early recovery rises linearly in the first 50 to 70 µs after CZ, as discussed in Section 4.1.3.2 for the example of air (Figure 4.18). The slope of rise of the C4-FN mixture is nearly twice as steep compared to the CO_2/O_2 mixture. After the fast rise of the early recovery, the recovery proceeds at a slower pace. This is indicated by the full lines in the insert of Figure 4.58(a). After 120 µs, for both the C4-FN and C5-FK mixtures, no breakdowns occurred; the arc zone withstood the maximum test voltage that could be applied in the laboratory. Such a level is close to the 'cold' breakdown level, which is the breakdown voltage without any current application at the same contact distance (measured prior to the power tests as given in the publication). The cold breakdown level is indicated by the dashed horizontal lines in Figure 4.58. Hence, after slightly more than 120 µs, the gap recovered approximately to the cold breakdown level when using C4-FN mixtures. This is similar for the C5-FK mixtures, when assuming that the cold breakdown level for this mixture is in between that of CO_2/O_2 (blue dashed line) and C4-FN mixtures (magenta dashed line). With CO_2/O_2 breakdowns were observed up to 1,000 µs after CZ at a level of about 350 kV, which is close to the cold breakdown level, i.e., the late dielectric recovery with CO_2/O_2 takes longer compared to the fluorinated mixtures.

For T30 currents, the dielectric recovery is slower compared to T10 for all gases investigated. Not only the rise of the early recovery is lower (see inset in Figure 4.58(b)) but also the late recovery, indicated by the full lines. The steepness of the rise is the highest for C4-FN mixtures and the lowest for the CO_2/O_2 mixture. The C5-FK recovery is only slightly below that of the C4-FN mixture. The differences in the T10 case can possibly be explained by clogging and moderate back-heating, which lead to higher temperatures in the arc zone after CZ. After 1,000 µs, the cold breakdown level is still not fully reached for CO_2/O_2. For the other gas mixtures, the maximum test voltage was too low to determine the late recovery characteristic after 120 µs.

The dielectric recovery depends on the electric field, the temperature and the pressure distribution in the arc zone, as explained in Section 4.1.3.2. The difference between the tests at low and medium currents shown above, which were done under nearly identical pressure conditions within one experiment can, therefore, be attributed to differences in plasma cooling in the early recovery and to differences in the cooling and the removal of hot gas in the late recovery. As discussed in Section 4.2.3 on material properties, the speed of sound in all CO_2-based gas mixtures is quite similar at the same gas temperature, leading to similar expected durations for the removal of hot gases from the arc zone. Differences are then expected to be explained by the temperature decay of the residual arc after CZ, the hot gas cooling properties (e.g., $\rho \cdot c_p$), and the differences in the critical field.

A strongly simplified comparison can be done for understanding the measurements shown in Figures 4.57 and 4.58. For simplicity, the influence of O_2

Table 4.3   *Ratio of particle density reduced critical fields of CO_2-based mixtures compared to SF_6 and CO_2*

Gas	Critical field ratio compared to SF_6	Critical field ratio compared to CO_2	Source
C5-FK/CO_2 (10%/90%)	0.6	2.2	Figure 4.36
C5-FK/CO_2 (1.5%/98.5%)	0.4	1.5	Figure 4.36
C4-FN/CO_2 (3%/97%)	0.55	2	Figure 4.39
CO_2	0.27	1	Figure 4.37

admixture on CO_2 and C5-FK mixtures is neglected here. As was shown in Figures 4.37 and 4.39, the particle density-reduced critical field of a C4-FN/CO_2/O_2 mixture with 3% C4-FN admixture is about 55% that of SF_6. Assuming the same pressure, temperature and electric field in the arc zone this would mean that the breakdown voltage for the C4-FN mixture would approximately be expected at about 55% that of SF_6. Such ratios are compared for C4-FN, C5-FK mixtures and CO_2 with SF_6, see Table 4.3.

Based on such reasoning, it can be deduced that mixtures with 10% C5-FK and pure CO_2 should show a breakdown voltage of about 60% and 27% that of SF_6, respectively. Within the scatter, this is roughly fulfilled in the late recovery, see e.g. at 100 µs after CZ in Figure 4.57. Note that the filling pressure in these tests was only 0.4 MPa, and the dielectric recovery performance for the typical filling pressures mentioned in Table 4.2 will be higher, approaching that of SF_6.

At high currents (Figure 4.57(c)), the breakdown voltages in the late recovery were similar considering the large scatter. It is likely that the pressure and temperature (and also composition) distributions are quite different for the different gases, i.e., breakdown voltage relations do not depend only on the critical field.

Similarly, we can compare the measurements of Figure 4.58 at low (T10) and medium (T30) currents to the critical field ratios of Table 4.3. Based on the critical field ratios, the C4-FN mixture should show a breakdown voltage about twice as high as CO_2 and about 33% higher than the C5-FK mixture used in these tests. Figure 4.58 confirms such ratios for the T10 currents during the early and late recovery. For the T30 currents, the ratios seem to be slightly less than these simple relations predict.

It can be concluded that for low (T10 like) and medium (T30 like) currents, the different dielectric recovery characteristics under the same pressure conditions are mainly determined by the differences in the critical field of the employed gas mixtures. At high currents, differences in pressure and temperature distributions in the arc zone will lead to deviations from such simple scaling relations.

At very high currents, as for example T100a/s conditions, little information on the dielectric recovery characteristics has been published to date. Due to gas decomposition and ablated nozzle material [49,133,134], increased concentrations

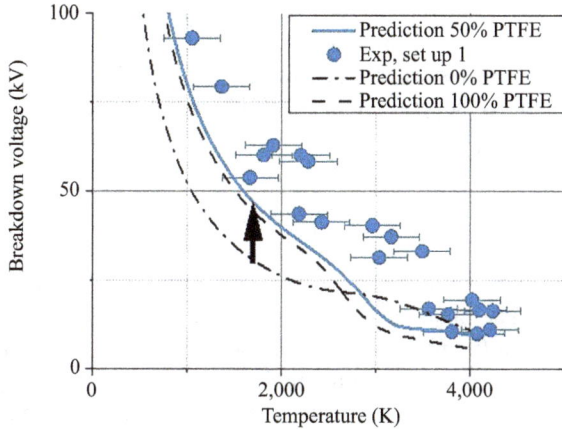

Figure 4.59 *Breakdown voltage in uniform field in CO_2/O_2 (90%/10%) in the*
presence of PTFE (C2F4) vapor at elevated temperatures, adapted
from [135]. The PTFE was not determined quantitatively but
estimated to be in the range above 50%. The curves show the
voltages at which the critical field is reached in the gap and the
arrow indicates the difference between CO_2/O_2 and CO_2/O_2 with
50% PTFE vapor admixture.

of CF_4 occur in the arc zone during the dielectric recovery. In [135], it was shown that PTFE (C_2F_4) vapor increases the critical field in CO_2/O_2 mixtures at temperatures up to 2,800 K. This is indicated by the arrow in Figure 4.59. Note that the main decomposition product of PTFE vapor is CF_4 under the assumption of LTE. Hence, at high currents, an increase in the critical field can be expected due to such decomposition effects. This improves the dielectric recovery.

For the absolute value of the critical field, the particle number density N is decisive. For an ideal gas $N=p/(k \cdot T)$ with pressure p and temperature T. Hence, for a given temperature and composition, particle number density is determined by the pressure. Increasing the filling pressure will therefore improve the dielectric recovery, as discussed in Section 4.1.3.2 and shown in [136] for SF_6 and [45] for air, see also Figure 4.18. The increased filling pressure in CO_2-based gas mixture which is needed to reach similar insulation performance as with SF_6 at ambient temperatures (Section 4.2.4) will also bring the dielectric recovery closer to that of SF_6.

4.4 Gas decomposition and lifetime

The lifetime of high-voltage circuit breakers in the field under service conditions is determined by various factors, including, for example, contact and nozzle erosion, leakage and condensation, generation of solid by-products and mechanical wear, as discussed in the detailed overview of [52]. These parameters can affect switching and insulation performance, mechanical behaviour and current carrying capabilities. Many of these issues are very similar in SF_6 and CO_2-based gases. In the following sub-sections, several factors that affect the lifetime of HVCBs will be addressed.

4.4.1   Contact and nozzle erosion

The arcing contacts in a circuit breaker need to fulfil mechanical endurance requirements, short-circuit current carrying and switching capabilities and insulation requirements. The main contacts need to carry nominal and short-circuit currents and need to fulfil mechanical endurance and insulation requirements. For sufficiently fast commutation from the main to the arcing contacts (Section 4.1.2) and for thermal requirements, the main contacts need to maintain a low contact resistance. Contact erosion depends on parameters such as material composition, manufacturing processes, contact design, arcing current and arcing time and polarity and leads to changes in shape, contact resistance and surface roughness [52,137].

Nozzles are usually made of PTFE with some fillers (Section 4.1.2) and are needed to confine the arc. They are important for pressure build up and gas flow distribution in the CB, which both strongly influence the switching performance. Nozzle erosion depends on the current magnitude and arcing time, design and filler content used and leads to change of shape and surface roughness [52,134]. For HV CB using SF_6 alternative gases, it is, therefore, important to know if new gases and mixtures could lead to a different contact and nozzle erosion, which might, in turn, impact the lifetime of the CB compared to that of an SF_6 CB.

Contact and nozzle erosion was investigated by [138], where it was shown that similar or lower erosion rates than for SF_6 were obtained with CO_2/O_2 mixtures. The statistical significance and measurement uncertainty of these experiments, however, is not known. In [52], it was concluded that contact and nozzle erosion does not significantly depend on the gas and that erosion estimates obtained with SF_6 could be used for lifetime estimates with SF_6 alternative gases. These findings are plausible since the arc roots at the contacts are embedded in metal vapor [137,139] and PTFE ablation is only dominant at high current densities where the arc burns in PTFE vapor [134]. As was discussed in [134], the radiation transport in the gas used in the CB (in this case SF_6) affects the ablation rate of PTFE nozzles only at low currents, below the onset of clogging and back-heating. In this current range, the ablation rate was less than half of that at high currents, i.e., the dominant ablation occurs in the back-heating mode. Hence, the blow gas is expected to influence the erosion of contacts and nozzles only at smaller currents when the arc is embedded in the gas flow from the compression or heating chamber. This possibly explains a reduced sensitivity of the contact and nozzle erosion on the quenching gas used in HVCB. From all these findings, it can be concluded that contact and nozzle erosion in CO_2-based gas mixtures is similar to SF_6 CB.

4.4.2 Gas decomposition

After the decomposition of fluorinated additives such as C4-FN and C5-FK, these molecules will not recombine, e.g. [52], in contrast to CO_2. Oxygen will recombine to O_2 if not consumed in reactions with metals, for example. In SF_6, gas decomposition will also occur, which is, however, mainly due to reactions of fluorine with metals to form metal fluorides and CF_4 from carbon freed by PTFE nozzle erosion [140]. Aside from such reactions, the SF_6 molecule itself, since it is highly stable (this is also one reason it has such a high GWP), will reform after decomposition in the arc.

The main gaseous decomposition products of C4-FN and C5-FK that are stable at room temperature (as noted above, for the most part, CO_2 and O_2 recombine) are predicted to be CO_2 and CF_4, while CO can also form. Solid graphite was predicted and observed in experiments in the absence of oxygen admixture [138,141]. Solid graphite and CO can be significantly reduced by the addition of O_2. In experiments also other compounds, such as, for example, C_2F_5CN, CF_3CN, C_2F_6, COF_2 and HF, were found at low concentrations after decomposition in a furnace [132,140]. In circuit breaker tests, the main final reaction products have been observed to be CO_2 and CF_4. CO may also form depending on oxygen admixture concentration. The CO concentration is important because to a large extent, it determines the toxicity of the gas mixture. Additional fluorinated compounds are produced in very low concentrations; see e.g., [52,115,132]. These results show that complex reaction paths can occur depending on the details of the decomposition and recombination processes. Generally, it was found that the decomposed CO_2-based gas mixtures can be classified as non-toxic and have the same GHS (Global Harmonised System of Classification and Labelling of Chemicals), health and safety classifications for new and arced gases as SF_6 [51,52]. A detailed recent discussion is given in [52]. Safety, handling and maintenance procedures remain, therefore, similar to those for existing SF_6 products. The GWP of arced gases is not significantly altered compared to new gas, as discussed, e.g., in [115], since the fluorinated compounds are produced only in low concentrations.

4.4.2.1 Decomposition rates due to arcing

Decomposition or consumption rates of quenching gas were quantified by various authors [120,128,132,142–147] and were partially summarised in [52]. Consumption rates are reported to depend on total CB volume, current and concentration of the considered compound. For SF_6, the consumption rate (order of magnitude) can be roughly estimated from the contact and nozzle erosion due to reactions with the SF_6 decomposition products. Hence, SF_6 consumption is characterized by the incomplete SF_6 recombination during cooling of the quenching gas from temperatures above decomposition to ambient temperature. An estimate is shown in the example below using numbers from Tables 3.9 and in [52].

Estimate of SF$_6$ consumption at SLF 90, 56.7 kA (numbers from [52])
Assumptions:

- Energy per shot: 0.5 MJ
- Current-time integral per shot: 810 g/As
- Ablated mass at fixed contact (W/Cu (80%/20%)): 1.25 g/shot, i.e. 0.25 g Cu and 1 g W
- Molar mass C: 63 g
- Molar mass W: 183 g
- Nozzle erosion rate: 17 g/MJ
- Molar mass PTFE (CF2): 50 g

Estimate of consumption rate:
The ablated moles per shot are: 4×10^{-3} mol Cu and 5.5×10^{-3} mol W. Considering that, stoichiometrically, one Cu atom and one W atom react with one SF$_6$ and three SF$_6$ molecules, respectively, it follows that per shot 0.02 mol SF$_6$ are consumed from the erosion of one contact in the given example. This gives a consumption rate of 0.04 mol SF$_6$/MJ for the fixed contact. Assuming a 25% higher ablation rate at the tulip [2], a value of 0.09 mol SF$_6$/MJ for the total contact erosion follows. From the nozzle erosion, 0.34 mol CF$_2$ are released per MJ. This reacts with SF$_6$ to CF$_4$ and SF$_4$, see e.g. [6], resulting in an SF$_6$ consumption rate of 0.34 mol/MJ.

→ Contact and nozzle erosion together consumes about 0.43 mol SF$_6$/MJ in the given example. This value is within the range 0.1–0.8 mol/MJ given in [52].

For CO$_2$ and CO$_2$/O$_2$ mixtures, values of 0.8 and 0.03 mol/MJ, respectively, are given in [52]. It seems that the addition of O$_2$ can significantly enhance the CO$_2$ recombination, which is in line with the observations of various authors that CO production is significantly lowered if sufficient O$_2$ is added. These numbers were based on a single publication only, however, and further studies are needed.

As discussed above, the C4-FN and C5-FK molecule, in contrast to SF$_6$, does not recombine when the arc plasma cools. The consumption rate of C4-FN and C5-FK is, therefore, determined by the decomposition of the molecules due to gas heating. A simple estimate of the consumption rate can be done on the example of C4-FN admixture.

Estimate of C4-FN consumption rate:
During arcing, the arc energy will heat gas from ambient temperature to plasma temperatures above 20,000 K in the arc. This energy is released into the circuit breaker and will lead to heating of gas and heat transfer to the walls by heat conduction and radiation. Additional energy is used for contact and nozzle erosion. As seen in many investigations, the arc energy used for heating of gas is only a fraction of the total arc energy, a fraction that is often denoted kp (e.g., [148,149]) in the literature and which is typically in the range of 0.2–0.8 depending on design, materials, fill pressure, current magnitude and gas. Hence, from the total energy

W_{tot}, only a fraction k_p is used for gas heating. Heating of the gas above the decomposition temperature of about 1,000 K will lead to the decomposition of the C4-FN in the gas. The temperature up to which the gas is heated T' is assumed in the following to be slightly above the decomposition temperature of $T'=1,500$ K, to consider limited reaction time scales. The amount of gas mass m which is heated to T' by a given energy $k_p \cdot W_{tot}$ can be estimated from $m = k_p \cdot W_{tot}/h'$ with the enthalpy increase h' from ambient temperature to T'. The mass per total input energy can then be expressed as $m/W_{tot} = k_p /h'$. With the molar mass M_{mol} of the gas mixture and molar fraction $C_{C4\text{-}FN}$ of C4-FN, this results in the moles of heated C4- FN per total energy W_{tot}:

$$\frac{\text{Mol}_{C4-FN}}{W_{tot}} = \frac{C_{C4-FN} \cdot k_p}{h' \cdot M_{mol}} \tag{4.8}$$

The molar mass of the mixture C4-FN/ CO_2/O_2 (10%/80%/10%) is $M_{mol} = 58$ g. From Figure 4.36, it can be deduced that h' is about 1.5 MJ/kg for a 10% C4-FN admixture to CO_2/O_2. With $C_{C4-FN} = 0.05$, which is a realistic value for practical applications, this results in 0.12–0.46 mol/MJ of decomposed C4-FN in this example, assuming k_p in the range 0.2–0.8. Hence, the consumption rate of C4-FN in CO_2-based mixtures is expected to depend via the k_p on the design, the heat dissipation to surfaces and on the concentration of the C4-FN admixture. The smaller the concentration, the lower will be consumption rate, i.e., the consumption rates cannot be expected to be constant over the lifetime of a CB when decomposition reduces the concentration of C4-FN. Similar estimates can be done for the C5-FK molecule resulting in similar numbers.

→ Consumption rates of C4-FN with a 5% C4-FN admixture to CO_2/O_2 are roughly expected to be in the range of 0.12–0.46 mol/MJ. This is similar to C5-FK.

These estimates can be compared to the numbers compiled in [52], see Figure 4.60.

A good agreement with the estimates from (4.8) can be seen. At low and high currents, the measured consumption rates agree with the lower and higher expected limits, respectively. This is plausible since at high currents a higher fraction of the total energy is used for gas heating compared to low currents, which is reflected in a higher value of k_p.

The dependence of the expected consumption rate on the C4-FN concentration is shown in Figure 4.61 for an average $k_p = 0.5$. Reduction of the C4-FN concentration by 20% leads to a reduction of consumption rate by 20%. Note that these estimated rates are independent of the volume of the CB. However, the absolute mol number of additives in the breaker will vary with volume, i.e., in a smaller volume, the molar concentration of C5-FK or C4-FN decreases faster.

The consumption rates discussed here can be compared to the requirements. This was done in [116] where it was concluded that even for an extreme case of 6 MJ (20 single phase rated short-circuit operations) of arc energy input into a GIS

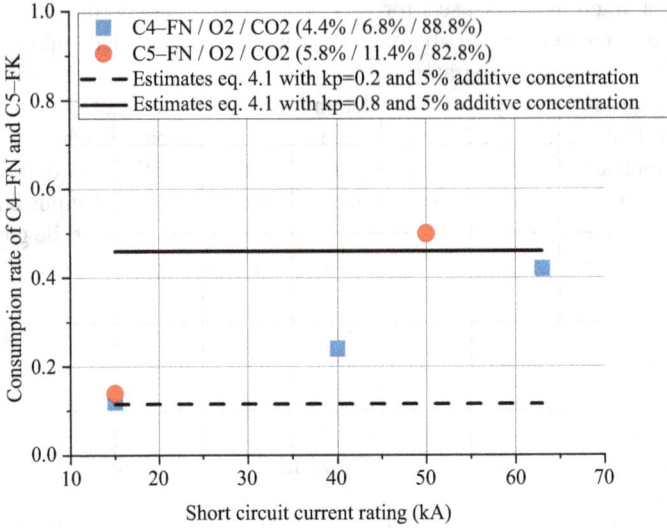

Figure 4.60   Maximum consumption rates of C4-FN and C5-FK in CO_2/O_2 mixtures adapted from [52]

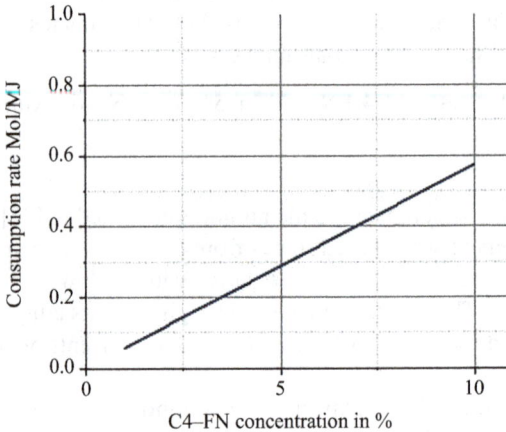

Figure 4.61   Consumption rate of C4-FN vs. C4-FN concentration for $k_p=0.5$

CB only about 6% of the total C4-FN concentration would be consumed. That is less than the reduction when going from nominal filling pressure to lock-out pressure, which corresponds to 10% pressure reduction. Hence, the C4-FN reduction even under these extreme assumptions is still within the limits of the performance shown in the type-test.

4.4.2.2 Decomposition rates due to partial discharge

Decomposition under partial discharges was investigated e.g., by [150–152]. A much lower decomposition rate compared to arcing was found, which can probably be explained by the different mechanisms involved. Partial discharges e.g. by streamers, are usually cold discharges and decompose molecules mainly by UV absorption and electron impact collisions. This is different from the thermal decomposition due to arc discharges. Mixtures with 4% C4-FN and C5-FK did not show significantly different decomposition per charge [23]. The main decomposition compounds were similar to those found under arcing, beside some compounds which are specifically formed under partial discharges. Zhou *et al.* [152] report that decomposition products with short carbon chains, such as CF_4, C_2F_4 and C_2F_6 are more likely produced under breakdown, i.e. arcing conditions, whereas partial discharges produce more likely long carbon-chain compounds such as C_3F_8, C_3F_6, and C_4F_{10}. Simka *et al.* [150] concluded that toxicity of the decomposed gas mixture due to partial discharges would be low (LC50 $\geq$ 50,000 ppm) throughout the lifetime. Hence, partial discharge activity will not significantly affect the lifetime or performance of a HV CB.

4.4.3 Change of dielectric strength due to decomposition

The reduction of concentration of C4-FN and C5-FK may lead to a reduction of dielectric strength as can be seen in Figures 4.41 and 4.45, respectively. From the previously shown consumption rates, the remaining additive concentration can be estimated or measured and, from this, the dielectric performance is estimated. In [143], it was shown that a combined SLF75 and T100s test with a cumulated energy of about 4 MJ in a 63 kA, 145 kV CB led to a reduction of C4-FN concentration from initially 3.8–2.6% after the tests. For such reduction, it was shown that the insulation level is expected to decrease by about 9%, see Figure 4.62. This heavy use of the CB is more than what is usually experienced during the service life of a circuit breaker. Hence, it was concluded that the reduction of dielectric strength in service will be less than 10%. A similar conclusion was reached in [52,116]. For circuit breaker designs with larger gas volume, as typical for GIS application, the reduction will be significantly smaller [153].

It should be noted that for Figure 4.62, it was assumed that the dielectric strength of the fluorinated decomposition products is similar to that of the CO_2/O_2 mixture. This might not be the case, as discussed in [70]. The fluorinated compounds produced by the decomposition of C4-FN and C5-FK mostly have a higher dielectric strength than CO_2, see Figure 4.63. Thus, the reduction of dielectric strength of decomposed CO_2 mixtures with fluorinated additives should be less than expected based on the reduction of C4-FN or C5-FK concentration alone, especially if there is a synergy of the fluorinated decomposition compounds with the background gas.

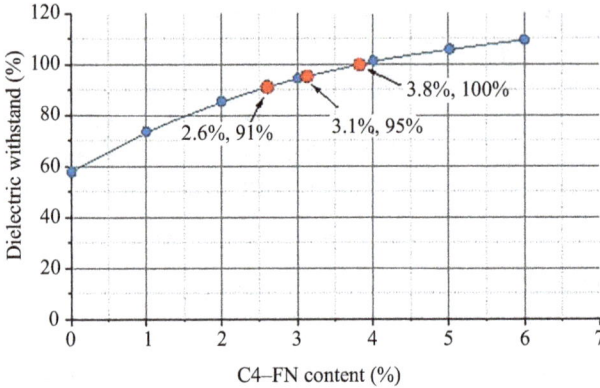

*Figure 4.62   Calculated insulation strength of C4-FN/CO$_2$/O$_2$ mixture with 3.8%/
83.2/13% (initial) vs. C4-FN content at a total pressure of 0.75 MPa,
reproduced from [143]*

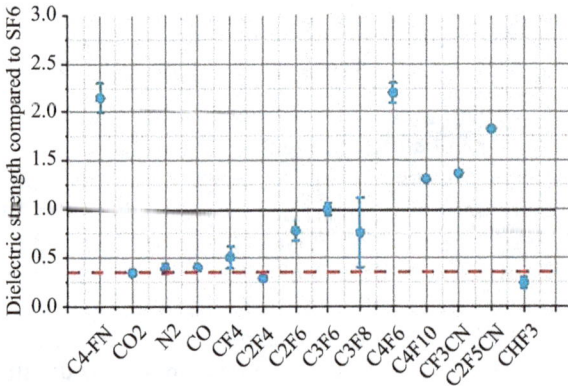

*Figure 4.63   Dielectric strength of potential C4-FN decomposition products
compared to SF$_6$ deduced from [70]. The dashed horizontal line
indicates the dielectric strength of CO$_2$.*

4.4.4   Change of interruption performance due to decomposition

As is shown in Section 4.3.2, the thermal interruption capability does not significantly depend on the concentration of fluorinated admixtures, i.e., it is not expected that with the consumption of C4-FN or C5-FK, a significant change in thermal interruption performance would occur.

For the dielectric recovery, it was shown that there is a significant effect due to the admixture of fluorinated additives in Section 4.3. Reduction of molar concentrations of

C4-FN and C5-FK might affect the dielectric recovery similar to the insulation performance, i.e., about 10% typically, if it is assumed that the decomposition compounds have a similar dielectric strength as CO_2/O_2 has. As shown in Figure 4.63, most of the fluorinated compounds produced have a higher critical field than CO_2 or O_2. Hence, as for the insulation strength, it can be expected that this also helps to improve the dielectric recovery and a reduction of dielectric recovery due to C4-FN or C5-FK consumption might be less than expected just based on the concentration of these additives.

It can be concluded that the effects of consumption of the fluorinated additives on interruption performance are less than due to change of contact and nozzle geometry by contact and nozzle erosion, as was shown also in [52].

4.4.5 Gas leakage

Gas leakage, which leads to pressure loss from the equipment, needs to be considered when using new gases. As discussed in the overview given by CIGRE WG A3.41 [52], 'interface' and 'permeation' leakage are of importance.

For interface leakage, gas can penetrate through a narrow channel in the interface between pressurized gas compartment and the outside. This is, for example, due to the presence of a hair, due to scratches on sealing surfaces or due to seal defects. Interface leakage can also result from faulty assembly. Under such conditions, the leakage channels are large compared to the molecule size and all components of the gas mixture will leak with the same relative rate, i.e., interface leakage does not affect the gas mixture composition.

This is different for permeation-induced leakage, which is due to the diffusion of molecules through the seals or the enclosure. Permeation depends on the materials and gas used and on the environmental conditions (e.g., temperature). Permeation can lead to inhomogeneous leakage, i.e., larger losses of some compounds compared to others. Through inhomogeneous leakage, the mixing ratios of CO_2-based mixtures can slightly change.

Permeation predominantly occurs through sealing materials, which are polymers, and is negligible through metallic and porcelain enclosures [52]. Since permeation of gases like CO_2 and O_2 through EPDM is enhanced compared to SF_6, different sealing materials need to be used, e.g., Butyl rubber as discussed in [52]. In this reference, permeation of CO_2 through Butyl rubber is reported to be higher than for O_2. This needs to be considered by sealing design, type, or material. The permeation of C4-FN and C5-FK through Butyl rubber is much less than for CO_2 and less than for O_2 [52]. The leakage ratio of CO_2 to C4-FN under identical conditions of partial pressure, temperature, material and volume is given to 21 to 1 [52]. Should a top up be necessary, the top-up is always to be done with the original gas mixture, independent whether the interface or permeation leakage occurred. In case the leakage is permeation dominated, a slight increase in C4-FN concentration might occur. However, this increase is small and does not negatively impact the performance of the equipment. Even though there might be some change in the gas composition over time, a regular monitoring of the gas composition is not necessary.

Acknowledgement

The author would like to express very special thanks to Patrick C. Stoller, Mahir Muratovic and Michael Gatzsche from Hitachi Energy Ltd. for the valuable contributions and extensive reviews of the chapter. Many thanks also to Martin Kriegel, Branimir Radisavljevic, Kunal Soni, Ullrich Straumann, Robert Voss and Markus Bujotzek from Hitachi Energy Ltd. for their inputs, discussions and reviews of parts of the manuscript and to Giulio Corsi from Hitachi Energy Switzerland Ltd. for the help in the reproduction of diagrams.

References

[1]   H. Ito (ed.), *Switching Equipment, CIGRE Green Book*, Cham: Springer, 2019.

[2]   M. Kapetanovic, *High-Voltage Circuit Breakers*, Sarajevo, Bosnia and Herzegovina: Faculty Elect. Eng., Univ. Sarajevo, 2011, p. 132–136.

[3]   H. M. Ryan and G. R. Jones, *SF6 Switchgear*, The Institution of Engineering and Technology, 1989.

[4]   Nakanishi and K. Switching, *Phenomena in High-Voltage Circuit Breakers*, CRC Press, 1991.

[5]   R. D. Garzon, *High Voltage Circuit Breakers: Design and Applications*, 2nd ed., CRC Press, 2002.

[6]   IEC, IEC Standard 62227-100: High-Voltage Switchgear and Control Gear – Part 100: Alternating-Current Circuit-Breakers, IEC (International Electrotechnical Commission), 2021.

[7]   IEEE Standard for Ratings and Requirements for AC High-Voltage Circuit Breakers with Rated Maximum Voltage Above 1000 V, 2018.

[8]   A. Kuechler, *High Voltage Engineering: Fundamentals-Technology – Applications*, Heidelberg, Germany: Springer Vieweg, 2017.

[9]   W. Boeck and W. Pfeiffer, *Conduction and Breakdown of Gases. Wiley Encyclopedia of Electrical and Electronics Engineering*, John Wiley & Sons, 1999.

[10]   W. Mosch and W. Hauschild, *Hochspannungsisolierungen Mit Schwefelhexafluorid*, Berlin, Germany: VEB Verlag Technik, 1979.

[11]   R. Arora and W. Mosch, *High Voltage and Electrical Insulation Engineering*, Piscataway, NJ: Wiley-IEEE Press, 2011.

[12]   M. Seeger, L. Niemeyer, and M. Bujotzek, "Leader propagation in uniform background fields in SF_6," *Journal of Physics D: Applied Physics*, vol. 42, p. 185205, 2009.

[13]   M. Seeger, Electric breakdown in high voltage gas circuit breakers, in *22nd International Conference on Gas Discharges and their Application*, Novi Sad, Serbia, 2018.

[14]   M. Seeger, L. Niemeyer, and M. Bujotzek, "Partial discharges and breakdown in SF_6 at protrusions in uniform background fields," *Journal of Physics D: Applied Physics*, vol. 41, p. 18520, 2008.

[15] M. Seeger, J. Avaheden, P. S., and T. Votteler, " Streamer parameters and breakdown in CO_2," *Journal of Physics D: Applied Physics*, vol. 50, p. 1–15, 2017.

[16] A. Pedersen, "Limitations of breakdown voltages in SF_6 caused by electrode surface roughness," in *Conference on Electrical Insulation & Dielectric Phenomena – Annual Report 1974*, Downingtown, PA, 1974.

[17] A. Pedersen, P. Karlsson, E. Bregnsbo, and T. Nielsen, "Anomalous breakdown in uniform field gaps in SF_6," *IEEE Transactions on Power Systems*, vol. 6, p. 1820–1826, 1974.

[18] S. Berger, "Onset or breakdown voltage reduction by electrode surface roughness in air and SF_6," *IEEE Transactions on Power Apparatus and Systems*, vol. 95, p. 1073–1079, 1976.

[19] I. McAllister, "A multiple protrusion model for surface roughness effects in compressed SF_6," *Elektrotechnische Zeitschrift*, vol. 99, p. 283–284, 1978.

[20] O. Feet, M. Seeger, D. Over, K. Niayesh, and F. Mauseth, "Breakdown at multiple protrusions in SF_6 and CO_2," *Energies*, vol. 13, p. 4449, 2020.

[21] M. Seeger, M. Schwinne, R. Bini, N. Mahdizadeh, and T. Votteler, "Dielectric recovery in a high-voltage circuit breaker in SF_6," *Journal of Physics D: Applied Physics*, vol. 45, p. 395204, 2012.

[22] W. Zängl and K. Petcharaks, "Application of streamer breakdown criterion for inhomogeneous fields in dry air ," in *Gaseous Dielectrics VII*, New York, NY: Plenum Press, 1994, p. 153.

[23] M. Yousfi, Ph. Robin-Jouan, and Z. Kanzari, "Breakdown electric field calculations of hot SF_6 for high voltage circuit breaker applications," *IEEE Transactions on Dielectrics and Electrical Insulation*, vol. 12, no. 6, p. 1192–1199, 2005.

[24] G. J. Cliteur, Y. Hayashi, E. Haginomo, and K. Suzki, "Calculation of the uniform breakdown field strength of SF_6 gas," *IEEE Transactions on Dielectrics and Electrical Insulation*, vol. 5, no. 6, p. 843–849, 1998.

[25] M. Seeger, M. Bujotzek, and G. V. Naidis, "Investigation on the temperature dependence of the critical electric field in SF_6/C_2F_4 mixtures," in *XVIIth Symposium on Physics of Switching Arc*, Brno, Czech Republic, 2007.

[26] L. Rothhardt, J. Mastovsky, and J. Blaha, "Dielectric strength of SF_6 at elevated temperatures," *Journal of Physics D: Applied Physics*, vol. 14, p. L205, 1981.

[27] C. B. Ruchti and L. Niemeyer, "Ablation controlled arcs," *IEEE Transactions on Plasma Science*, vol. 14, no. 4, p. 423–434, 1986.

[28] J. J. Lowke and H. C. Ludwig, "A simple model for high-current arcs stabilized by forced convection," *Journal of Applied Physics*, vol. 46, no. 8, p. 3352–3360, 1975.

[29] W. Hermann, U. Kogelschatz, K. Ragaller, and E. Schade, "Investigation of a cylindrical, axially blown, high-pressure arc," *Journal of Physics D: Applied Physics*, vol. 7, p. 607–619, 1974.

[30] L. Niemeyer and A. Plessl, "The influence of flow geometry on gas blast arc interruption," in *6th International Conference on Gas Discharges and Their Applications*, 1980.

[31] C. B. Ruchti, "Cylindricity of ablation-controlled arcs," *IEEE Transactions on Plasma Science*, vol. 16, no. 1, p. 47–49, 1988.

[32] J. Liu, Q. Zhang, J. D. Yan, J. Zhong, and M. T. C. Fang, "Analysis of the characteristics of DC nozzle arcs in air and guidance for the search of SF_6 replacement gas," *Journal of Physics D: Applied Physics*, vol. 49, p. 435201, 2016.

[33] W. Wang, J. Yan, M. Rong, A. Murphy, and J. W. Spencer, "Theoretical investigation of the decay of an SF_6 gas-blast arc using a two-temperature hydrodynamic model," *Journal of Physics D: Applied Physics*, vol. 46, no. 6, p. 065203, 2013.

[34] C. M. Franck and M. Seeger, "Application of high current and current zero simulations of high-voltage circuit breakers," *Contributions to Plasma Physics*, vol. 46, no. 10, p. 787–797, 2006.

[35] M. Seeger, B. Galletti, R. Bini, *et al.*, "Some aspects of current interruption physics in high voltage circuit breakers," *Contributions to Plasma Physics*, vol. 54, no. 2, p. 225–234, 2014.

[36] Q. Zhang, J. Liu, J. D. Yan, and M. T. C. Fang, "The modelling of an SF_6 arc in a supersonic nozzle: II. Current zero behaviour of the nozzle arc," *Journal Physics D: Applied Physics*, vol. 49, p. 335501, 2016.

[37] U. Habedank and H. Knobloch, "Zero-crossing measurements as a tool in the development of high-voltage circuit breakers," *IEE Proceedings Science, Measurement and Technology*, vol. 148, no. 6, p. 268–272, 2001.

[38] M. Seeger, "Perspectives on research on high voltage gas circuit breakers" *Plasma Chemistry and Plasma Processing*, vol. 35, p. 527–541, 2015.

[39] Y. Tanaka and K. Suzuki, "Development of a chemically nonequilibrium model on decaying SF_6 arc plasmas," *IEEE Transactions on Power Delivery*, vol. 28, Art. no. 4, 2013.

[40] H. Sun, Y. Tanaka, K. Tomita, *et al.*, "Computational non-chemically equilibrium model on the current zero simulation in a model N2 circuit breaker under the free recovery condition," *Journal of Physics: Applied Physics*, vol. 49, no. 5, p. 55204, 2016.

[41] Y. Wu, H. Sun, Y. Tanaka, *et al.*, "Influence of the gas flow rate on the nonchemical equilibrium N_2 arc behavior in a model nozzle circuit breaker," *Journal of Physics D: Applied Physics*, vol. 49, p. 425202, 2016.

[42] T. Nakano, Y. Tanaka, K. Murai, *et al.*, "Evaluation of arc quenching characteristics of various gases using power semiconductors," *Journal of Physics D: Applied Physics*, vol. 50, p. 485602, 2017.

[43] T. Christen and M. Seeger, "Current interruption limit and resistance of the self-similar electric arc," *Journal of Applied Physics*, vol. 97, p. 1–3, 2005.

[44] W. Hermann and K. Ragaller, "Theoretical description of the current interruption in HV gas blast breakers," *IEEE Transactions on Power Apparatus and Systems*, vols. PAS-96, p. 1546–1555, 1977.

[45] M. Seeger, G. Naidis, A. Steffens, H. Nordborg, and M. Claessens, "Investigation of the dielectric recovery in synthetic air in a high voltage circuit breaker," *Journal of Physics D: Applied Physics*, vol. 38, p. 1795, 2005.

[46] M. T. Dhotre, X. Ye, M. Seeger, M. Schwinne, and S. Kotilainen, "CFD simulation and prediction of breakdown voltage in high voltage circuit breakers," in *IEEE Electrical Insulation Conference (EIC)*, Baltimore, MD, 2017.

[47] Y. Kieffel and F. Biquez, "SF6 alternative development for high voltage switchgears," in *IEEE Electrical Insulation Conference (EIC)*, Seattle, WA, 2015.

[48] J. G. Owens, "Greenhouse gas emission reductions through use of a sustainable alternative to SF_6," in *IEEE Electrical Insulation Conference (EIC)*, Montreal, QC, Canada, 2016.

[49] P. Stoller, C. Doiron, D. Tehlar, P. Simka, and N. Ranjan, "Mixtures of CO_2 and $C_5F_{10}O$ perfluoroketone for high voltage applications," *IEEE Transactions on Dielectrics and Electrical Insulation*, vol. 24, no. 5, p. 2712–2721, 2017.

[50] T. Uchii, Y. Hoshina, H. Kawano, K. Suzuki, T. Nakamoto, and M. Toyoda, "Fundamental research on SF_6-free gas insulated switchgear adopting CO_2 gas and its mixtures," in *International Symposium on Eco Topia Science ISETS07*, 2007.

[51] M. Seeger, R. Smeets, J. Yan, *et al.*, "Recent development of SF_6 alternative gases for switching applications," *Electra*, vol. 291, 2017.

[52] N. S. Støa-Aanensen, E. Attar, M. Claessens, *et al.*, "Current interruption in SF_6-free switchgear, Technical Brochure 871," in *Cigre – International Council on Large Electric Systems*, 2022.

[53] Intergovernmental Panel on Climate Change, Climate Change 2022 – Impacts, Adaption and Vulnerability Sixth Assessment Report of the Intergovernmental Panel on Climate Change, 2022.

[54] A. Chachereau and C. Franck, "Characterization of HFO1234ze mixtures with N_2 and CO_2 for use as gaseous electrical insulation media," in *22nd International Symposium on High Voltage Engineering*, Buenos Aires, Argentina, 2017.

[55] H. Kasuya, Y. Kawamura, H. Mizoguchi, Y. Nakamura, S. Yanabu, and N. Nagasaki, "Interruption capability and decomposed gas density of CF_3I as a substitute for SF6 gas," *IEEE Transactions on Dielectrics and Electrical Insulation*, vol. 17, no. 4, p. 1196–1203, 2010.

[56] H. Katagiri, H. Kasuya, H. Mizoguchi, and S. Yanabu, "Investigation of the performance of CF_3I gas as a possible substitute for SF_6," *IEEE Transactions on Dielectrics and Electrical Insulation*, vol. 15, no. 5, p. 1424–1429, 2008.

[57] 3M^{TM} Novec^{TM} 4710 Dielectric Fluid, Technical Data Sheet, 2021.

[58] 3M^{TM} Novec^{TM} 5110 Dielectric Fluid, Technical Data Sheet, 2021.

[59] J. D. Mantilla, N. Gariboldi, S. Grob, and M. Claessens, "Investigation of the insulation performance of a new gas mixture with extremely low GWP," *2014 IEEE Electrical Insulation Conference (EIC)*, Philadelphia, PA, USA, 2014, p. 469–473, doi: 10.1109/EIC.2014.6869432.

[60] L. Niemeyer, A systematic search for insulation gases and their environmental evaluation. In: Christophorou, L.G., Olthoff, J.K. (eds) (eds.) *Gaseous Dielectrics VIII*. Springer, Boston, MA, 1998.

[61] C. Preve, R. Maladen, D. Piccoz, and J.-M. Biasse, "Validation method for SF_6 alternative gas," *Cigre*, 2016.

[62] K. Juhre and E. Kynast "High Pressure N_2, N_2/CO_2 and CO_2 Gas Insulation in Comparison to SF_6 in GIS Applications", *14th International Symposium on High Voltage Engineering (ISH)*, Paper C-01, p. 1–6, 2005.

[63] L. Christophorou, J. K. Olthoff, and D. S. Green, "Gases for Electrical Insulation and Arc Interruption: Possible Present and Future Alternatives to Pure SF6," US Department of Commerce, Technology Administration, National Institute of Standards and Technology, 1997.

[64] D. Tehlar, T. Diggelmann, P. Müller, R. Buehler, N. Ranjan, and C Doiron, "Ketone Based Alternative Insulation Medium in a 170 kV Pilot Installation", *CIGRE Colloquium*, Nagoya, Japan, 2015.

[65] E. Laruelle, A. Ficheux, Y. Kie el, and M. Waldron, "Reduction of Greenhouse Gases in GIS Pilot Project in UK", *Cigré*, Paper C3–304, 2016.

[66] H. Zhao, X. Li, S. Jia, and A. B. Murphy. "Prediction of the critical reduced electric field strength for carbon dioxide and its mixtures with 50% O2 and 50% H2 from boltzmann analysis for gas temperatures up to 3500 K at atmospheric pressure," *Journal of Physics D: Applied Physics*, vol. 47, no. 32, p. 325203, 2014.

[67] J. C. Kim, H. C. Kim, Z. Tanasic, X. Ye, and J. Mantilla, "Experience in the development of a 170kV/50kA/60Hz HVCB using a $CO_2+C_4F_7N$ mixture," in *CIGRE Session*, 2022.

[68] P. Stoller, M. Seeger, A. Iordanidis, and G. V. Naidis, "CO_2 as an arc interruption medium in gas circuit breakers," *IEEE Transactions on Plasma Science*, vol. 41, no. 8, p. 2359–2369, 2013.

[69] P. Stengard, P. Stoller, S. Buffoni-Scheel, *et al.*, "SF_6-alternative 145 kV live-tank circuit breaker," in *CIGRE Session*, 2022.

[70] M. Gatzsche, V. Tilliette, U. Straumann, *et al.*, "Moving towards carbon-neutral high-voltage switchgear by combining eco-efficient technologies," in *CIGRE Session*, 2022.

[71] Y. Kieffel, T. Irwin, P. Ponchon, and J. Owens, "Green gas to replace SF_6 in electrical grids," *IEEE Power and Energy Magazine*, vol. 14, no. 2, p. 32–39, 2016.

[72] J. Park, M. Ha, K. Seo, H. Kim, and J. Lee, "Experimental and numerical analysis of the interruption capability of SF_6-free 245kV 63kA GCB," in *CIGRE Session*, 2022.

[73] D. Peng and D. Robinson, "A new two-constant equation of state," *Industrial and Engineering Chemistry Fundamentals*, vol. 15, no. 1, p. 59–64, 1976.

[74] Y. Zhikang, T. Youping, W. Cong, Q. Sichen, and C. Geng, "Research on liquefaction characteristics of SF_6 substitute gases," *Journal of Electrical Engineering and Technology*, vol. 13, no. 6, p. 2545–2552, 2018.

[75] E. Lemmon, I. Bell, M. Huber, and M. McLinden, "NIST Standard Reference Database 23: Reference fluid thermodynamic and transport properties-REFPROP, version 10.0.," in *National Institute of Standards and Technology*, Gaithersburg, 2018.

[76] C. Guder and W. Wagner, "A reference equation of state for the thermo-dynamic properties of sulfur hexafluoride (SF_6) for temperatures from the melting line to 625K and pressures up to 150 MPa," *Journal of Physical and Chemical Reference Data*, vol. 38, no. 1, p. 33–94, 2009.

[77] "Fluid Workbench highlights," Kintech Laboratory, https://www.kintechlab.com/products/fluid-workbench/. [Accessed 11 04 2023].

[78] X. Lin, J. Zhang, J. Xu, J. Zhong, Y. Song, and Y. Zhang, "Dynamic dielectric strength of C_3F_7CN/CO_2 and C_3F_7CN/N_2 gas mixtures in high voltage circuit breakers," *IEEE Transactions on Power Delivery*, vol. 37, no. 5, p. 4032–4041, 2022.

[79] Y. Tu, G. Chen, C. Wang, *et al.*, "Feasibility of C_3F_7CN/CO_2 gas mixtures in high-voltage DC GIL: a review on recent advances," *High Voltage*, vol. 5, p. 377–386, 2020.

[80] M. Walter, "Low temperature behaviour and dielectric performance of Fluoronitrile/CO_2/O_2 mixture," in *CIGRE Centennial Session*, 2021.

[81] Y. Wu, C. Wang, H. Sun, *et al.*, "Properties of C_4F_7N–CO_2 thermal plasmas: thermodynamic properties, transport coefficients and emission coefficients," *Journal of Physics D: Applied Physics*, vol. 51, no. 15, p. 4003, 2016.

[82] L. Zhong, J. Wang, J. Xu, *et al.*, "Effects of buffer gases on plasma prop-erties and arc decaying characteristics of C_4F_7N–N_2 and C_4F_7N–CO_2 arc plasmas," *Plasma Chemistry and Plasma Processing*, vol. 39, p. 1379–1396, 2019.

[83] L. Chen, B. Zhang, J. Xiong, X. Li, and A. B. Murphy, "Decomposition mechanism and kinetics of iso-C4 perfluoronitrile (C_4F_7N) plasmas," *Journal of Applied Physics*, vol. 126, p. 163303, 2019.

[84] X. Li, X. Guo, A. B. Murphy, H. Zhao, J. Wu, and Z. Guo, "Calculation of thermodynamic properties and transport coefficients of $C_5F_{10}O$-CO_2 thermal plasmas," *Journal of Applied Physics*, vol. 122, p. 143302, 2017.

[85] L. Chen X. Li, J. Xiong, A. B. Murphy, M. Fu, and R. Zhuo, "Chemical kinetics analysis of two C5-perfluorinated ketone (C5 PFK) thermal decomposition products: C_4F_7O and C_3F_4O," *Journal of Physics D: Applied Physics*, vols. 51, p. 435202, 2018.

[86] Y. Wu, C. Wang, H. Sun, *et al.*, "Evaluation of SF_6-alternative gas C5-PFK based on arc extinguishing performance and electric strength," *Journal of Physics D: Applied Physics*, vol. 50, no. 38, p. 385202, 2017.

[87] V. R. T. Narayanan, M. Gnybida, and C. Rümpler, "Transport and radiation properties of C_4F_7N-CO_2 gas mixtures with added oxygen," *Journal of Physics D: Applied Physics*, vol. 55, p. 295502, 2022.

[88] X. Li, H. Zhao, and A. Murphy, "SF$_6$-alternative gases for application in gas-insulated switchgear," *Journal of Physics D: Applied Physics*, vol. 51, p. 153001, 2018.

[89] E. André-Maouhoub, P. André, S. Makhlouf, and S. Nichele, "Production of graphite during the extinguishing arc with new SF$_6$ alternative gases," *Plasma Chemistry and Plasma Processing*, vol. 40, no. 3, pp. 795–808, 2020.

[90] W. Wang, M. Rong, Y. Wu, and J. D. Yan, "Fundamental properties of high-temperature SF$_6$ mixed with CO$_2$ as a replacement for SF$_6$ in high-voltage circuit breakers," *Journal of Physics D: Applied Physics*, vol. 47, no. 25, p. 255201, 2014.

[91] H. R. Griem, *Plasma Spectroscopy*, New York: McGraw-Hill, 1964.

[92] H. R. Griem, *Spectral Line Broadening by Plasmas*, New York: Academic Press, 1974.

[93] R. Liebermann and J. Lowke, "Radiation emission coefficients for sulfur hexafluoride arc plasmas," *Journal of Quantitative Spectroscopy and Radiative Transfer*, vol. 16, no. 3, p. 253–264, 1976.

[94] A. Gleizes, J. Gonzalez, and P. Freton, "Thermal plasma modelling," *Journal of Physics D: Applied Physics*, vol. 38, 2005.

[95] H. Nordborg and A. A. Iordanidis, "Self-consistent radiation based modelling of electric arcs: I. efficient radiation approximations," *Journal of Physics D: Applied Physics*, vol. 41, 2008.

[96] WG D1.51, "Dry air, N$_2$, CO$_2$, and N$_2$/SF$_6$ mixtures for gas-insulated systems, Technical Brochure 730," CIGRE, 2018.

[97] C. Franck, "Electric performance of new non-SF$_6$ gases and gas mixtures for gas-insulated systems, Technical Brochure 849," *CIGRE*, 2021.

[98] B. Zhang, J. Xiong, L. Chen, X. Li, and A. B. Murphy, "Fundamental physicochemical properties of SF$_6$-alternative gases: a review of recent progress," *Journal of Physics D: Applied Physics*, vol. 53, p. 173001, 2020.

[99] H. E. Nechmi, A. Beroual, A. Girodet, and P. Vinson, "Fluoronitriles/CO$_2$ gas mixture as promising substitute to SF$_6$ for insulation in high voltage applications," *IEEE Transactions on Dielectrics and Electrical Insulation*, vol. 23, no. 5, p. 2587–2593, 2016.

[100] H. Nechmi, A. Beroual, A. Girodet, and P. Vinson, "Effective ionization coefficients and limiting field strength of fluoronitriles-CO$_2$ mixtures," *IEEE Transactions on Dielectrics and Electrical Insulation*, vol. 24, no. 2, p. 886–892, 2017.

[101] Y. Long, L. Guo, C. Chen, *et al.*, "Measurement of ionization and attachment coefficients in C$_4$F$_7$N/CO$_2$ gas mixture as substitute gas to SF$_6$," *IEEE Access*, vol. 8, p. 76790, 2020.

[102] A. Chachereau, A. Hösl, and C. M. Franck, "Electrical insulation properties of the perfluoroketone C$_5$F$_{10}$O," *Journal of Physics D: Applied Physics*, vol. 51, no. 33, p. 335204, 2018.

[103] "UNAM database data," 27 01 2017. www.lxcat.net/UNAM. [Accessed 12 04 2023].

[104] A. Hösl, A. Chachereau, J. Pachin, and C. Franck, "Identification of the discharge kinetics in the perfluoro-nitrile C_4F_7N with swarm and breakdown experiments," *Journal of Physics D: Applied Physics*, vol. 52, no. 23, p. 235201, 2019.

[105] B. Zhang, Y. Yao, M. Hao, X. Li, J. Xiong, and A. B. Murphy, "Study of the dielectric breakdown strength of CO_2–O_2 mixtures by considering ion kinetics in a spatial–temporal growth avalanche model," *Journal of Applied Physics*, vol. 132, p. 093302, 2022.

[106] T. Uchii, D. Yoshida, S. Tsukao, *et al.*, "Recent development of SF_6 alternative switchgear using natural-origin gases in Japan," in *CIGRE Session*, Paris, France, 2022.

[107] M. Rong, H. Sun, F. Yang, *et al.*, "Influence of O_2 on the dielectric properties of CO_2 at the elevated temperatures," *Physics of Plasmas*, vol. 21, no. 11, p. 112117, 2014.

[108] Y. Li, X. Zhang, F. Ye, D. Chen, S. Tian, and Z. Cui, "Influence regularity of O_2 on dielectric and decomposition properties of C_4F_7N–CO_2–O_2 gas mixture for medium-voltage equipment," *High Voltage*, vol. 5, p. 256–263, 2020.

[109] M. Aints, I. Jõgi, M. Laan, P. Paris, and J. Raud, "Effective ionization coefficient of C5 perfluorinated ketone and its mixtures with air," *Journal of Physics D: Applied Physics*, vol. 51, no. 13, p. 135205, 2018.

[110] M. Seeger, P. Stoller, and A. Garyfallos, "Breakdown fields in synthetic air, CO_2, a CO_2/O_2 mixture, and CF_4 in the pressure range 0.5–10 MPa," *IEEE Transactions on Dielectrics and Electrical Insulation*, vol. 24, no. 3, p. 1582–1591, 2017.

[111] P. Simka and N. Ranjan, "Dielectric strength of C5 Perfluoroketone," in *19th International Symposium on High Voltage Engineering*, Pilsen, Czech Republic, 2015.

[112] Y. Li, W. Ding, Z. Li, Z. Zheng, and Y. Wang, "Breakdown characteristics of $C_5F_{10}O/N_2$ mixtures in a uniform field under AC voltage," *IEEE Transactions on Dielectrics and Electrical Insulation*, vol. 7, p. 73854, 2019.

[113] E. A. Egüz, A. Chachereau, A. Hösl, and C. Franck, "Measurements of swarm parameters in $C_4F_7N:O_2:CO_2$, $C_5F_{10}O:O_2:CO_2$ and $C_5F_{10}O:O_2:N_2$ mixtures," in *The 21st International Symposium on High Voltage Engineering*, Xi'an, China, 2021.

[114] C. Franck, J. T. Engelbrecht, M. Muratovic, P. Pietrzak, P. Simka, and Current Zero Club, "Comparative test program framework for non-SF_6 switching gases," *BH Electrical Engineering*, 2021.

[115] J. Mantilla, M. Claessens, and M. Kriegel, "Environmentally friendly perfluoroketones-based mixture as switching medium in high voltage circuit breakers," in *CIGRE Session*, Paris, France, 2016.

[116] P. C. Stoller, T.H.D. Braun, J. Korbel, and M. Richter, "SF_6 alternative circuit breaker for 145 kV gas insulated switchgear," in *CIGRE Session*, Paris, France, 2022.

[117] H. Jung, J. U. Yeun, H. S. Ahn, *et al.*, "Breaking and switching performance with fluoronitrile (C_4F_7N) mixtures on 170kV 50kA 60Hz GIS," in *5th International Conference on Electric Power Equipment – Switching Technology*, Kitakyushu, Japan, 2019.

[118] T. Uchii, Y. Hoshina, T. Mori, H. Kawano, T. Nakamoto, and H. Mizoguchi, "Investigations on SF_6-free gas circuit breaker adopting CO_2 gas as an alternative arc-quenching and insulating medium," in *Gaseous Dielectrics X*, Springer, 2004, p. 205–210.

[119] H. Kotsuji, H. Urai, T. Sakuyama, K. Shiraishi, and N. Yaginuma, "Measurement of gas temperature in self-blast chamber of model gas circuit breaker at high current interruption," *IEEJ Transactions on Electrical and Electronic Engineering*, vol. 13, no. 10, p. 1440–1445, 2018.

[120] J. Mantilla, M. Kriegel, and E. Panousis, " Switching interruption performance comparison between SF_6, CO_2 and fluoroketones-based mixtures in HVCB," in *CIGRÉ Colloquium Study Committees A3, B4 & D1*, Winnipeg, Canada, 2017.

[121] S. Kosse, P. Nikolic, and G. Kachelriess, "Holistic evaluation of the performance of today's SF_6 alternatives proposals," in *CIRED 24th International Conference on Electricity Distribution*, Glasgow, 2017.

[122] S. Kosse, "Development of CB with SF_6 alternatives," in *Presentation at CIGRE SC A3 - SF6 Gas Alternatives Workshop*, Paris, France, 2016.

[123] T. Uchii, A. Majima, T. Koshizuka, and H. Kawano, "Thermal interruption capabilities of CO_2 gas and CO_2-based gas mixtures," in *18th International Conference on Gas Discharges and their Applications (GD2010)*, Greifswald, Germany, 2010.

[124] D. Schiffbauer, A. Majima, T. Uchii, *et al.*, "High voltage F-gas free switchgear applying CO_2/O_2 sequestration with a variable pressure scheme," in *CIGRE-IEC 2019 Conference on EHV UHV AC DC*, 2019.

[125] W. Y. Lee, J.-U. Jun, H.-S. Oh, *et al.*, "Comparison of the interrupting capability of gas circuit breaker according to SF_6, g3 and CO_2/O_2 mixture," *Energies*, vol. 13, p. 6388, 2020.

[126] R. Smeets and V. Kertesz, "A new arc parameter database for characterization of short-line fault interruption capability of high-voltage circuit breakers," in *CIGRE SEssion*, Paris, France, 2006.

[127] R. Smeets, " High-voltage circuit breaker test statistics 2011–2016 and test analysis tools," in *CIGRE Session*, Paris, France, 2018.

[128] C. Grégoire, D. Leguizamon, J. Ozil, and L. Darles, "60Hz breaking capability of g3," *Matpost*, 2020.

[129] B. Blez and C. Guilloux, "Post-arc current in high voltage SF_6 circuit-breakers when breaking at up to 63 kA," *IEEE Power Engineering Review*, vol. 9, no. 4, p. 64–65, 1989.

[130] V. Hermosillo, C. Gregoire, D. Vancell, J. Ozil, Y. Kieffel, and E. Pierres, "Performance evaluation of CO_2/fluoronitrile mixture at high short circuit current level in GIS and dead-tank high-voltage circuit breakers," in *CIGRE Session*, Paris, France, 2018.

[131]   K. Bousoltane, Y. Vigoroux, Y. Kieffel and P. Robin-Jouan, "Performance evaluation of CO_2 and fluoronitrile mixture in comparison with SF_6," in *CIGRE Session*, Paris, France, 2018.

[132]   B. Radisavljevic, P. C. Stoller, C. B. Doiron, D. Over, A. Di-Gianni, and S. Scheel, "Switching performance of alternative gaseous mixtures in high-voltage circuit breakers," in *20th International Symposium on High Voltage Engineering*, Buenos Aires, Argentina, 2017.

[133]   P. Stoller, C. B. Doiron, S. Scheel, *et al.*, "Behavior of eco-efficient insulation mixtures under internal-arc-like conditions," *IEEE Transactions on Plasma Science*, vol. 49, no. 10, p. 3200–3211, 2021.

[134]   M. Seeger, J. Tepper, T. Christen, and J. Abrahamson, "Experimental study on PTFE ablation in high voltage circuit-breakers," *Journal of Physics D: Applied Physics*, vol. 39, p. 5016, 2006.

[135]   M. Seeger, T. Votteler, S. Pancheshnyi, J. Carstensen, A. Garyfallos, and M. Schwinne, "Breakdown in CO_2 and CO_2/C_2F_4 mixtures at elevated temperatures in the range 1000–4000 K," *Plasma Physics and Technology*, vol. 6, no. 1, p. 39–42, 2019.

[136]   E. Schade, "Similarity of the dielectric recovery characteristic of axially blown arcs in SF_6," in 8th *International Conference on Gas Discharges and their Applications*, Oxford, UK, 1985.

[137]   J. Tepper, M. Seeger, T. Votteler, V. Behrens, and T. Honig, "Investigation on erosion of Cu/W contacts in high-voltage circuit breakers," *IEEE Transactions on Components and Packaging Technologies*, vol. 29, no. 3, p. 658–665, 2006.

[138]   A. Majima, U. T. T. Yasuoka, T. Inoue, and D. Schiffbauer, "Properties of CO_2/O_2 gas mixture as an alternative medium for gas circuit breakers," in 22nd *International Conference on Gas Discharges and their Applications (GD2018)*, Novi Sad, Serbia, 2018.

[139]   S. Franke, R. Methling, D. Uhrlandt1, R. Bianchetti, R. Gati, and M. Schwinne, "Temperature determination in copper-dominated free-burning arcs," *Journal of Physics D: Applied Physics*, vol. 47, p. 015202, 2014.

[140]   K. Pohlink, Y. Kieffel, and J. Owens, "Characteristics of fluoronitrile/ CO_2 mixture – an alternative to SF_6," in *CIGRE*, Paris, 2016.

[141]   K. Bousoltane, Y. Kieffel, L. Maksoud, *et al.*, "Investigation on the influence of the O_2 content in fluoronitrile/CO_2/O_2 (g3) mixtures on the breaking in high voltage circuit breakers," in 22nd *International Conference on Gas Discharges and their Applications*, Novi Sad, Serbia, 2018.

[142]   F. Y. Chu, "SF_6 decomposition in gas-insulated equipment," *IEEE Transactions on Electrical Insulation*, vols. EI-21, no. 5, p. 693–725, 1986.

[143]   J. Ozil, C. Gregoire, A. Ficheux, Y. Kieffel, F. Biquez, and L. Drews, "Return of experience of the SF_6-free solution by the use of fluoronitrile gas mixture and progress on coverage of full range of transmission equipment', Cigre Sess. 48, vol. A3-11," in *CIGRE Session*, Paris, France, 2020.

[144] S. Brynda, P. Y. H. Sohn, T. H. Song, X. Ye, and J. D. Mantilla, "Theoretical and practical behaviour of eco-friendly SF_6 alternatives in high voltage switchgear," in *CIGRE Session*, Paris, France, 2020.

[145] J. D. Mantilla, M. Kriegel, and M. Claessens, "Physical aspects of arc interruption in CO_2/O_2/fluoroketones gas mixtures," in *CIGRE Session*, Paris, France, 2018.

[146] M. Sato, "Interruption performance of circuit breakers with SF_6 and with SF_6 alternatives," in *GICRE – AORC Technical Meeting*, Japan, 2020.

[147] D. Schiffbauer, "The effects of oxygen on the decarbonization and detoxification of arc byproducts," in *CIGRE Session*, Paris, France, 2018.

[148] N. Uzelac, M. Glinkowski, L. del Rio, *et al.*, "Tools for the simulation of the effects of the internal arc in transmission and distribution switchgear," *Electra*, 2015.

[149] X. Zhang, G. Pietsch, J. Zhang, and E. Gockenbach, "Fundamental investigation on the thermal transfer coefficient due to arc faults," *IEEE Transactions on Plasma Science*, vol. 34, no. 3, p. 1038–1045, 2006.

[150] P. Simka, C. Doiron, S. Scheel, and A. Di-Gianni, "Decomposition of alternative gaseous insulation under partial discharge," in *20th International Symposium on High Voltage Engineering*, Buenos Aires, Argentina, 2017.

[151] T. Hammer, M. Ise, T. Kishomoto, and F. Kessler, "Decomposition of low GWP gaseous dielectrics caused by partial discharge," in *21st International Conference on Gas Discharges and their Applications*, Nagoya, Japan, 2016.

[152] Y. Zhou, C. Li, N. Tang, *et al.*, "Comparison of decomposition by-products of C_4F_7N/CO_2 mixed gas under AC discharge breakdown and partial discharge," in *5th International Conference on Electric Power Equipment – Switching Technology (ICEPE-ST)*, Kitakyushu, Japan, 2019.

[153] P. Stoller, "C4-FN-based gas mixtures – electrical endurance not limited by decomposition," in *CIGRE Session*, 2021.

[154] B. Chervy and A. Gleizes, "Electrical conductivity in SF_6 thermal plasma at low temperature (1000–5000 K)," *Journal of Physics D: Applied Physics*, vol. 31, p. 2557–2565, 1998.

[155] L. Zhong, M. Rong, X. Wang, *et al.*, "Compositions, thermodynamic properties and transport coefficients of high-temperature $C_5F_{10}O$ mixed with CO_2 and O_2 as substitutes for SF_6 to reduce global warming potential," *AIP Advances*, vol. 7, p. 075003, 2017.

[156] A. Chachereau, A. Hösl and C. M. Franck, "Electrical insulation properties of the perfluoronitrileC_4F_7N," *Journal of Physics D: Applied Physics*, vol. 51, no. 49, 2018.

Chapter 5

DC switching technology

Nadew Belda[1]

5.1 Basics of DC current interruption principle

In the first part of this chapter, the fundamentals of DC current interruption and the main differences between AC and DC current breaking principles are discussed in detail. A mathematical analysis of DC current interruption at different stages of the current breaking process is provided.

The requirements of DC current breaking and a few DC circuit breaker (CB) concepts have been extensively investigated in 1970–1990. The fundamentals of DC current breaking remain the same, but the requirements on High Voltage Direct Current (HVDC) CBs have changed due to the progress in the HVDC transmission technology.

5.1.1 Brief review of terminologies and definitions

Although the actual operation of various HVDC CB technologies varies, the general interruption process is similar – creation of a counter-voltage that exceeds the system voltage and absorption of system inductive energy. As such, it is possible to define a common terminology regarding the operational modes and timings. CIGRÉ JWG A3/B4.34 developed terminology associated with voltage and current waveform, and timing definitions related to fault current interruption by a generic HVDC CB [1]. The most important timing and wave trace definitions are described with reference to Figure 5.1.

5.1.1.1 Timing-related terminology

1. **Relay time (T_{RI})** – the time interval between fault inception and the sending of a trip order to the HVDC CB (from t_0–t_1, in Figure 5.1). Although this is a characteristic of the protection system and is not part of the test requirements of the HVDC CB, it has a significant impact on the design (requirements).
2. **Breaker operation time (T_{Br})** – the time interval between the reception of the trip order and the beginning of the rise of the TIV (from t_1–t_2, in Figure 5.1). This determines the speed of operation of the HVDC CB, which is a critical parameter for the application in the envisaged multi-terminal DC (MTDC) grids.

[1]TenneT TSO B.V., Arnhem, the Netherlands

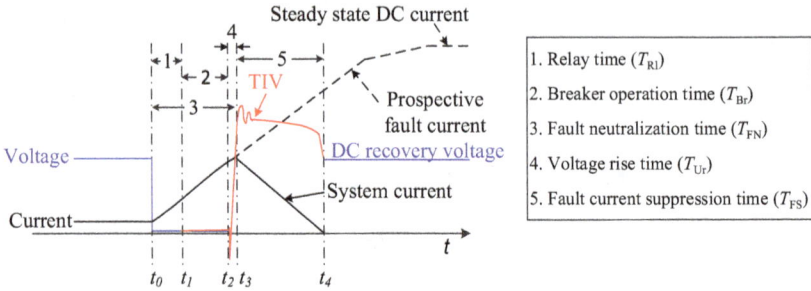

Figure 5.1   Diagram for illustration of timing definitions and terminology related to HVDC CB operation, reproduced after [2]

3. **Fault neutralization time (T_{FN})** – the time interval between fault inception and the instant when the fault current starts to decrease (from t_0–t_3, in Figure 5.1). This is important from the perspective of the system since at the end of this time, the system voltage starts to recover.

4. **Voltage rise time (T_{Ur})** – the time required to build the transient interruption voltage (TIV) (from t_2–t_3, in Figure 5.1). The rate-of-rise of the TIV is important because the dielectric recovery of the mechanical switching device (s) must be coordinated with this.

5. **Fault current suppression time (T_{FS})** – the time interval between the peak of the interrupted fault current and the instant when the current has been suppressed to zero (arrester leakage) current level (from t_3–t_4, in Figure 5.1). This time depends on several system and CB-related parameters. From the CB perspective, two events take place during this period. The HVDC CB absorbs magnetic energy stored in the system's inductance and the TIV sustained by the metal oxide surge arrester (MOSA) appears across the HVDC CB during this period. From the system perspective, the voltage recovers during this period.

6. **Break time** – the duration from the trip order until the fault current is reduced to the residual current level.

5.1.1.2   Wave trace-related terminology

1. **Prospective fault current** – the fault current that results at a CB location when no action to clear the fault is made. The rate-of-rise and the peak value of the current are essential parameters.

2. **Peak interruption current** – the maximum value of the fault current that flows during the current interruption process. The peak value of the interruption current occurs at the moment the TIV equals the system voltage.

3. **Transient interruption voltage (TIV)** – the voltage that the HVDC CB generates[*] during current interruption, see the red trace in Figure 5.1. During a test,

---

[*]In fact, the HVDC CB does not have an active voltage source that generates a TIV, rather it is a sheer action and provision of the HVDC CB itself combined with the system inductance that leads to the generation of this voltage.

the HVDC CB must clearly demonstrate that this voltage is higher than the rated operation voltage of the CB with sufficient margin. About 50% margin has been suggested as a typical compromise between system insulation coordination and CB performance. The initial value, the rate-of-rise, the peak value, and the duration of the TIV are crucial parameters.

4. **Rated short-circuit breaking current** – the maximum current that the HVDC CB can interrupt within specified breaker operation time.

5. **Leakage (residual) current** – the current that flows through the MOSA after fault current interruption.

The above and other terminologies are described in detail in [1]. Interruption of load/nominal current follows the same procedure except in this case the relay time is not critical since the trip order can be directly sent to the appropriate CB. This can be performed, for example, for maintenance and/or reconfiguration purposes without de-energizing the converters or the other DC side of the system.

5.1.2   DC current breaking

The critical aspect that makes DC current breaking challenging is the fact that the DC current does not cross zero level. This means there is magnetic energy stored in the system inductance ($\frac{1}{2}LI^2$) at all times, where L is the equivalent inductance of the system and I is the DC current flowing, see Figure 5.2. Under a short-circuit condition, this energy can be very large as the short-circuit current increases. When a breaker attempts to stop a DC current (breaking), the system tries to get rid of the magnetic energy in its inductance. Initially, this energy is transformed into electric energy by charging any capacitive element (C) that appears in parallel to the breaker as shown in the figure. The ideal (lossless) relationship is as follows:

$$\frac{1}{2}LI^2 \Leftrightarrow \frac{1}{2}CU_{CB}^2 \tag{5.1}$$

Equation (5.1) shows that when an interrupter stops DC current, a capacitor (C – stray or including a lumped element) is charged to a voltage (U_{CB}) that is proportional to the square root of the magnetic energy in the system. It is charged in such a way that its voltage is in opposite polarity to the source supplying the current, i.e., countering the source. It is the creation and management of this counter voltage that is the main principle in any DC current break.

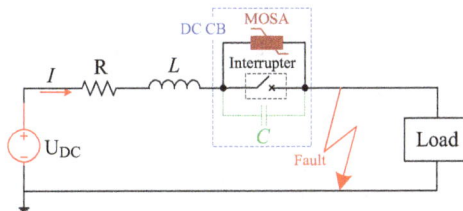

Figure 5.2   A simplified DC circuit during a fault reproduced after [3]

In addition, the smaller the capacitor, the higher the charging voltage that appears across the interrupter. The interrupter can withstand only limited voltage stress within a certain time of operation. Thus, if not somehow limited, the capacitor will be charged to an extremely high voltage, which may exceed the dielectric insulation strength of the interrupter and cause damage to the system components.[†] A possible solution to reduce the charging voltage is by increasing the capacitance of the capacitor appearing in parallel to the interrupter. In fact, this becomes impractical as the energy in the system increases. A practical solution is to use a voltage-limiting component, for example, MOSA, in parallel with the interrupter. In the latter case, the excess magnetic energy in the system is transformed into thermal energy during the current breaking process.

An important conclusion here is that to achieve DC current breaking, the counter voltage, henceforth termed as the transient interruption voltage (TIV), is essential and must be generated, limited to a safe level and maintained across the terminals of the breaker. Nevertheless, to produce and maintain the TIV, the interrupter must be able to stop (locally) the DC current in the first place in order to force the current into the capacitor and MOSA. For this, a local current zero must be achieved in the interrupter.

Moreover, to suppress the system current, the TIV must be sufficiently higher than the source voltage, i.e., $U_{CB} > U_{DC}$ in Figure 5.2. This is mathematically illustrated in a later section. At low voltage, DC current interruption and suppression are achieved by an arc voltage that makes sufficient TIV. In this case, the interrupter has to dissipate the system's magnetic energy in the form of arc thermal energy. In order to enhance the arc voltage, intensive arc cooling mechanisms such as arc elongation and arc splitting have been used. However, the same approach cannot be applied at high voltage since the arc voltage that is produced in such a way cannot exceed a few kilovolts.

In general, a DC CB performs the following fundamental functions to achieve DC current interruption:

1. Local current interruption (internal current commutation)
2. TIV generation
3. Energy absorption

It was quite clear from the early stages of HVDC fault current-breaking investigations that these functions should be separated in a DC CB [4]. There is no way that a single component can achieve these functions concurrently.

5.1.3   Comparison of DC vs. AC current interruption

DC current interruption is fundamentally different from AC current interruption. Figure 5.3 shows current (black) and voltage (blue and red) waveforms during AC

[†]Even if the breaker could sustain infinitely high voltage across its terminals, such interruption results in a very high voltage due to $L\frac{di}{dt}$ that results in the system. Thus endangering system insulation coordination.

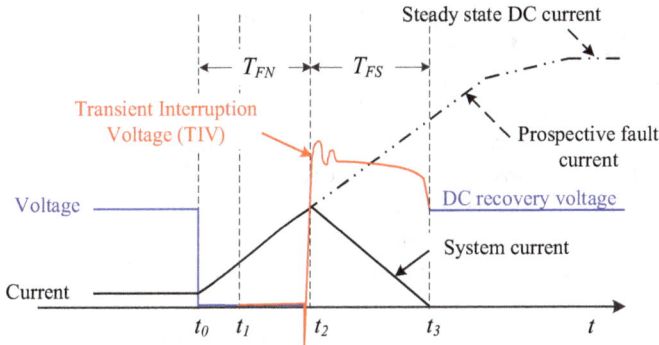

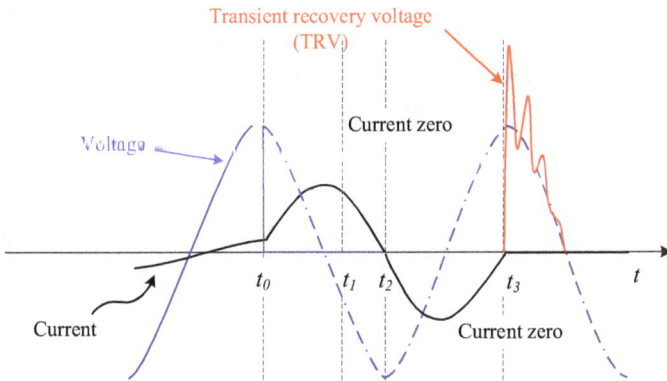

Figure 5.3 AC vs. DC current interruption principles – waveform comparison from [2]. (a) DC current interruption waveform. (b) AC current interruption waveform.

and DC current interruptions. The current in an AC system is sinusoidal, crossing the zero level at a rate of twice the power frequency. At each current zero, the system inherently de-energizes itself i.e., there is no magnetic energy stored in the system inductance.

Figure 5.3(a) shows typical current and voltage waveforms during DC fault current interruption. At t_0, a short-circuit occurs in the system. This is followed by a rapid rise of fault current, the rate-of-rise of which is limited by the system inductance until steady-state DC current is reached. Unlike an AC fault current, a DC fault current does not have naturally recurring current zero crossings. The protection relay detects the fault and trips a DC CB at t_1. The DC CB performs its internal operation (involving local current interruption and commutation) and starts to generate the TIV at t_2. The actual internal actions of the DC CB, which precede the generation of the TIV, vary depending on the technology of DC CB. This is explained in Section 5.4. Then, the generated TIV is limited and maintained until the system short-circuit current is suppressed at t_3. The magnitude of the TIV is a design parameter of the DC CB, which must exceed the system voltage for which the breaker is intended. While maintaining the TIV during the interval between t_2 and t_3, the DC CB absorbs the magnetic energy from the system. Furthermore, the faster the DC CB produces the TIV following a trip command, the lower the peak value of the short-circuit current in the system. Thus, a DC CB has a short-circuit current limiting feature. After current suppression at t_3, the DC CB is subjected to the DC recovery voltage from a system.

On the other hand, if no fast action is taken, the steady-state DC short-circuit current, the magnitude of which is determined (on the DC side) only by inherent resistance in the circuit, continuously flows as shown by the black dash-dotted trace in Figure 5.3(a). This can subsequently damage the system components which are predominantly power electronic semiconductors with a limited safe operation area (SOA[‡]). The power electronic components in HVDC systems are protected well before the steady-state DC short-circuit current is reached – for example by system control actions (such as converter blocking). However, to ensure continuous operation of these components (a converter as a whole), a fast-acting DC CB is indispensable so that the impact of a fault on the remaining healthy DC and AC systems is limited.

Generic current and voltage waveforms during AC fault current interruption are depicted in Figure 5.3(b). When a short-circuit occurs at t_0, the current starts to rise with its peak value determined by the system voltage and circuit impedance up to the fault location. After a short while, the protection relays detect the fault and send trip commands to the appropriate CBs. The AC CB operates its mechanism and opens its contacts at t_1 in Figure 5.3(b). After contact separation, the system current flows through an arc established between the parting contacts. Current zero is reached naturally at t_2, where the arc is momentarily extinguished. This provides an opportunity for a CB to stop the continuation of the current flow by ensuring the arc between the contacts remains extinguished from that moment onward. For an AC CB to achieve current interruption at the first (natural) current zero (t_2), the contacts must have reached sufficient distance or fulfill a minimum arcing duration to withstand the transient recovery voltage (TRV). The latter is necessary for SF_6-based CB, in which sufficient SF_6 pressure must be built to extinguish the arc. If the arcing duration is too short or the contact gap is too small, the CB fails to clear, and the arc re-ignites to re-establish current flow until the next current zero crossing which is at t_3 in Figure 5.3(b). At t_3, the CB could withstand the TRV and subsequently the system voltage.

Table 5.1 summarizes the crucial features that distinguish a DC CB from an AC CB.

[‡]The voltage and current conditions over which the device can be expected to operate without damage.

Table 5.1 Summary of comparison of AC CB vs. DC CB

DC CB	AC CB
- Active – limits the peak value of the short-circuit current	- Passive – does not limit the peak value of short-circuit current
- Produces, limits and maintains a TIV and imposes it on the system	- The system determines the TRV and imposes it on the CB
- Absorbs magnetic energy stored in the system inductance	- Does not need to absorb the magnetic energy stored in the system inductance
- Requires active power (Megawatts) source for testing – (traditional) synthetic testing method cannot be applied	- Requires reactive power (Mega VArs) source for testing – synthetic testing method can be applied

5.2 Mathematical model of DC current interruption

Figure 5.2 shows a simplified DC system where an AC to DC converter is represented by an ideal DC source. The part shown as a load can be another DC to AC converter on the receiving end and is not relevant for the analysis in this section since it does not affect the operation of a DC CB after a short-circuit. For simplicity, it is assumed that a bolted short-circuit fault occurs near a converter terminal. Otherwise, the effects of traveling waves shall be considered. Applying Kirchhoff's voltage law (KVL) to the simplified circuit in Figure 5.2,

$$U_{DC} - Ri(t) - L\frac{d(i(t))}{dt} - U_{CB} = 0 \tag{5.2}$$

where $i(t)$ is the fault current, U_{DC} is the converter terminal voltage, R is the inherent resistance in the circuit up to the fault location, and U_{CB} is the voltage across a DC CB. Rearranging (5.2), the rate-of-change of the short-circuit current can be described as follows:

$$\frac{di(t)}{dt} = \frac{U_{DC} - Ri(t) - U_{CB}}{L} \tag{5.3}$$

For mathematical convenience, the analysis is split into two consecutive time periods; fault current neutralization period ($T_{FN} = t_0 \rightarrow t_2$) and fault current suppression (energy absorption) period ($T_{FS} = t_2 \rightarrow t_3$) as discussed in [5], see Figure 5.3 (a). In some cases, a DC CB might have started to build the TIV a little earlier than at t_2. However, the period in which the voltage across the breaker rises to the desired level of TIV is assumed to be negligible for mathematical simplicity.

5.2.1 Fault current neutralization period – T_{FN}

This is the period from the fault inception at t_0 until the moment the TIV generated by a DC CB equals the system voltage; for example, at t_2 in Figure 5.3(a). In this period, a DC CB does not influence the fault current since it is not inserting a counter voltage yet except for a negligible arc voltage in some cases or a small voltage drop in the intermediate branches of some DC CBs. Referring to (5.3), when the voltage across a CB is neglected i.e., $U_{CB} \approx 0$, the rate-of-change of current remains positive until $U_{DC} = Ri(t)$. Therefore, during this period, the fault current rises with its rate-of-rise limited mainly by the equivalent inductance of the system represented by L. Hence, neglecting the voltage across a CB and assuming initially steady state load current I_0 flowing before the short-circuit occurs, the solution of (5.3) yields the prospective fault current described as,

$$i(t) = \frac{U_{DC}}{R}(1 - \exp(-t/\tau)) + I_0 \exp(-t/\tau) \qquad (5.4)$$

where $\tau = \frac{L}{R}$ is the time constant of the system. From (5.4), it can be seen that the prospective fault current increases to a steady-state value of $\frac{U_{DC}}{R}$ if not quickly interrupted by a DC CB. Hence, on the DC side,[§] the magnitude of the prospective fault current in DC systems is limited only by the intrinsic resistance R of the system.

5.2.2 Fault current suppression period – T_{FS}

This is an interval from the moment that the TIV equals the system voltage (at t_2) until the system current ceases to flow or falls below a residual value (at t_3), see Figure 5.4. Assuming the U_{CB} is constant in this period and I_P is the value of fault current at the end of the fault neutralization period ($I_P = i(T_{FN})$ in (5.5), the sign of rate-of-change of the short-circuit current ($d(i)/dt$ in (5.3) can be changed only if $U_{CB} > U_{DC}$. In other words, for a DC CB to achieve current suppression, it must produce a TIV higher than the system voltage. With this assumption, (5.3) has the following solution [5]:

$$i(t) = \frac{U_{DC} - U_{CB}}{R}[1 - \exp(-t/\tau)] + I_P \exp(-t/\tau) \qquad (5.5)$$

Moreover, from (5.5), the fault current suppression time (T_{FS}) can be determined by setting $i(t)$ to zero. In reality, there could be a small leakage current after current

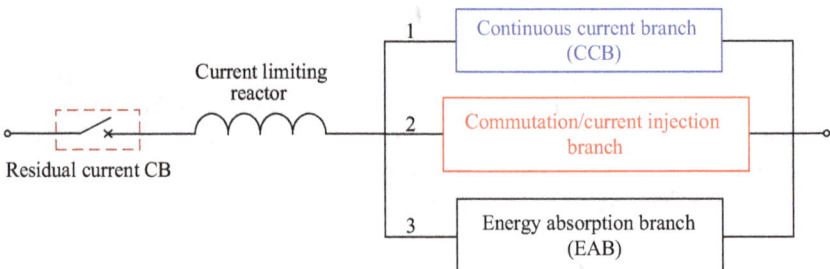

Figure 5.4   Generic DC CB showing multiple current branches

[§]If the DC source is an AC-to-DC converter, the fault current is limited by AC side impedance as well.

suppression and this is neglected for simplicity. Accordingly, the fault current suppression time T_{FS} can be determined from the following expression [5–7]:

$$T_{FS} = \tau \ln\left[1 + \frac{R}{U_{CB} - U_{DC}}I_P\right]$$ (5.6)

Using Maclaurin series expansion of the natural logarithm, (5.6) can be simplified as follows:

$$T_{FS} = \tau \ln\left[\frac{RI_P}{U_{CB} - U_{DC}}\left(1 - \frac{1}{2}\left(\frac{RI_P}{U_{CB} - U_{DC}}\right) + \frac{1}{3}\left(\frac{RI_P}{U_{CB} - U_{DC}}\right)^2 - \cdots\right)\right] \approx \frac{LI_P}{U_{CB} - U_{DC}}$$ (5.7)

The approximation on the right-hand side is obtained assuming an ideal situation where the total resistance in the circuit is neglected. This assumption makes the terms in parentheses of the middle equation converge to unity. Thus, (5.7) shows that the fault current suppression time is directly proportional to the system inductance L, and the value of interruption current I_P, and inversely proportional to the voltage difference between the TIV and the system voltage.

Combining (5.5) and (5.6), the fault current suppression time T_{FS} can be expressed as a function of a system and DC CB parameters as follows:

$$T_{FS} = \tau \ln\left[1 + \frac{U_{DC}}{U_{CB} - U_{DC}}[(1 - \exp(-T_{FN}/\tau)) + I_0\exp(-T_{FN}/\tau)]\right]$$ (5.8)

From (5.6) and (5.8), it can be seen that the time required by a DC CB to suppress current I_P to zero depends on the system as well as DC CB parameters. For example,

$$\lim_{U_{CB}\to\infty} T_{FS} \to 0$$ (5.9)

Equation (5.9) shows that the higher the TIV of a DC CB, the shorter the fault current suppression duration. However, the TIV is practically limited by the system as well as the DC CB insulation coordination. In addition, the larger the inductance L in the system, the longer the T_{FS}.

The energy absorbed from the system during the current suppression period can as well be described mathematically. Multiplying both sides of the KVL (5.2) by $i(t)$ dt and integrating over a period of T_{FS},[**] the following expression is obtained [3]:

$$\int_0^{T_{FS}} U_{DC}i(t)\mathrm{d}t - \int_0^{T_{FS}} Ri^2(t)\mathrm{d}t - \int_{I_P}^0 Li(t)\mathrm{d}i(t) - \int_0^{T_{FS}} U_{CB}i(t)\mathrm{d}t = 0$$ (5.10)

After some mathematical operation and rearrangement, (5.10) results in the energy balance equation corresponding to the energy dissipation phase of the current

---

[**]$i(t)$ changes from I_P to zero (negligible value) within the same duration.

interruption process. Accordingly, the energy balance in the system is given by the following expression:

$$\frac{1}{2}LI_P^2 + \int_0^{T_{FS}} U_{DC}i(t)\mathrm{d}t = \int_0^{T_{FS}} U_{CB}i(t)\mathrm{d}t + \int_0^{T_{FS}} Ri^2(t)\mathrm{d}t \tag{5.11}$$

The first term on the left-hand side of (5.11) represents the magnetic energy stored in the system inductance prior to the start of the fault current suppression whereas the second term on the same side represents the electrical energy injected by a DC voltage source (the rest of the system) during the fault current suppression period. The two terms on the right-hand side of (5.11) represent the energy absorbed by the DC CB and the energy dissipated in the circuit resistance, respectively. The latter can be ignored compared to the former since the energy dissipated in the circuit resistance R over a short T_{FS} is negligible. Thus, (5.11) implies that a DC CB absorbs not only the energy stored in the system inductance but also the energy contributed by a DC source (system) during fault current suppression. The latter depends on the magnitude of the DC system voltage as well as the duration of T_{FS} as shown in (5.13).

Assuming the ideal situation as in (5.7) and solving the right-hand side of (5.11), the total energy absorbed by a DC CB can be computed as follows:

$$\int_0^{T_{FS}} U_{CB}i(t)\mathrm{d}t = \frac{1}{2}LI_P^2\left(\frac{U_{CB}}{U_{CB} - U_{DC}}\right) = \frac{1}{2}LI_P^2 + \frac{1}{2}LI_P^2\left(\frac{U_{DC}}{U_{CB} - U_{DC}}\right)$$

$$\tag{5.12}$$

Equation (5.12) can be further simplified as follows:

$$\int_0^{T_{FS}} U_{CB}i(t)\mathrm{d}t = \frac{1}{2}LI_P^2 + \frac{1}{2}U_{DC}I_P T_{FS} \tag{5.13}$$

Equations (5.12) and (5.13) show that the portion of energy that is contributed by a source during current suppression is proportional to the source voltage magnitude and inversely proportional to the voltage difference between the TIV and the source voltage. The higher the TIV relative to U_{DC}, the smaller the energy contributed by a source. In practical HVDC transmission systems, the TIV is limited to 1.7 p.u. by the insulation level of the transmission equipment [8]. The minimum TIV considering the minimum interruption current is suggested to be at least 1.3 p.u.

To manage the energy absorption requirement of a DC CB, the contribution of the DC source in (5.11) can be controlled. For example, by actively reducing the magnitude of the DC source voltage during fault condition (fault suppression) [9]. This is normally a common procedure in LCC HVDC transmission systems where the converter voltage is reduced by increasing the thyristor firing angle. It is even possible to make the source absorb the magnetic energy in the circuit by changing the polarity of the DC source. That is, by converting a rectifier into an inverter [10], however, this action affects the normal operation of the healthy part of the system in a multi-terminal environment. The preferred operation mode is indeed a "constant voltage" mode similar to AC systems [9].

Assuming at TIV (U_{CB}) is 1.50 p.u. of the system voltage ($1.5U_{DC}$), which is the currently recommended practice, the energy contribution from the source is approximately twice the energy stored in the system inductance i.e. neglecting resistive losses during fault current suppression. Combining (5.9) and (5.11), it can be seen that the magnitude of the TIV also has an impact on the energy absorbed by a DC CB:

$$\lim_{U_{CB}\to\infty}\left(\int_0^{T_{FS}} U_{CB}i(t)\mathrm{d}t\right) = \frac{1}{2}LI_P^2 \tag{5.14}$$

Equation (5.14) shows the minimum energy absorption requirement of a DC CB cannot be influenced even if the DC source can be shut off during fault current suppression.

5.3 DC current breaking techniques

The above fundamental discussion shows that to achieve DC current interruption, a DC CB must produce a TIV of sufficient magnitude and be able to dissipate the magnetic energy in the system inductance. In addition, it ideally needs to produce the TIV as quickly as possible after receiving a trip command to prevent the fault current from further increasing.

Therefore, in order to achieve DC current interruption, HVDC CBs employ multiple parallel branches as shown in Figure 5.4, each branch having components that serve various purposes during the current interruption process. The main branches are:

- **Continuous current branch (CCB)**: A branch for nominal/load current conduction and local current interruption (to commutate current to another parallel branch).
- **Commutation/current injection branch**: Branch(es) for reverse current injection, temporary current conduction, commutation and/or interruption and TIV generation.
- **Energy absorption branch (EAB)**: A branch in which the TIV is limited and maintained to a desired level and system energy is absorbed.

In addition, there is a residual current CB in series (see Figure 5.4), which is an additional provision to isolate the HVDC CB from the system voltage after current suppression. This prevents leakage current through the EAB and/or through semiconductor switches (if they exist) when subjected to system voltage which, otherwise, could lead to thermal overload of these components.

The type and configuration of components in the first two branches, and especially the function and/or the number of current paths in the commutation branch vary from one CB technology to another. Nevertheless, the main objective in any HVDC CB is to systematically commutate the DC current into an EAB under sufficient TIV condition.

In a nutshell, an HVDC CB must fulfill the following basic conditions to successfully interrupt a DC fault current [11]:

1. Achieve local current interruption in the continuous current commutation branch – internal current commutation.
2. Gain sufficient dielectric strength to withstand the voltage across its terminals during the current interruption process as well as the response from the network at a later stage.
3. Build up and sustain a TIV higher than the system voltage to drive the fault current to zero.
4. Dissipate the magnetic energy stored in the circuit inductance and supplied by the system meanwhile absorbing the magnetic energy.

Hence, depending on the type and function of the main component(s) in the continuous current commutation branch and in the commutation branch, the state-of-the-art technologies of HVDC CBs can be classified into three major categories. These are:

1. **Solid-state (power electronic DC CB)** – PE switches interrupt system current, produce and sustain TIV.
2. **Mechanical HVDC CB** – mechanically opening contact system achieves local current interruption and sustains the subsequent TIV.
3. **Hybrid HVDC CB** – combined operation of mechanical and PE switches achieves local current interruption, produce and sustain the TIV.

Each of the above categories has its associated pros and cons. For example, operating speed, maximum current interruption capability, cost, footprint/volume, complexity and operating losses as well as re-closing capability are the main distinguishing features among these technologies. The design and configuration of internal components, operation principles, status of development and finally a critical look at pros and cons of some of the state-of-the-art technologies of HVDC CBs are discussed in the following sections.

5.4   Review of the state-of-the-art of DC circuit breakers

In this section, an extended, in-depth discussion of HVDC circuit breaker technologies, operation principles, high-voltage realizations, development and application status, major pros and cons associated with different technologies and realizations is presented. The discussion of this section is taken from the work in [3].

Like HVDC converter valves, HVDC CBs are also designed in a modular approach to achieve a high-voltage rating. A module (breaker unit) of an HVDC CB is the smallest part that contains all the required functionality of the full-scale HVDC CB. However, modular design of HVDC CBs is not necessarily a series connection of independent modules. Different modular design approaches are considered for extra high-voltage ratings. The most common approaches are depicted in Figure 5.5. Each of these approaches has been considered in the recently implemented HVDC CBs where each approach has its own technical pros and cons.

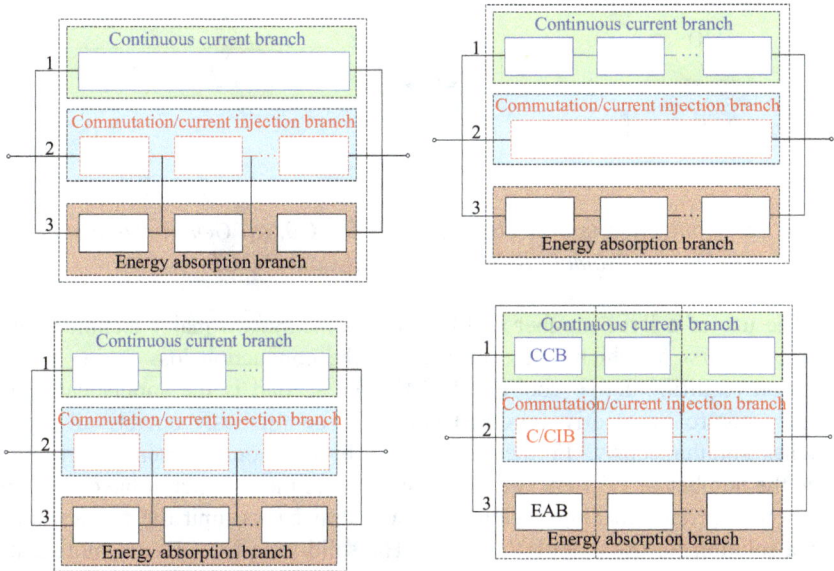

Figure 5.5 *Four different modular design approaches of HVDC CBs [3]. Top left: Modular current commutation branch. Top right: Modular continuous current branch. Bottom left: Modular continuous current commutation branch and commutation branch. Bottom right: Cascaded independent DC CB modules.*

5.4.1 Solid-state (power electronic) HVDC CBs

Static power electronic (PE) devices such as insulated-gate bipolar transistors (IGBTs), injection-enhanced gate transistors (IEGTs), bi-mode insulated gate transistors (BIGTs), insulated gate commutated thyristors (IGCTs) and gate turned-off thyristors (GTOs) are capable of changing state from conducting to insulating within a few to tens of microseconds using their gate controls. A generic model of a solid-state DC CB is shown in Figure 5.6(a). Essentially, there are two current branches: (1) a continuous current branch (CCB) and (2) an energy absorption branch (EAB) although there are additional snubber circuits for smooth switching of the semiconductors – the equivalent of which is shown by the green shadows in Figure 5.6(b). Since there are no mechanically moving parts in these devices, they can stop DC current flow nearly instantly forcing the current to flow through the EAB made of MOSA. However, the PE switches have limited voltage and current rating, which means a large number must be used in series to withstand the TIV, and in parallel to cope with the large current interruption requirements [12].

Figure 5.6(b) depicts an example of a bidirectional solid-state DC CB composed of a series connection of IGBTs in parallel with MOSA. Each IGBT has a snubber circuit for smoothing du/dt and di/dt stresses during turn OFF and ON, respectively.

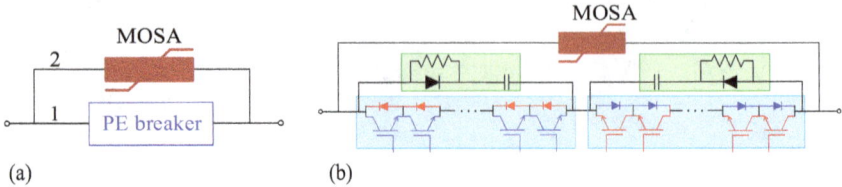

*Figure 5.6   Solid state (power electronic) HVDC CB. (a) Generic solid state DC
CB. (b) Example IGBT-based solid state DC CB.*

The use of a large number of PE switches incurs two major technical challenges, the first of which is the increase in the conduction loss during normal
operation requiring active cooling [13,14]. The second is the sophisticated and
complex control (gate drive circuits) needed to continuously operate all the PE
switches simultaneously. Furthermore, for a bi-directional current interruption,
twice the number of PE switches are required since these switches have only unidirectional current blocking capability. These drawbacks limit the application of
such a solution for use in an HVDC grid. The solid-state DC CBs are more suited
for low- to medium-voltage applications [12].

5.4.2   Mechanical (resonant current zero creating) HVDC CB

The mechanical HVDC CBs consist of only mechanical CB(s) and passive components like saturable reactors in the CCB, and auxiliary circuit needed for local
current zero creation in the commutation branch. Hence, under normal operation
condition, current flows through closed mechanical contacts of a CB. These types
of HVDC CBs exhibit low conduction loss and are naturally capable of bidirectional current breaking without additional component(s). With the help of auxiliary
circuits, a mechanical CB is responsible for local current interruption. This leads to
current commutation into a parallel TIV generating branch. For successful local
current interruption, local current zero(s) must be created in the mechanical interrupter which is achieved using an auxiliary circuit in the commutation branch.

Based on current zero creation schemes, two categories of mechanical HVDC
CBs exist. These are:

1. Passive oscillation current zero creation scheme
2. Active current injection current zero creation scheme – many variants exist

Especially in the active current injection category, several concepts have been
proposed, developed and prototype tested, and a few are put in service. The most
prominent developments are discussed below.

5.4.2.1   Passive oscillation – negative dynamic arc resistance

This is the earliest, the least complex, the most cost-effective construction and the
most exhaustively studied and prevalent DC CB concept to date. The electrical
diagram of a passive oscillation DC CB is shown in Figure 5.7. It consists of a
mechanical interrupter (AC CB) in the CCB, and a series-connected capacitor and
inductor (L–C) in the commutation branch as shown in the figure. In addition, it has

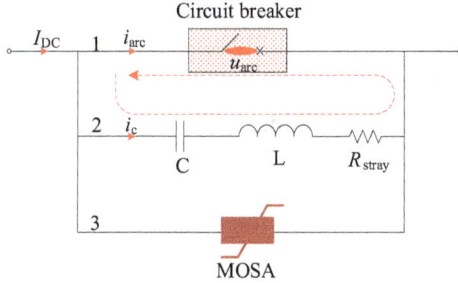

Figure 5.7   Passive resonant HVDC circuit breaker [3]

an EAB composed of a MOSA in the third parallel branch. The interrupter is usually an AC gas CB, for example air-blast CB and mostly SF_6 CB.

A gas CB is necessary because it exhibits negative dynamic arc (resistance) behaviour which is essential during the current interruption process. The interaction of the L–C with the arc voltage of the interrupter excites a negatively damped resonant current that superimposes onto the arc current through the interrupter.

Referring to Figure 5.7, during the current interruption process, the characteristics of the arc voltage (u_{arc}) as a function of arc current (i_{arc}) follow a dynamic pattern dictated by the parameters of the main interrupter as well as the frequency of the oscillation branch. Thus, $\frac{du_{arc}}{di_{arc}}$ represents the arc's dynamic resistance. The arc current i_{arc}, when interrupting current I_{DC}, is given as

$$i_{arc} = I_{DC}\left(1 + e^{-\left(\frac{du_{arc}}{di_{arc}} + R_{stray}\right)\frac{t}{2L}}\sin\left(\omega t\right)\right) \tag{5.15}$$

where $\omega = \frac{1}{\sqrt{LC}}$ is the angular frequency of the superimposed oscillating current, R_{stray} is the parasitic resistance of the circuit. The second term on the right-hand side of (5.15) represents the oscillating part of the arc current which is superimposed. If the $\frac{du_{arc}}{di_{arc}} < 0$ and its magnitude is larger than the R_{stray}, then the amplitude of the self-excited oscillatory current increases exponentially. When the amplitude of the oscillating current equals (or exceeds) the I_{DC}, then current zero is achieved in the interrupter and hence, an opportunity for local current interruption is created. After a local current interruption, the system current commutates, first to the commutation (L–C) branch and then, to the EAB after the capacitor in the commutation branch is charged to the TIV level determined by the MOSA, see Figure 5.8.

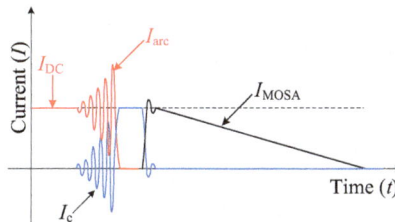

Figure 5.8   Illustration of passive current oscillation

Considering the recent advances in HVDC transmission technology, two main challenges prohibit this HVDC CB from consideration for a fault current interruption in MTDC grids:

1. **Very slow operation speed** – there are two main factors contributing to the slow operation. The first is the opening time[††] of a gas CB which could be as long as 20 ms. The second is the speed at which a current zero is achieved through the main interrupter once the arcing contacts separate. This depends on the magnitude of the interruption current as well as on the magnitude of the arc voltage and its characteristics.
2. **Limited maximum current interruption capability** – The maximum current interruption capability of passive oscillation HVDC CB is limited to a few kiloamperes, whereas the fault current levels in HVDC grids are expected to be significantly higher.

To meet the second challenge, some efforts have been made to improve current interruption capability by using series interrupters with double blast interrupter nozzle arrangement. Using such arrangement, a maximum current interruption capability of 5,500 A was achieved [15]. This was sufficient at the time, since it was expected that a DC CB with maximum current interruption capability of 4,000 A would be adequate for the then envisaged LCC-based multi-terminal HVDC grid. For instance, such DC CB is operated after an LCC converter controls reduce current to within the interruption capability (e.g., to ≈2,000 A) of a DC CB. However, the same technique cannot be applied in VSC HVDC multi-terminal systems since these converters do not have current control capability during a DC side fault. Hence, nowadays the passive oscillation-based HVDC CB is used mainly as a transfer switch rather than a CB. Thus, it is used for commutating DC current from low impedance path to high-impedance, parallel current path. This includes neutral bus switch (NBS), earth return transfer switch (ERTS) and metallic return transfer switch (MRTS). For these applications, recently, current transfer in excess of 6,500 A has been achieved using double-blast interrupter nozzle arrangement [16,17].

Although both functions require local current interruption, fault current interruption is a far more onerous duty compared to current commutation in terms of current level to be dealt with, the magnitude of counter voltage to be generated and the energy that the device has to absorb [18]. Another critical difference is operation time. A commutation does not need to be performed with acute urgency, whereas short-circuit currents in HVDC systems need to be cleared extremely rapidly. Figure 5.9 illustrates the difference between current commutation and current interruption.

A typical application of a transfer switch, for example, a MRTS, is the reconfiguration of a bipolar transmission scheme to a monopolar metallic return scheme without de-energizing the system. When one of the pole converters in a bipolar operation needs to be taken out of service, for example, for a scheduled

---

[††]From the trip command until the contacts separate.

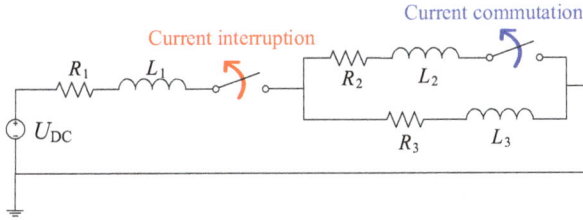

Figure 5.9   Illustration of current interruption (red) versus current commutation (blue)

maintenance of a fault in the converter pole, first such a system is configured to operate as a monopolar system with earth return. Then, the transmission line/cable of the unused pole can be configured as a return path, to avoid continuous current flowing into the ground (with all associated problems, e.g., corrosion of nearby metallic structure). In this case, the earth current needs to be transferred from earth return to pole line return using an MRTS. However, the earth return offers a low-resistance path and a significant proportion (reportedly up to 90%) of current flows through the earth path. Thus, an MRTS must transfer current from the earth (low impedance) path to the transmission line (high impedance) path. Therefore, using an MRTS, continued power transmission at 50% capacity is ensured without de-energization of the DC side.

To achieve current transfer in HVDC system, a TIV by arc voltage is not sufficient; however, a TIV in the order of 100 kV is necessary instead[‡‡] [16,17]. Thus, in principle, the TIV of a MRTS is a trade-off between the over voltage on a neutral bus, the energy absorption requirement, and the commutation time. Normally, the energy absorption takes re-closing into account without any cooling in between. Thus, double the rated energy absorption is normally specified.

5.4.2.2   Active current injection

The main drawbacks of the passive oscillation HVDC CB – slow speed of operation, dependency on arc behaviour and limited maximum current interruption capability are addressed by using energy stored in a pre-charged capacitor in the commutation/current injection branch as shown in Figure 5.10. The corresponding voltage and current waveforms are shown in Figure 5.11. The pre-charged capacitor is discharged to inject counter current which, when superimposed onto the system current, creates local current zero in the interrupter. In such a way, the necessity for slow and bulky AC gas CBs has disappeared and instead, fast, and lightweight interrupters such as vacuum interrupters (VIs) can be used. Unlike in the passive oscillation DC CB, the capacitor, in this case, remains charged and is isolated from the circuit by a high-speed making device which, when required,

---

[‡‡]At the same time, the TIV cannot be extremely high since it must comply with over-voltage withstand of a neutral bus. Usually, a neutral bus is protected against over-voltage by surge arresters. Thus, the TIV of an MRTS must conform to the insulation coordination of DC neutral bus with sufficient margin from DC neutral bus over-voltage.

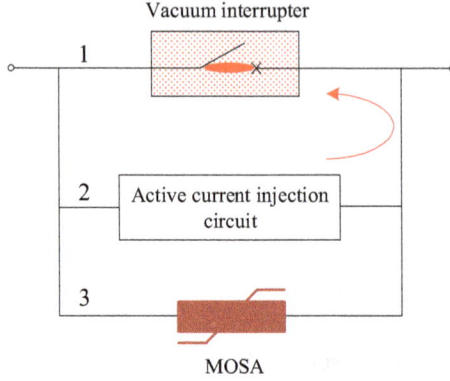

Figure 5.10   *Active current injection HVDC circuit breaker*

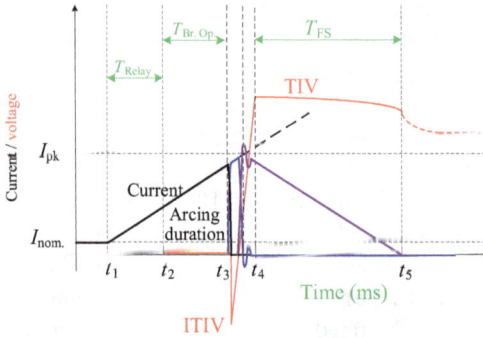

Figure 5.11   *Generic mode of operation of a direct capacitor discharge active injection type HVDC CB. VI current (black), L–C injection current (blue), MOSA current (purple), TIV (voltage across the breaker) (red).*

closes to discharge the capacitor. Various realizations of active current injection HVDC circuit breakers have been proposed. The most practical ones are discussed below (Figure 5.11).

5.4.2.2.1   *Active current injection – direct discharge of pre-charged capacitor*

The concept is introduced for the first time in [4] where a VI is used as a main interrupter because of its superior performance at short contact separation. The local current zero(s) can be created at any desired moment to enable local current interruption by a VI, provided that its contacts have reached sufficient separation to, subsequently, withstand the TIV. This is achieved by discharging the pre-charged capacitor by closing the high-speed mechanical making switch (HSMS), see Figure 5.12. In fact, the local current interruption does not stop the fault current, rather it forces the current to commutate out of the interrupter into a branch that generates a TIV. At this stage, the system current remains unaffected. However, from that instant on, the fault current commutates first into the parallel L–C branch which causes the voltage to rise (by charging a capacitor) until a level is reached that leads to the commutation to a MOSA in the EAB.

Normally current flows through the mechanical interrupter (VI in path 1) shown in Figure 5.12. In order to demonstrate the operation of the direct capacitor discharge active injection type HVDC CB, a short circuit is introduced at time t_1 – in reference to Figure 5.11. Then, a circuit breaker receives a trip command after which at t_2 the VI separates its contacts. This is followed by arcing between the contacts of the VI(s) as shown in the figure. When the contact gap reaches sufficient distance (at t_3) to withstand the transient interruption voltage (TIV) at later stage, an oscillating current from a pre-charged capacitor bank (C_{inj}) is injected into an arcing gap to create current zero and interruption in the VI. From that instant on, the circuit current commutates first into the parallel L–C branch which causes the voltage to rise until a level is reached that will lead to the commutation to a parallel metal oxide surge arrester designed to absorb the energy in the circuit (from t_3–t_4).

In addition, precise control and accurate timing of current injection are essential. In the 1980s, an HSMS that can close within 3–4 ms was inconceivable. Therefore, a simple closing switch like an ignitron or a triggered vacuum gap was used [4]. Triggered spark gaps are fast and precise, however, the continuous voltage stress from the pre-charged capacitor under normal operation conditions poses a reliability issue. Recently, HSMSs operated by electromagnetic repulsion mechanisms are used instead. The latter can also suffer from pre-strike when closing under high voltage, which causes current injection at an undesirable moment in the current interruption process. To mitigate this, high-speed, high-voltage vacuum-making switches are used. Nevertheless, several HSMSs are needed in series connection to cope with the continuous stress from the pre-charge voltage.

To address the above issues, power-electronic-based making switches are proposed. PE (power electronics) making switches are the most accurate and reliable given their track record in other applications. However, a large number of PE

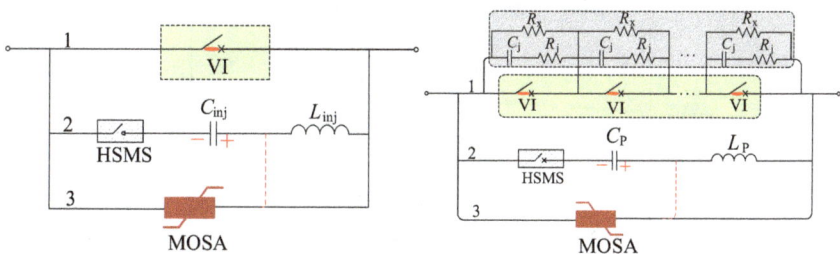

Figure 5.12   Electrical diagram of active current injection HVDC CB – direct discharge of pre-charged capacitor

devices must be used in series to withstand the continuous voltage stress from the pre-charged capacitor. An example of a PE making switch, in this case based on IGCTs, is shown in Figure 5.13. A 500 kV active current injection HVDC CB using this PE-making switch is realized for the Zhangbei HVDC grid in China [19].

The success of the overall DC current interruption is determined not only by the creation of current zeros but also by the capability of a CB's components to withstand the magnitude, the duration and the rate-of-rise (du/dt) of the TIV and, especially for mechanical interrupters, the current history prior to the current zero (di/dt) [4]. The product of di/dt and du/dt is usually considered as a figure of merit of performance. The higher this product, the lower the probability of current interruption. However, the exact value demarcating the success and failure of interruption depends on many other factors of the interrupter. Compared to AC current interruption, a vacuum interrupter is subjected to much higher di/dt in DC current interruption.

Methods to reduce the di/dt prior to current zero and the du/dt after current zero are introduced in [4]. Saturable reactors are used in series with the main interrupter to limit the di/dt prior to the current zero crossing and the du/dt for a few microseconds. Prior to current zero, a saturable reactor comes out of saturation and its inductance increases rapidly. In order to reduce the du/dt snubber circuits are used; for example, resistor–capacitor (RC) circuits are connected in parallel with the breakers – see Figure 5.12. Moreover, in an active current injection HVDC CB, multiple current zeros can be created in quick succession. It was identified that every even numbered zero crossing provides better opportunity for current interruption since current is held low during this period [4]. In other words, the residual plasma density at even numbered zero crossing is lower because the high-frequency peak current prior to the even numbered zero crossings is lower than prior to odd numbered zero crossings.

Recently, there is significant progress in the fast mechanical switching devices based on electromagnetic repulsion mechanisms. As part of the European union funded PROMOTioN project, a prototype of an active current injection HVDC CB, rated at 160/200 kV, 16 kA, was tested at KEMA Labs, see Figure 5.14. It has two series-connected high-voltage VIs in the CCB. The counter current is injected from a pre-charged capacitor by closing two HSMSs (also vacuum) [20]. This breaker

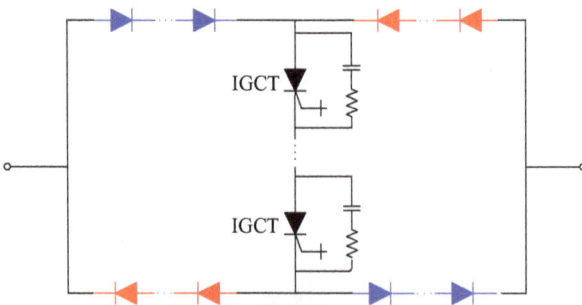

Figure 5.13   Example of PE-making switch

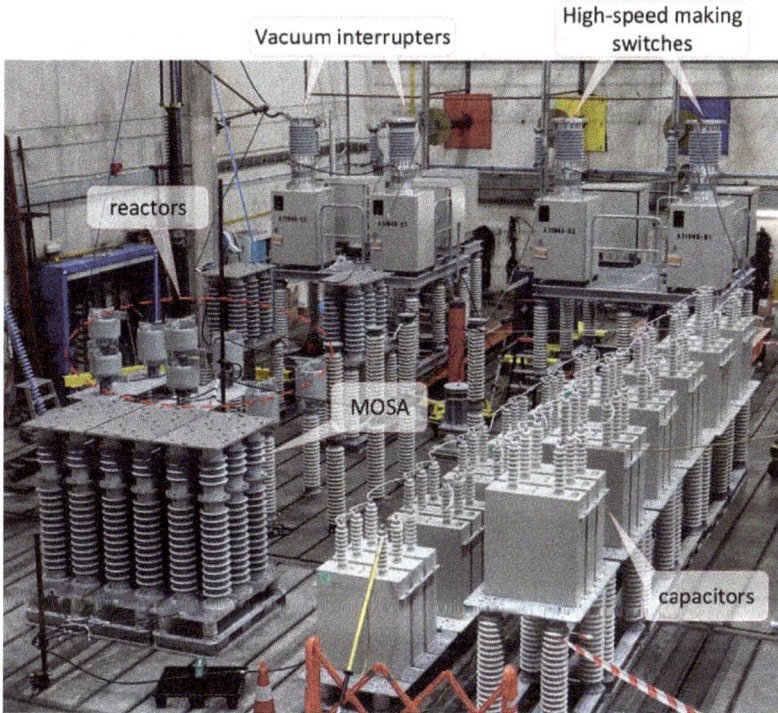

Figure 5.14 Active current injection HVDC CB – 160/200 kV prototype tested at KEMA Labs [2]. Source*: EU PROMOTioN project.*

could demonstrate 16 kA current interruption within a breaker operation time of 7 ms while producing a TIV well over 300 kV.

A 500 kV active current injection HVDC CB using a power electronic making switch is realized for the Zhangbei HVDC grid in China [19]. A photo of such a breaker, installed at a converter station, is shown in Figure 5.15 although the detailed internal (sub-)components cannot be seen. In this case, 12 vacuum interrupters are connected in series to sustain a TIV of over 800 kV within 3 ms after trip command.

5.4.2.2.2 Active current injection – pulse transformer (coupled inductors)

During the early developments of an active current injection HVDC CB, the capacitor is charged to the line voltage from the DC bus. Some practical issues of such a design have been identified. The major issues are:

1. Spontaneous breakdown of gaps due to continuous high-voltage stress from the pre-charged capacitor over service lifetime.
2. Long-time storage of charges at high voltage on the capacitor.
3. Unavailability of HVDC CB during system energization because the capacitor cannot be sufficiently charged yet.

Figure 5.15   A 500 kV active current injection HVDC CB installed in the Zhangbei project. Courtesy of Sieyuan Electric Company [2].

Especially, the triggered spark gaps become sensitive at high voltage and must be carefully designed considering the basic insulation level (BIL) of the HVDC CB as a whole. For example, for a 500 kV HVDC CB, the required BIL is 1,550 kV. This means, the triggered spark gaps must be designed to meet the BIL and, it should operate immediately upon reception of the trigger signals. To manufacture spark gaps that can fulfill these two requirements at the same time is very challenging. Therefore, a design based on a pulse transformer (coupled inductor) was introduced. An electrical diagram of a recently revised pulse transformer-based design is shown in the Figure 5.16 here below [21]. In this design, a pre-charged capacitor is placed on the low voltage (LV) side of a pulse transformer to mitigate the above issues. The counter current injection is applied to the interrupter through a pulse transformer so that the HVDC CB can conveniently be used in the system. In addition, a re-closing feature can be supplemented relatively easily. For example, two pre-charged parallel capacitors, each having its own closing device, can be placed on the LV side of the pulse transformer for quick re-closing functionality. A 250 kV, 3,500 A HVDC CB having the above features has been designed, manufactured and tested in [8].

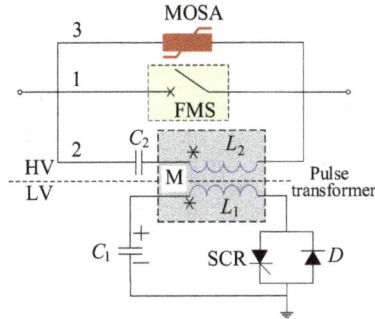

Figure 5.16 Active injection HVDC CB based on coupled inductors (pulse transformer)

In the latest design, a thyristor valve (silicon-controlled rectifier (SCR) in Figure 5.16) with anti-parallel diode is used as a making switch instead of a triggered spark gap. Thus, the LV side has an SCR with anti-parallel diodes (D), a pre-charged capacitor (C_1) and LV inductance of the pulse transformer (L_1). Since thyristors are subjected to LV stress in this case, only a few series-connected valves are needed. The HV side has capacitor C_2 (not pre-charged) and pulse transformer inductance (L_2). M is the mutual inductance of the pulse transformer. Thus, when the current in the CCB needs to be interrupted, the thyristors are turned ON with a continuous trigger mode. This ensures, together with the anti-parallel diodes, an oscillation creating multiple current zeros.

A sufficient number of current zero crossings are essential for the reliable operation of this concept. This requires proper design of the parameters on both LV and HV sides of the injection circuit. Especially, the pulse transformer must have a good coupling coefficient for efficient energy transfer from LV side to HV side during the current interruption process [21]. To enhance the coupling coefficient as well as the insulation level, optimized oil-immersed air core coupling reactors are proposed. In fact, sufficient energy must be stored on the LV capacitor to create the necessary injection current peak on the HV side.

A 160 kV, 9 kA prototype based on this concept has been developed, tested and put in service in the Nana'o radial HVDC pilot project in China [22,23]. The electrical diagram of this breaker is shown in Figure 5.17. The photo of the in-service HVDC CB is shown in Figure 5.18. In this case, four series-connected fast VIs, each rated at 40 kV having own R-C//R grading elements, are used as the main interrupter in the CCB.

5.4.2.2.3 *Active current injection – voltage source-assisted oscillation*

This is similar to the passive oscillation DC CB described in Section 5.4.2.1, in the sense that an oscillating current with increasing amplitude is superimposed on the current through the main interrupter by an auxiliary L–C circuit. Similar to the passive oscillation DC CB, the resonant capacitor (C_R) remains uncharged during

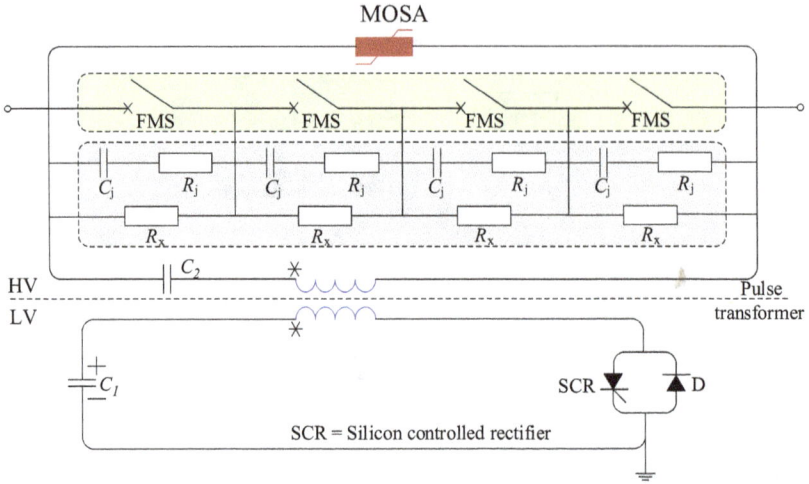

Figure 5.17   Modular approach for 160 kV system voltage

Figure 5.18   Active injection HVDC CB based on pulse transformer – on-site operation [2]

normal operation. However, the main difference is that the L–C resonant circuit, in this case, has a voltage source in series. Thus, instead of self-excited oscillation based on the negative dynamic arc resistance of the main interrupter, the excitation of the oscillating current is driven by a voltage source in the commutation branch. The power frequency of the voltage source must be tuned to the resonant frequency of the L–C circuit. This approach alleviates many of the main challenges of the passive resonant DC CB, such as for example:

1. The maximum current interruption capability is not limited to a few kilo-amperes since an oscillating current of large amplitude can be excited to create current zero.
2. Since the excitation is driven by an external voltage source, there is no requirement for the main interrupter exhibiting negative dynamic arc resistance behaviour nor high arc voltage.
3. Resonant current with rapidly increasing amplitude can be achieved by proper design of the resonance frequency and source voltage magnitude. Thus, faster current zero creation (within a few cycles) is ensured.

Hence, lightweight VIs operated by fast actuators can be used as the main interrupter instead of bulky gas CBs. An innovative design based on this concept is proposed in [24,25] and is shown in Figure 5.19 here below. This design uses a full bridge voltage source converter (VSC) with a low-to-medium voltage (pre-)charged[§§] DC link capacitor (C_{DC}) to excite the oscillating current in the L–C circuit. This principle is called VSC-assisted resonant current (VARC) DC CB. Compared to the resonant capacitor (C_R), C_{DC} is very large i.e., $C_R \ll C_{DC}$ for sufficient energy storage.[***] The resonant components (C_R and L_R) are dimensioned at a frequency of a few tens of kHz in order to achieve rapidly increasing oscillation current. The frequency of the resonant circuit determines the switching frequency of the IGBTs in the VSC.

The operation of the VARC HVDC CB during the current interruption process is as follows [24,25]: upon reception of a trip command, the VI opens its contacts which in turn is followed by arcing between the contacts. When the contacts of the VI reach sufficient separation, the VSC initiates the excitation of the resonance circuit to create current of increasing amplitude at high frequency. Referring to Figure 5.19, the VSC is operated in such a way that $IGBT_1$ and $IGBT_4$ are switched ON/OFF simultaneously whereas $IGBT_2$ and $IGBT_3$ are switched OFF/ON simultaneously at each current zero of the resonant current so that, at these switching instants, the polarity of C_{DC} voltage is in phase with C_R voltage – see Figure 5.20 (a) and (b). The resonant current is superimposed onto the main current through the VI. After a number of cycles, see Figure 5.20(c), the resonant current grows to a value higher than (or equal to) the main current, thus creating a local current zero

---

[§§]The charging voltage of C_{DC} determines the number of series-connected IGBTs in each arm of the VSC.

[***]Ideally, the VSC is a stiff voltage source for the creation of the oscillating current with sufficiently increasing amplitude during the current interruption process.

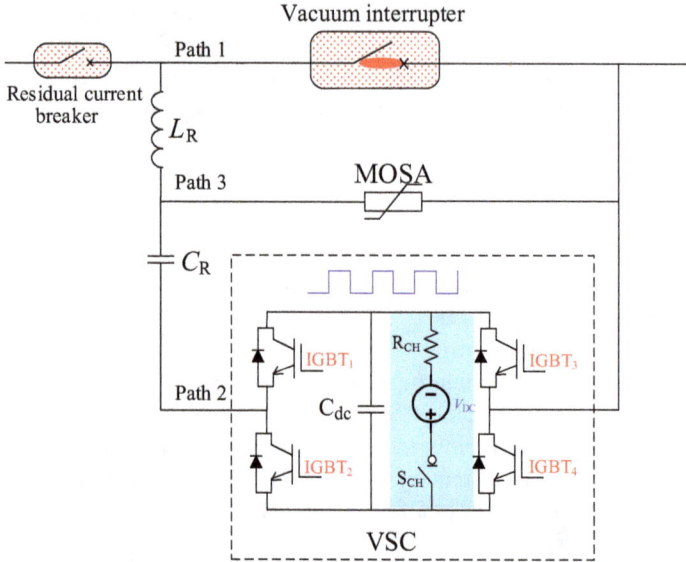

Figure 5.19   *Electrical diagram of VSC-assisted resonant current (VARC) DC CB*

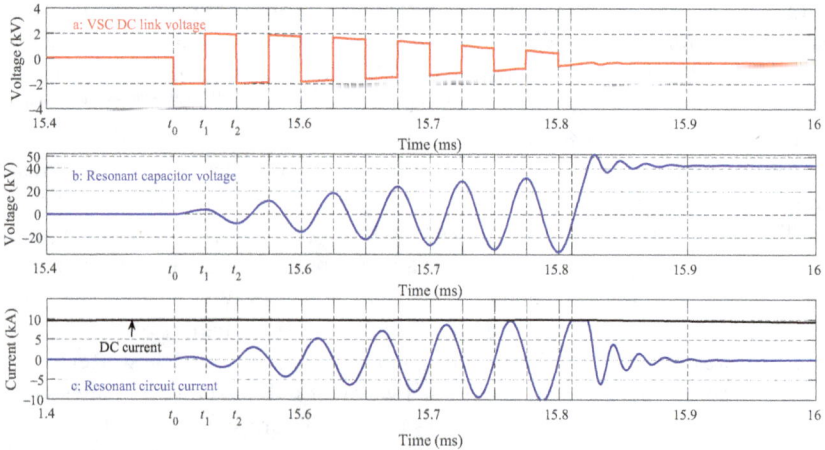

Figure 5.20   *Operation principle of VARC DC CB – resonant current excitation*

through the VI at which point local current interruption in the VI occurs. This forces the main current to commutate, first, to the commutation branch (path 2 in Figure 5.4) thus charging C_R until the clamping voltage of the MOSA is reached, subsequently leading to the system current suppression.

An important feature of the VARC HVDC CB is that the current interruption is achieved when the resonant current is near its peak value despite the magnitude of

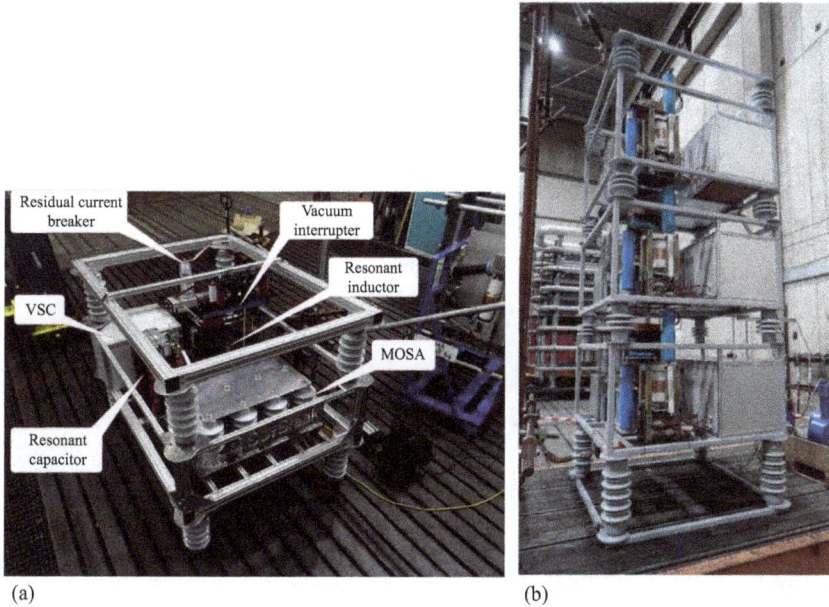

Figure 5.21 *Photos of VARC HVDC circuit breaker prototypes tested at KEMA Labs' high-power laboratory: (a) 27 kV single-module VARC DC circuit breaker; (b) three modules 80 kV VARC DC CB*

the interruption current. Hence, the di/dt is not so severe, and since the C_R is almost entirely discharged near the peak value of the excitation current, the initial TIV is very low. Thus, both di/dt and the initial TIV which are crucial success factors for current interruption adjust adaptively.

A 27 kV, 10 kA prototype unit of this concept, see Figure 5.21, has been developed and tested within the framework of the PROMOTioN project at KEMA Labs [24]. The first prototype could achieve 10 kA current interruption within the breaker operation time of 3 ms while producing 40 kV TIV.

Later, an 80 kV, 15 kA VARC HVDC CB consisting of three independent modules connected in series has been tested. In this case, the breaker operation time of <2 ms has been achieved [3].

For extra high-voltage application (EHV) application, many such units shall be connected in cascade. For example, 12 such modules are needed for 320 kV system voltage while 16 modules are needed for 500 kV system [26].

5.4.2.2.4 Magnetic field controlled gas discharge tube-based commutation branch

A different approach that provides current blockade without mechanical action is a gas discharge tube [27]. In fact, gas discharge tubes cannot conduct current continuously, hence can only form a commutation branch. Therefore, in the continuous current commutation branch, a parallel AC CB with a fast opening speed having a

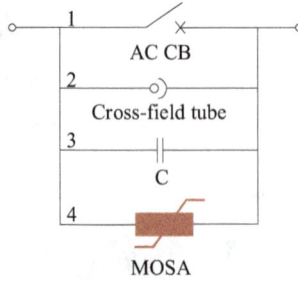

Figure 5.22   Cross-field tube HVDC CB, redrawn after [27]

few hundred volts of arc voltage is used. An electrical diagram of a cross-field gas discharge tube-based HVDC CB is shown in Figure 5.22. A DC CB based on such a concept has been developed and field tested in HVDC projects [27–29].

During current interruption, the AC CB opens its contacts and at the same time the gas discharge tube is turned ON by applying cross-field (perpendicular to the electric field) magnetic field. The arc voltage of the AC CB commutates current into the gas discharge tube, which conducts current until the mechanical CB opens sufficiently. Once the mechanical gap gains sufficient dielectric strength, the magnetic field is removed which results in current quenching and a drastic increase in the arc voltage. This in turn commutates current from the gas discharge tube to a parallel capacitor which charges until the conduction voltage of the MOSA is reached.

At the time, it was found that the main limitation to the current interruption capability of this concept was the AC CB used in the continuous current commutation branch [30]. The gas discharge tube could sustain a TIV of about 100 kV after current quenching although it is very sensitive to high du/dt. Further development of DC CB based on gas discharge tubes was halted for many challenging reasons [30]. Nevertheless, recently the use of gas-discharge tubes has revived especially for use in hybrid HVDC CBs [31,32]. In this case, the gas discharge tube could be used in the commutation branch of hybrid HVDC CBs instead of a large number of PE switches, see Section 5.4.1.

5.4.3   Hybrid HVDC CB

Hybrid HVDC CBs combine the desirable features of mechanical switches with those of PE switches. The desirable features of a mechanical switch are low conduction loss in a closed position and high-voltage withstand in an open position. However, the mechanical switches suffer from relatively long operation time (both in opening and in closing operations). The desirable feature of a PE switch is high-speed operation which can be achieved within a few tens of microseconds. The PE switches, however, suffer from high conduction losses when carrying current, and large numbers are required to withstand any practical high voltage in open (blocked) position.

Therefore, a hybrid HVDC CB is proposed in such a way that the continuous current commutation branch consists of a fast mechanical switch(es) in series with a (continuous) current commutation switch (CCS)[†††] as shown in Figure 5.23. The CCS is needed for commutating the continuous/fast rising fault current into the commutation branch. The commutation branch is a PE breaker. This branch conducts the fault current until the (ultra-)fast mechanical switch(es) reaches sufficient dielectric withstand to protect the CCS against the subsequent overvoltage. Then,

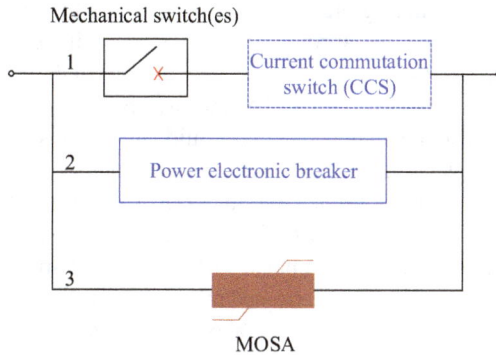

Figure 5.23   Generic hybrid HVDC CB

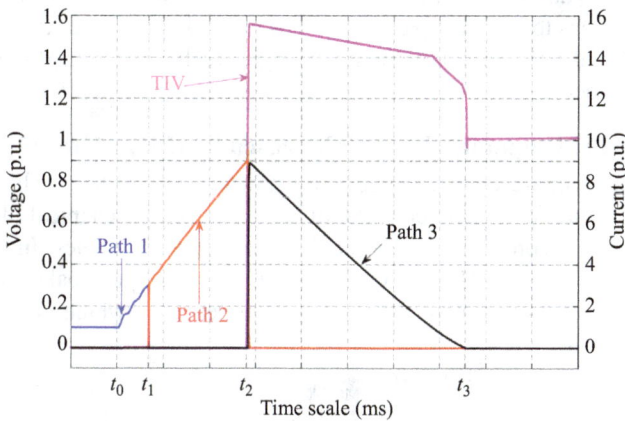

Figure 5.24   Generic mode of operation of a hybrid HVDC CB. Dashed lines for current through various branches (right side scale), solid lines for voltage (left side scale) [5].

[†††]In some cases, the CCS is absent (hence indicated by dashed blue box in Figure 5.24) where current transfer to the commutation branch is achieved by arc voltage of the mechanical switch or a counter current injection feature. If the CCS is available, the mechanical switch does not interrupt current except a small residual and/or transient current (<5 A) caused by stray inductance during commutation.

the PE breaker interrupts the fault current, and this is followed by a steep voltage rise across the breaker, which is limited by the MOSA serving the same purpose as in the mechanical type HVDC CBs.

The CCS is usually a matrix of several PE switches arranged in series and parallel (for example, 3×3) in order to reduce the on-state losses and increase reliability [33]. Moreover, these switches have a (water) cooling system since they conduct continuous current during normal operation. However, no cooling system is required for the PE breaker in the commutation branch since, under normal operation conditions, no/negligible current flow through this branch.

Depending on the type and arrangement of the PE switches in the CCS as well as in the commutation branch, many variants of hybrid HVDC CB have been proposed, some are prototype tested and a few have been commissioned in service [34–36]. Also, the fast mechanical switch could be SF_6 insulated ultra-fast disconnector(s) or a series connection of vacuum interrupters.

The generic mode of operation of a hybrid HVDC CB during a fault is illustrated in Figure 5.24. The fault current starts at t_0, then at t_1 current is interrupted in path 1 (see Figure 5.23) and the fault current commutates into path 2 (the main breaker for hybrid type CB). Next, at t_2, the current is interrupted in path 2. This leads to a steep rise of the TIV, until the MOSA limits that voltage to about 1.5 p.u. as shown in the figure. It is important to realize that, from this point on the system starts to recover, although the current in the fault location has not been interrupted yet. From this point on, the TIV steadily suppresses the current to zero; while the system's inductive energy is dissipated in the MOSA of path 3. Current suppression is completed at t_3. A small residual current remains, which has to be interrupted by a series residual current breaker.

Six different and most prominent topologies/realizations are discussed in the following subsections.

5.4.3.1   Hybrid topology I – serially cascade of IGBT topology

The electrical diagram of this topology is shown in Figure 5.25. The continuous current commutation branch consists of an ultra-fast mechanical disconnector (UFD) and a CCS consisting of a limited number of PE switches. In the commutation branch, it consists of a PE CB, which is divided into breaker units (cells), sometimes known as the main breaker modules, each of which can be operated

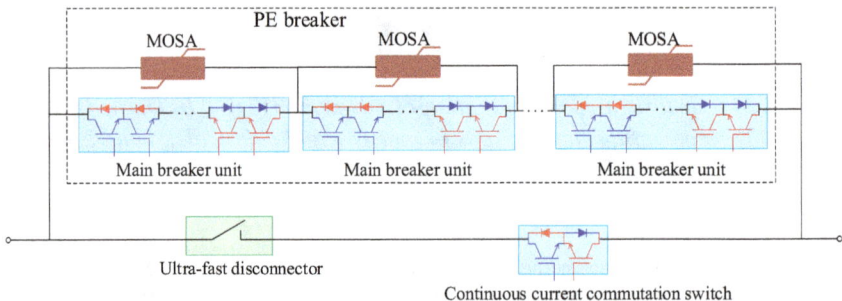

Figure 5.25   Bi-directional hybrid HVDC CB – original topology

independently. Each breaker unit consists of a cascade of PE switches, which operate simultaneously and a parallel EAB (MOSA).

Since under normal operation it continuously conducts load current, it is equipped with a water-cooling system [34]. The commutation branch is a power electronic-based CB. It is divided into breaker units (cells), sometimes known as main breaker modules, each of which can be operated independently. Each breaker unit consists of a cascade of power electronic switches which operate simultaneously, and a parallel energy absorber branch (MOSA). The fact that the breaker units have their own MOSA ensures equal voltage distribution across the modules. Furthermore, this can be manipulated to produce a controlled TIV (staircase waveform) by successively switching these breaker units one after the other so that the overall du/dt stress on the UFD is reduced. In addition, the fact that the breaker units can be switched independently enables the HVDC CB to be operated in a current limiting mode as discussed in [34]. To increase current breaking capability, the power electronic switches (IGBTs) can be connected in parallel.

The UFD provides a low-loss path during normal operation and protects the continuous current commutation switch during and after current interruption. It has been reported that the UFD can open within 2–3 ms and close within less than 10 ms after receiving a trip command [34]. In order to quickly build up its insulation level, multiple overlap contacts with double motion and dual current paths are designed [37,38]. In the closed current-carrying state of the switch, the conducting elements align to form parallel current paths between terminals of the switch along an axial direction. For opening the switch, the contact elements are mutually displaced by means of two drives along a direction perpendicular to the axial direction [37].

Each of the IGBTs within the breaker units has their own resistor-capacitor-diode (RCD) snubber circuit, see Figure 5.26, for controlling du/dt during turn OFF and equal voltage grading thereafter. The snubber circuit limits the du/dt to within the range acceptable to the IGBTs, for example 300 V/μs [35]. Simultaneous turn OFF of the power electronic switches in a breaker cell of the commutation branch is essential. There is a slight dispersion (μs) in the turn OFF time of semiconductors in the commutation branch and proper snubber circuit must be used to ensure acceptable voltage stress at the turn OFF [39]. In addition, the snubber circuits limit the di/dt during the turn ON of the power electronic switches. There are two resistors shown in the RCD circuit shown in Figure 5.26. The resistor R_1 is the discharge resistor during turn ON (when the IGBTs are turned ON) whereas R_2 is a relatively large resistor intended for discharge of the snubber capacitor after a current interruption for safety purpose (or make it ready for the next switching if re-closing is involved) [33]. Hence, R_2 assists in

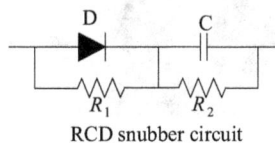

RCD snubber circuit

Figure 5.26   RCD snubber circuit based on the concept presented in [33]

preventing surge current through the IGBTs during re-closing by discharging the snubber capacitor during the de-ionization period of the system – in case re-closing is required. In addition, R_2 enhances voltage grading across IGBTs during the current suppression period. During the current interruption process, R_1 is bypassed by the diode ensuring rapid charging of the snubber capacitor while generating the TIV.

Recently, a newly developed power electronic semiconductor that integrates the anti-parallel freewheeling diode into the IGBT was proposed and used for HVDC CB application. The new development resulted in a single component, a reverse conducting IGBT, often referred to as bi-mode insulated gate transistor (BIGT). The main claimed advantage of a BIGT is that it is compact and has higher current rating with up to double the interruption capability of the traditional IGBTs. The BIGT is a reverse conducting IGBT device that combines the functionality of a high-power IGBT and fast diode utilizing the same silicon volume. The BIGTs have a failsafe (short-circuit failure mode) feature – this is the ability to fail into a stable low-impedance state.

A prototype of unidirectional hybrid HVDC CB, rated at 350 kV (480 kV TIV), 20 kA, tested at KEMA Labs' high-power laboratory and is shown in Figure 5.27. Power supply at high potential to the gate units of the power electronic

Figure 5.27   Prototype of 350 kV hybrid HVDC CB based on original topology [2]. Source: PROMOTioN Project.

switches as well as to the driving mechanism of the disconnector is challenging. In order to address this optically powered gate drives are used.

5.4.3.2 Hybrid topology II – full-wave diode rectifier bridge

One of the major drawbacks of hybrid HVDC CB topology I is the fact that twice the number of power electronic switches required for unidirectional current interruption are needed for bidirectional current interruption capability. This is because power electronic switches are naturally unidirectional. This has a significant impact on the cost and footprint of the HVDC CB.

Some optimization efforts have been made to minimize the number of IGBTs required for bidirectional current interruption capability [36]. An electrical diagram of one of the optimized designs is depicted in Figure 5.28. In this case, only a unidirectional power electronic breaker is required in the commutation branch while bidirectional current interruption capability is achieved by using a full-wave diode rectifier bridge as shown in figure.

The figure shows that the power electronic breaker (divided into several breaker units) is placed across the diode rectifier bridge. Hence, compared to the bi-directional hybrid HVDC CB shown in Figure 5.25, only half of the total IGBTs are needed in this case at the price of a full-bridge diode rectifier. During the current interruption process, the diode rectifier bridge ensures unidirectional current flow through the power electronic breaker despite the system current direction. The components of the continuous current branch remain the same as for the topology described in Section 5.4.3.1 except in this case the UFD is built from a number of series-connected vacuum interrupters [40]. When using a large number of series-connected vacuum interrupters, redundancy can be introduced in order to enhance

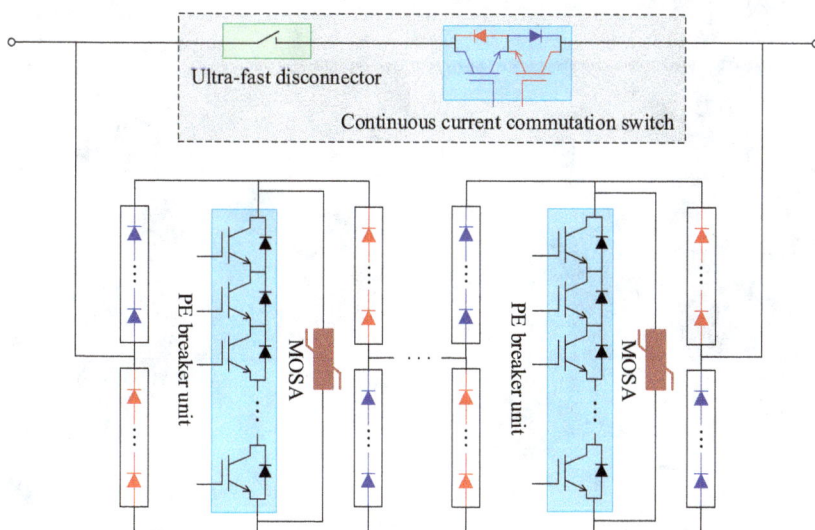

Figure 5.28   Hybrid HVDC CB – modified topology from [3]

reliability compared to a single high-voltage UFD. However, this comes at the expense of additional complexities such as voltage grading, synchronized operation and power supply to different units at high potential.

The full-wave diode rectifier bridge introduces extra stray inductance which affects the current commutation time from the continuous current branch to the commutation branch. In addition, the UFD can only be operated when the current in the continuous current branch falls below a certain threshold, for example, below 5 A. Stray inductance introduces the current oscillation between the continuous current branch and the commutation branch, thus prolonging the current commutation process. The latter affects the overall performance of the breaker. Therefore, compact design is essential.

A 500 kV realization of the optimized hybrid HVDC CB (topology II) is developed for application in the Zhangbei multi-terminal HVDC grid [40]. A photo of two such hybrid HVDC CBs installed in one of the converter stations is shown in Figure 5.29. In this case, the continuous current commutation switch is composed of 14 series X 6 parallel PressPack IGBTs with a commutation of time less than 200 μs [40]. A UFD is composed of 10 series-connected vacuum switches each of which are operated simultaneously by electro-magnetic repulsion mechanisms. It is reported that within 2 ms, these vacuum switches can achieve a total insulation level of 1,000 kV with a voltage unbalance of less than 5%.

Figure 5.29   500 kV hybrid HVDC CBs based on modified topology. Courtesy of
NR Electric.

The commutation branch is divided into modules where each module is a full-wave diode rectifier bridge consisting of 32 series-connected IGBTs and 32 series-connected diodes in a valve. Ten such modules are connected in series for a 500 kV rated voltage. Each module has a 15 MJ energy absorber MOSA at a clamping voltage of 80 kV. This makes the total energy absorption capability of the entire hybrid HVDC CB 150 MJ. Each MOSA module has 40 parallel columns where each column is composed of 12 series stacked MO varistor blocks.

5.4.3.3 Hybrid topology III –full-bridge IGBT submodule-based topology

Another hybrid HVDC CB topology based on full-bridge sub-module is depicted in Figure 5.30. The operation principle is similar to the topology II described above. However, current interruption is achieved not by a cascade of single IGBTs/BIGTs (IGBT valve), rather by a series connection of full-bridge sub-modules as shown in the figure. A full-bridge sub-module consists of a capacitor and four IGBTs each having anti-parallel diodes. The continuous current branch is composed of a matrix of full-bridge sub-modules (e.g., 2×3) connected in series with the ultra-fast mechanical switch(es). Likewise, the transfer branch consists of a large number of series-connected full-bridge sub-modules designed to interrupt fault current and withstand the TIV. In any direction, current in a full-bridge sub-module flows through two parallel paths (blue and red coloured paths) – each path being a series connection of an IGBT and a diode. Hence, the current breaking capability of such a design is twice the interruption performance of the hybrid HVDC CB original topology built from the same type of IGBTs. In fact, such performance is achieved at double the number of power electronic switches compared to the topology I. Finally, energy absorber units (MOSA modules) are connected across a few series-connected sub-modules making one breaker unit. For example, one breaker unit is rated for 50 kV (80 kV TIV) and consists of 32 sub-modules [36,41]. Multiple breaker units are connected in series to cope with the desired system voltage, for instance, four units are stacked for 200 kV system voltage. The modular design makes maintenance and replacement of sub-modules easier.

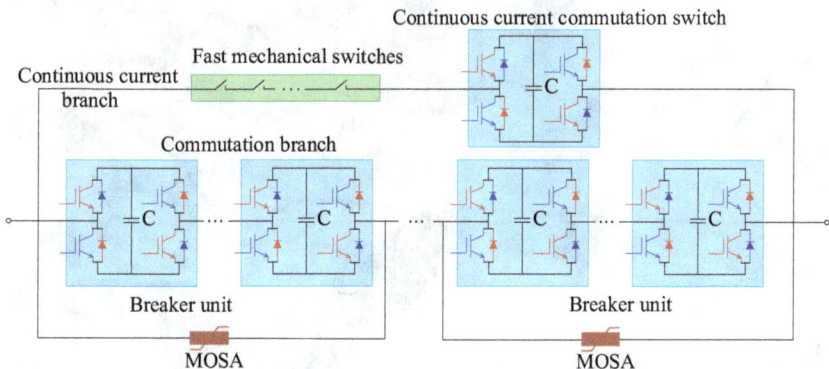

Figure 5.30   Hybrid HVDC CB – full-bridge IGBT-based topology based on the method presented in [36,41]

The matrix arrangement of sub-modules in the continuous current branch reduces on state losses and increases reliability. Moreover, similar to the other hybrid HVDC CB topologies described above, the power electronic switches in the continuous current branch have a water-cooling system since they conduct load current during normal operation. However, no cooling system is required for the power electronic switches in the commutation branch since, under normal operation condition, no/negligible current flow through this branch.

Another important difference worth noting is that unlike in the topologies described above, the IGBTs in the full-bridge IGBT-based topology do not require special snubber circuits since this function is provided by the sub-module capacitor. These capacitors are not normally charged like in a VSC, so a different form of isolated power supply to the gate drive units has to be developed which is a challenge. The capacitors have associated discharging resistors connected in parallel (not shown in Figure 5.30).

The fast mechanical switch (FMS) is made of series connection of vacuum switches driven by electromagnetic repulsion mechanism (Thomson coil actuators) shown in Figure 5.40 [36]. Repulsive force produced by the discharge of precharged capacitors through a coil is used to open the vacuum switches. One of the main challenges of using such a large number of mechanical devices is synchronous separation of contacts as well as proper grading of the TIV during the current interruption process.

For example, six vacuum switches (of which one is redundant) each rated for 40.5 kV, 2,000 A, constructed in sets of two interrupters are used in 200 kV hybrid HVDC CBs shown in [41]. Figure 5.31 shows two in-service hybrid HVDC CBs (of

Figure 5.31   Hybrid HVDC CB on-site – Zhoushan Five-Terminal MTDC grid [41,43]

full-bridge IGBT-based topology in series with current limiting reactors) installed at the 200 kV Zhoushan five terminal HVDC pilot project in China [42,43]. The six series-connected fast mechanical switches must achieve sufficient dielectric distance to withstand the TIV (within 2 ms) before switching of current in the commutation branch. Nevertheless, the vacuum switches in hybrid HVDC CBs open under no current – thus arc-less operation is ensured since current interruption is performed by power electronic switches in the continuous current branch.

Dual power supply based on a high-frequency electromagnetic energy system (with a high-frequency isolation transformer) is used together with a laser power system which serves as an auxiliary power supply in case the former fails to energize the IGBTs [40]. The latter is put into operation several milliseconds after the failure of the electromagnetic power supply is detected.

The overall test of such a breaker consisting of one breaker unit in the commutation branch has been performed in a laboratory where 15 kA current interruption is reportedly achieved within 3 ms [36].

5.4.3.4 Hybrid topology IV – modular diode full-bridge based

The hybrid HVDC CB described in the previous subsection is further optimized in a similar way as the original hybrid HVDC CB is optimized to the modified topology. In this case, the IGBT-based full-bridge sub-modules in the commutation branch are replaced by diode-based full-bridge sub-modules having two parallel IGBTs connected across the bridge as shown in Figure 5.32 [40]. The sub-module capacitor has a series-connected diode across the bridge to prevent the IGBTs from stress during and after current interruption. Such an arrangement results in the reduction of the required number of IGBTs by half while achieving the same current interruption performance as a full-bridge IGBT-based topology. The diode bridge redirects current in such a way that it flows through the IGBTs in the

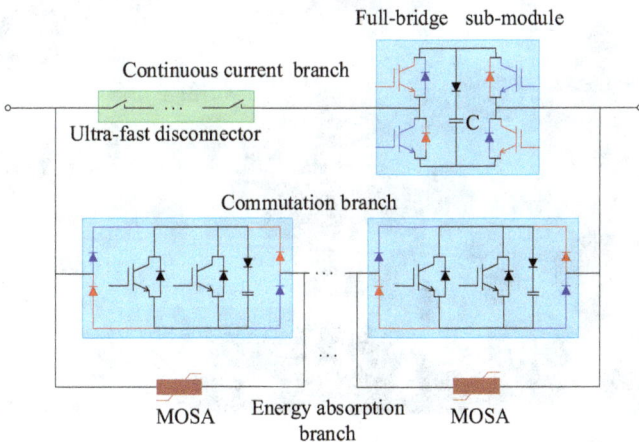

Figure 5.32   Hybrid HVDC CB – diode full-bridge-based topology

forward direction despite the system current direction. In such a way, the size and the cost of hybrid HVDC CB are further optimized. However, the continuous current branch remains the same as described in Section 5.4.3.3.

A 500 kV (800 kV TIV) hybrid HVDC CB based on this topology has been developed and functionally tested [40]. A photo of the prototype is depicted in Figure 5.33 [40]. In this prototype, all the components are integrated onto a single high-potential platform as displayed by the photo. In this case, a 2×5 matrix of IGBT-based full-bridge sub-modules is used in the continuous current branch. However, in the commutation branch, diode-based full-bridge sub-modules are used as can be seen in the diagram of Figure 5.32. The commutation branch is composed of five breaker cells (modular units) each rated for 100 kV system voltage (160 kV TIV). Each breaker cell is placed on a separate layer, thus the prototype shown in Figure 5.33 has six layers of which the bottom layer is used for the continuous current commutation switch.

The UFD is composed of 10 series-connected vacuum interrupters rated for 60 kV and 3,300 A [40]. Compared to the vacuum interrupters used in the 200 kV, 2,000 A prototypes shown in Figure 5.31 bigger contact system (thus heavier) is used for the increased current rating. Nevertheless, a similar operating mechanism as well as arrangement of vacuum interrupters, as illustrated in Figure 5.40, is used. The opening speed of the vacuum interrupters is verified per interrupter, and it is reported that a single vacuum interrupter reaches 10 mm contact separation after

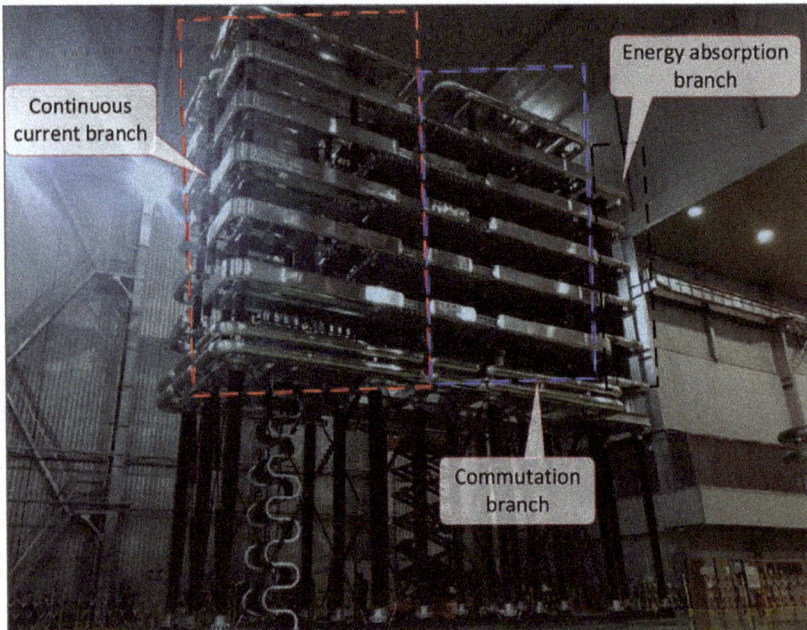

Figure 5.33   500 kV realization of hybrid HVDC CB – diode full-bridge-based topology [40]

Figure 5.35   A 500 kV prototype of negative voltage coupled hybrid HVDC CB [2]

It consists of eight 100 kV FMSs and 320 bidirectional submodules in the commutation branch [45]. The sub-modules in the commutation branch are divided into five breaker units each with a common MOSA branch.

(B) Parallel current commutation switches

In the second approach, a power electronic-based commutation switch is placed in parallel with a mechanical current commutation switch [46], see Figure 5.36. In the continuous current branch, there are two mechanical switches, an ultra-fast disconnector and a current commutation. The former is required to withstand the full TIV while the latter is required to commutate the load current under arcing condition. The current commutation switch is in parallel with an injection-enhanced gate transistor (IEGT) based full-bridge sub-module with pre-charged capacitor across the bridge. Under normal operation condition, the

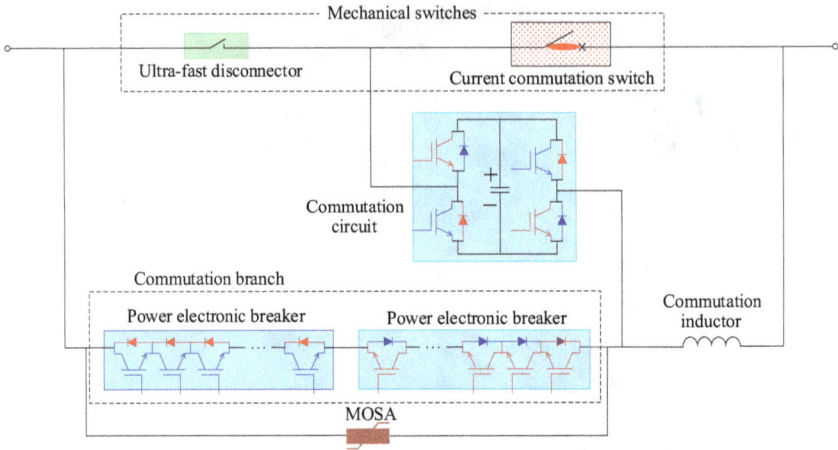

Figure 5.36   Arc voltage-based current commutation hybrid HVDC CB

full-bridge IEGTs remain in the OFF-state; hence, no ON-state loss. The characteristics and performance of IEGTs in comparison with other power electronic switches is described in [46]. Compared to the NVC hybrid HVDC CB described above, an added value of the latter approach is that it still has proactive breaking feature.

When the current interruption is required following a fault, a trip signal is sent to the breaker after which the breaker opens its mechanical switches. This is followed by arcing in both the fast disconnector and the current commutation switch. Then, the full-bridge sub-module designated as commutation circuit in Figure 5.36 turns its IEGTs ON.

Depending on the current direction either bottom-left and top-right (coloured blue) or top- left and bottom-right (coloured red), IEGTs are turned ON to commutate forward or reverse current, respectively. This discharges the pre-charged capacitor of the sub-module through the commutation inductor shown in the figure to create a current zero in the mechanical commutation switch. This results in the current commutation from the commutation switch to the commutation circuit. Meanwhile the (ultra-)fast disconnector switch is opening under arcing condition. Once current is fully commutated to the commutation circuit, the semiconductor breaker in the commutation branch is turned ON and the IEGTs in the full-bridge sub-module are turned OFF. This results in the charging of sub-module capacitor through the freewheeling diodes of the sub-module IEGTs. The voltage across the sub-module capacitor thus facilitates the commutation of current from the arcing (ultra-)fast disconnector to the power electronic breaker in the commutation branch. Following this step, the disconnector continues to open under no current/arcing condition. Note that in the latter stage, the system current needs to commutate from the commutation inductor. The commutation inductor is required to limit the counter injection current peak value and frequency. After dielectric recovery of the UFD, the power electronic breaker in the commutation branch turns its IEGTs OFF to produce the TIV after which current suppression starts in the MOSA.

One of the main differences between current injection-based hybrid HVDC CBs and the active current injection-based mechanical HVDC CBs is that the counter current injection in the former case occurs well before the mechanical switches are fully open. This is because the TIV in the former case is delayed due to the current conduction of the commutation branch unlike in the latter case where TIV generation follows immediately after current injection. This means that the current injection circuits can be optimized since a high-peak counter current pulse may not be required to create a current zero at such an early stage of fault current development. In addition, components (capacitors and inductors) do not need to be designed for rated current and voltage.

5.4.3.6   Hybrid topology VI – thyristor-based topology

This is a thyristor-based realization of hybrid HVDC CB [35] which consists of several current branches (paths): namely, the continuous current branch, time delaying branches, arming branch and extinguishing branch as shown in Figure 5.37. The figure depicts an electrical diagram a bidirectional thyristor based HVDC CB while Figure 5.38 shows the artist impression design.

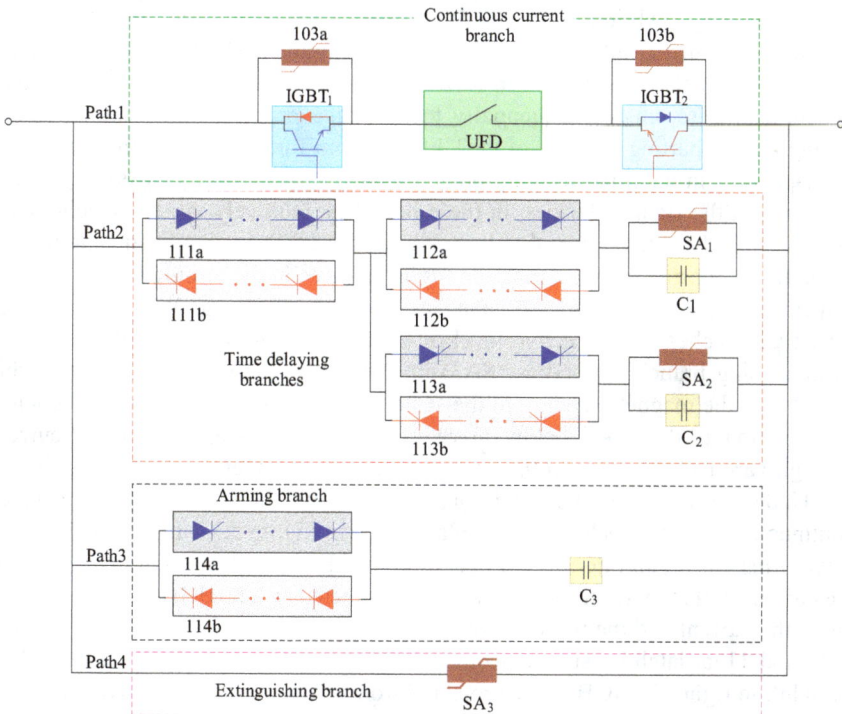

Figure 5.37   Electrical diagram of thyristor-based hybrid HVDC CB

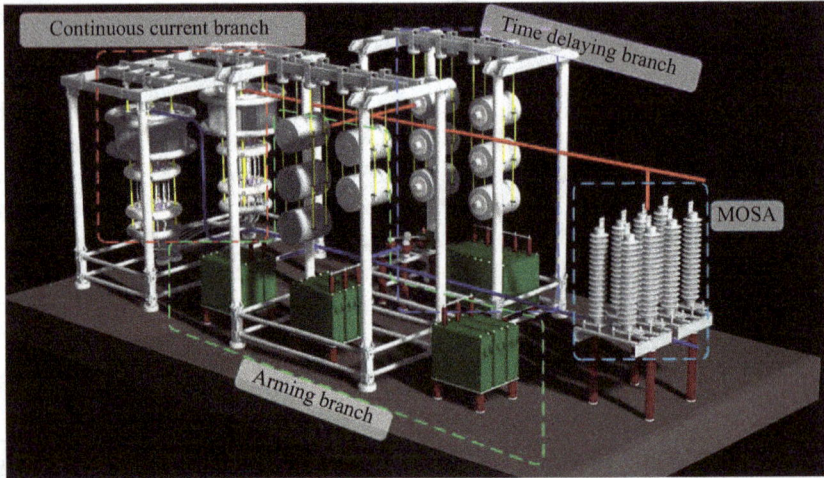

*Figure 5.38   Artist impression (rendering) of hybrid HVDC CB – thyristor-based
topology. Source: EU TWENTIES Project.*

The main components in each branch are the following.

Path1 is a low-impedance branch like the continuous current branch of the original topology hybrid HVDC CB. It consists of a UFD in series with IGBTs for commutating current into the first time-delaying branch. The time delaying branches consist of thyristor valves in series with capacitors which, during the commutation process, charge sequentially to a voltage proportional to the dielectric strength of the parting contacts of the UFD. To limit further increase in the voltage, the capacitors in the time-delaying branches have surge arresters (SAs) in parallel, as shown in the figure. Besides, in these branches, relatively larger capacitors are used to limit the rate-of-rise of voltage. When the UFD reaches sufficient contact separation (reported within 2 ms), the thyristor valve in the arming branch (shown in path3) is latched on and the capacitor C3 starts charging. When the voltage across C3 reaches the protective level of the main energy absorber MOSA in the extinguishing branch (named as (SA3) in path4), the current commutates to this branch and the magnetic energy in the system is dissipated. The detailed sequence of operation is described below in reference to the waveforms of the current through various branches and the voltage across the breaker.

Under normal operating condition, current flows through the low-impedance continuous current branch shown by black trace in Figure 5.37. In Figure 5.39, the waveforms during an interruption operation are shown. When a fault occurs at t_1, the current starts to rise. As soon as a fault is detected (t_2), the IGBTs in this branch block the current and the thyristor valves in the first time-delaying branch, namely 111a and 112a, latch on simultaneously – assuming the short-circuit current flow from left to right. The IGBTs are protected from a small voltage surge by the surge arrester 103a. Hence, current flows through the surge arresters until the thyristors 111a and 112a are fully turned on.

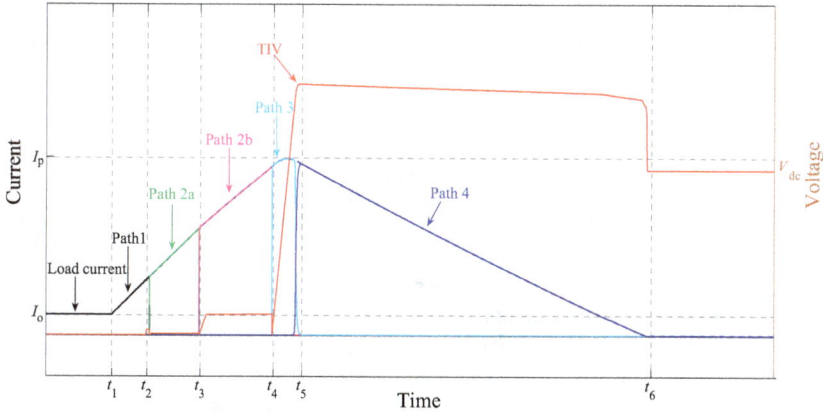

Figure 5.39 *Waveforms describing the operation principle of hybrid HVDC CB – thyristor-based topology*

As soon as the current is commuted from path1 to path2 (green trace), the TIV starts to build up across the breaker since capacitor C_1 starts to charge. The voltage across C_1 is limited from further rise by surge arrester SA1 – between t_2 and t_3. At t_3 the thyristor valve 113a latches on and similarly, the current commutates to this branch (the magenta trace). Since the capacitor C_1 is charged it attempts to inject current in the opposite direction to the fault current as a result which the thyristor valve 112a turns off naturally. In a similar way, the thyristor valve 114a latches on at $t4$ and subsequently, the current is commutated to path3 (cyan trace) charging the capacitor C_3.

The UFD which starts opening by the time the current is fully commutated to the time-delaying branch must have reached the full TIV withstand distance by now. After current commutates to path3, the full TIV is produced by charging the C_3 up to the protection level of the MOSA designated by SA_3. This leads to the current commutation into path4 (at I_5), where the MOSA suppresses the fault current while absorbing the energy in the system – the blue trace. The commutations in the several auxiliary branches are, therefore, important so that the breaker can handle the fault current until the TIV withstand level of the UFD builds to a required level in a controlled manner. The protection level of surge arresters (and the voltage across capacitors) at each branch is designed to exceed the protection level in the previous branch. Further details can be found in [35].

5.5 Sub-components of HVDC CBs and critical issues

Unlike an HVAC CB, an HVDC CB is no longer a single device, rather it is a system of components designed and arranged in several current branches operating in a prescribed sequence to achieve DC current interruption as discussed in Section 5.4. These components include PE switches like IGBTs, BIGTs, IGCTs, IEGTs, thyristors, components such as MOSA and mechanical switching devices like UFD, VCB and HSMS including their newly developed high-speed actuators.

Most of these are standard components used in non-standard applications, and therefore, have to face non-standard stresses [2,47]. Some are newly developed, for example, the UFDs and the high-speed actuators on all other mechanical switching devices. The main components of an HVDC CB are discussed below.

5.5.1   Mechanical switching devices

Every practical HVDC circuit breaker is equipped with a mechanical switching device. Its function is to enable low losses in continuous operation and to alleviate (when in the open state) dielectric stresses on the power electronic components. In HVDC circuit breakers, every mechanical switching device has to achieve contact separation very fast, which is achieved by electromagnetic repulsion drives. Such drives are electronically controlled, which implies a certain susceptibility to EM interference from transients of the primary sources (arcing, re-ignition, fast switching, high di/dt-current, high du/dt, etc.). In most designs, a (considerable) number of mechanical switching devices are put in series. This implies that power to the individual drives cannot be supplied through galvanic connections. Usually, transformers that have enough insulation capability are used. For example, several isolation transformers are stacked in series to achieve adequate insulation from earth for 500 kV HVDC circuit breakers [48].

High-speed drives and their isolated power supply are not used in such a way before in power equipment and service experience is very limited or non-existent. Due attention needs to be paid to the verification of the mechanical endurance of the total kinematic chain. In addition, the proper functioning of a stack of a larger number of smaller interrupters needs a well-synchronized contact separation as well as a built-in redundancy to overcome the functional loss of one or more individual interrupters. Differences of ± 0.1 ms in switch opening times are reported in a stack of 10 vacuum fast disconnecting switches which achieves 1,000 kV isolation in 2 ms [2].

The fast mechanical switch (FMS) is made of a series connection of vacuum switches driven by electromagnetic repulsion mechanism (Thomson coil actuators) shown in Figure 5.40 [36]. Repulsive force produced by the discharge of

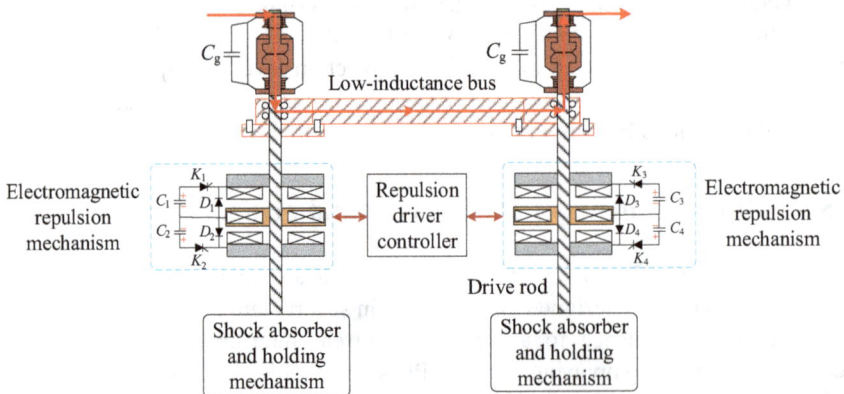

Figure 5.40   Driving mechanism of FMS created based on the method presented in [49]

pre-charged capacitors through a coil is used to open the vacuum switches. One of the main challenges of using such a large number of mechanical devices is synchronous separation of contacts as well as proper grading of the TIV during the current interruption process.

The special high-speed drives cause huge impact forces on the contact systems of vacuum or SF_6 interrupters. Care must be taken when applying standard AC vacuum interrupters in combination with fast electromagnetic (EM) drives. Especially the bellows of standard spring-actuated vacuum interrupters are not designed to withstand a certain number of high-impact force opening operations.

The series combination of many interrupters has its challenges, not only mechanically but also electrically. Sharing (grading) of voltage needs to be considered seriously, not only regarding DC voltage but also during transients. For AC applications, niche products like capacitor bank switches consisting of up to 9 series vacuum interrupters do exist but have a poor service record. In general, for AC, high-voltage vacuum circuit breakers are developed with as few as possible series interrupters, never more than two at present.

Grading capacitors (Cg) are used for voltage distribution as shown in Figure 5.40. A bi-stable spring device is utilized as a holding mechanism (also provides sufficient contact force under closed condition) and a special hydraulic buffering device is adopted as a shock absorber (come into play at the last phase of opening operation), see Figures 5.40 and 5.41, the design detail of which is described in [49].

The use of a limited number of HV vacuum breakers is an advantage compared to a much higher number of MV vacuum breakers, though it is a challenge to realize a high enough opening speed with the heavier HV vacuum interrupter contact systems. Vacuum is a very good "medium" regarding interruption of high-frequency current and very fast recovery of the gap against steep rising recovery voltage. Nevertheless, the application of vacuum interruption in active current injection type of HVDC circuit breakers may approach performance limits. Re-ignition and restrikes become apparent as shown in Figure 5.42.

Mechanical gaps (vacuum/SF_6) breakdown electrically when they are not able to withstand voltage. Most critical is the fault current suppression phase, where the overvoltage is around 1.5 p.u. whereas at the same time, the gaps are recovering

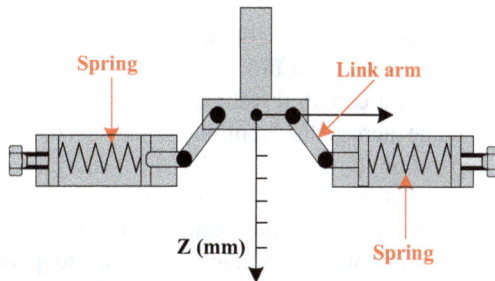

Figure 5.41   Bi-stable Spring Mechanism (holding unit) redrawn after [49]

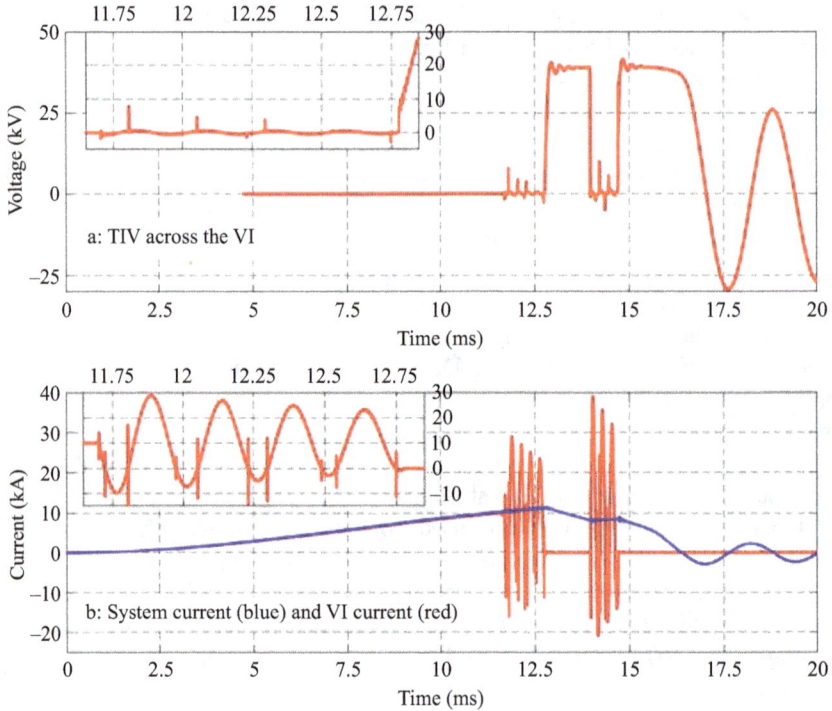

Figure 5.42   Re-ignitions and a late restrike in the vacuum interrupter during the current suppression period [47]

from interruption and/or switching. During (after) fault current suppression, there is a much longer exposure to the recovering system voltage and its (slow) transients, until the residual current switch takes over that voltage stress. Breakdown of (a) vacuum gap(s) is less critical than of an SF_6 gap, since the vacuum is able to restore its insulation state very fast (see Figure 5.42), while SF_6 generally cannot. In addition to the damage of its contacts, the impact on SF_6 switchgear is that power electronic switches in the continuous current path will be damaged which results in complete malfunctioning.

In HVDC circuit breakers, SF_6 disconnectors need to open with a very low current and with very low voltage in order to avoid arcing. Once the current (at contact separation) exceeds a certain threshold, the arc will persist during a time, depending on the voltage across the commutation branch. The current at disconnector contact separation is determined by the leakage current through the snubbers/grading elements of the series (semiconductor) switches in the continuous current branch (up to a few amperes). The voltage against which the disconnector is opening is determined by the on-state voltage of the semiconductors in the commutation branch (main breaker), which can be several hundreds of volts to a few kilovolts. The latter could be sufficient to keep current conduction in the disconnector if opened before the current is fully commutated. Therefore, the design of (ultra-)fast disconnectors is extremely critical. Opening of the disconnector needs to be synchronized carefully, after current transients in the continuous current branch have decayed sufficiently.

Once the main breaker has interrupted the fault current, soon the full TIV appears across the switching gap which must have been sufficiently open to isolate. Breakdown of this gap would lead to dielectric overload of the continuous current commutation switch and a free burning arc in the SF_6 disconnector. Therefore, the dielectric coordination of this disconnector allows only a very small variation in its opening time over service life. Requirements for mechanical stability are more severe than for controlled (capacitor bank) SF_6 CBs for HVAC applications [50].

In many cases, standard AC vacuum CBs are used as ultra-fast disconnector. Such devices are not optimized for DC voltage withstand, so proper verification is necessary. Moreover, standard vacuum breakers contain arc control devices (axial or radial magnetic field arc control devices, see Chapter 3). They add unnecessary weight to the interrupters and additional limits to the actuator, given the fact that currents that DC vacuum breakers have to deal with are far below those in AC systems. In addition, the standard contact material, a CuCr alloy, may not be the optimum choice for DC conduction and insulation. From dedicated studies, it was found that different designs of commercially available 36–38 kV AC vacuum circuit breakers acted very differently regarding the interruption of HF counter current and TIV withstand capability [3].

In the active current injection schemes using capacitor discharge as the counter current source, an ultra-fast switching device must be used to start the discharge. This may be a mechanical switch (vacuum-making switch, triggered spark gap) or a semiconductor stack (IGCTs, thyristors). In both applications, the very large di/dt and peak current need to be evaluated as a non-standard stress. In the case of vacuum making switch, provisions must be made to avoid contact welding, originating from the pre-strike arc that is comparable to the back-to-back capacitor bank-making function in AC application. For extra high voltage (EHV) applications, several of these switches need to be connected in series. Synchronous operation of these making switches is essential to avoid premature current injection as a result of pre-strike. In the case of semiconductor making switches, di/dt and short-time thermal and dynamic stresses can be extreme and far from standard. In addition, during normal operations the making devices in the open position are subjected to continuous DC voltage stress.

5.5.2   Power electronic switches

The application of large stacks of semiconductors is not new and ample experience exists in AC/DC converters.

In the continuous current commutation switch, that needs to commutate the current into the commutation (main breaker) branch, a large current is ramped down to zero with a very high di/dt. This switch consists of a limited number of power-electronic switches. The proper choice of the number of elements is critical from thermal point of view and continuous cooling is necessary.

Continuous current commutation switches are mostly made from the state-of-the-art IGBTs slabs which consist of IGBTs and diodes at different sections of the semiconductor package. These packages have IGBTs and diodes to be able to conduct current in the reverse direction. Since the continuous current commutation switch is conducting DC current continuously in normal operation only the IGBT parts of the semiconductor volume are heated due to on-state loss. The results of this unequal heating must get sufficient attention as there is not much operational experiences. During functional tests of HVDC circuit breakers which focus on the proof of concept, steady-state effects like unequal heating cannot be observed. The latest developments of IGBTs, known as bi-mode insulated gate transistor (BIGT), which uses the same semiconductor volume for both IGBT and diode are not susceptible to this situation [51].

In specification, one needs to make sure what the thermal limit in the case of a full-through short-circuit current (short-time current) is. Power electronics are sensitive, and contrary to converters, HVDC breakers may have no protection from power electronics (like blocking and bypassing current into freewheeling diodes) in case of an unexpected short-time current exceeding the rated breaking capability.

In hybrid HVDC breakers, the continuous current commutation switch is conducting current continuously (the main breaker may be as well, but at very low current) without switching. This is a different operation than in "normal" operation, like in converters, where semiconductors are switching continuously. In such a way, semiconductors conduct during on-state while diodes might conduct during off-state. This results in a lower thermal gradient across the package volume. Alternatively, the breaker in the commutation branch is idle, or conducting small current while under "standard converter operation" its semiconductors are operating continuously. It is not known how the "one time only" activation to switch off all semiconductors might have an impact on a possible conditioning of the semiconductor junction and/or package [2].

5.5.3 MOSA

The other essential component of all HVDC CBs, which is subjected to unique stresses during DC breaking, is the energy-absorbing component, i.e., the MOSA. In HVDC CBs, the MOSA serves two main functions: clamp and maintain the TIV to a desired level and absorb the system energy during current suppression. The MOSA must be designed to meet the requirements of both functions simultaneously. The desired level of the TIV must be sufficiently higher than the system operation voltage while the energy absorption is dependent on the system as well as on the circuit breaker parameters [5]. A TIV about 50% higher than the system voltage is considered as a reasonable value for system insulation coordination as well as current interruption duration [1].

Conventionally, MOSA is used for overvoltage protection in power systems. A few major differences between the use of MOSA for overvoltage protection (both in AC and DC applications) and for HVDC CB applications are highlighted in [52] and are presented in Table 5.2.

*Table 5.2   Comparison of MOSA use for overvoltage protection and for HVDC
circuit breaker application*

MOSA for overvoltage protection in AC and DC systems	MOSA for energy absorption in HVDC circuit breaker application
- Active all the time/conducts small leakage current at system operation voltage	- Passive during normal system operation – becomes involved only during DC CB operation
- Subject to system voltage under normal operation	- Is not subject to any voltage stress during normal operation – it is bypassed by the main current path of DC CB
- One or a few columns are sufficient (low energy absorption requirement)	- Large number of parallel columns are needed for high energy absorption – careful column matching is essential
- Normally conducts short duration pulses (far less than a millisecond up to a few milliseconds)	- Long duration conduction up to 10 or even more milliseconds
- Suppresses surge voltage before reaching its prospective peak value and diverts charge to ground	- Suppresses fault current before reaching its prospective peak value

A large volume of MOSA is needed for absorbing the energy from the faulted system and maintaining the TIV. Many columns are needed in parallel to cope with large energy absorption. This means the individual zinc-oxide (ZnO) varistor discs composing each column need to be carefully selected to have an equal current flowing through the column. Given the high non-linearity of the u–i characteristic, equal stacking and paralleling can change its characteristic unfavourably [52,53]. Therefore, very careful selection of individual varistor disks and matching of the parallel columns is essential. A small mismatch in conduction voltage would lead to a large current difference. This, in turn, would heat the columns.

It is reported in the literature that the energy handling capability of an MO varistor depends on several factors such as the diameter and volume, impulse duration, and current density [54–57]. Moreover, different makes of MO varistors exhibit considerably varying energy absorption capabilities [56–58]. Nevertheless, the optimum energy absorption per volume, considering a large number of MO varistors in a MOSA module of an HVDC CB, might need further investigation.

When designing a MOSA module for HVDC CB application, the desired TIV determines the residual voltage (and thus the height) of the active part whereas the expected energy in the system determines the total volume required. Since a large amount of energy is absorbed during DC fault current interruption, several parallel columns of MOSA are required to cope with the volumetric requirement. However, this requires a column-matching procedure: a crucial design step when constructing a multi-column MOSA. This column-matching procedure is necessary to ensure equal current sharing during the fault current suppression, which otherwise would lead to unequal energy distribution and hence, possible thermal overloading of one or more columns.

Note that typically after manufacturing, all MO varistors are screened by applying 8/20 μs impulses to check the *u–i* characteristics, specifically by measuring the residual voltage at 10 kA impulse current. Even after this, not all the MO varistors have identical *u–i* characteristics and hence, the MOSA columns built from the same batch of MO varistors do not necessarily have matched *u–i* characteristics. This is due to inherent imperfections in the manufacturing process that do not always lead to MO varistors with an identical distribution of microscopic ZnO grains. The voltage of the MO varistor is determined by the number of ZnO grain boundaries conducting along the current path. Thus, it is difficult to ensure a homogeneous distribution of grain sizes and/or impurities along all current paths even within the same MO varistor batch. Current initially flows in the path that results in the fewer number of grain boundaries until it is distributed across the entire cross-section. This is what results in localized current conduction especially at low current densities [59].

The issue of MOSA column matching has been identified in the early developments of HVDC CBs [60]. Different column matching techniques exist. During the column matching procedure, current impulses are applied (10–20 times) successively to the parallel arrangement of MOSA columns – see an example of impulse measurement through six columns in Figure 5.43. One among these columns serves as a reference column. The arithmetic mean of the peak values of the current through each column is computed, and compared against the arithmetic mean of the peak values of the current measured through the reference column. Columns with current measurements within an acceptable margin, for example 3% from the reference column current, are accepted as matched.[§§§] The procedure is repeated until the required number of columns are obtained.

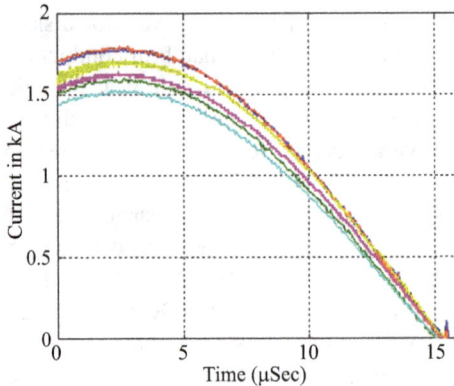

Figure 5.43   An example of impulse current measurements during MOSA column matching procedure

[§§§]Three percent in current sharing typically means <<1% in residual voltage difference, due to the extremely high degree of non-linearity in the considered current range of only a few hundred amperes per column in many cases.

Another similar routine procedure of column matching is by applying an impulse current of around 500 A per column [61]. In this case, by measuring an impulse current through each column, the deviation from the arithmetic mean is calculated after 20 impulses. The selection criterion is that those MO arrester columns with a maximum deviation of +7% from the arithmetic mean are accepted as matched.[****] In fact, the columns are constructed from initially pre-selected MO varistors. The MO varistors are pre-selected by applying an impulse current of 500 A and 10 kA of 8/20 μs.

The total mass of ZnO material in HVDC circuit breakers can be over one thousand kilograms, which implies that cooling down (after interruption of a significant fault current) is very time consuming, in the order of several hours after rated fault current interruption. The consequence of this is that when a reclosure and re-open function is required, the design should be able to absorb at least double the energy that is associated with a single interruption (and proportionally more counter voltage creation, etc. when more than two reclosures are expected). Moreover, the other functions of the breaker (local current zero creation), should be accommodated for quickly repeated operation.

Multi-reclosure of HVDC breakers is required in overhead line (OHL) systems. OHL arcing faults most often disappear after a reclosure and a subsequent opening (O-CO sequence) of the breaker. HVDC circuit breakers for the Zhangbei (overhead line) project in China have been specified to deal with total energy absorption exceeding 150 MJ. When the actual short circuit, carrying the large fault current is removed, in many cases a low-current secondary arc to earth persists, which is fed through the stray impedances of the transmission system. Reclosure should then be delayed until the secondary arc ceases, mostly by natural reasons, like wind or by thermal elongation. CIGRE studies have indicated that in HVAC overhead line systems, after a single open-close action, the fault is removed in the vast majority of the cases. Further study needs to reveal the persistence of secondary arcs in HVDC OHL systems.

In cable systems, reclosure does not seem to be a suitable action since faults in cable systems are normally destructive and need repair.

When designing a MOSA module for HVDC CB application, the desired TIV determines the height of the active part whereas the expected energy in the system determines the total volume. Since a large amount of energy is absorbed during DC fault current interruption, several parallel columns of MOSA are required to cope with the volumetric requirement. However, this requires a column matching procedure: a crucial design step when constructing a multi-column MOSA. This column matching procedure is necessary to ensure equal current sharing during the fault current suppression which otherwise would lead to unequal energy distribution and hence, thermal overloading of one or more columns.

[****]In this case, there is no reference column for comparison.

5.5.4 Summary of critical issues

Table 5.3 HVDC circuit breaker components facing "non-standard" stresses and potential issue [2]

	Subcomponent	Characteristic in standard operation	Non-standard stress in HVDC circuit breaker application	Potential issue(s)
Mechanical switching devices	(Multiple) actuator(s) for mechanical switching device	Speed 1 – few m/s	• Ultra-high speed • High impact forces • Control electronics on board	• Mechanical reliability • Compatibility with equipment attached • Electromagnetic sensitivity • Synchronicity
	Power supply to actuator	At earth potential	At high potential	Non-galvanic power supply
	Vacuum interrupters for interruption	Power frequency AC current interruption	• HF current interruption at high di/dt • Recovery at small gap length against very high du/dt • Arc stays at the same polarity on the contacts	• Interruption of high-frequency current • Very high du/dt recovery • Uneven contact conditioning
	Vacuum interrupters for insulation	AC voltage insulation	DC voltage applied after interruption	DC voltage withstand capability
	High-speed making switches (vacuum, power electronics, triggered gap)	Capacitor bank inrush current making	Injection current making above highest IEC standardized value	• Contact welding by pre-strike arc • Mechanical synchronicity

	Multiple vacuum breakers in series	Usually single break	Several to many interrupters in series	• Grading for transients and DC • Redundancy • Mechanical synchronicity
	SF$_6$ gap(s) for insulation	Very low opening speed in GIS AC application	Ultra-high contact separation speed	• Shall not switch current • Dynamic DC voltage withstand capability • Extreme mechanical consistency over time
Semiconductor switching devices	Semiconductors in the continuous current branch (continuous current commutation switch function)	Power electronics switch with high frequency	Conduct continuously and switch only occasionally	• Thermal stability. • Unequal thermal distribution
	Semiconductors in the commutation branch (main breaker function)	Power electronics switch with high frequency	Never/hardly conduct and switch only occasionally	Reliability after long idle time
MOSA	MOSA consisting of multiple columns	Overvoltage protection	Significant energy absorption (similar application in series capacitor bank protection)	• Thermal overload, runaway • Current sharing between columns
	MOSA columns	Always under (AC) voltage	Occasionally stressed by voltage	• Conditioning • Stability of u–i characteristic • DC stability

References

[1]   CIGRE Joint Working Group A3/B4.34, "Technical requirements and specifications of state-of-the-art HVDC Switching Equipment," Cigre Technical Brochure 683, Paris, 2017.

[2]   R. P. P. Smeets and N. A. Belda, "High-voltage direct current fault current interruption: a technology review," *High Voltage*, vol. 6, pp. 171–192, 2021.

[3]   N. A. Belda, *HVDC Circuit Breakers – Test Requirements, Methods and Circuits*, Darmstadt: TU Darmstadt, 2021.

[4]   A. N. Greenwood, P. Barkan, and W. C. Kracht, "HVDC vacuum circuit breakers," *IEEE Transactions on Power Apparatus and Systems*, vol. PAS-91, no. 4, pp. 1575–1588, 1972.

[5]   N. A. Belda and R. P. P. Smeets, "Test circuits for HVDC circuit breakers," *IEEE Transactions on Power Delivery*, vol. 32, no. 1, pp. 285–293, 2017, doi:10.1109/TPWRD.2016.2567783.

[6]   W. Grieshaber, J.-P. Dupraz, and M. Collet, "DC current during commutation and interruption: relation between interruption time, insulation level and energy in the transmission means," in Matpost, 2011.

[7]   D. Kind, E. Marx, K. Moellenhoff, and J. Salge, "Circuit-breakers for HVDC transmission," in Cigre Session, Paris, 1968.

[8]   S. Yanabu, T. Tamagawa, S. Irokawa, T. Horiuchi, and S. Tomimuro, "Development of HVDC circuit breaker and its interrupting test," *IEEE Transactions on Power Apparatus and Systems*, vol. PAS-101, no. 7, pp. 1958–1965, 1982.

[9]   A. N. Greenwood and T. H. Lee, "Theory and application of the commutation principle for HVDC circuit breakers," *IEEE Transactions on Power Apparatus and Systems*, vol. 91, no. 4, pp. 1570–1574, 1972.

[10] EPRI, *EPRI High Voltage Direct Current (HVDC) Transmission Reference Book*, Palo Alto: Electric Power Research Institute, 2018.

[11] C. M. Franck, "HVDC circuit breakers: a review identifying future research needs," *IEEE Transactions on Power Delivery*, vol. 26, no. 2, pp. 998–1007, 2011.

[12] K. Sano and M. Takasaki, "A surgeless solid-state DC circuit breaker for voltage-source-converter-based HVDC systems," *IEEE Transactions on Industry Applications*, vol. 50, no. 4, pp. 2690–2699, 2013.

[13] L. Feng, R. Gou, F. Zhuo, X. Yang, and F. Zhang, "Development of a 10kV solid-state DC circuit breaker based on press-pack IGBT for VSC-HVDC system," in *2016 IEEE 8th International Power Electronics and Motion Control Conference (IPEMC-ECCE Asia)*, Hefei, 2016.

[14] T. Wei, Z. Yu, Z. Chen, *et al.*, "Design and test of the bidirectional solid-state switch for an 160kV/9kA hybrid DC circuit breaker," in *2018 IEEE Applied Power Electronics Conference and Exposition (APEC)*, San Antonio, TX, USA, 2018.

[15] B. Pauli, G. Mauthe, E. Ruoss, G. Ecklin, J. Porter, and J. Vithayathil, "Development of a high current HVDC circuit breaker with fast fault

clearing capability," *IEEE Transactions on Power Delivery*, vol. 3, no. 4, pp. 2072–2080, 1998.

[16] L.-R. Jänicke, J. Teichmann, S. B. Weidong Han, and X. Wang, "New two chamber transfer switch for 6500 A commutation current," *Global Energy Interconnection*, vol. 1, no. 3, p. 344–351, 2018.

[17] M. Backman, L. Liljestrand, F. Rafatnia, and S. Nyberg, " Advances in DC neutral breaker performances for bipolar HVDC schemes," in *Cigre Paris Session*, Paris, 2018.

[18] K. Kanngieser, "The current commutation function of HVDC switching devices," *Cigre Electra*, vol. 124, pp. 33–39, 1989.

[19] W. Gu, D. Feng, and Z. Guo, "Development of high-speed mechanical switchgear with vacuum interruption technology and application in HVDC circuit breaker," in *2019 IEEE 3rd International Electrical and Energy Conference (CIEEC)*, 2019.

[20] S. Tokoyoda, T. Inagaki, H. Sadakuni, T. Minagawa, D. Yoshida, and H. Ito, "Development and testing of EHV mechanical DC circuit breaker," in *2019 5th International Conference on Electric Power Equipment - Switching Technology (ICEPE-ST)*, Kitakyushu, Japan, 2019.

[21] L. Liu, Z. Yuan, H. Xu, *et al.*, "Design and test of a new kind of coupling mechanical HVDC circuit breaker," *IET Generation, Transmission & Distribution*, vol. 13, no. 9, pp. 1555–1562, 2019.

[22] W. Wen, Y. Wang, B. Li, Y. Huang, R. Li. and Q. Wang, "Transient current interruption characteristics of a novel mechanical DC circuit breaker," *IEEE Transactions on Power Electronics*, vol. 33, no. 11, pp. 9424–9431, 2018.

[23] X. Leishi, S. Chao, and L. Qifu, "Research on short-circuit test and simulation of CSG first mechanical HVDC circuit breaker in VSC- HVDC," in *2018 International Conference on Power System Technology (POWERCON)*, 2018.

[24] L. Ängquist, S. Norrga, and T. Modeer, "A new DC breaker with reduced need for semiconductors," in *2016 18th European Conference on Power Electronics and Applications (EPE'16 ECCE Europe)*, IEEE, September 2016.

[25] L. Ängquist, S. Nee, T. Modeer, A. Baudoin, S. Norrga, and Belda, "Design and test of VSC assisted resonant current (VARC) DC circuit breaker," in *15th IET International Conference on AC and DC Power Transmission (ACDC 2019)*, Birmingham, 2019.

[26] S. Liu, M. Popov, S. S. Mirhosseini, *et al.*, "Modeling, experimental validation, and application of VARC HVDC circuit breakers," *IEEE Transactions on Power Delivery*, vol. 35, no. 3, pp. 1515–1526, 2020.

[27] G. Hofmann, G. L. Barbera, N. Reed, and L. Shillong, "A high speed HVDC circuit breaker with crossed-field interrupters," *IEEE Transactions on Power Apparatus and Systems*, vol. 95, no. 4, pp. 1182–1193, 1976.

[28] G. A. Hofmann, W. F. Long, and W. Knauer, "Inductive test circuit for a fast acting HVDC interrupter," *IEEE Transactions on Power Apparatus and Systems*, vols. PAS-92, no. 5, pp. 1605–1614, 1973.

[29] J. G. Gorman, C. W. Kimblin, R. E. Voshall, R. E. Wien, and P. G. Slade, "The interaction of vacuum arcs with magnetic fields and applications,"

IEEE Transactions on Power Apparatus and Systems, vols. PAS-102, no. 2, pp. 257–266, 1983.

[30]   A. L. Courts, J. J. Vithayathil, N. G. Hingorani, J. W. Porter, J. G. Gorman, and C. W. Kimblin, "A new DC breaker used as metallic return transfer breaker," *IEEE Transactions on Power Apparatus and Systems*, vols. PAS-101, no. 10, pp. 4112–4121, 1982.

[31]   C. C. Davidson, C. D. Barker, J. M. De Bedout, W. Grieshaber, J. W. Bray, and T. J. Sommerer, "Hybrid DC circuit breakers using gas-discharge tubes for high-voltage switching," in *CIGRÉ Colloquium*, Winnipeg, 2017.

[32]   T. Augustin, M. Becerra, and H. P. Nee, "Enhanced active resonant DC circuit breakers based on discharge closing switches," *IEEE Transactions on Power Delivery*, vol. 36, no. 3, pp. 1735–1743, 2021.

[33]   A. Hassanpoor, J. Häfner, and B. Jacobson, "Technical assessment of load commutation switch in hybrid HVDC breaker," *IEEE Transactions on Power Electronics*, vol. 30, no. 10, pp. 5393–5400, 2014.

[34]   J. Häfner and B. Jacobson, "Proactive hybrid HVDC breakers – a key innovation for reliable HVDC grids," in *CIGRE Symposium on the Electric Power System of the Future: Integrating Supergrids and Microgrids*, Bologna, 2011.

[35]   W. Grieshaber, J. P. Dupraz, and D. L. P. a. L. Violleau, "Development and test of a 120 kV direct current circuit breaker," in *Cigre Session*, Paris, 2014.

[36]   W. Zhou, X. Wei, S. Zhang, *et al.*, "Development and test of a 200 kV full-bridge based hybrid HVDC breaker," in *2015 17th European Conference on Power Electronics and Applications (EPE'15 ECCE-Europe)*, 2015.

[37]   L. Liljestrand, L. E. Jonsson, P. Skarby, and R. Chladny, "Switch having two sets of contact elements and two drives," United States Patent US8797128B2, 5 August 2014.

[38]   R. Derakhshanfar, T. U. Jonsson, U. Steiger, and M. Habert, "Hybrid HVDC breaker – a solution for future HVDC system," in *Cigre Session 45*, Paris, 2014.

[39]   W. Wen, Y. Huang, B. Li, Y. Wang, and T. Cheng, "Technical assessment of hybrid DCCB with improved current commutation drive circuit," *IEEE Transactions on Industry Applications*, vol. 54, no. 5, pp. 5456–5464, 2018.

[40]   G. Tang, "Research on key technology and equipment for Zhangbei 500 kV DC grid," *High Voltage Engineering*, vol. 44, no. 7, pp. 2097–2106, 2018.

[41]   G. F. Tang, X. G. Wei, W. D. Zhou, and S. Zhang, "Research and development of a full-bridge cascaded hybrid HVDC breaker for VSC-HVDC applications," in *Cigre Session Paper A3–117*, Paris, 2016.

[42]   G. Tang, Z. He, H. Pang, X. Huang, and X. p. Zhang, "Basic topology and key devices of the five-terminal DC grid," *CSEE Journal of Power and Energy Systems*, vol. 1, no. 2, pp. 22–35, 2015.

[43]   G. Tang, W. Zhou, Z. He, X. Wei, and C. Gao, "Development of 500kV modular cascaded hybrid HVDC breaker for DC grid applications," in Cigre Session, Paris, 2018.

[44] X. Zhang, Z. Yu, and R. Zeng, "State-of-the-art 500 kV hybrid circuit breaker for DC grid: design, development, and experiment—the world's largest capacity HVDC circuit breaker," *IEEE Industrial Electronics Magazine*, 2019.

[45] S. Zhang, G. Zou, C. Xu, and W. Sun, "A reclosing scheme of hybrid DC circuit breaker for MMC-HVDC systems," *IEEE Journal of Emerging and Selected Topics in Power Electronics*, vol. 9, no. 6, pp. 7126–7137, 2020.

[46] Y. Koyama, R. Hasegawa, K. Kanaya, T. Matsumoto, and T. Ishiguro, "Operation and experimentation of a current commutated hybrid DC circuit breaker for HVDC transmission grids," *IEEJ Journal of Industry Applications*, vol. 8, no. 5, pp. 835–842, 2019.

[47] N. A. Belda, R. P. P. Smeets, and R. M. Nijman, "Experimental investigation of electrical stresses on the main components of HVDC circuit breakers," *IEEE Transactions on Power Delivery,* vol. 35, no. 6, pp. 2762–2771, 2020.

[48] X. Zhang, Z. Yu, R. Zeng, *et al.*, "High voltage isolated power supply system for complex multiple electrical potential equipments in 500 kV hybrid DC breaker," *High Voltage*, vol. 5, no. 4, pp. 425–433, 2020.

[49] W. Wen, Y. Huang, M. Al-Dweikat, *et al.*, "Research on operating mechanism for ultra-fast 40.5-kV vacuum switches," *IEEE Transactions on Power Delivery*, vol. 30, no. 6, pp. 2553–2560, 2015.

[50] IEC TR 62271-302: *High-Voltage Switchgear and Controlgear – Part 302: Alternating Current Circuit-Breakers with Intentionally Non-simultaneous Pole Operation*, IEC, 2010.

[51] M. Rahimo, L. Storasta, and F. Dugal, "The Bimode Insulated Transistor (BIGT), an ideal power semiconductor for power electronics based DC breaker applications," in *CIGRE Session, Paper B4–302 (2014)*, Paris, 2014.

[52] P. Hock, N. Belda, V. Hinrichsen, and R. Smeets, "Investigations on metal-oxide surge arresters for HVDC circuit breaker applications," in *INMR World Conference*, Tucson, USA, 2019.

[53] R. Le Roux, R. Smeets, and N. Belda, "Utilization of metal oxide surge arresters in HVDC circuit breakers and similar application," in *International Conference on Condition Monitoring, Diagnosis and Maintenance*, Bucharest, 2019.

[54] K. Eda, "Destruction mechanism of ZnO varistors due to high currents," *Journal of Applied Physics*, vol. 56, no. 10, pp. 2948–2955, 1984.

[55] K. Ringler, P. Kirkby, C. Erven, M. Lat, and T. Malkiewicz, "The energy absorption capability and time-to-failure of varistors used in station-class metal-oxide surge arresters," *IEEE Transactions on Power Delivery*, vol. 12, no. 1, pp. 203–212, 1997.

[56] {Cigre WG A3.17}, "Technical Brochure 544 – MO surge arresters: stresses and test procedures," Cigre, 2013.

[57] {WG A3.25}, "Technical Brochure 696: MO surge arresters – Metal oxide resistors and surge arresters for emerging system conditions," Cigre, 2017.

[58] M. N. Tuczek, "Recent experimental findings on the single and multi-impulse energy handling capability of metal–oxide varistors for use in high-

voltage surge arresters," *IEEE Transactions on Power Delivery*, vol. 29, no. 5, pp. 2197–2205, 2014.

[59]   M. Bartkowiak, M. Comber, and G. Mahan, "Failure modes and energy absorption capability of ZnO varistors," *IEEE Transactions on Power Delivery,* vol. 14, no. 1, pp. 152–162, 1999.

[60]   S. Yanabu, T. Tamagawa, S. Irokawa, T. Horiuchi, and S. Tomimuro, "Development of HVDC circuit breaker and its interrupting test," *IEEE Transactions on Power Apparatus and Systems*, vols. PAS-101, no. 7, pp. 1958–1965, 1982.

[61]   ABB HV Switchgear AB, "Zinc oxide surge arrester – Current sharing consideration," ABB, 1990.

[62]   N. A. Belda, C. A. Plet, and R. P. P. Smeets, "Analysis of faults in multi-terminal HVDC grid for definition of test requirements of HVDC circuit breakers," *IEEE Transactions on Power Delivery*, vol. 33, no. 1, pp. 403–411, 2018.

Chapter 6

Open research questions and future trends

Kaveh Niayesh[1]

In the first five chapters of this book, most important concepts for the high voltage switching devices within the context of more environmentally friendly switching technologies suitable for new requirements forced by the emerging developments of the power systems have been thoroughly discussed.

In this chapter, some of the other relevant topics will be briefly addressed, which have not been resulted in matured products or are on the way to be introduced as standard commercial products. The main idea here is to give an overview on the possible interesting research questions for the interested reader which would eventually gain in relevance for the field of high voltage switching technologies in future. Each section of this chapter is written in a form of compact technology review with reference to many recent investigations on different research fields highlighting the potential impact and research directions.

6.1 Areas for future research

6.1.1 HVDC switchgear

The switching technologies related to HVDC have been extensively covered in Chapter 5. However, there are still several open issues to be addressed. In a comprehensive review paper of 2011 [1], future research needs have been listed. It would be interesting to evaluate which of those research needs have been addressed during the last 12 years, and which are still relevant.

The focus on the switching arc to commutate the current in HVDC CB schemes has been now shifted towards other commutating technologies involving more power semiconductors [2], also to some extent in the main current path, for more information on the topology of HVDC CBs, refer to Section 5.3. Nevertheless, understanding of the behavior of the DC arc, especially if it is used for arc current commutation, and development of mathematical models and multiphysics simulations are of essential importance for realization of a cost-efficient DC CB with minimum use of power semiconductor switches.

---

[1]Norwegian University of Science and Technology (NTNU), Trondheim, Norway

Development of fast mechanical switches or disconnectors with high recovery voltage withstand has been successfully demonstrated in several prototypes [3], but there is still no commercial out-of-shelf product available. Some of the research needs are still highly relevant, the optimization of the size of elements with the goal of reduction of the size, interruption time and most importantly the cost of an HVDC CB is still an active research field. This also comprises new schemes and configurations for HVDC CB as detailed in Chapter 5. In an interesting paper [4], the cost drivers of HVDC CBs are considered in four different categories, including commodities, labor, engineering as well as energy, and a cost reduction of 38% by 2050 compared to the 2017 cost level for DC CBs is predicted.

In addition, the research on power semiconductor switches, also for HVDC CB applications, with the goal of reduction of their on-state losses is a highly active research field. The details are discussed in Section 6.1.5.

Technologies for fault current limiters could also be used in order to interrupt faults in HVDC networks. This has been the motivation to list the development of fault current limiters for medium voltage and high voltage applications, as one of the research needs for HVDC CBs. The development trends of fault current limiters are presented in detail in Section 6.1.4.

Another important topic mentioned in the review paper from 2011 [1] has been the need to develop new test methods for HVDC CBs which should follow the standards for multiterminal HVDC networks. Meanwhile, several test circuits [5] have been developed for short circuit test of HVDC circuit breakers. However, there is a lack of field experience for HVDC CBs in multi-terminal HVDC networks, as the number of installations is still limited. Furthermore, there are still no standardized test methods and standards for multiterminal HVDC networks at the time of writing of this book, but several active working groups in different technical bodies are working on standardization of DC CBs in medium and high voltage networks. Reference [6] details the efforts on standardization of DC circuit breaker specifications initiated in CIGRE, IEC and also on the national level. China has been pioneering in application of HVDC circuit breakers, as there are a few installations and a number of future projects, so it would not be a big surprise to have already some Chinese standards available for DC switches as well as for HVDC circuit breaker [7]. Joint WG B4/A3.80 of CIGRE has finalized a study of HVDC circuit breakers and its test requirements [8] and the first IEC standards are drafted [9] and expected to be made available by 2023/2024.

6.1.2   High-voltage vacuum circuit breakers

In Chapter 3, the application of vacuum interrupter technology for transmission level voltages has been covered extensively. In this section, a summary of the most important open issues and research trends will be presented.

In vacuum circuit breaker, magnetic fields are used to control the vacuum arc and ensure a quasi-homogeneous heat distribution over the whole vacuum contact to avoid anode melting (spot formation) which eventually leads to an interruption failure. As explained in Section 3.4, this can be accomplished either using AMF or RMF contact designs. In AMF designs, the magnitude of the axial magnetic field

In Chapter 4, it is explained that the nozzle material usually used in high-voltage circuit breakers is PTFE, another substance which contains fluorine, among the so-called fluoropolymers also known under "perfluorinated and polyfluorinated alkyl substances (PFAS)." As discussed in Section 4.3, the high current arc burns in PTFE vapor, and its superior characteristics help the switching device to interrupt the current. There are, however, some activities to restrict the use of fluorine containing materials [24] where 119 European organizations urge EU to ban the production of PFAS by 2030. European Chemical Agency (ECHA) recently published a proposal for PFAS regulations [25], it is at the time of writing this book not clear when and to what extent the use of PTFE in high-voltage switchgear will be affected. Strict restrictions could impose significant influence on switching arc characteristics and switching performance of gas circuit breakers. This may in turn open another interesting research field to investigate other nozzle materials, both their compatibility to the interruption medium (gas) employed in the switching devices and their impact on the switching arc characteristics and switching performance of the gas circuit breaker.

Another emerging application is the so-called HVDC gas-insulated switchgear, (HVDC GIS) following the idea of creating an emission free power network, these should also be realized using SF_6 – alternatives, and most preferably using natural gases. From the physical point of view, not all knowledge gained for the AC insulation and breakdown characteristics of the gases, explained in Section 4.1.3.1 or gas insulated switchgear [26], can be transferred to DC application because some other physical mechanisms such as the accumulation of space charges under DC voltage stresses as well as increased significance of current conduction mechanisms on the electric field distribution, and the temperature dependence of the conductivities could play an important role. Recently, some HVDC GIS solutions for voltages up to $\pm$ 320 kV [27] has been installed, and designs for voltages up to $\pm$ 550 kV [28] are presented. Like the HVDC CBs, there is still no IEC standard for HVDC GIS, but a draft is prepared and commented [29], and most probably will be made available by 2023/2024.

6.1.4 Fault current limiters

Fault current limitation has been of interest since many decades because of its potential to improve the performance of the power networks and to reduce the overall cost of power system apparatus by reducing the maximum short circuit current levels. Higher short circuit current level is a consequence of more interconnectivity in the power networks that is normally utilized to increase the reliability and availability in the system, or because of newly installed power generation units in response to the increasing demand for electric power, a good example is given for extension of the grid in Java-Bali power system [30]. One of the main application fields of FCL in conventional power networks is the so-called network coupling or busbar coupling where two different parts of a network are connected to each other leading to much higher short circuit current levels; placing an FCL in between may isolate the network parts during fault condition resulting in a reduced short circuit current level. Another important application could be using an FCL in networks with increased short circuit current levels beyond the rated short circuit current of switching components, e.g., because of an extension of the network, as a cost-competitive alternative to upgrade all switching equipment. In addition, splitting buses and using bus-tie breakers and series reactors can be avoided, and the voltage dips reduced. In modern power networks with high penetration of renewables (or distributed generation, DG) and solid-state devices, some more applications have been identified for FCLs. Some of the DGs would increase the short circuit current in the distribution systems. One expensive solution would be to connect them through a transformer, but if a cost-effective solution for FCLs exists, these could be also connected through an FCL [31]. Another interesting application of FCL would be to enhance the fault ride through capability and to improve the transient stability of grid-connected PV/wind power system [32].

The fault current limitation can be performed actively (by realizing a current controlled impedance), so that the impedance of the device is near zero under normal conditions for rated currents and is very high for short circuit currents, or passively so that a constant impedance is inserted in series in the power network and in this way the short circuit impedance is increased. The latter one is, however, associated with steady power losses and lowers voltage regulation. The so-called passive fault current limiters, e.g. current limiting reactors, and active one-shot fault current limiters, such as fuses or Is-limiter (or FC protector) [33] for medium voltage networks have been in service for many decades. Multi-shot active FCLs, however, have been considered as a nice-to-have device for certain application fields. There has been a lot of effort with many innovative ideas to use different nonlinear material properties to realize a current controlled impedance. A few examples include employing temperature dependence of resistivity, the so-called PTC materials [34], or utilization of Lorentz forces at short circuit currents either to compress (pinch) the current carrying conductor [35], or to move part of the current carrying conductor in order to force the current to flow through a high resistive path [36], as well as using the non-linear permeability in ferromagnetic materials [37], or strongly non-linear temperature and magnetic field dependent resistivity of superconducting materials in many different configurations [38]. In some other concepts, the nonlinear materials are combined with fast opening mechanical switches forming a hybrid solution, where fast mechanical (current commutating) switches, nonlinear materials, e.g., PTC materials, and solid-state switches with appropriate overvoltage control (or energy absorption) devices are combined [39]. From the physical point of view, the same technologies with slight modifications can be applied for the limitation of fault currents and DC circuit breakers as forcing the current in HVDC circuits to zero requires a fault current limitation principle. This is the reason why many solid-state-based concepts, which were originally developed for HVDC current interruption, can also be applied as FCL.

Out of many different research ideas, it seems that three concepts could establish themselves as technically matured solutions for high-voltage applications, namely saturable core inductors, superconductor-based fault current limiters, and solid-state fault current limiters. There are also several combinations of these three concepts to add features or reduce the cost.

In the saturable core fault current limiters, the nonlinearity of B–H curve of magnetic materials is used to realize a current controlled inductance. In this type of FCLs, the core is kept in saturation under normal operation, when a fault occurs it comes out from the saturation region entering the linear region resulting in a sudden increase of the inductance as the relative permeability in the linear B–H region is much larger than unity. This concept is simple in operation and robust as no moving parts are necessary but results in very large and heavy apparatus. The idea has been commercialized by few manufacturers [40] with few installations around the globe [41].

Another concept which attracted a lot of attention in the last few years has been using superconducting materials for fault current limiting purposes. Superconductor-based FCLs are either resistive or inductive. One of the first superconducting fault current limiters (a 10.5 kV/1.2 MVA demonstrator) [42] was installed in Switzerland in 1996. Another pilot installation in Germany (10 kV/ 10 MVA demonstrator) was reported in 2005 [43]. Recently, a few large European projects [44,45] with participation of leading manufacturers have been focused on improvement of this type of FCLs. In another recent paper, a high-voltage SC FCL was proposed to limit fault current in Russian grid [46]. The main problem here is the need for cryogenic systems to reduce the operational temperature as the available superconducting materials possess the desired properties at very low temperatures. This makes the FCLs based on superconductors very complex with several subsystems and therefore less reliable and expensive. Moreover, the cooling system consumes significant amount of energy, which can be considered as dissipation losses of the whole fault current limiting system.

Power semiconductor devices can also be used in combination with some other passive components to limit the fault currents. Advances in power semiconductor technologies, allowing higher current and voltage ratings, enable applications of these technologies in FCLs. This type of FCLs is called solid-state fault current limiters. As the power semiconductors are used in realizing other applications in transmission and distribution networks, the solid-state FCLs would be easily integrated in other parts. The main idea in many concepts is to realize a modular design which can be scaled up to higher voltage and current levels. In an interesting report, a 15 kV module which can be used for FCLs in the range of up to 69 kV was introduced [47]. An extensive list of studies on different configurations and designs of solid-state FCLs is collected in [48], which shows the large activities by many different research groups in this field. The main challenge will, however, remain the rather large power dissipations related to the carrying of rated current under normal operational conditions, which leads to the need having an enhanced cooling system and an increasing number of components.

There have been many efforts to reduce the total cost of the above-mentioned concepts but not many installations can be found world-wide. In the author's opinion, the reason has been that no cost-competitive reliable solution for multi-shot active FCLs have been so far available for medium- or high-voltage applications. Cost-competitiveness remains one of the most important challenges in future research.

6.1.5   *Switching in supercritical fluid*

In previous chapters of this book, it has been discussed how very low pressure medium (vacuum) as well as different gases at pressures up to a 1 MPa are used as current interruption dielectric medium in different switching concepts. A gas pressure increase could eventually help to improve the switching performance of gas circuit breakers. Some of the gases may, however, experience a transition to a new phase, called *supercritical fluid*, by a further pressure increase beyond a critical point. In the supercritical phase, the good properties of liquid, e.g., high density, high thermal conductivity and large heat capacity, and gas phases, e.g., low viscosity and no bubble formation, coexist. This combination may be beneficial for current interruption. Furthermore, an increase of the gas pressure also improves the dielectric breakdown strength [49].

With this background, a research question has recently attracted a lot of attention, namely if supercritical fluids can be employed as a current interruption medium in power switching devices.

Several researchers have taken different approaches: a large number of investigations, e.g., [50–54], focused on the dielectric breakdown strength of supercritical fluids. Table 1 of [52], summarizes many of the experiments performed by different researchers. The main conclusion has been that the transition to supercritical state does not change the breakdown strength, as the breakdown strength is controlled by the particle number density [50]. Some anomalies have been reported for the pressures near the critical point, e.g., decrease of the breakdown voltage in micrometer-scale gaps [55,56]. In addition, a saturation of the breakdown voltage in high-density regions [50] is observed, which can be explained considering the cluster formation at higher densities which results in a lower density-reduced critical electric field of the supercritical fluid [57].

A first step towards the application of supercritical fluids for switching purposes has been linked to pulse power switching applications, where a breakdown in supercritical is used to realize a pulsed power closing switch with high repetition rate [58,59]. There is, strictly seen, no current interruption as such in those switches as the current is a damping oscillation to zero. The current magnitudes (of maximum few tens of ampere) and arcing times of tens to hundreds of nanoseconds are far from the conventional switching arcs in gas circuit breakers.

In another comprehensive study [60], current interruption in supercritical N_2 has been investigated. The arc currents correspond the rated currents of load breaking switches and the arcing times have been in the range of few milliseconds. Several arrangements including free-burning arcs [61], arcs in ablative and non-ablative cylindrical tubes [62], as well as arcs in some self-blast and puffer type model switches [63], have been studied. The conclusion has been that there is no change in arcing voltage, or current interruption performance by the transition from the gas to the supercritical state. Furthermore, a supercritical fluid is only beneficial if there is an active flow of gas.

In another endeavor, an ultra-fast switch in a supercritical fluid (CO_2) with piezoelectric drive is used in a hybrid switching device on the main path in series with a saturable inductor [64]. The contact separation occurs just before the current zero, so that there is almost no arcing in the supercritical fluid switch [65], no detailed information on the arc behavior in this switch was available. The main role of an increasing impedance in the main path is performed in this design by the variable inductor, and the supercritical fluid switch plays the role of a superior insulating switch.

In the author's opinion, the characteristics of switching arcs in supercritical fluids and their limitations have not been fully understood, especially at larger currents that correspond to typical current ranges of power switching equipment. There is active research on characterization of high-density plasmas, i.e., plasma in supercritical fluids as well as cryogenic plasmas, see e.g., [66], which can be eventually used to characterize the switching arc behavior in supercritical fluids. With the current level of knowledge, it is hardly imaginable that the supercritical fluids would be employed in gas circuit breakers in general purpose CBs in the transmission system as the current interruption medium, because the need for efficient gas flow, and additional challenges related to realization of and operation under the necessary high pressures, e.g., 3.4 MPa for N_2 or more than 7.2 MPa for CO_2 would bring no decisive advantage. However, if further studies confirm good current interruption capability of switches with supercritical fluids especially at high currents, supercritical fluid switches may become attractive for some special applications, e.g., subsea applications, because despite the vacuum or gas CBs there would be no need for pressure compensation leading to cost-efficient enclosures of subsea switchgear. In addition, it will be possible that the supercritical fluid switch as part of a hybrid CB as proposed in [64] finds application, where it is merely used as a superior dielectric medium with high breakdown strength.

6.1.6 Hybrid switchgear

Depending on the context, the expression "Hybrid Switchgear" can have different meanings. In general, if two different technologies are combined together to realize a switchgear, then the term *hybrid switchgear* is used. The idea would be to take advantage of the strengths of the technologies to realize a solution with much better performance. One example has already discussed and presented in Chapter 5, where HVDC circuit breakers employing fast mechanical switches and power electronic switching technologies have been combined together to realize the so-called hybrid HVDC CBs [67], where the low dissipation of mechanical switches for rated current carrying is combined with good current interruption performance of power semiconductor switches, see e.g., [68].

Another possibility would be to combine different arc-based switching technologies, e.g., vacuum and gas switching in one circuit breaker. Here, very fast recovery of vacuum can be combined with high-voltage dielectric insulation withstand of gas switching devices. This idea has been first invented in [69], then its different aspects have been studied by several researchers, see e.g., [70,71]. In those studies, the vacuum has been used in series with SF_6 gas circuit breaker to improve its short line fault performance. The same concept can also be applied to combine SF_6-alternative gas circuit breakers, e.g., using CO_2, with vacuum interrupters [72,73]. This concept of hybrid switchgear is interesting from the scientific point of view, but it struggles with cost competitiveness. In the author's opinion, this is the reason why it was not the preferred solution of the leading manufacturers so far.

6.1.7　High power solid-state switching technologies

Power semiconductor switching devices are used in many different applications in modern power networks. The question which is going to be discussed in this subsection is, however, if this type of switching devices can take over the role of conventional switching arc-based electromechanical circuit breakers. There are several advantages using solid-state circuit breakers, including their very fast response resulting in significantly reduced switching times, and having no arc what can be considered as a desired characteristic especially for those applications where arc hazard exposure is of concern. In addition, the solid-state circuit breakers have no moving parts, and therefore much higher number of operations can be realized. The downside of the coin is very large losses of the solid-state circuit breakers when they are in closed position and carry current, because the solid-state devices have a certain forward voltage drop. This becomes even more critical when higher rated voltages are targeted, because many modules are to be connected in series in order to increase the reverse blocking voltage of the device, leading to much higher turn on voltage drops over the circuit breaker in closed position. This makes a cooling system of the solid-state circuit breakers necessary. Moreover, the solid-state devices are vulnerable to high rates of changes of current and tolerating large short-circuit currents, in one of the recently launched products a series inductance if foreseen to reduce the rate of rise of current [74]. As this type of switches does not interrupt the current at its zero crossing, the energy stored in the stray inductance of the system may cause large overvoltages and therefore some parallel components known as voltage-clamping and snubber circuitry are required to control the overvoltages across the switch itself. Therefore, besides power semiconductor device and its fault sensing and tripping system, many other subcomponents, such as gate driver, cooling system, voltage clamping/snubber circuitry and in some cases di/dt limiting components, are required for solid-state circuit breakers [75]. So, they become hardly cost competitive compared to conventional switching-arc based circuit breakers.

There are, however, several application fields, such as DC microgrids [76] and (dc) power distribution systems for marine [77] and aviation [78] applications, where the pros overweigh contras for employing solid-state circuit breakers, and this has been the reason for recent increased research activities and also new product launches [74] by leading manufacturers of power system equipment. The recent research activities are addressing solutions for voltage ratings in the low-voltage range or in few cases related to lower medium voltage range.

On the other hand, there is active research on different semiconductor materials as well as semiconductor device designs with the aim of increasing the current density and the breakdown voltage of the devices, targeting application where higher power densities are necessary. One of the promising research fields has been investigations on wide band gap materials, such as Silicon Carbide (SiC) and Gallium Nitride (GaN). Reference [79] details opportunities and limitations of SiC for power switching device applications. More details on current status and future trends of GaN devices, especially those based on high electron mobility transistors, can be found elsewhere [80]. These materials show much higher electrical breakdown fields, which translates in higher voltage ratings, as well as larger conductivities in on-state and enhanced thermal properties, e.g., larger thermal conductivities, enabling higher current ratings of the devices.

The recent improvements of the semiconductor technology, e.g. the transition from Insulated Gate Bipolar Transistor (IGBT) technology to reverse blocking integrated gate-commutated thyristor (RB-IGCT) technology has resulted in reduction of conduction losses by a factor of four [74], but the reduced losses (e.g. 3,500 W for a current of 2,500 A in a device with rated voltage of 1 kV) are still too high to be able to use this technology for higher voltage ratings. To illustrate this, note that tens (or hundreds) of this module are necessary to realize a MV (or HV) circuit breaker, resulting in on-losses of several orders of magnitude higher than the losses of an electromechanical circuit breaker with the same ratings. Some other technologies seem to be promising in increasing the voltage rating of the devices; in an interesting investigation [81], a 15 kV/200 A solid-state circuit breaker is presented based on parallel operation of the 15 kV SiC emitter turn-off (ETO) thyristors. The reported forward voltage drop of the proposed switch has been about 5.6 V at 80 A. In another endeavor [82], super cascode structure of multiple low voltage, high current SiC JFETs in series has been employed to make a medium voltage solid-state circuit breaker with an efficiency of more than 98% at a nominal power of 10 MW. This means that although the losses have been substantially reduced through the recent improvements in material and design of semiconductor devices, a thermal management (cooling) system is still necessary, even for LV applications. The main reason would be to develop topologies and designs to reduce both the losses and enhance the transfer of the generated heat out of the semiconductor junctions in order to keep the temperature of the junctions under maximum allowable temperature. The thermal management system may contain heat sinks, forced air cooling system [74] or even water-cooling system [83], depending on the losses. Another important aspect related to design of solid-state circuit breakers for higher voltages would be the increased dielectric stresses at high voltage and high temperatures. This makes new fabrication and packaging designs necessary for solid-state modules. If the standards for testing of circuit breakers are followed strictly, the solid-state switch should also be able to withstand the impulse voltage tests. The solid-state devices, however, are not able to handle the impulse voltage levels, and therefore a mechanical disconnecting switch in series with the solid-state breaker (the so-called isolator switch) is still necessary.

To the author's opinion, there is not likely that solid-state circuit breakers replace high voltage switching arc-based circuit breakers in near future if no quantum leap jumps occur in the semiconductor technology, or their cooling technologies. However, as mentioned in Chapter 5, the solid-state switching devices will be inevitable components of the high-voltage DC circuit breakers in combination with mechanical switches.

6.1.8   Controlled switching

Conventional switching devices are not synchronized, i.e., they open or close at a random phase of the voltage (or current) when the command is received. Energization (or de-energization) of power system (or equipment) may create very large transient voltages or currents in the system stressing the power system equipment. As the magnitude of the overvoltage or overcurrent is dependent on the phase of the voltage, where the power equipment or system is energized, the switching overvoltage or overcurrent can be significantly reduced if the switching operation is synchronized with the phase of the voltage. This is called synchronized switching, also called controlled (or point-on-wave) switching in literature. To make a synchronized switching possible, the stochastic change in delay of the operating mechanism, i.e., jitter, should be minimized and the dependence of the delay of the operating mechanism on different operational parameters such as temperature and idle time of the circuit breaker should be understood and accordingly compensated. In addition, a control and measurement system responsible for synchronizing the operation of the circuit breaker is required. Furthermore, it is necessary to have separate operations for all three poles of a circuit breaker because an intentional non-simultaneous pole operation is necessary. As the standard for conventional high-voltage switchgear [84] excludes non-simultaneous pole operated circuit breakers, a new standard for circuit breakers with controlled switching [85] has been recently provided based on the guidelines developed by CIGRE WG A3.35 [86].

Synchronized closing could be beneficial for switching of different equipment, including overhead lines, capacitors, shunt reactors, and transformers. A good overview of the details can be found in [87–90].

Controlled switching with the intention to control the transient overvoltage and overcurrent has been implemented since more than two decades and leading manufacturers have developed appropriate control units [91,92], where energizing of capacitors, transformers, over-headlines and power cables, as well as de-energizing of shunt reactors are implemented. The number of circuit breaker installations with controlled switching has been increasing within the last two decades, reaching a level of more than 15,000 aggregated controlled switching installations in 2015, where more than 50% cover capacitor switching [86]. The controlled switching is, however, mainly related to closing operation to reduce the inrush current, as explained in Section 3.4.3.6, it would be conceivable to extend the control switching to opening operation by capacitive current switching to ensure a sufficiently large gap at the time of current zero crossing to reduce the probability of re-strikes.

Besides reduction of switching transients, a synchronized opening operation may serve minimizing the stresses applied to the circuit breaker, e.g., minimizing the arcing time, and optimally adapting the stresses applied to different parts during the switching operation. This is normally not the case in conventional circuit breakers, but as already mentioned in previous subsections, is a "must" for many hybrid circuit breakers employing power semiconductors which also used for HVDC switching.

It seems that the controlled switching targeting reduction of switching transients has reached a matured level from the technical point of view and has found acceptance within several utilities. Nevertheless, a failure in synchronization could lead to catastrophic consequences as illustrated in a case in [93].

In order to evaluate the optimum closing time in case of transformers to limit the inrush current, it is necessary to estimate the residual flux of the transformer, see e.g., [94]. As indicated in [95], estimation of the residual flux in transformers could be challenging for the controller units of controlled switching devices; detailed electromagnetic transient simulations of hysteresis and remanence may help to improve the estimations. In addition to controlled closing, controlled opening can also be used to prevent large overvoltages, e.g., because of multiple re-ignitions when de-energizing shunt reactors.

Another important aspect is related to the random nature of the breakdown (pre-strike) by decreased gap length between the contacts that causes the exact time of the circuit breaker closure to be a stochastic variable. The time of pre-strike depends on the rate of decrease of dielectric strength (RDDS) of the switching device [96] that is dependent on the insulation gas, contact roughness, etc. [97]. In order to be able to apply the controlled switching to SF_6-alternative high voltage circuit breakers, it is required to study their RDDS. The first studies on a circuit breaker with an alternative gas [98] report much higher dispersion in the dynamic breakdown compared to SF_6. This could be an interesting research topic to investigate their behavior and find appropriate measures to adapt the controlled switching algorithms to SF_6-free HVAC CBs.

A recent report [99] reveals interesting details on the experience of 11 European transmission system operators (TSO) with application, use and maintenance problems of their controlled switching devices: The controlled switching devices have been used for almost 20 years in average by almost all TSOs, where the majority (88%) are installed in EHV (230–420 kV) networks mostly for transformer energization purpose, but surprisingly the majority of TSOs do not use the residual flux calculation function for transformer energization control. In the same way, the compensation of delay time for idle time of the circuit breakers, and environmental parameters like temperature is performed by few TSOs, instead of this, many TSOs vary the delay time of the CB based on its last mechanical operation. Half of the TSOs report occasional lack of complete functionality, in seven cases resulting in power equipment damage. These results show the potential to improve the reliability of operation of controlled switching devices by better understanding the interaction between controlled switching devices and the rest of the network.

6.1.9 Condition monitoring and diagnostics

As discussed in Chapters 3 and 4, proper functionality of circuit breakers is only provided if the opening and closing operations are performed according to the design of circuit breakers and the contact separation (arc initiation) occurs in a dielectric medium with sufficient current interruption properties. There are, however, several ageing and degradation mechanisms of high-voltage arc-based switching circuit breakers that would eventually cause that the minimum requirements for the proper functionality of circuit breakers are not met. Some are linked to the mechanical wear caused by mechanical operations under no-load or by clearing short circuit current faults, and affect mostly the mechanical operating mechanism, and some are linked to the interruption chamber conditions which could also be degraded because of arcing, e.g., contact and nozzle degradation, or other non-arc dependent damages, e.g., loss of vacuum in VIs or pressure decrease of the gas (gas leakage) in gas CBs.

To ensure full functionality of CBs as one of the most critical components of power networks, it is imperative to select an appropriate maintenance strategy. The time-based (periodical) maintenance strategy is becoming obsolete, and a predictive or reliability centered maintenance (RCM) program based on condition monitoring, data analysis and spare parts inventory management, is a more effective methodology for cost effectively managing switching equipment. In [100], the components or operating conditions suggested to be accessible for condition monitoring of circuit breakers, are listed in four different categories, namely (1) interrupter: insulation medium status, temperature rise and interrupter condition, mainly contact and nozzle degradation; (2) control circuits: trip/close circuit integrity, AC/DC control voltage; (3) operating mechanism: opening-, closing-, travel- and recharge time, stored energy, total number of operations and motor current; (4) accessories: control/mechanism cabinet heater operation, SF_6 tank heater operation, cabinet door status. Some of the items mentioned here are directly accessible/measurable; these include most of the items related to the second, third and fourth categories. A very simple way would be comparing the measured values with a reference measurement of new devices, but defining the acceptable deviation from the reference measurements is not straightforward, and most importantly it is not easily possible to precisely predict the future degradation trend.

Understanding the interrelations between different measurable physical quantities and the state of the circuit breaker or its degradation mechanisms has been subject of research in a large number of studies during the last three decades, a good overview of existing methods is given in [101]. Several extensive surveys, see e.g., [102,103], of the root cause of different major and minor failures of high voltage SF_6 gas circuit breakers revealed that almost half of the failures are related to the operating mechanism and control circuitry of the circuit breakers. There is no reason to believe that this would be different for the SF_6-free circuit breakers. Many physical quantities used to detect any possible failure in operating mechanism of the circuit breakers, including vibration signals [104,105], coil current [106], travel curve [107], and spring charging motor current [108].

In addition to the operating mechanism, almost one-fourth of the major failures are linked to the interruption chamber, where the switching arc burns and is extinguished; these are mostly linked to the degradation of contacts, nozzle as well as loss of the required gas pressure or vacuum. The gas pressure in gas circuit breakers is usually monitored using a gas density gauge, but detection of the vacuum level is not directly possible. As mentioned in Section 3.5, there are several physical mechanisms which can be used to estimate the vacuum pressure level, but none of them can be considered as a matured online diagnostics method for circuit breakers, even though some recent studies [109] claim to make it possible.

Moreover, detection of any wear in interrupter is not easily possible as no direct access to the interruption chamber is available. Therefore, several researchers have studied how to relate the degradation of contacts [110–112], and nozzle [113,114] to the measurable physical quantities like the current and voltage. Measurement of dynamic resistance of circuit breakers [115,116] has also been used to estimate the state of the arcing contacts in gas circuit breaker. Moreover, other physical quantities like transient electric fields [117] have been used to detect early problems related to interrupter insulation system, e.g., partial discharges. Some of the proposed methods are only applicable when the circuit breaker is out of the network, i.e., off-line, and some can be adapted to on-line operation of circuit breakers. A good overview of existing methods is given in [101].

Besides measuring appropriate diagnostics signal (i.e., physical quantities linked to the state of CB), it is also important to interpret the measured signals correctly. The ultimate goal here would be to understand the degradation development mechanisms and to predict any failure occurrence in advance. In many different investigations, application of various digital transformative technologies and advance mathematical methods is demonstrated for the analysis of different signals. This brings, in particular for some complex signals, e.g., vibration signals, great advantages. It will be also possible to combine different diagnostic signals to enhance the quality of the decision-making process.

Leading manufacturers of circuit breakers have developed circuit breaker monitoring systems, see e.g., [118], where many of the diagnostic methods including the coil current measurement, travel curve, arc contact and nozzle degradation for specific types of interrupters are integrated. This type of condition monitoring equipment is usually part of the new trend of digitalization of high voltage equipment, e.g., switchgear, to increase their utilization in a smart grid context, which will be detailed in Section 6.3.

There are still some topics within this subject which deserve more investigation: understanding the correlation between measurable signals and contact and nozzle degradation in SF_6-free circuit breakers is important to be able to apply the degradation models developed for SF_6 CBs to the new SF_6-free CBs. In addition, working on interpretation of the measured signals in terms of degradation of CB is still an active field of research. In some recent studies, ideas related to development of digital twins for circuit breakers are presented; these are however in their early stages and there is a large potential to further develop them. Another interesting research field would be investigation of noninvasive diagnostic methods based on new sets of physical quantities. There has been some preliminary work on switching transient electric fields [117] and acoustic signals [119] for detecting of different failure types. As mentioned earlier, continued research on application of methods to estimate the vacuum level in vacuum interrupters will remain among the interesting research topics.

6.2 Switching in networks with high penetration of renewables

As mentioned in Chapter 1, in future power grids, renewable energy will play a more prominent role and a high penetration of renewable energy sources into the power networks will be expected. Distributed generation, also partly inverter interfaced, such as offshore (and onshore) wind and photovoltaic will gain in importance, and it will impact the electric power system from different perspectives [120]. Also, the network configurations around the circuit breakers used in transmission and distribution systems will substantially change. As detailed in Chapter 2, the surrounding network components of a circuit breaker are decisive for the shape and magnitude of the short circuit current flowing through the CB, as well as for the transient recovery voltage applied to the CB just after current zero crossing. With this introduction, the legitimate question here would be if a high penetration of renewable energy sources would lead to much higher stresses to circuit breakers compared with what they are designed for.

This research question has started to attract some attention from different researchers during the last few years. Many simulations, see e.g., [121], and few experimental studies, e.g., [122], indicate that some increased stresses are to be expected under certain circumstances.

6.2.1 Short circuit current

Estimation of the short circuit current in networks with renewable energy sources demands a lot of modeling effort and could be a delicate task. Many researchers have developed ideas to accurately calculate the fault currents in networks with renewables [123–127]. The magnitude of the short circuit current as well as its DC component can be significantly affected in power networks with many distributed generators. In general, adding the distributed generation to the system leads to a decrease in short circuit impedance and consequently to higher magnitudes of short circuit currents. Furthermore, depending on the type of the renewable energy source, a severe increase in the DC component of the short circuit current can be observed.

When it comes to wind turbine technologies, wind turbine generators are categorized in different types [125]; type I: squirrel case induction generator connected directly through a step-up transformer, type II: wound rotor induction generator with a variable resistor in the rotor circuit connected through a step-up transformer, type III: doubly fed induction generator with partial power electronics conversion (variable AC frequency excitation in the rotor circuit), type IV: variable speed with full power electronics conversion. In few comprehensive studies [121,125,128], short circuit current of all different types of wind turbine generators for certain network configurations have been investigated. The conclusion has been that in case of type I, II and III wind turbine generator configurations, the short circuit current has a very large DC component, which results in no current zero crossing for several hundreds of milliseconds. This means an eventual CB opening in this period would lead to a failure of the circuit breaker, and its consequent damage. The reason for the increased DC component and missing current zeros is the time-dependent short circuit reactance of the rotating machines like the phenomenon already detailed in Section 2.7.4. for generator circuit breakers.

Even though the type IV, i.e., wind turbines connected through a full power electronics converter, is the modern topology mostly used for new installations, the majority of installed wind turbines seems to be of the first three types. In an interesting report of ENTSO-E [129], distribution of different types of wind turbines among 17 European TSOs is presented. Seven TSOs declared that more than 70% of their installations are of type III, and three TSOs stated that 40–70% of their installations are of type III, whereas type IV amounted only 15–40% of installations for six TSOs, and 40–70% for two TSOs. The conclusion here would be that the type III is widely available in the European power networks, and most probably elsewhere. Combining this information with the already mentioned conclusion of [125,126], at least under some certain circumstances, stresses beyond the capabilities of CBs are expected. This calls for special measures for example a delayed opening of circuit breaker or insertion of a series resistance to accelerate the damping of the DC component [128].

Another scenario, already detailed in Section 2.7.1, is related to the missing current zeros when switching long AC cables with shunt reactor compensations, see e.g., [130]. In the same way, where large DC components of the short circuit current and a risk of CB failure is given. Long AC cables are also associated with one of the possibilities for the transmission link between the offshore wind park and the onshore grid.

There are also some studies on the short circuit current in networks with high penetration of photovoltaic (PV) energy sources. In several studies [127,131,132], the conclusion based on a large number of simulation cases has been that the fault current contribution is insignificant even for large-scale PV power plants. Due to the fast response of inverter control systems the behavior of large-scale PV applications is controllable even during the worst-case fault scenarios [133]. However, some problems related to the coordination of the power system protection are reported [134].

6.2.2 *Transient recovery voltage*

Besides short circuit current, the waveform, peak and rate of rise of the transient recovery voltage are of essential importance for successful current interruption. The estimation of the transient recovery voltage waveform is, however, much more complicated than the short circuit current as besides the lumped network components, the stray circuit elements as well as wave propagation phenomena become relevant.

There are just few studies [135–137] which focused on transient recovery voltage characteristics under different fault scenarios in presence of renewable energy sources, and all are simulation-based considering some special circuit configurations. The conclusion of some of the studies [135,137] has been that the

TRV remains in the manageable range of the CB, but in another study [136], TRV exceeds the capabilities of CB at least for some circuit configurations in case of wind generation units. In addition, in another study [138], TRV by capacitive switching in the presence of wind turbines in a network has been investigated and shown that it could lead to some restrikes of the circuit breaker. It is not possible to draw convincing conclusions based on these scarce studies. Understanding the impact of high penetrating renewables on TRV will be a potential research topic.

Another impact of high penetration of renewables would be the increased risk of switching transients (e.g., overvoltages) applied to different network components. This has been investigated in many studies and as it is not directly related to the stresses on the circuit breaker which could eventually impact the switching technology; it is, therefore, not going to be detailed here.

6.3 Digitalization: how power switching devices are impacted?

Application of digital transformative technologies is a hot and emerging topic within many disciplines for several years, including electric power grids [139]. The main idea has been to utilize the technologies, such as artificial intelligence, Internet of Things (IoT), and digital twins to realize sustainable solutions at higher reliability, availability, and lower cost [140]. The idea has also been extended to high voltage equipment within the last few years [141]. The main urgent challenges which are hoped to be addressed by digitalization of high-voltage equipment are: ageing infrastructure, retiring know how, increasing reliability demand as well as increasing power demand. These general concepts can be more concretized in the form of shorter project time, less physical changes, easy extensions, higher working safety, less traveling time, lower equipment stress, lower stress to the grid, less outages, less unproductive time, and remote inspection [142].

It is important to emphasize that the physical concepts applied to high-voltage switching technologies, discussed in the preceding chapters, will not be affected. Switchgear itself is, however, not a standalone component in the network. It needs to be controlled and linked appropriately to the power system protection and is vulnerable to degradation (ageing). Even though the failures are quite rare, but the consequences are catastrophic and therefore a very high level of reliability is necessary. Digitalization of switchgear mainly focuses on the systems around a physical switching device and aims at improving the switchgear operation at higher reliability, imposing less stress on the system and at a lower cost. The digitalization of switchgear has two different facets; the hardware around the switching device will be changed, e.g., non-conventional instrument transformers are used, and more sensors are utilized, but also the way to handle the data and make decisions is also changed (the software part). This enables some new functionalities or features which could have not been realized in analog conventional circuit breakers.

In the following, digitization movement in switchgear is considered from three different perspectives, namely, enabling hardware technologies, enabling digital technologies and new functionalities or features.

6.3.1 Enabling hardware technologies

In order to improve the operation of a switching device, it is essential to gain more information on its state (or condition). For this purpose, different measurement techniques (or sensors) are needed to be utilized which can be easily integrated to the protection, monitoring, and control system.

6.3.1.1 Non-conventional instrument transformers

Conventional current and voltage transformers are based on magnetic induction in ferromagnetic cored transformers, which are heavy, dissipate significant power, and suffer from saturation of their magnetic cores. Beside magnetic core instrument transformers, there have been several other methods studied for many decades, and now reached a level of maturity and performance that enables their implementation in high-voltage substations. The main physical mechanisms and concepts are Rogowski coil-based current transformer [143,144], and fiber optics instrument transformer both for current [145] and voltage measurement [146], as well as electric field probe (EFP) [144] for voltage measurements, which are all low power instrument transformer concepts.

Optical transformers and Rogowski coils have been around for a quite long time, but as their output signals are at very low power, they could not be directly used for activation of electromagnetic protection relays. Development of digital protection systems (relays) makes it possible to potentially replace the conventional instrument transformers with low power non-conventional ones. Relevant standard guidelines [147] and calibration methods [148] for testing of these new instrument transformers have been provided to make the integration in the power system protection possible.

The integration of non-conventional instrument transformers to commercial medium voltage switchgear products started couple of years ago, see e.g., [149], and leading high-voltage equipment manufacturers introduced their concepts and products to integrate non-conventional instrument transformers to high-voltage switchgear.

6.3.1.2 Other sensors

Measurement of voltage and current is essential for protection and electricity cost evaluation, but these do not contain sufficient information on the condition of the high-voltage switchgear. To be able to assess the functional state of switchgear, other physical quantities need to be measured. These include gas density, temperature, humidity, breaker counter, travel curve, breaker position, vibrations, as well as partial discharge or arc activities. The sensors are an integral part of the switchgear which provide the measured data through a communication platform. The measured quantities can be used for different purposes, including condition monitoring of switching devices, optimized utilization of switchgear (dynamic rating) as well as for control and protection. A short description of the possible physical parameters which can be measured and are of relevance for condition assessment of the switchgear has been covered in Section 6.1.9.

6.3.1.3   Communication networks and systems

In addition to appropriate sensors, a communication platform compatible to digital systems is also necessary. In conventional switchgear, it has been a bunch of copper wires connecting every output to every input establishing many different physical signal communication lines, but a smarter way would be to use one common physical communication platform and utilize it in such a way that different devices can send and receive their signals through the common communication platform. Appropriate standard protocols and a new communication system architecture known as bus architecture (e.g., process bus, station bus) have been introduced in an IEC standard, namely the standard IEC 61850 [150]. Even though the primary goal of introducing the standard has not been directly related to switchgear, but it can also be used as a platform for transmission of measured signals. For this purpose, some more digital hardware become necessary such as the merging units. Applications of IEC 61850 in substation automation and power system protection have been subject of significant number of papers, see for example [151], and this has been increasingly employed in new so-called digital substations.

6.3.2   Enabling digital technologies

6.3.2.1   IoT

The IoT is one of the digital transformative technologies and is generally defined as a network of connected objects and devices that are equipped with sensors (and other technologies) allowing them to transmit and receive data – to and from other things and systems. With progress of sensors – and communication technologies, it would be possible to consider the switchgear as one of the "things" in a larger network of many other equipment which are all connected together.

6.3.2.2   Digital twins

The digital twin is the real projection of all components in the product life cycle using physical data, virtual data, and data in between [152]. The digital twin includes data on the characteristics of the object in a detailed mathematical model, the parameters of which are refined using the actual data [141]. In a very simplified description, digital twin can be considered as a continuously improving dynamic model connected to the real object, e.g., switchgear, getting its state (e.g., the sensor data) and eventually giving control action commands. One of the possible applications of the digital twin in high-voltage equipment like high-voltage switchgear is for condition assessment purposes, where the degradation mechanism of switchgear may be replicated enabling a better estimation of the condition of the switchgear, see e.g., [153].

6.3.2.3   Artificial intelligence

Artificial intelligence is a general term covering many different intelligent methods such as machine learning, neural networks, support vector machine, which are used for postprocessing and feature extraction out of sensor data. One of the most important applications of artificial intelligence in switchgear would be diagnostics based on its condition, where symptoms of abnormalities can be detected in an early stage and with high accuracy [154]. In several studies, different AI-based techniques and machine learning have been used for this purpose; Table I of Ref. [101] gives a good summary of different mathematical methods applied for various diagnostic signals.

6.3.3 New functionalities and features

There are several reasons why the electric power equipment manufacturers, see e.g., [140], are moving toward digital switchgear. New functionalities and features resulting to increased safety [149], increased availability [140,155], reduced operational cost [142]. Leading power switchgear manufacturers have introduced their own ideas and products related to digitalization of switchgear, see e.g., [156,157]. In digital switchgear, features and functionalities such as digital control and protection, remote access, condition monitoring, and controlled (or point-on-wave) switching can be realized.

In Sections 6.1.8 and 6.1.9, the importance of controlled switching and condition monitoring to increase reliability, availability and reduced life cycle cost of the switchgear by reduction of failures as a consequence of reduced stresses applied to CB, and other network equipment, as well as early detection of the faults in switchgear, has been explained. These functionalities can also be integrated to a digital switchgear.

Different aspects in a digital switchgear contribute to the increased safety. Application of non-conventional instrument transformers eliminates the danger of ferroresonance and the danger of high voltage across the secondary terminal of an open CT. Self-supervision and error detection, a simplified wiring (i.e., replacement of point-to-point wire connections by bus architecture) facilitate the troubleshooting of relays and exposure of the personnel to risks during the troubleshooting. Remote access makes it possible to operate the switchgear from a safe distance eliminating the exposure of operators to arc hazards in case of a failure in the switchgear. According to the data presented by one of the manufacturers [142], the risk exposure both in air-insulated substations and gas-insulated substations can be reduced by 98%.

In digital switchgear, substantial cost reduction can be realized by using compact, lightweight, and cost-efficient non-conventional instrument transformers, bus architecture, and minimizing the risk exposure of equipment and operators, reducing the failure rates. These can be because of reduced footprint for relay and switchgear, reduced wiring costs, reduced operational and maintenance costs, shorter outages (faster times to energization) and increased productivity (reduced manufacturing and testing times). In an example [149], the manufacturing time has been reduced by 82% in the case of a digital switchgear.

6.4 Concluding remarks

The present book is a narrative on a piece of a long and continuing story on evolution of high-voltage switching devices as a response to various drivers that have been changed over time in a context of given constraints and boundary conditions. In the early days, the need for an equipment with sufficiently high-current interruption capability (i.e., increasing rated voltages and increasing short circuit currents) has been the main driver for the development of switching equipment. This was then later amended by the requirement to reduce the cost, which was meant at first only manufacturing costs, and at a later stage included the maintenance and operational costs.

With the increased awareness on environmental aspects, reducing the environmental footprint of the power system has become the utmost driver of the recent development of the switching technology. This has impacted the technologies in two different ways: changing the boundary conditions for example by limitation of the range of eligible materials, most prominently the limitations on SF_6, or imposing new generation, transmission, and distribution trends resulting in new specifications and stresses for the switching devices, e.g., high penetration of renewables, and emerging applications of multi-terminal HVDC networks.

The technologies presented in this book are intended to address the recent developments in response to the drivers towards more environmentally friendly switching technologies, where the Chapters 3 and 4 presented new solutions in the light of limitation on materials, and Chapter 5 had a focus on solutions for HVDC switching in modern power networks. As explained in the preceding sections of this chapter, there are however several potential areas for further improvement of switching technologies, which call for continued research in this field.

References

[1]   C. M. Franck, "HVDC circuit breakers: a review identifying future research needs," *IEEE Transactions on Power Delivery*, vol. 26, no. 2, pp. 998–1007, 2011.

[2]   F. Mohammadi, K. Rouzbehi, M. Hajian, *et al.*, "HVDC circuit breakers: a comprehensive review," *IEEE Transactions on Power Electronics*, vol. 36, no. 12, pp. 13726–13739, 2021.

[3]   P. Skarby and U. Steiger, "An ultra-fast disconnecting switch for a hybrid HVDC breaker – a technical breakthrough," in *Cigre International Symposium*, Alberta, Canada, 2013.

[4]   D. Jovcic, G. Chaffey, W. Leterme, *et al.*, "HVDC grid rollout in the European context: current status and challenges of HVDC grid protection," *Cigre Science and Engineering*, no. 20, pp. 106–124, 2021.

[5]   N. A. Belda and R. P. P. Smeets, "Test circuits for HVDC circuit breakers," *IEEE Transactions on Power Delivery*, vol. 32, no. 1, pp. 285–293, 2017.

[6]   R. P. P. Smeets and N. A. Belda, "High-voltage direct current fault current interruption: a technology review," *High Voltage*, no. 6, pp. 171–192, 2021.

[7]   GB/T 38328-2019: Common Specifications of High-Voltage Direct Current Circuit-Breakers for High-Voltage Direct Current Transmission Using

Voltage Sourced Converters (VSC-HVDC), The Standardization Administration of the People's Republic of China, 2019.

[8] J. Cao, P. Tuennerhoff, C. Gao, *et al.*, Technical Brochure 873: Design, Test and Application of HVDC Circuit Breakers, 2022.

[9] IEC TC 17 WG6, IEC TS 62271-5 ED1: High-voltage Switchgear and Controlgear – Part 5: Common Specifications for Direct Current Switchgear, 2023.

[10] S. Paul, *The Vacuum Interrupter: Theory, Design, and Application*, 2nd ed., Boca Raton, FL: CRC Press, 2021.

[11] R. Renz, "On criteria of optimized application of AMF and RMF contact systems in vacuum interrupters," in *19th ISDEIV International Symposium on Discharges and Electrical Insulation in Vacuum*, 2000.

[12] H. Li, Z. Wang, Y. Geng, Z. Liu, and J. Wang, "Arcing contact gap of a 126-kV horseshoe-type bipolar axial magnetic field vacuum interrupters," *IEEE Transactions on Plasma Science*, vol. 46, no. 10, pp. 3713–3721, 2018.

[13] R. Smeets, A. Derviškadić, A. B. Hofstee, R. M. Nijman, N. A. Belda and B. Baum, "The Impact of the Application of Vacuum Switchgear at Transmission Voltages, Cigre Technical Brochure 589," Cigre Working Group A3.27, 2014.

[14] H. Komatsu, K. Saito, and T. Furuhata, "Progress of Vacuum Interrupter (VI) and recent technical trends," *Meiden Review*, vol. 163, no. 1, p. 8, 2015.

[15] L. Min-fu, D. Xiong-ying, Z. Ji-yan, F. Xing-ming, and S. Hui, "Dielectric strength and statistical property of single and triple-break vacuum interrupters in series," *IEEE Transactions on Dielectrics and Electrical Insulation*, vol. 14, no. 3, pp. 600–605, 2007.

[16] H. Schellekens and G. Gaudart, "Compact high-voltage vacuum circuit breaker, a feasibility study," *IEEE Transactions on Dielectrics and Electrical Insulation*, vol. 14, no. 3, pp. 613–619, 2007.

[17] Z. Xiang, X. Zhang, R. Huang, Z. Wu, M. Liao, and X. Zhang, "Research on voltage distribution of double-break vacuum circuit breakers with different interrupters in series," in *28th International Symposium on Discharges and Electrical Insulation in Vacuum (ISDEIV)*, Greifswald, Germany, 2018.

[18] L. Baron, F. Graskowski, A. Lawall, and C. Stiehler, "Overpressure-Resistant Vacuum Interrupter Tube". US: Patent 11,289,292, 29 March 2022.

[19] L. Baron, F. Graskowski, U. Jahnke, C. Stiehler, A. Lawall, and K. Schachtschneider, "Vacuum Interrupter". U.S. Patent 2021/0074494A1, 11 March 2021.

[20] T. Rokunohe, Y. Yagihashi, K. Aoyagi, T. Oomori, and F. Endo, "Development of SF 6-free 72.5 kV GIS," *IEEE Transactions on Power Delivery*, vol. 22, no. 3, pp. 1869–1876, 2007.

[21] Joint Working Group A3.32/CIRED, "Non-intrusive Methods for Condition Assessment of Distribution and Transmission Switchgear, Cigre Technical Report 737," CIGRE, 2018.

[22] P. G. Slade, "The application of vacuum interrupters in HVDC circuit breakers," *IEEE Transactions on Plasma Science*, vol. 50, no. 11, pp. 4675–4682, 2022.

[23] "Regulation of the European Parliament and of the Council: On Fluorinated Greenhouse Gases, Amending Directive (EU) 2019/1937 and Repealing Regulation (EU) No. 517/2014, 05 04 2022. https://eur-lex.europa.eu/legal-content/EN/TXT/?uri=CELEX%3A52022PC0150 [Accessed 8 April 2023].

[24] "Manifesto for an Urgent Ban of 'forever chemicals' PFAS," 10 2022. https://banpfasmanifesto.org/en/ [Accessed 8 April 2023].

[25] European Chemical Agency (ECHA), "ECHA Publishes PFAS Restriction Proposal: ECHA/NR/23/04," 07 February 2023. https://echa.europa.eu/-/echa-publishes-pfas-restriction-proposal [Accessed 8 April 2023].

[26] H. Koch (ed.) *Gas Insulated Substations*, 2nd ed., Wiley-IEEE Press, 2022.

[27] M. Kosse, M. Tuczek, C. Klein, and M. Claus, "Experiences with on-site dielectric testing during commissioning tests of world's first offshore HVDC GIS ±320 kV," in *VDE Hochspannungstechnik*, Berlin, 2022.

[28] M. Kosse, L. Dejun, K. Juhre, and M. Kuschel, "Overview of development, design, testing and application of compact gas-insulated DC systems up to ±550 kV," *Global Energy Interconnection*, vol. 2, no. 6, pp. 567–577, 2019.

[29] IEC SC 17C, "IEC TS 62271-318 ED1: High-Voltage Switchgear and Control Gear – Part 318-DC Gas-Insulated Switchgear Assemblies," International Electrotechnical Commission (IEC), 2023.

[30] Z. A. Siahaan, A. Wijaya, A. J. Chandra, K. Banjar-Nahor, and N. Hariyanto, "Fault current limitation roadmap to anticipate the problem of high fault currents in Indonesian Java-Bali Power System," in *2022 IEEE International Conference on Power Systems Technology (POWERCON)*, Kuala Lumpur, Malaysia, 2022.

[31] M. R. Barzegar-Bafrooei, J. Dehghani-Ashkezari, A. Akbari Foroud, and H. Haes Alhelou, *Fault Current Limiters Concepts and Applications*, Singapore: Springer, 2022.

[32] S. B. Naderi, M. Negnevistky, A. Jalilian, and M. T. Hagh, "Non-controlled fault current limiter to improve fault ride through capability of DFIG-based wind turbine," in *2016 IEEE Power and Energy Society General Meeting (PESGM)*, Boston, MA, 2016.

[33] "Fault Current Limiter – FC Protector," ABB, https://new.abb.com/medium-voltage/apparatus/fault-current-limiters/fc-protector [Accessed 04 April 2023].

[34] R. Strumpler, J. Skindhoj, J. Glatz-Reichenbach, J. H. W. Kuhlefelt, and F. Perdoncin, "Novel medium voltage fault current limiter based on polymer PTC resistors," *IEEE Transactions on Power Delivery*, vol. 14, no. 2, pp. 425–430, 1999.

[35] C. Niu, B. Wang, H. He, *et al.*, "A novel liquid metal fault current limiter based on active trigger method," *IEEE Transactions on Power Delivery*, vol. 36, no. 6, pp. 3619–3628, 2021.

[36]   K. Niayesh, J. Tepper, and F. Konig, "A novel current limitation principle based on application of liquid metals," *IEEE Transactions on Components and Packaging Technologies*, vol. 29, no. 2, pp. 303–309, 2006.

[37]   Z. Zhang, J. Yuan, Y. Hong, H. Chen, C. Zou, and H. Zhou, "Hybrid multifunctional saturated-core fault current limiter," *IEEE Transactions on Power Delivery*, vol. 37, no. 6, pp. 4690–4699, 2022.

[38]   W. Song, X. Pei, H. Alafnan, *et al.*, "Experimental and simulation study of resistive helical HTS fault current limiters: quench and recovery characteristics," *IEEE Transactions on Applied Superconductivity*, vol. 31, no. 5, pp. 1–6, 2021.

[39]   M. Steurer, K. Frohlich, W. Holaus, and K. Kaltenegger, "A novel hybrid current-limiting circuit breaker for medium voltage: principle and test results," *IEEE Transactions on Power Delivery*, vol. 18, no. 2, pp. 460–467, 2003.

[40]   F. Moriconi, F. De La Rosa, F. Darmann, A. Nelson and M. L., "Development and deployment of saturated-core fault current limiters in distribution and transmission substations," *IEEE Transactions on Applied Superconductivity, vol. 21, no. 3, pp. 1288–1293, 2011.

[41]   A. Heidary, H. Radmanesh, K. Rouzbehi, A. Mehrızı-Sani, and G. Gharehpetian, "Inductive fault current limiters: a review," *Electric Power Systems Research*, vol. 20, p. 106499, 2020.

[42]   W. Paul and W. Chen, "Superconducting control for surge currents," *IEEE Spectrum*, vol. 35, no. 5, pp. 49–54, 1998.

[43]   J. Bock, S. Elschner, F. Breuer, *et al.*, "Field demonstration of worldwide largest superconducting fault current limiter and novel concepts," in *CIRED 2005 – 18th International Conference and Exhibition on Electricity Distribution*, Turin, Italy, 2005.

[44]   H. P. Kraemer, A. Bauer, and M. Frank, *et al.*, "ASSiST – a superconducting fault current limiter in a public electric power grid," *IEEE Transactions on Power Delivery*, vol. 37, no. 1, pp. 612–618, 2022.

[45]   P. Tixador, A. Akbar, M. Bauer, *et al.*, "Some results of the EU Project FASTGRID," *IEEE Transactions on Applied Superconductivity*, vol. 32, no. 4, pp. 1–6, 2022.

[46]   M. Moyzykh, D. Gorbunova, P. Ustyuzhanin, *et al.*, "First Russian 220 kV superconducting fault current limiter (SFCL) for application in city grid," *IEEE Transactions on Applied Superconductivity*, vol. 31, no. 5, pp. 1–7, 2021.

[47]   R. Adapa and D. Piccone, *Solid-State Fault Current Limiter Development: Design and Testing Update of a 15 kV SSCL Power Stack*, Electric Power Research Institute and Silicon Power Corporation, 2012.

[48]   J. Xu, L. Gao and H. Zhang, "Design of self-powered solid-state fault current limiters for VSC DC grids," *Frontiers in Energy Research*, vol. 9, p. 760105, 2021.

[49]   A. Küchler, *Hochspannungstechnik: Grundlagen – Technologie -Anwendungen*, Berlin, Heidelberg: Springer, 2009.

[50] M. Seeger, P. Stoller, and A. Garyfallos, "Breakdown fields in synthetic air, CO_2, a CO_2/O_2 mixture, and CF_4 in the pressure range 0.5–10 MPa," *IEEE Transactions on Dielectrics and Electrical Insulation*, vol. 24, no. 3, pp. 1582–1591, 2017.

[51] T. Kiyan, T. Ihara, S. Kameda, T. Furusato, M. Hara and H. Akiyama, "Weibull statistical analysis of pulsed breakdown voltages in high-pressure carbon dioxide including supercritical phase," *IEEE Transactions on Plasma Science*, vol. 39, no. 8, pp. 1729–1735, 2011.

[52] J. Wei, A. Cruz, C. Xu, F. Haque, C. Park and L. Graber, "A review on dielectric properties of supercritical fluids," in *2020 IEEE Electrical Insulation Conference (EIC)*, Knoxville, TN, 2020.

[53] D. Young, "Electric breakdown in CO_2 from low pressures to the liquid state," *Journal of Applied Physics*, vol. 21, no. 222, pp. 222–231, 1950.

[54] F. Haque, J. Wei, L. Graber, *et al.*, "Modeling the dielectric strength variation of supercritical fluids driven by cluster formation near critical point," *Physics of Fluids*, vol. 32, p. 077101, 2020.

[55] T. Ito, H. Fujiwara and K. Terashima, "Decrease of breakdown voltages for micrometer-scale gap electrodes for carbon dioxide near the critical point: temperature and pressure dependences," *Journal of Applied Physics*, vol. 94, no. 8, pp. 5411–5413, 2003.

[56] T. Ito and K. Terashima, "Generation of micrometer-scale discharge in a supercritical fluid environment," *Applied Physics Letters*, vol. 80, no. 16, pp. 2854–2856, 2002.

[57] Y. Tian, J. Wei, C. Park, Z. Wang, and L. Graber, "Modelling of electrical breakdown in supercritical CO_2 with molecular clusters formation," in *12th IEEE International Conference on the Properties and Applications of Dielectric Materials, Xi'an, China*, 2018.

[58] J. Zhang, E. J. M. van Heesch, F. J. C. M. Beckers, *et al.*, "Breakdown strength and dielectric recovery in a high pressure supercritical nitrogen switch," *IEEE Transactions Dielectrics and Electrical Insulation*, vol. 22, no. 4, pp. 1823–1832, 2015.

[59] J. Zhang, B. van Heesch, F. Beckers, T. Huiskamp, and G. Pemen, "Breakdown voltage and recovery rate estimation of a supercritical nitrogen plasma switch," *IEEE Transactions on Plasma Science*, vol. 42, no. 2, pp. 376–383, 2014.

[60] F. Abid, *Characteristics of Switching Arc in Ultrahigh-pressure Nitrogen*, PhD Thesis, Norwegian University of Science and Technology (NTNU), 2020.

[61] F. Abid, K. Niayesh, E. Jonsson, N. S. Støa-Aanensen, and M. Runde, "Arc voltage characteristics in ultrahigh-pressure nitrogen including supercritical region," *IEEE Transactions on Plasma Science*, vol. 46, no. 1, pp. 187–193, 2018.

[62] F. Abid, K. Niayesh, and N. S. Støa-Aanensen, "Ultrahigh-pressure nitrogen arcs burning inside cylindrical tubes," *IEEE Transactions on Plasma Science*, vol. 47, no. 1, pp. 754–761, 2019.

[63] F. Abid, K. Niayesh, E. Viken, N. S. Støa-Aanensen, E. Jonsson, and H. K. Meyer, "Effect of filling pressure on post-arc gap recovery of N2," *IEEE Transactions on Dielectrics and Electrical Insulation*, vol. 27, no. 4, pp. 1339–1347, 2020.

[64] T. Damle, C. Xu, J. Wei, *et al.*, "EDISON: a new generation DC circuit breaker," in *CIGRE Session*, Paris, France, 2020.

[65] Y. He, Q. Yang, Y. Li, and L. Graber, "Control development and fault current commutation test for the EDISON hybrid circuit breaker," *IEEE Transactions on Power Electronics*, vol. 99, pp. 1–14, 2023.

[66] S. Stauss, H. Muneoka, and K. Terashima, "Review on plasmas in extraordinary media: plasmas in cryogenic conditions and plasmas in supercritical fluids," *Plasma Sources Science and Technology*, vol. 27, no. 2, p. 023003, 2018.

[67] M. Callavik, A. Blomberg, J. Häfner, and B. Jacobson, "The hybrid HVDC breaker: an innovation breakthrough enabling reliable HVDC grids," *ABB Grid Systems*, 2012.

[68] G. Demetriades and A. Shukla, "Hybrid circuit breaker". U.S. Patent US 20120218676A1, 30 08 2012.

[69] J. Stechbarth, K. Kaltenegger, W. Hofbauer, L. Niemeyer, M. Claessens, and K. D. Weltmann, "Hybridleistungsschalter". Patent DE19958645A1, 06 12 1999.

[70] D. Dufournet and C. Lindner, "Hybrid chamber with vacuum and gas interrupters for high-voltage circuit-breakers," in *Proceedings of Cigre Session*, 2004.

[71] R. P. P. Smeets, V. Kertesz, D. Dufournet, D. Penache, and M. Schlaug, "Interaction of a vacuum arc with an SF6 arc in a hybrid circuit breaker during high-current interruption," *IEEE Transactions on Plasma Science*, vol. 35, no. 4, pp. 933–938, 2007.

[72] X. Cheng, Z. Chen, G. Ge, Y. Wang, M. Liao, and L. Jiao, "Dynamic dielectric recovery synergy of hybrid circuit breaker with CO2 gas and vacuum interrupters in series," *IEEE Transactions on Plasma Science*, vol. 45, no. 10, pp. 2885–2892, 2017.

[73] N. Götte, T. Krampert, and P. G. Nikolic, "Series connection of gas and vacuum circuit breakers as a hybrid circuit breaker in high-voltage applications," *IEEE Transactions on Plasma Science*, vol. 48, no. 7, pp. 2577–2584, 2020.

[74] A. Antoniazzi, T. Masper, P. Cairoli, and T. Strassel, "One of a kind: SACE infinitus for the future of electrical distribution," *ABB Review*, no. 4, pp. 14–19, 2022.

[75] R. Rodrigues, Y. Du, A. Antoniazzi, and P. Cairoli, "A review of solid-state circuit breakers," *IEEE Transactions on Power Electronics*, vol. 36, no. 1, pp. 364–377, 2021.

[76] W. A. Martin, C. Deng, D. Fiddiansyah, and J. C. Balda, "Investigation of low-voltage solid-state DC breaker configurations for DC microgrid applications," in *IEEE International Telecommunications Energy Conference (INTELEC)*, Austin, TX, USA, 2016.

[77]   R. Rodrigues, Y. Zhang, U. Raheja, P. Cairoli, L. Raciti, and A. Antoniazzi, "Robust 5 kA, 1 kV solid-state DC circuit breaker for next generation marine power systems," in *2021 IEEE Electric Ship Technologies Symposium (ESTS)*, Arlington, VA, USA, 2021.

[78]   A. Barzkar and M. Ghassemi, "Components of electrical power systems in more and all-electric aircraft: a review," *IEEE Transactions on Transportation Electrification*, vol. 8, no. 4, pp. 4037–4053, 2022.

[79]   X. She, A. Q. Huang, Ó. Lucía, and B. Ozpineci, "Review of silicon carbide power devices and their applications," *IEEE Transactions on Industrial Electronics*, vol. 64, no. 10, pp. 8193–8205, 2017.

[80]   N. Keshmiri, D. Wang, B. Agrawal, R. Hou, and A. Emadi, "Current status and future trends of GaN HEMTs in electrified transportation," *IEEE Access*, vol. 8, pp. 70553–70571, 2020.

[81]   L. Zhang, R. Woodley, X. Song, S. Sen, X. Zhao, and A. Q. Huang, "High current medium voltage solid state circuit breaker using paralleled 15kV SiC ETO," in *2018 IEEE Applied Power Electronics Conference and Exposition (APEC)*, San Antonio, TX, USA, 2018.

[82]   U. Mehrotra, B. Ballard, and D. C. Hopkins, "High current medium voltage bidirectional solid state circuit breaker using SiC JFET super cascode," in *2020 IEEE Energy Conversion Congress and Exposition (ECCE)*, Detroit, MI, USA, 2020.

[83]   S. Wang, Z. Song, F. Peng, *et al.*, "Thermal analysis of water-cooled heat sink for solid-state circuit breaker based on IGCTs in parallel," *IEEE Transactions on Components, Packaging and Manufacturing Technology*, vol. 9, no. 3, pp. 483–488, 2019.

[84]   IEC, *IEC Standard 62227-100: High-Voltage Switchgear and Controlgear – Part 100: Alternating-CURRENT CIRCUIT-BREAKERS*, IEC (International Electrotechnical Commission), 2021.

[85]   IEC, "IEC 62271-113 ED1: High-Voltage Switchgear and Controlgear – Alternating Current Circuit-Breakers Intended for Controlled Switching," 2019.

[86]   CIGRE WG A3.35, "Technical Brochure 757: Guidelines and Best Practices for the Commissioning and Operation of Controlled Switching Projects," CIGRE, 2019.

[87]   CIGRE WG 13.07, "Controlled switching of HVAC circuit breakers. Guide for application lines, reactors, capacitors, transformers. 1st part," *Electra*, vol. 183, p. 43–73, 1999.

[88]   C. W. 13.07, "Controlled switching of HVAC circuit breakers. Guide for application lines – reactors–capacitors–transformers. 2nd Part," *Electra*, vol. 185, p. 37–57, 1999.

[89]   D. Goldsworthy, T. Roseburg, D. Tziouvaras, and J. Pope, "Controlled switching of HVAC circuit breakers: application examples and benefits," in *61st Annual Conference for Protective Relay Engineers*, College Station, TX, USA, 2008.

[90]   CIGRE WG A3.07, "Technical Brochure 263: Controlled Switching of HVAC CBs – Guidance for Further Applications Including Unloaded Transformer Switching, Load and Fault Interruption and Circuit-Breaker Uprating," CIGRE, 2004.

[91]   Siemens, "SIPROTEC 5 Point-on-Wave Switching, V9.50 and Higher, Manual," SIEMENS, 2023.

[92]   Hitachi Energy, "PWC600 – Point-on-Wave Controller Switchsync^{TM}," https://www.hitachienergy.com/products-and-solutions/substation-automa-tion-protection-and-control/products/protection-and-control/breaker-protec-tion/pwc600. [Accessed 14 04 2023].

[93]   M. Nagpal, Z. Jiao, S. Manuel, T. Martinich, S. Merriman, and D. Sydor, "Sub-synchronous ferroresonance causes catastrophic failure of a line shunt reactor – a post mortem investigation," *IEEE Transactions on Power Delivery*, vol. 37, no. 1, pp. 374–382, 2022.

[94]   J. H. Brunke and K. J. K. J. Frohlich, "Elimination of transformer inrush currents by controlled switching. I. Theoretical considerations," *IEEE Transactions on Power Delivery*, vol. 16, no. 2, pp. 276–280, 2001.

[95]   W. Chandrasena, D. Jacobson and P. Wang, "Controlled switching of a 1200 MVA transformer in Manitoba," *IEEE Transactions on Power Delivery*, vol. 31, no. 5, pp. 2390–2400, 2016.

[96]   U. Parikh and B. R. Bhalja, "Challenges in field implementation of controlled energization for various equipment loads with circuit breakers considering diversified dielectric and mechanical characteristics," *International Journal of Electrical Power & Energy Systems*, vol. 87, pp. 99–108, 2017.

[97]   H. Hamada, A. Eto, T. Maekawa, *et al.*, "RDDS (rate of decrease of dielectric strength) measurement for gas circuit breaker," in *IEEE/PES Transmission and Distribution Conference and Exhibition*, Yokohama, Japan, 2002.

[98]   X. Lin, J. Zhang, J. Xu, J. Zhong, Y. Song, and Y. Zhang, "Dynamic dielectric strength of C3F7CN/CO2 and C3F7CN/N2 gas mixtures in high voltage circuit breakers," *IEEE Transactions on Power Delivery*, vol. 37, no. 5, pp. 4032–4041, 2022.

[99]   ENTSO-E, "Controlled Switching Devices: Applications, Use and Maintenance Problems," European Network of Transmission System Operators for Electricity, 16 112022. https://eepublicdownloads.entsoe.eu/clean-documents/SOC%20documents/CONTROLLED_SWITCHING_DEVICE_APPLICATION__USE_AND_MAINTENANCE_PROBLEMS.pdf. [Accessed 14 04 2023].

[100]   "IEEE Std C37.10.1^{TM}-2018: IEEE Guide for the Selection of Monitoring for Circuit Breakers," IEEE Power and Energy Society, 2018.

[101]   A. Razi Kazami and K. Niayesh, "Condition monitoring of high voltage circuit breakers: past to future," *IEEE Transaction on Power Delivery*, vol. 36, no. 2, pp. 740–750, 2021.

[102]   CIGRE Working Group A3.06, "Final Report of 2004–2007 International Enquiry on Reliability of High Voltage Equipment, Part 2—Reliability of High Voltage Circuit Breakers, Tech. Brochure 510," CIGRE, October 2012.

[103]   CIGRE Working Group 13.06, "Final Report of the Second International Enquiry on High Voltage Circuit-Breaker Failures and Defects in Service, Tech. Brochure 83," CIGRE, June 1994.

[104]   Q. Yang, J. Ruan, Z. Zhuang, D. Huang, and Z. Qiu, "A new vibration analysis approach for detecting mechanical anomalies on power circuit breakers," *IEEE Access*, vol. 7, pp. 14070–14080, 2019.

[105]   H. K. Hoidalen and M. Runde, "Continuous monitoring of circuit breakers using vibration analysis," *IEEE Transactions on Power Delivery*, vol. 20, no. 4, pp. 2458–2465, 2005.

[106]   A. Razi-Kazemi, M. Vakilian, K. Niayesh, and M. Lehtonen, "Circuit-breaker automated failure tracking based on coil current signature," *IEEE Transactions on Power Delivery*, vol. 29, no. 1, pp. 283–290, 2014.

[107]   F. N. Rudsari, A. A. Razi-Kazemi, and M. A. Shoorehdeli, "Fault analysis of high-voltage circuit breakers based on coil current and contact travel waveforms through modified SVM classifier," *IEEE Transactions on Power Delivery*, vol. 34, no. 4, p. 1608–1618, 2019.

[108]   A. Pöltl and M. Lane, "Field experiences with HV circuit breaker condition monitoring," in ABB, June 2011.

[109]   F. Zhang, H. Yuan, A. Yang, X. Wang, D. Liu and M. Rong, "On-line vacuum degree monitoring of vacuum circuit breaker based on laser-induced breakdown spectroscopy combined with random forest algorithm," in *EEE International Conference on High Voltage Engineering and Applications (ICHVE)*, Chongqing, China, 2022.

[110]   M. Mohammadhosein, K. Niayesh, A. A. Shayegani-Akmal, and H. Mohseni, "Online assessment of contact erosion in high voltage gas circuit breakers based on different physical quantities," *IEEE Transactions on Power Delivery*, vol. 34, no. 2, pp. 580–587, 2019.

[111]   A. Bagherpoor, S. Rahimi-Pordanjani, A. A. Razi-Kazemi, and K. Niayesh, "Online condition assessment of interruption chamber of gas circuit breakers using arc voltage measurement," *IEEE Transactions on Power Delivery*, vol. 32, no. 4, pp. 1776–1783, 2017.

[112]   J. Tepper, M. Seeger, T. Votteler, V. Behrens, and T. Honig, "Investigation on erosion of Cu/W contacts in high-voltage circuit breakers," *IEEE Transactions on Components and Packaging Technologies*, vol. 29, no. 3, pp. 658–665, 2006.

[113]   X. Yang, C. Zhang, W. Li, Z. Huang, J. Chu, and Y. Li, "Condition assessment on arc nozzle of high voltage circuit breaker through SF6 decomposition products," in *IEEE 8th International Conference on Advanced Power System Automation and Protection (APAP)*, Xi'an, China, 2019.

[114]   T. Kampert, S. Wetzeler, P. G. Nikolic, and A. Schnettler, "Minimum-intrusive diagnostic system for SF6 high voltage selfblast circuit breaker

nozzles," in *IEEE International Power Modulator and High Voltage Conference (IPMHVC)*, San Francisco, CA, USA, 2016.

[115] M. Landry, O. Turcotte, and F. Brikci, "A complete strategy for conducting dynamic contact resistance measurements on HV circuit breakers," *IEEE Transactions on Power Delivery*, vol. 23, no. 2, pp. 710–716, 2008.

[116] M. Mohammadhosein, K. Niayesh, A. A. Shayegan-Akmal, and H. Mohseni, "Sensitivity of dynamic resistance of gas circuit breakers to the arc-induced contact erosion," in *5th International Conference on Electric Power Equipment–Switching Technology (ICEPE-ST)*, Kitakyushu, Japan, 2019.

[117] H.-Y. Zhang, Y.-Z. Xie, T.-Q. Yi, X. Kong, L. Cheng, and H.-J. Liu, "Fault detection for high-voltage circuit breakers based on time–frequency analysis of switching transient E-fields," *IEEE Transactions on Instrumentation and Measurement*, vol. 69, no. 4, pp. 1620–1631, 2020.

[118] Hitachi Energy, "Circuit Breaker Sentinel^{TM} (CBS)," Hitachi Energy, https://www.hitachienergy.com/products-and-solutions/high-voltage-switch-gear-and-breakers/monitoring-and-controlled-switching/circuit-breaker-sen-tinel [Accessed 20 04 2023].

[119] T. Iwata, T. Endo, J. Nukaga, Y. Takahashi, and T. Nishimura, "Development of acoustic diagnostics for opening and closing operations of gas circuit breakers,," in *6th International Conference on Electric Power Equipment – Switching Technology (ICEPE-ST)*, Seoul, Korea, 2022.

[120] R. Ufa, Y. Malkova, V. Rudnik, M. Andreev, and V. Borisov, "A review on distributed generation impacts on electric power system," *International Journal of Hydrogen Energy*, vol. 47, pp. 20347–20361, 2022.

[121] E. Muljadi, N. Samaan, V. Gevorgian, J. Li, and S. Pasupulati, "Short circuit current contribution for different wind turbine generator types," in *IEEE PES General Meeting*, Minneapolis, MN, USA, 2010.

[122] N. Leonov, G. C. Cho, and A. Poluektov, "Short-circuit currents from wind turbine generators study in laboratory workshop," in *International Conference on Information Technologies in Engineering Education (Inforino)*, Moscow, Russia, 2020.

[123] R. Aljarrah, M. Al-omary, D. Alshabi, *et al.*, "Application of artificial neural network-based tool for short circuit currents estimation in power systems with high penetration of power electronics-based renewables," *IEEE Access*, vol. 11, pp. 20051–20062, 2023.

[124] J. Li, T. Zheng, and Z. Wang, "Short-circuit current calculation and harmonic characteristic analysis for a doubly-fed induction generator wind turbine under converter control," *Energies*, vol. 11, no. 9, p. 2471, 2018.

[125] D. F. Howard, "Short-circuit currents in wind-turbine generator networks," Ph.D. Thesis, Georgia Institute of Technology, December 2013.

[126] R. Aljarrah, "Assessment of fault level in power systems with high penetration of non-synchronous generation," Ph.D. Thesis, The University of Manchester, 2020.

[127] N. Mourad and B. Mohamed, "Short circuit current contribution of distributed photovoltaic integration on radial distribution networks, Boumerdes," in *4th International Conference on Electrical Engineering (ICEE)*, Boumerdes, Algeria, 2015.

[128] R. Walling, R. Harley, D. Miller, *et al.*, "Fault current contributions from wind plants," in *68th Annual Conference for Protective Relay Engineers*, College Station, TX, USA, 2015.

[129] ENTSO-E, "Short Circuit Contribution of New Generating Units Connected with Power Electronics and Protection Behaviour," European Network of Transmission System Operators for Electricity, April 2019.

[130] F. Ait-Abdelmalek, "How controlled switching technique supports integration of offshore wind to the grid," in *CIGRE A3 Gen. Disc. Meeting, CIGRE Conference*, 2022.

[131] B. Wang, R. Burgos, Y. Tang, and B. Wen, "Fault characteristics analysis on 56-bus distribution system with penetration of utility-scale PV generation," in *2021 6th IEEE Workshop on the Electronic Grid (eGRID)*, New Orleans, LA, USA, 2021, pp. 01–08, in *6th IEEE Workshop on the Electronic Grid (eGRID)*, New Orleans, LA, USA, 2021.

[132] S. Mihaes and M. Istrate, "Influence of photovoltaic power plants on single-phase faults in medium voltage electrical networks," in *2018 International Conference and Exposition on Electrical And Power Engineering (EPE)*, Iasi, Romania, 2018.

[133] T. Neumann and I. Erlich, "Short circuit current contribution of a photovoltaic power plant," *IFAC Proceedings Volumes*, vol. 45, no. 21, pp. 343–348, 2012.

[134] H. A. Álvarez-Macías, R. Peña-Gallardo, C. Soubervielle-Montalvo, E. D. De León-Mendoza, and J. A. Pecina-Sánchez, "Impact of high penetration of photovoltaic distributed generation on the protection coordination in distribution networks," in *IEEE International Autumn Meeting on Power, Electronics and Computing (ROPEC)*, Ixtapa, Mexico, 2022.

[135] Z. Zhou, X. Wang, and P. Wilson, "Transient recovery voltage assessment for 138 kV breakers with the new addition of a wind farm," in *International Conference on Power System Technology*, Chongqing, China, 2006.

[136] S. K. Wetjen, "Investigation of the transient recovery voltage across circuit breakers in networks with distributed energy resources," Master Thesis, Norwegian University of Science and Technology, 2019.

[137] M. Heydari and A. A. Razi-Kazemi, "Impacts of various wind turbine generators on transient recovery voltage in a medium voltage power network," in *30th International Conference on Electrical Engineering (ICEE)*, Tehran, Iran, 2022.

[138] B. Badrzadeh, "Transient recovery voltages caused by capacitor switching in wind power plants," in *IEEE Industry Applications Society Annual Meeting*, Las Vegas, NV, USA, 2012.

[139] M. Mhamud Hussen Sifat, S. M. Choudhury, S. K. Das, *et al.*, "Towards electric digital twin grid: Technology and framework review," *Energy and AI*, vol. 11, p. 100213, 2023.

[140] D. Helbig, P. Singh, and E. Gomez, "Transmission products and systems for utilities of the future – IoT connected, digital twin based, intelligent," in *CIGRE Canada Conference*, Toronto, Ontario, Canada, October 2020.

[141] A. I. Khalyasmaa, A. I. Stepanova, S. Eroshenko, and P. V. Matrenin, "Review of the digital twin technology applications for electrical equipment lifecycle management," *Mathematics*, vol. 11, no. 6, 2023, p. 1315, https://doi.org/10.3390/math11061315.

[142] Hitachi Energy, "Digitalization of high-voltage switchgear webinar," Hitachi Energy, https://go.hitachienergy.com/Digital-SWG-webinar-2022? utm_source=website&utm_medium=referral. [Accessed 21 04 2023].

[143] R. Thomas, A. Vujanic, D. Z. Xu, *et al.*, "Non-conventional instrument transformers enabling digital substations for future grid," in *IEEE/PES Transmission and Distribution Conference and Exposition (T&D)*, Dallas, TX, USA, 2016.

[144] W. Olszewski and M. Kuschel, "New smart approach for a U/I-measuring system integrated in a GIS cast resin partition (NCIT) – design, manufacturing, qualification and operational experience," in *International ETG Congress*, Bonn, Germany, 2017.

[145] K. Bohnert, A. Frank, L. Yang, *et al.*, "Fiber-optic current sensor in 420 kV circuit breaker," in *Conference on Lasers and Electro-Optics (CLEO)*, San Jose, CA, USA, 2016.

[146] S. Wildermuth, K. Bohnert, O. Steiger, *et al.*, "Electro-optic high voltage sensor for utility application," in *Conference on Lasers & Electro-Optics Europe & International Quantum Electronics Conference CLEO EUROPE/ IQEC*, Munich, Germany, 2013.

[147] IEEE C37.235-2021, "IEEE Guide for the Application of Rogowski Coils used for Protective Relaying Purposes," IEEE Standards Association, 2021.

[148] Y. Chen, G. Crotti, A. Dubowik, *et al.*, "Novel Calibration systems for the dynamic and steady-state testing of digital instrument transformers," in *IEEE 11th International Workshop on Applied Measurements for Power Systems (AMPS)*, Cagliari, Italy, 2021.

[149] H. Karandikar, T. Neighbours, and R. Pate, "Digital switchgear: the next phase in the evolution of safety by design," *IEEE Industry Applications Magazine*, vol. 27, no. 3, pp. pp. 23–30, May-June 2021.

[150] IEC 61850 Communication Networks and Systems In Substations, International Electrotechnical Commission (IEC).

[151] S. Hodder, B. Kasztenny, D. McGinn, and R. Hunt, "IEC 61850 process bus solution addressing business needs of today's utilities," in *Power Systems Conference*, Clemson, SC, USA, 2009.

[152] F. Tao, F. Sui, A. Liu, *et al.*, "Digital twin-driven product design framework," *International Journal of Production Research*, vol. 57, no. 12, pp. 3935–3953, 2019.

[153]   Y. Zhu, Z. Qian, S. Yuan, and H. Yu, "Fault diagnosis of high-voltage circuit breaker based on digital twin," in *International Conference on Advanced Electrical Equipment and Reliable Operation (AEERO)*, Beijing, China, 2021.

[154]   A. Yamaguchi, K. K. Ueno, K. Uchida, E. Matsumoto, and T. Saida, "Development of advanced AI technologies for condition diagnosis of high voltage switchgear in substations," *CIGRE Science and Technology*, no. 26, November 2022.

[155]   Cigre Study Committee B3, "Expected impact on substation management from future grids. Cigre Technical Brochure 764," CIGRE, 2019.

[156]   "Sensproducts^{TM} – From Products to System Intelligence," Siemens Energy, https://www.siemens-energy.com/global/en/offerings/power-transmission/innovation/sensproducts.html [Accessed 22 04 2023].

[157]   "Digital Switching and Components," Hitachi Energy, https://www.hitachienergy.com/products-and-solutions/digitalization/digital-switching-and-components [Accessed 22 04 2023].

Index

www.ingramcontent.com/pod-product-compliance
Lightning Source LLC
Chambersburg PA
CBHW050126240326
41458CB00124B/1456